WARFARE SINCE THE SECOND WORLD WAR

Klaus Jürgen Gantzel
Torsten Schwinghammer

WARFARE SINCE THE SECOND WORLD WAR

with an afterword by Dietrich Jung, Klaus Schlichte, and Jens Siegelberg

Translated by Jonathan P.G. Bach

Routledge
Taylor & Francis Group
LONDON AND NEW YORK

First published 2000 by Transaction Publishers

Published 2017 by Routledge
2 Park Square, Milton Park, Abingdon, Oxon OX14 4RN
711 Third Avenue, New York, NY 10017, USA

First issued in paperback 2018

Routledge is an imprint of the Taylor & Francis Group, an informa business

Library of Congress Catalog Number: 99-043673

Library of Congress Cataloging-in-Publication Data

Gantzel, Klaus Jürgen.
[Kriege nach dem Zweiten Welkrieg, 1945 bis 1992. English]
Warfare since the Second World War / Klaus Jürgen Ganztel and Torsten Schwinghammer; with an afterword by Dietrich Jung, Klaus Schlichte, and Jens Siegelberg; translated by Jonathan P.G. Bach
p. cm.
Includes bibliographical references and index.
ISBN 1-56000-413-4 (cloth: alk. paper)
1. War--History--20th century. 2. World politics--1945- I. Title. II. Schwinghammer, Torsten.

U21.2.G36413 1999
355'.009'1045 21--dc21

99-043673

ISBN 13: 978-1-138-51779-0 (pbk)
ISBN 13: 978-1-56000-413-4 (hbk)

Dedicated in Memory of

Professor Dr. iur., Ph.D., Dr. Mult. h.c. Karl W. Deutsch

July 21, 1912–November 1, 1992

Contents

List of Tables and Figures

List of Abbreviations

AKUF	Working Group for Research into the Causes of War
CIA	Central Intelligence Agency
EA	East Asia
EAF	East Africa
ECOWAS	Economic Community of West African States
EU	European Union
EUR	Europe
CAF	Central Africa
CAM	Central America and the Caribbean
CSCE	Conference on Security and Cooperation in Europe
NAF	North Africa
NAM	North America
NMO	Near and Middle East
OAS	Organization of American States
OAU	Organization of African Unity
OECD	Organization for Economic Cooperation and Development
SAF	Southern Africa
SAM	South America
SSOA	South/Southeast Asia
UN	United Nations
UNO	United Nations Organization
WAF	West Africa

Note: Abbreviations and acronyms not listed here are explained in the text. Abbreviations of periodicals are explained in the bibliography.

Foreword

This translation is based on the revised and updated edition of *Wars Since the Second World War—Data and First Analyses* published in 1986 by the Weltforum Verlag (Munich/Cologne/London), supported by the German Overseas Institute in Hamburg and edited by Klaus Jürgen Gantzel and Jörg Meyer-Stamer. The list of wars and their characteristics have been reworked many times since then. The data set presented and analyzed here extends until December 31, 1992.

An afterword written by my colleagues Dr. Dietrich Jung, Dr. Klaus Schlichte, and Dr. Jens Siegelberg has been added to bring us closer to the current situation, including the impact of developments in Eastern Europe. The Transaction edition also contains a list of wars waged each year from 1993 to 1995. An update through 1997 is in preparation and can be requested directly from the Institute in Hamburg (see note 1). A web site containing this information is also underway.

In addition to the detailed war register, upon which all our statistical data and analyses are based, the reader will find appendices with directories useful for locating specific wars, as well as several supplementary lists offering additional information. Should the reader discover any mistakes or questionable findings, we would be glad to receive corrections and critical responses.

We limit ourselves to the presentation, in tables or figures, of those data we find most important. Extensive theoretical reflections are not included here, and on this point we refer the reader to Jens Siegelberg's book *Capitalism and War—A Theory of War in World Society* (1994), as well as Siegelberg (1990), Jung (1992), and Schlichte (1992) (see bibliography). We restrict ourselves solely to explanatory commentary. The reason for this restriction lies in the particularly high level of abstraction necessitated by a social-theoretical explanatory approach to contemporary wars. This level of abstraction does not presently permit this approach to be operationalized in all cases at the empirical and typological level.

Klaus Jürgen Gantzel is primarily responsible for the descriptive and interpretive findings. Torsten Schwinghammer is responsible for the generation of tables and diagrams on the basis of the computer-based war databank and with the help of SPSS (Statistical Program for the Social Sciences) and Harvard Graphics. He conceived this databank and has been in charge of its maintenance and improvement. Ulrike Borchardt and Jens Siegelberg critically read the manuscript. Wilhelm Nolte attended to the final editing of the German version.

As with the 1986 book of wars, this new edition is a common product of the Working Group for Research into the Causes of War (AKUF). Our institutional home is located in the Research Center for War, Armament and Development, founded in 1986 as part of the Institute of Political Science at the University of Hamburg.[1] AKUF was founded by Klaus Jürgen Gantzel in 1978 and continues under his direction. It consists of a small number of scientific colleagues financed mostly through third-party resources and a larger number of advanced students from various disciplines. They work without special remuneration, voluntarily, and on the basis of their respective scientific motivation. Currently AKUF has approximately 40 members. This form of research-driven learning has been honored by the University of Hamburg through its conferment of the newly founded Fischer Appelt Prize to Gantzel.

AKUF received a noteworthy lift from the presence in 1982 of the Hungarian colleague István Kende (1917-1988), who was the Carl von Ossietzky Guest Professor, made possible by the German Society for Peace and Conflict Research (DGFK) at the University of Hamburg. Using methodology which was extremely unusual for research in Eastern Europe at the time, Kende took it on his own, without a staff or institutional support, to record and classify wars and to analyze and interpret them in view of the distribution of their frequency and trends (Kende 1971, 1978, 1982a). His guest lectures resulted in, among other works, a preliminary realization of his war directory (Kende/Gantzel/Fabig 1982) which drew much interest and was reprinted multiple times. AKUF continues his work as his expressed legacy, even if from a different theoretical perspective.[2] The volume edited by Jens Siegelberg (1991) is gratefully dedicated to the memory of István Kende.

The current book is dedicated to the memory of Karl W. Deutsch. Born in Prague, he was forced to emigrate under the national socialist dictatorship. He held professorships in political science at many well-known universities in the United States; his last position was at Harvard University. He was also a guest professor at the University of Hamburg during the 1983 summer semester where he gave an important lecture on the theories and analytical models of political evolution. As perhaps the last truly universal scholar of social science Deutsch focused his intellectual concerns above all on two connected questions: how can societies peacefully control their conflicts both internally and with other national societies, and which processes of social transformation—he spoke of "social mobilization"—lead to the inner and intrastate integration of humanity? Methodologically, Deutsch united modern social research methods, as developed primarily in the United States, with the tradition of European thought on long-term, formation-building historical processes. Together with the development of the communicative-theoretical model of "political cybernetics" he promoted the intensive collection and the comparative analysis of social-scientific data sets.

AKUF is indebted to Karl Deutsch precisely for the following methodology, which works empirically and theoretically. The empirical work consists in the analysis—comparatively when possible—of the causes of individual wars,[3] in case studies and annotated documentations,[4] in the directorship of a press archive as well as in the regular realization and revision of the war databank. The individual wars /armed conflicts and the countries of the subregions are under continuous observation by the Working Group's members. This leads unavoidably to categorization problems, since such complex societal processes as wars do not lend themselves to classification in crystal clear operationalization schemas. Such problems are discussed together in biweekly meetings where it is decided whether an armed conflict should be regarded as a war in the sense of our definition, how a particular definition or typological characteristic is generally or specifically interpretable, how a specific set of circumstances is to be interpreted in context, and so forth. In addition, the periodical preparation of the data for publication insures that external persons are involved in the critical evaluation process. In this way a "team spirit" emerges which guarantees the use of uniform criteria in the collection of information and in the coding of entries for the data bank. New AKUF members are "socialized" through these tasks. They are, however, in no way "domesticated," and the style and scope of their contributions to the war register are completely individual. This does not compromise the statistical analysis in any way.

Furthermore, unlike many other research subjects, the delineation and classification of violent conflict can change with improved information. Unlike other similar projects, AKUF does not base its definition of war on the number of victims, and thereby does not make the classification of war dependent on one variable which relies on a question of degree. Rather, in addition to conventional definitions, AKUF uses a complex set of information. Of course AKUF's definition criteria, the continuation and direction of hostilities, as well as the evaluation of the cause of war and the goals of the warring parties depends on fluctuations in the availability of this information.

These changes in analysis are also mirrored in the regular surveys edited by AKUF on ongoing wars. With new information an ongoing war can be reclassified as a terminated war or as an armed conflict below the threshold of war. Naturally new information can also lead to reclassifying a previous armed conflict as a war, or breaking down one war into several independent wars.

Specifically, there have been several changes of the above nature which were unable to be accounted for in this edition's data and analysis. These include the hostilities between the FLEC and the Angolan central government, which can now be seen as a war, and the war in Yugoslavia, after long and controversial discussion, was divided into three wars (Slovenia, Croatia, and Bosnia-Herzegovina).

As a correlate to the empirical studies, AKUF pursues a social-theoretical framework for the comparative analysis of causes of war. The essential design for this approach stems from Jens Siegelberger (1990, 1994), using developed bourgeois society as a standard.[5] AKUF's future work seeks to differentiate and classify this approach with regard to the ways in which capitalist modernization functions as war-provoking in its encounter with traditional forms of socialization.

At this point a comment about our research strategy is called for. We are often asked why we do not concentrate our exploratory energy on possibilities and forms of war termination or war prevention. In fact our research program stems directly from this goal. But how can we convey systematic questions of peace-creation and peace-securing for the real world without a systematic knowledge of the structural and situational conditions from whence war emerges? The lack of such knowledge would imply that the causes of war, as well as the objects over which wars are fought, are irrelevant for war prevention and war termination, and that consequently war is an insubstantial, ahistorical and, in principle, eternally recurring mechanism of collective interaction.[6] Unfortunately, modern research into the causes of war has extremely little to show for its efforts.[7] Furthermore, the analytical processing of wide-ranging societal problems cannot be done without a division of labor among the social sciences.

This book aims to contribute to the comparative study of the causes of war through the adaptation and presentation of empirical data. A comprehensive, theoretically informed systematic analysis of the material contained herein would be over hasty at the present time.[8] In order to maintain the character of a handbook we have often relegated ancillary explanations or suggestions for further research to the footnotes, since we felt they would disturb the flow of the text. This necessitates more extensive footnotes than we ourselves would have preferred.

This book would not have been possible without the empirical efforts of the AKUF members, whose biographies are included in the author's index. The abbreviations at the end of each war in the war register correspond to the list of authors, so the reader may identify the author of each particular contribution. The involvement of the AKUF members deserves special praise in light of the unfavorable and deteriorating conditions for scholarship and research at German universities in general, and at Hamburg in particular, and the evaporation of funding from the German federal government for peace and conflict studies.

We are grateful to the German Research Society (DFG) in Bonn for financial support. They provided personnel support for the more than two years necessary for the creation of the AKUF war databank and the input of the data.[9] We also wish to thank the Berghof Foundation for Conflict Research in Berlin who supported AKUF with personnel support and the updating and ordering of the press archive. [10] Special recognition is due to Rudolf Augstein, former editor of the news magazine Der Spiegel and honorary senator of the city of Hamburg. His financial donation in 1991 allowed AKUF to overcome a tight situation. Finally we wish to thank the University of Hamburg, which created the necessary infrastructure through the provision of space and supplies and awarded the above-mentioned Fischer-Appelt Prize, with a financial contribution to AKUF's work. The translation was supported by a grant from the German Foreign Ministry through *Internationes* in Bonn. It was our good fortune to find, through Lit Verlag, a competent and conscientious translator in Dr. Jonathan P.G. Bach. We regard the appearance of this book with Transaction Publishers as a mark of distinction.

With this foray into the English-language realm we hope our work will be received as an important contribution to the international discussion of the causes of war and their relevance for peace research and policy.

Hamburg, January 1998
Klaus Jürgen Gantzel
Torsten Schwinghammer

Notes

1. Mailing address for the Research Center: Allende Platz 1, D-20146, Hamburg, Germany. Visiting address: Sedanstr. 19, 3rd floor, D-20146, Hamburg, Germany. telephone (49)(40)4123-3689, fax (49)(40)4123-2460.

2. See the detailed dedication with biographical references in Siegelberg (1991:V).

3. See, for example, above all Siegelberg (1991), Kurtenbach (1991), Niebling (1992), Jung (1992) and Schlichte (1992).

4. For a list of Working Papers please contact the Research Center at the address in footnote 1.

5. See also the summary in Gantzel/Siegelberg (1990a).

6. Such a position—without evidence and limited to international wars—is found, for example, in Singer (1981:7), one of the most influential representatives of the modern comparative study of war. Here it is not deemed worthwhile to study the internal structure of states and their decision-makers. The intellectual style of foreign and security policymakers since the nineteenth century is seen as similar and constant over decades and across nations, as are the decision-making mechanisms of states, especially in conflict situations, regardless of the governing system, national culture, ideology and economic development status, such that one can hardly explain anything from such data; bilateral relations are decisive.

7. On a completely different set of attempts to assess research into the causes of war see Mendler/Schwelger-Rohmeis (1989).

8. For the first, preliminary results see Gantzel (1988a) and Gantzel/Siegelberg (1990a).

9. The core of the databank contains AKUF's data collection of wars since 1945. Additionally, it contains data from a set of other lists of wars from international research. These other lists are based upon different criteria, definitions and examination time periods which defy easy summaries and reach back in time, for example into the 18th Century.

10. In order to bridge the growing gap between the original timeframe of the first edition—1945-84, in Gantzel/Meyer-Stamer (1986)—we published short extensions of our work thanks to the interest of the Foundation for Development and Peace (SEF, Bonn) and the Institute for Development and Peace (INEF) at the University of Duisburg. See Gantzel/Siegelberg (1994) as well as Global Trends 1991: 183-193 and Global Trends 93/94:177-188.

Glossary

We use specialized terms to simplify the language dealing with the descriptive characteristics and content analysis of specific wars and general war development. They are briefly presented in alphabetical order in this glossary. For more detailed discussion of the terms please see chapter two.

Armed Conflict: Organized collective violent confrontation which does not (yet) fulfill or no longer fulfills the definition criteria for war. See the definition of war.

Third-Party Intervention / Foreign Intervention: Participation which directly affects the warring parties by military units, guerrillas, etc., whose highest commander (state leadership, supreme command, etc.) does not belong to the country in question or to one of the countries participating in the war. In the same sense we also refer to such participation as "from outside" or "from a third party." This does not include arms transfers or similar support.

Intervention: See Third-Party Intervention.

Conflict / Object of Conflict means the object or the issue which generates a clash of interests (e.g., societal system, type of government, autonomy, territory, etc.) between opponents who go to war in its name. Unlike common linguistic usage we differentiate between conflict and war. War is a special, extreme form of conflict.

War is a violent mass conflict which exhibits all of the following characteristics:

a. two or more armed forces take part in the fighting, where *at least one side* consists of regular *government* troops (military, paramilitary groups, armed police units);

b. on both sides a minimum amount of *centrally directed* organization of the warring parties must be present, even when this means nothing more than organized armed defense or strategic/tactical methodical assaults (e.g., guerrilla operations, partisan war, etc.);

c. the armed operations occur with a certain *continuity* and not only as occasional, spontaneous clashes. This means that both sides operate following a planned strategy, regardless of whether the fighting takes place on the territory of one or several societies and how long they last.

For a more detailed definition see section 2.1.

War List is the compilation, in table form, of all individual wars observed in this time frame. It is ordered by region, and within each region chronologically by the date of each war's outbreak.

War Register is the documentation of all of the wars observed here and their characteristics including a sketch of their background and development. The register is more detailed than the "war list" as it contains basic information for the list and for statistical analysis.

Warring Parties are the states, governments, movements, groups, including the opponents, who participate with troops of any sort in the fighting of the war.

War Susceptibility means the frequency (number) of wars, which are ***begun*** in one region or country within a particular time period worldwide. Also referred to as "war-proneness."

Forms of War: This indicates the difference between intranational wars (inner-societal, inner-state, or simply internal wars), international wars (between states, as well as colonial or decolonialization wars) and intranational/international mixed wars. We want to avoid the term "civil war," which is widely used in the existing literature, media, and daily language for all intra-national wars, in order to leave the term free to specify a particular instantiation of war. Likewise we find it unhelpful that war between states is generally referred to as "border war" (e.g., Kende 1982a:28). We understand border wars to refer only to those wars which are specifically fought over a territorial border.

War Burden means the frequency (number) of *ongoing* wars in a particular region or in a particular time period worldwide.

War Participation (national war participation) is the direct participation of an actor (a state, a movement, party, guerrilla, etc.) in war activity, either as the one of the direct opponents or in the form of a third party intervention, even — if not otherwise characterized — as UN forces. It is irrelevant for this term whether the actor is the aggressor/initiator of war or the aggressed/ defender, or whether the actor is unwillingly drawn into someone else's war.

War Zone depicts the land or territory of the respective region where the war is fought. Colloquially this is referred to as "theater of war." This usage has a macabre aftertaste, as if organized murder is carried out as spectacle! Critically used, however, this usage is not completely off the mark: the leaders look on as if from box seats, and we watch the collective killing regions near and far from our "first-row seats" in front of the TV screen.

War Events (Kriegsgeschehen) is a general term for war-like developments.

War Frequency depicts the number of wars fought.

War Persistence describes the duration of war or participation in a war as the measure of the time a society expends on it. The expression "persistence" is oriented to lengthy wars. It is to be used sparingly if the war is relatively short.

War Intensity is determined, above all, by the number of persons killed in the war.

War Complexity describes the extent to which different types of war (see below) and/or conflicts overlap or run into one another.

War Tendency: The amount of participation by an actor (usually a state) in wars as an indicator of its degree of readiness to carry through a war, independent of the actor's status as aggressor or aggressed.

Types of War are special instantiations of forms of wars (see above and section 2.5). "Pure" types correspond only to the four types A, B, C or D. A "mixed" type occurs when two or more types are present at the same time in the same war ("simultaneous mixed type") or in phases one after the other ("successive mixed type"). Occasionally we also refer to "complex types" or "complex wars," although this seldom occurs due to the small numbers of such wars. These cases are a sub-type of mixed types and occur when more than two types are present.

Scope of War measures the number of warring parties in a war, including the armed forces.

1

Purposes and Problems for the Statistical Analysis of Wars

1.1 Enlightenment and Peace Research

At first glance the recording and analyzing of wars seems a rather thankless task of cataloging information. More than bookkeeping, however, it serves the dual goals of clarifying crucial social processes and developing practical research programs for the study of war and peace.

The first goal extends the Enlightenment legacy of intellectualization, rationalization and the de-fetishizing of human world-views (Max Weber 1919:317). The end of the murderous Second World War released an outpouring of hope for world peace. The end of the East-West conflict in 1990 raised similar euphoric hopes for a world without war. This optimism, however, has encountered the very obstacles to action described by Goethe in his "Easter Stroll" (*Faust*, part 1, scene "Before the Gate"), where good people indulge in idle, distant sympathy for "far away Turkey, where the people battle amongst each other. One fellow stands at the window, empties his wine and watches the colorful ships glide along the river. Upon cheerfully returning home for the evening he then blesses peace and peaceful times."

Except for the Korean and Vietnam wars most of us in the West have hardly noticed the steady rise of wars in the world. Of the 44 wars taking place at the end of 1992 many already belong to the "forgotten wars." We gloss over the fact that there has been nary a single day without war somewhere in the world since World War II. It was thus a rather rude awakening when television brought the Gulf War into our living rooms, when the slaughter in ex-Yugoslavia began sending new streams of refugees toward the center of Europe and when "humanitarian intervention" in Somalia soured into failure.

Post-Cold War optimism rested initially on the dual assumption that most wars were "proxy wars," whose actors could be controlled by the Great Powers like marionettes, and that the great East-West conflict was more axial than historical, its end signifying a qualitative shift toward a more peaceful world. Yet careful observation and counting of wars, especially over time, disproves or at least severely relativizes (in Weberian language "disenchants") these assumptions upon which political decisions are often based.

The second purpose of our study of war stems directly from empirical scientific necessity and indirectly from the interests of peace research: A systematic theory of the causes of war is imperative for moving beyond the short-term risky crisis management of brinksmanship.

Instead we must promptly recognize war potential in conflict situations to enable successful peace-securing and peace-management strategies (assuming the requisite political will!). Such a theory is sought in the spirit of Christa Wolf's *Cassandra* (1983:76f): "You can tell when a war starts, but when does the pre-war start? If there are rules about that, we should pass them on, we should hand them down etched into stone. What would be written there? It would state: Let not your own people deceive you." There is, however, no such reliable theory (Mendler/Schwegler-Rohmeis 1989), for the necessary research has been—and continues to be—universally neglected. [11]

There are serious scientists who claim that general statements about the causes of war are impossible due to the uniqueness of each particular case. [12] Consequently, war prevention becomes the sole province of disarmament. Against this we find claims that the human species is naturally prone to war-like violence, such that war-fighting is an inherent part of the political creature (see Lorenz 1963 and, somewhat more differentiated, Eibl-Eibesfeldt 1975; Morgenthau 1963). From this perspective it is not possible to discover and address the causes of war. Instead one merely creates barriers to war through political acts such as power balancing or by redirecting violent instincts toward substitute rituals. Both of these extreme positions remain mere assertions. Their empirical evidence remains untested and insulated from attempts to prove their claims in real situations.

The development of a theory of the causes of war and the testing of corresponding hypotheses in real situations assumes that wars can be classified following uniform qualitative criteria and quantitative measurements—that wars can be "listed" and compared. The cataloging and analysis of wars over time is indispensable for achieving the even narrower goal of spotting trends following the outbreak of war and for finding out where ourselves stand in the historical development of war.

1.2 Theoretical and Methodological Problems

Our project to create and update a directory of wars since the Second World War is neither new nor unique. We take our cue from Kende's prior work (1971, 1978, 1982a, 1982b) and follow exemplary research by the sociologist Sorokin (1937), the legal scholar Wright (1965), the meteorologist Richardson (1960), [13] and the sociopolitical team of Small and Singer (1982 [14]) among others. [15] Important recent lists of war are to be found in Butterworth and Scranton (1976), in Kaye, Grant, and Emond (1985—the so-called ORAE list [16]), in Luard (1986: 421-447 [17]) and in K.J. Holsti (1991 [18]). The five-volume compendium from Pfetsch (1991) is a useful reference work for individual conflicts, as is the earlier but far less detailed book by Dingemann (1983). Because of its lack of conceptual and systematic precision, however, it is not useful as a war register. The same problem applies to Arnold (1991). Only the ongoing state-supported project of our colleagues in Uppsala (which began after ours but enjoys more extensive personnel resources) is of current significance. [19]

All of the lists vary from one to another. They not only cover different time periods, but diverge substantially, since authors' categories proceed from different qualitative and quantitative assumptions even in lists which concern the same or overlapping time periods. Definitions of war differ depending on the goal of the research, the specificity of the approach, the availability of data, and the theoretical and methodological assumptions employed. We cannot expand further here on all the differences and justifications for divergent definitions. [20] Several typical difficulties in settling on definitions, however, merit closer attention.

How to determine the start and end of a war, and thereby its duration? Since official declarations of war seem almost completely out of fashion[21] along with official peace treaties, and since even cease fire agreements are often ignored, we are forced to focus on those moments when hostilities actually begin or end. However there are usually scant specific details, especially if one aims to pinpoint an exact day.

This difficulty is connected to more theoretical questions: can we consider individual acts of sabotage, terrorism and turbulent disruptions (often appearing as violent expressions of liberatory, revolutionary and secessionist conflicts in the "third world"[22]) as war or as a part of a "prewar" phase? Similar questions arise regarding massacres, "cleansings," expulsions and like events when undertaken by a warring party against the population after the end of regular military operations.

A further problem results from wars of long duration, where fighting is not continuous and there are often lengthy pauses between battles, advances, etc., as during the Greek Civil War or the wars in Colombia. Occasionally temporary cease-fire arrangements cause pauses only until they are broken and the war resumes. How short must a cease-fire be to count as a pause in a war, or how long must it be to signify two separate wars? Should such interruptions be wholly ignored, regardless of how long they last? This problem becomes even more complicated when the character of the war changes during its course.[23]

Should we determine a *threshold of war activity* which must be surpassed in order for violent hostile actions to be defined as war? How could we characterize this threshold? Should it include the proportion of armed forces involved? If so then we must ascertain the size of the units involved in the hostilities and separate them from those which are used by the warring party outside of the war zone as reserves or for other purposes. Few exact figures exist for such information. The simple number of soldiers involved directly in hostilities is a problematic figure, for a thousand air force personnel with a fleet of bombers can destroy and kill far more than a troop of a thousand infantry men in the same amount of time. Even measuring firepower itself is not the best method, because its efficacy depends on the specifics of the war. It would be rashly one-sided to compare, for example, guerrilla wars with modern conventional warfare, since high-technology armed forces have a far greater share of logistical and support personnel than guerrilla forces.

Occasionally the means of war themselves make distinctions difficult. Should one count the stones which Palestinian youths threw against the Israeli military? Or does it have to be at least bows and arrows, with which, for example, the Bougainville Revolutionary Army began its resistance against the Australian tin mines on their island and against the troops of Papau New Guinea?

Ethical considerations tend to focus on the intensity of war and therefore on the victims of wars, yet this too encounters immense difficulties.[24] First, data about war victims are usually highly unreliable. Insofar as they even keep count of losses the warring parties have an interest to slant the numbers: to exaggerate the number of "successfully exterminated" opponents and to underestimate the number of their own fallen, or to exaggerate the number of their own civilian victims and to minimize the numbers of those on the opposing side. The registration of losses also assumes a basic degree of efficient bureaucratic organization which is often lacking in Third World countries.

Second, who counts as a victim? Only those killed by weapons or also those who perished from war's companion elements of hunger, cold and disease? Should the wounded and injured count alongside the dead as victims? If only the dead are considered, should this include those who died after the war ended from war-related injuries or illnesses, such as the late victims of Hiroshima and Nagasaki?

Should only the victims of the fighting units count (combatants such as soldiers and quasi-combatants such as guerrillas and partisans[25]), or also the civilian population? Can we always differentiate between combatants and non-combatants, or, as for most of the so-called covert wars, is this distinction not relevant, leading only to the cynical "body counting" as during the Vietnam War? The question of whether "fallen" soldiers or all the dead (including civilians) should be considered touches on ethical aspects of epistemology as well as theoretical aspects connected with changes in modern military strategy and technology. If our suspicion is correct that the number of civilians killed in internal and international wars is growing relative to combatants (Eckardt 1992:254), then limiting the number of victims to military personnel as the basis for a definition or as a characteristic of intensity as Small and Singer (1982) do is more than questionable.

Assuming that all of these definition and measurement problems could somehow be resolved gives rise to a third question: how many victims does it take to turn a violent conflict into a war? After only one victim, or after at least ten victims, or one hundred, or one thousand, or even more? Should the number of the total victims of the war be decisive, or the minimum number of those killed per year? And were one to settle on the number of 1,000 dead as the basis for differentiating between war and other forms of violence, as Small and Singer or the Uppsala group do in the SIPRI Yearbook, what happens with cases whose numbers are 990 or 995 or 998? Such arbitrary lines of demarcation are allowable for statistical analysis of mass phenomena, but war is a phenomenon which cannot be seen as a mass event even in a statistical sense (despite the approximately 190 cases which we have established since 1945). Should one then follow the Richardson's rigorous analysis (1960:4 ff.) and register all "deadly quarrels," from individual murder to world war with megatons, and then arbitrarily decide that certain quarrels whose victims amount to a logarhythmic scale of 317 (=log 2.5) or more can be called war?

Perhaps it is simply insufficient to use war dead as the basis for the definition of war. Why not also include financial, economic and social costs of war, not to mention the cultural costs?

It is also not clear who should be considered a "warring party." Should one consider only those directly involved in the fighting (and perhaps make this even more precise by establishing a minimum number of soldiers and/or victims involved, as do Small and Singer)? Or should one also include indirect political, financial, material (e.g., weapons delivery) and logistical support for one of the warring sides in the calculations? [26]

What minimum amount of organization, strategic planning or fighting ability must be present in order to distinguish war from tumultuous disorders or other forms of violence? How can we differentiate between marauding bands on the one side (even if they wear uniforms), and war-fighting units on the other, and among the latter between government troops, police units and similar security forces, secret paramilitary organizations, mercenaries and "volunteers," guerrillas and partisans, terrorist groups and so forth?

What is the minimum amount of military activity which must occur for a war? Should a military occupation which encounters no significant resistance count as a war, such as the 1968 Warsaw Pact invasion in Czechoslovakia? This is not merely a pragmatic question, for behind it lies fundamentally divergent theoretical approaches. While behavioralists (ethnologists) and some psychologists of aggression "terminologically dissolve war *qua* aggression into areas normally designated as peaceful, Carl von Clausewitz, the most important theoretician of war, curtly summarized the concept: War begins not with attack, but with defense (Münkler 1985: 281)."[27]

Further problems arise with the definition of "state" and "government" as war-fighting actors. Does simply declaring political independence from a foreign power turn a warring party into a state? Must sovereignty be internationally recognized, and if yes, then by whom (the U.N.? the "Great Powers"?)? Must a warring party control the institutions in its territory (legally or illegally)? Political science and international law have difficulty answering these questions, as evidenced in Cyprus, Western Sahara, Eritrea, Rhodesia, Namibia, Kurdistan, East Timor, Korea, Yugoslavia, Moldavia, Abkasia, etc. Conclusions about these conflicts vary depending on the evaluation criteria. Opinions also differ about what constitutes an international war on the one hand and an "internationalized" civil war on the other, or for transitions from one form of war to another during a war.[28]

We have only touched upon the most common problems in compiling a comprehensive war directory. It is hardly surprising to find almost as many war lists as there are researchers. Much intensive research and discussion is needed before standard definitions emerge and are accepted as a basis for scientific understanding. Until now the opposing approaches of Small and Singer's (1982) Correlates of War project and that of Kende (1982a) present the most likely contenders.

We base our work on Kende's criteria because we find it the most theoretically and empirically open and profitable approach. In the next chapter we will explain and justify our use of Kende. For reasons of space we forgo in-depth engagement with other authors' criteria. Only Small and Singer will be dealt with in some detail, more as justification for our own decisions than as an independent evaluation of their approach.

Our preference for Kende does not mean that his work is free of problems or in no need of improvement, especially in questions of precision and differentiation. In our earlier book and in this present one we attempt several improvements and have expanded the information on wars found in Kende's database. To the best of our knowledge no other list of wars for the period after 1945 is so complete, so precise and presents such a rich trove of information. Yet our work is not solely an extension of existing war lists. We offer an ongoing contribution to expand scientific knowledge about the development and background of wars since World War II. Whether our claims are just will be determined by our critics. In any event we hope that our data collection and analysis will be widely acknowledged and—above all—put to use.[29] Our data are not only the empirical basis for our own research, but are conceived as a service for science, the media, higher and secondary education, and not least of all as grist for political discussion.

One last methodological comment: This book does not concern itself with detailed case studies as Siegelberg (1991) does. We focus rather on the indicators of trends common to most or all wars over time. This book is therefore concerned primarily with quantitative research. Our definitions, typologies and decisions are constructed with this purpose in mind. Many readers, even among the sciences, castigate such an approach as mere bean counting and consider only qualitative approaches valid for the study of war. Aside from the fact that our Center uses different methods depending on the goals of the research, presenting one or the other approach as absolute is absurd. We want to spare ourselves and our readers the equivalent of the "methodology debates" of the 1960s and 1970s. At very least quantification has two advantages: it simplifies and in some cases makes possible comparative research, thereby allowing for the analysis of events over time. Secondly, it compels precision. Scientific language, like our quotidian vocabulary, is full of descriptive terms for quantities ("large/small," etc.), full of comparisons ("larger/smaller," etc.) and full of stochastic expressions ("as a rule," etc.) all of which contain quantitative statements in imprecise and often unverifiable forms. Even sceptics are therefore invited to engage themselves with the statistical research in our book.

Notes

1. The relevance of Barbara Tuchman's dictum of the "Rule of Folly" is evidenced by the fact that now, when wars are unfolding on our doorstep, so to speak, and when the German federal army intervenes outside of NATO territory, that Bonn has eliminated the already meager funds for peace research. But it is not only politicians who are foolish, but also prominent intellectuals. Their excitement, theses and declarations during

the Second Gulf War of 1991 (against Iraq) and afterwards in journals and paperbacks are impressive evidence for this judgment.

2. See Anatol Rapoport in Lider (1983:48 f.).

3. See also Wilkinson (1980).

4. Since then this project has become inactive, apparently because of theoretically grounded reorganization of the entire data set and because of political-financial difficulties.

5. A series of further precursors is presented and discussed by Small and Singer (1982:25 ff.).

6. This list of wars in the period 1720-1985 was compiled from other existing lists.

7. Keep in mind, however, Luard's important reservations in 1986:7!

8. As with Luard, Holsti's focus is more on historical analysis than "datamaking." Holsti restricts himself to international wars from 1648-1989. He wants to explain the following aspects using historical periodization: the type of conflict materials and their significance for the generation of interstate war; the changes in the sociohistorical milieu that take place during these wars; and the relationship between peace arrangements and war.

9. The project is directed by Peter Wallensteen, who similar to Gantzel (1972) appeared with a systematic analysis of war in 1973. Since 1987 his research group, under different editors, has published annually an overview of armed conflict in the world entitled *Major Armed Conflicts* in the Department of Peace and Conflict Research at Uppsala university. They make use in part of our data. To the extent that the registered conflicts involve more than 1000 dead, the list is also included (since 1988) in the yearbook "World Armament and Disarmament" of the Stockholm International Peace Research Institute (SIPRI).

10. See the compendium in Mendler and Schwelger-Rohmeis (1989:7-19). Regarding fundamental types of definitions see Gantzel 1985 or 1991.

11. The Second Gulf War (of 1991) against Iraq with an international coalition, authorized by the U.N. Security Council and under the leadership of the United States was an exception.

12. Since the end of the Cold War the term "Third World" has dispensed with its typological sense, which originally was to differentiate the block-free

states from either the "first world" of the Western camp or the "second world" of the Eastern. It became, however, a general term for developing countries. We agree with the proposals of Nohlen and Nuscheler (1992:7, 29 f.) to continue to use its naturalized meaning.

13. For example, our 1986 book was criticized for not considering the Israeli incursion into Lebanon in 1982 as its own war, but rather as part of the conflict between Israel and the Palestine Liberation Organization.

14. AKUF is preparing a publication which addresses this particularly difficult aspect, which systematically analyzes the reasons why victim numbers diverge so starkly. Incidentally, the study of the *consequences* of war suffers more from neglect than the study of the causes of war.

15. These terms are not used here in the sense of international law.

16. In our earlier book and in every following discussion about our data critics have found fault with our decision not to count cost-free deliveries, sales of weapons and military training as participation in war or intervention. We continue to not use these as part of the definition criteria, because then there would only be wars which involved third parties, and the distinction between wars with and without third party participation would no longer make sense.

17. "When we think philosophically about the origins of war, then the concept of war arises not with the attack, since this does not have both fighting and possession as the absolute goal, rather war emerges from the defense, for this has fighting as its direct goal, because defense and fighting apparently are identical. (von Clausewitz 1832/34:384)."

18. Thus Small and Singer (1982), in compiling their data bank for interstate and so-called civil wars in the long era between 1816-1980 use intellectual acrobatics to settle on definitions for "states," shlep ideological balast with them (in making the determination of sovereignty directly and indirectly dependent on the interests and power calculations of Great Powers) and tolerate partially absurd consequences of all this (e.g., when only the colonial powers, but not the anti-colonial liberation movements, appear as warring parties). For a summary and critique of Small and Singer's definitions see Gantzel (1987:29-32, 79 and 82).

19. AKUF would be grateful for a copy, or at least for the citation, of any research which uses our data set.

2

Definitions and Inclusion Criteria for the List of Wars Since the Second World War

This chapter defines war and sets out the inclusion criteria upon which we base our list of wars from the end of the Second World War through 1992.

Although the first of these wars—the first Greek Civil War—began on December 3, 1944, and thus technically predates 1945, it nonetheless belongs to the new epoch emerging out of the Second World War and accordingly shares its symbolic significance. In order to avoid statistical bias and to maintain the symbolic caesura of "1945" we portray the first Greek Civil War unless otherwise noted as beginning in January of 1945 and extend it fictively 29 days after its actual end on February 12, 1945.

2.1 Definition of "War"

Following István Kende (1982a:5f) we regard any violent mass conflict which includes *all* the following characteristics as war:

- Two or more armed forces take part in the fighting and *at least one side* is composed of regular *government* troops (military, paramilitary groups, armed police units).
- A minimum of *centrally directed* organization of both warring parties must be present, if only as organized armed defense or strategic/tactical attacks (guerrilla operations, partisan war, etc.).
- The armed operations must have *continuity* and not be only occasional, spontaneous clashes. This means both sides must follow a planned strategy

whether the fighting takes place on the territory of one or several societies and regardless of the length of the conflict.

Since the same questions and critiques arise from almost every discussion concerning our definition, let us briefly elucidate it further.

In contrast to Small and Singer (1982), among others, our definition does not initially differentiate between *international* and *intranational* wars.[1] It includes both kinds of wars and submits them to the same inclusion criteria. This does not exclude treating them separately for analytical or comparative purposes. The following points are of fundamental importance:

First, by submitting international and intranational wars to the same criteria we do not assume *ex definitionem* that war as "the most violent expression of politics and the most political expression of violence (Kende 1982b:6)" stems principally from different causes, whether external or internal to a state, whether within a territorially-demarcated society or between several societies. Furthermore, we avoid preempting the question of whether the source of war is located in the very existence of territorial borders as much as in state power. We also keep open the question of whether war results from societal causes and the substantive content of politics, or perhaps even more, with social processes. This does not, however, imply that war has nothing to do with the state. We believe that it does, and will argue this point here and in later passages.

Second, we avoid engaging in the theoretically questionable evasion of definitions of the sort which Small and Singer (1982:38 ff., 51 ff., 210 ff.) undertake so that their category of warring parties includes both formal state units and those societal units which wholly or partially lack internationally recognized state sovereignty (colonies; protectorates; annexed and occupied territories; divided nations, etc.).

On the other hand our definition does *not* include *every type of collective violence*: The requirement that at least one side consist of *government* forces excludes violent private feuds such as conflicts between criminal gangs, cattle-breeding nomads, or local latifundia. This exclusion is based on a historical distinction between private violence and, respectively, public or state violence.[2] In the interests of providing security for its citizens a state can overcome or at least control private and arbitrary violence by claiming a monopoly over the use of physical force.[3] The next historical step is the public, democratic control of state power. Here we follow Max Weber's (1964:39, 1043; 1916:453) concept of state which consists of the successful societal claim to the legitimate monopoly of physical violence within a certain territory. Weber saw no other possibility of defining the "state."[4]

The political dimension of the collective use of violence needs also be recognized if an armed conflict counts as war, even when concrete material interests of individual persons or groups form the background of the conflict. We are aware that such a claim sets a standard oriented toward European history.[5] We are also aware that until now only a small number of states in the world approach this standard. We assume, however, that the largest number of wars since 1945 are directly or indirectly fought about state monopolization of physical force and the consolidation of power, and that concrete reasons, interests and motives for each individual case are, analytically at least, of secondary importance.

Accordingly the type of government involved remains open. Whether the government in question came to power legally or illegally, whether it holds to its constitution or not, whether recognized by international law or not, whether it represents a central state government, a provincial government or colonial government, whether it sits "straight in the saddle" and actually holds sway over the population and essential institutions or whether its power is limited is of no significance here.[6] Whether the rulers abide by international law or human rights or whether the rule of law is even operative in the respective state is equally inconsequential.

Various other forms of violent conflict are also excluded by the above condition that both sides have a minimum of organization and centrally directed, planned and armed war-fighting capability and that the fighting must entail an obvious sense of continuity. These excluded forms include:

- coup d'état, assassinations and similar terrorist actions (e.g., hostage taking, airplane hijacking), acts of sabotage, etc., unless they lead to a planned, armed mass conflict. [Anti-regime wars and anti-colonial liberation wars in the third world often start with sabotage and terrorist acts, however they become war only when they and the state reaction which they provoke take on a planned character and occur with regularity. The occasional attacks of the Armenian liberation movement which have been observable for years do not count as war, whereas the reoccurring attacks by commandos of the

African National Congress (ANC) since 1976 against industrial targets, police stations etc., do count as war, as do the strategically planned attacks and murders by the fundamentalist Islamic opposition in Algeria.]

- spontaneous, tumultuous uprisings such as in the German Democratic Republic in 1953 or in the black ghettos of Los Angeles in 1965 and 1992;
- massacres against a population group which is unarmed or not prepared to defend itself (as in Indonesia in 1965 or periodically in Haiti) unless such mass murder or exterminations occur during a war or in immediate connection with one, such as recently in Bosnia-Herzegovina or Rwanda;
- the military occupation of a territory which does not meet with organized, armed resistance, such as the 1968 Warsaw Pact invasion of Czechoslovakia;
- border incidents or isolated skirmishes as in 1969 on the Soviet-Chinese border or in May 1985 between the police and the MOVE organization in Philadelphia.

Additionally, our definition does not require that the leadership of a state exert total control over its own armed forces. In wars of our epoch individual commanders frequently call their own shots or fail to obey certain instructions from their political leadership. The war in former Yugoslavia provides numerous examples of this. Bear in mind that the primacy of politics over the military is after all a recent achievement of civilization.

The subject of our examination is thus "war" and can only be analytically delimited in reference to the historical function of the "state." We are not dealing with violence in general, nor with group violence. Nonetheless we hope that our approach will contain insights useful for the understanding of other forms of violence, including private violence.

Our definition also makes do without precise *quantitative* characteristics and delimiters such as the number of casualties, the size of the participating forces, the length of the conflict, and so forth. As we have already pointed out Small and Singer (1982) proceed in an exactly opposite fashion.[7] We find the uneven way in which they apply their standards of measurement according to the type of war especially problematic. Even more unsatisfactory are the increasing number of studies which rely on Small and Singer's data sets, among others, which fail to openly detail their definition and delimitation criteria let alone critically assess them. Their war statistics tend to take on, so to speak, an uncontrollable life of their own.

We decline to use these types of quantitative characteristics in our definition for the following reasons:

- The assignment of any minimum size or threshold of intensity can only be arbitrary and ahistorical.
- The numbers of casualties or the size of the armed forces are determined, *inter alia*, by the individual dynamic of the particular war and through the "peculiarity of its means" (Clausewitz 1980:34). These figures depend on changes within military strategy and weapons technology, which are not necessarily connected to the causes of war or political goals.
- We do not find the distinction between military and civilian casualties justified for purposes of the definition.
- The available figures for participating combatants and casualties are consistently inexact, especially regarding the civil population.

Naturally our definition includes quantitative aspects as an integral part of concepts such as "mass" conflict, "minimum amount" or "continuity." Since we do not assign number values to these concepts, our war list occasionally contains subjective evaluations. However, even a seemingly precise operationalization such as Small and Singers' cannot escape such subjective evaluations (1982:53, 58 ff., 73ff). Furthermore, we thoroughly discuss uncertain cases in order to exclude, to the extent possible, any arbitrariness, negligence or errors attributable to individual persons. With the complex topic of war it is unavoidable that every definition, no matter how exact, will face limitations (see Kende 1982a:6 f). The claim to exactitude rests on one's respective epistemological approach and the goal of the investigation. Arbitrary statements, however, are problematic when they grossly distort the results.[8]

Certain figures (or close estimates) are important and even indispensable for the description and analysis of wars. These include the size of the armed forces, the number of casualties, and even the extent of the destruction and financial costs. Even our data set contains significant gaps and imprecisions. For the *definition* of war, however, we do not claim any explicitly quantitative characteristics.

2.2 The Wagers of War: "Aggressors" and the "Aggressed"

"Aggressors," according to our definition, are only those actors whose armed forces participate directly in armed conflicts.

This includes third-party (usually foreign) states or groups who intervene for the military benefit of one side or the other, as well as fighting contingents who belong to neither the state nor military leadership of the respective country nor come from the theater of war. So-called volunteers also count, while mercenaries do not. Unlike Kende we do not consider the war-fighting activities of troops belonging to a colonial power as foreign participation as long as they are active only in their own colony.

When relevant we note whether foreign participation takes place within the framework of the United Nations, e.g., either directly at the U.N.'s behest or authorized as part of an internationally created U.N. force (see the list of countries involved in the wars, ordered by region and by type in the register). In order to be registered as a warring party such military units must participate directly in the fighting. U.N.-related third-party participation is worthy of inclusion for several reasons: 1. War, according to our definition, does not depend on the stated intentions or justifications of the warring parties for its analytical status. 2. We cannot exclude the possibility that a state which supplies troops for a U.N. contingent or U.N. action is not acting for its own selfish purposes. This is especially true since the options and decisions of the United Nations (above all the Security Council) depend not only on universally accepted norms for peace-keeping and peace-creating, but also on interest groups and power relations. The most recent evidence is provided by the 1991 Gulf War and the 1993 Somalia intervention. Note that we do not consider U.N. peace-keeping or peace-creating as normal interventions. They are not included in the statistical portrayal and evaluation of third-party participation and intervention (see section 4.10 on interventions). Third-party participation and intervention refer only to partisan intervention on a particular side of a warring party to enable military or political success or to avoid defeat.

Third-party participation also excludes the few cases of armed action by state groups (e.g., the OAS in the Dominican Republic in 1965 or the ECOWAS troops in Liberia since August 1990) or individual states performing a policing role to either end hostilities or force a cease-fire or negotiations (e.g., India in Sri Lanka or the Russian Federation in Moldavia in 1992). (Keep in mind, however, that the formal neutrality of a "pacifying power" may mask less neutral designs.) These types of interventions are also treated separately in section 4.10.

We wish to emphasize that the following do not count as participation in war: the mere presence of military advisors or U.N. peace-keeping troops as long as they do not directly take part in the fighting; the indirect support of a warring party through diplomatic or political actions; military assistance pacts or cooperation agreements; payment of subsidiaries; military training; weapons sales and gifts; supply, transport and intelligence assistance; establishment of logistical infrastructure; sharing information on armament technology, intelligence agency cooperation, etc.

We designating the warring parties as follows:

- As a rule states are referred to by their colloquial names, e.g., "Great Britain" or "England" instead of "the United Kingdom of Great Britain and Northern Ireland," or "Egypt" instead of "the Arab Republic of Egypt."
- The purposely broad term "lands" refers to territorially demarcated societies which do not possess their own statehood, for example colonies, overseas departments, mandate areas, or areas where statehood has been denied or usurped, such as Eritrea or East Timor. This allows us to include societies which are not formal states.
- The term "government troops" in the context of intrastate conflicts refers to the state-run military, police, and other similar units, including paramilitary ones, which fight on the side of the government. If troops from another government take part then their origin is duly noted (e.g., Soviet or French troops).
- All other groups in intrastate conflicts are referred to in general by their official or self-designated name (e.g., the Palestinian Liberation Organization—PLO; the Eritrean Peoples Liberation Front—EPLF). Occasionally we characterize these groups with additional attributes such as conservative, liberal, republican, nationalist, national-revolutionary, socialist, communist, and so forth, even when this does not necessarily coincide with the respective group's self-image. We also use the term "oppositional" when the political direction of a given group is either unknown or unclear. We do not conceptually differentiate terms such as "uprising," "rebels," "guerrilla," "revolutionary," "movement," etc.
- We refer occasionally to the participation of "colonial troops" in wars of colonial conquest. This definition includes troops belonging to a colonial power and recruited from the population of one or more colonies (at least the regular soldiers). The colonial troops do not have to consist of recruits native to the colony where they are engaged.

We have tried to distinguish between aggressor and aggressed, though this is often a subjective judgment. In the register and lists the aggressor is always listed first, and the aggressor is designated "side A" and the aggressed "side B." Perhaps the term "war-initiator" would have been more accurate than aggressor, since sometimes one side provokes a violent response from the other. For this reason our criteria for aggressor consists of who initiates the use of weaponry, rather than which side fires the first shot. Obviously there are instances where this cannot be determined. In cases where military action does not provide a clear indication of who attacked whom we assess from which side the decisive political direction toward war emanated.[9]

2.3 Regional Categorization of Countries at War

By regions we mean "world regions," political-geographic areas which encompass a number of lands. We will make clear from the context if a "region" is used to indicate a local area within a land (e.g., southern Sudan) or an ethnographically or culturally demarcated area which can extend over the territory of several formal states (e.g., Kurdistan).

In light of material and temporal differences in the process of global development it is important to determine regional differences concerning susceptibility to war. It is often difficult to demarcate the regions and allocate a particular land to this or the other region. Cartographers, physical and social geographers, historians, political scientists, and ethnologists, among others, address these issues from fundamentally different perspectives.

We have approached our classifications practically, focusing primarily on areas of political conflict, interconnections, and comparisons. We differentiate the following regions as follows (Africa is substantially subdivided more than the other continents because of its relatively large number of states):[10]

- *Europe* (abbr. EUR) , encompasses North, West, South, Southeast, Middle and Eastern Europe, the entire former Soviet Union and the present Russian Federation, the Baltic states, Belarus, Ukraine and Moldova. Cyprus and Turkey count as part of Europe, however, we classify the war in Eastern Turkey against the Kurdish PKK as well as the war in North Ossetia as belonging to the "Near/Middle East" region.
- *Africa* (AF), subdivided into
- North Africa (NAF), includes Mauritania, Western Sahara, Morocco, Algeria, Tunisia, Libya, Egypt and Sudan;
- West Africa (WAF) includes Senegal, Gambia, Guinea Bissao, Guinea, Sierra Leone, Liberia, Ivory Coast, Mali, Burkina Faso, Ghana, Togo, Benin, Niger and Nigeria;
- Central Africa (CAF) includes Chad, Cameroon, Central African Republic, Gabon, Congo, the Democratic Republic of Congo (formerly Zaire), Rwanda and Burundi;
- East Africa (EAF) includes Ethiopia, Djibouti, Somalia, Kenya, Uganda, Tanzania and Madagascar;
- Southern Africa (SAF)[11] includes Angola, Zambia, Malawi, Mozambique,

Namibia, Botswana, Zimbabwe, the Republic of South Africa (RSA), Swaziland and Lesotho.

- *Near and Middle East* (NME): This region extends from the Arabian peninsula, Israel and Palestine, Iraq and Iran until Afghanistan, and after the fall of the Soviet Union encompasses in the northern region the Caucasus countries of Azerbaidjan, Kazakstan, Uzbekistan, Turkmenistan, Tadjikistan, Kirgizhia and Georgia.
- *South and Southeast Asia* (SSEA), begins in the west with Pakistan and ends in the east with the Philippines, further south it includes Oceania, which also includes Australia and New Zealand.
- *East Asia* (EA), encompasses Mongolia, the People's Republic of China,[12] Taiwan, North and South Korea, and Japan.
- *North America* (NAM) includes Canada and the USA. Conventionally Mexico is also considered part of this region, especially since the creation of NAFTA. However, because of the culture, language, and history of Mexico it seems more appropriate to include this country in the Central America region. The 1994 Zapatista Indian uprising in the Mexican province of Chiapas provides the most recent evidence for this.
- *Central America* (CAM) extends from Mexico in the north to Panama in the south.
- *South America* (SAM) includes all of the countries south of the Panamanian border (in other words beginning with Colombia).

2.4 Duration of the Wars

As noted it is difficult to pinpoint the exact beginning and end of a war. We have recorded the beginnings and endings according to the day, month and year whenever possible. Where it is impossible to determine a particular day, we use the middle of the month in order to calculate the duration of the war. In cases where the month is unclear, we have used the middle of the year. We assume that any errors arising from these seemingly arbitrary determinations will be mitigated by a large number of wars resulting eventually in a suitable mean.

It was more difficult for us to classify multiple wars within the same territorial area or long wars with extended interruptions, such as those in Colombia or the Palestinian conflict. We have already referred to this problem in section 1.2. In each case we proceed accordingly: We are dealing with one continuous war if, after the end of hostilities, the main opponents and the substance of the conflict remain unchanged. If, however, an essential actor was significantly transformed or replaced by another, even if the war-zone remains the same or if the point of dispute has shifted, then a new war has begun. We consider wars in the recent past to be terminated when no continuing hostilities have been reported for one year. We "re-activate" the conflict as a war, however, if after this one-year war breaks out under the same conditions as before. It happens frequently enough that cease-fires are negotiated or one party pulls back temporarily only in order to fortify their forces for further battle.

2.5 Types of War

Kende (1982a:27 f.) differentiates between three types of war:

- *Anti-regime wars* (type A), are those which aim, respectively, to maintain or overthrow the government. In other words, the object is political power. The government itself is often the immediate point of dispute, especially in the Third World, if only because they control the access to graft and sinecures. In numerous cases the government is only the superficial goal, and the real object of the conflict is the societal system and the corresponding social order (e.g., with revolution, fundamental reform, or counter-revolution). Examples include the Vietnam War, the "Shining Path" campaign in Peru, the "New People's Army" (NPA) in the Philippines or the "African National Congress" (ANC) in South Africa.
- *Other internal wars* (type B), are those in which the direct object of the conflict is not the government itself or the system it controls, although the government is involved in the war. These wars are waged primarily for the accomplishment of (or defense against) particular ethnopolitical interests, limited territorial claims, political hegemonial ambitions, religious supremacy and similar interests (as long as they do not fundamentally alter the society as did the "Islamic Revolution" in Iran from 1978 to 1982). This category includes the numerous wars in ethnoculturally, economically or power-politically demarcated provinces and wars led by national minorities or repressed national majorities for the elimination of discrimination, greater autonomy within a state structure, or even for secession or separation from the state. The latter can be connected with irredentist striving for annexation to a neighboring state. Examples include the wars in Cyprus, Kashmir, Sri Lanka (Tamil Rebels), South Sudan, in Spanish Basque territory and in Northern Ireland.
- *Interstate war* (Type C), consists of the "classic" war type, so to speak, in which the armed forces of an established government preside over a conflict involving at least two state-organized territories without regard for their status under international law. Examples include the armed conflict between the People's Republic of China and "Nationalist China" (Taiwan) in the 1950s, the 1982 Falkland War or the Iran-Iraq War (1980-1988) and the (second) Gulf War of 1991. Note that a war does not necessarily become interstate solely when a foreign government intervenes in an internal conflict by supporting a rebellious opposition fighting against its own government.

This typology is not wholly sufficient for our purposes. First, the class of "other cases" (type B) is impractical, for it encompasses too many different sorts of cases. Second, factual sovereignty is definitive for some type B and all type C cases, while it plays no role at all for type A cases. These problems notwithstanding, we have deferred any improvement of Kende until our own theoretical work is advanced enough to enable a meaningful and stringent typology.

However, we have expanded upon Kende's schema on one particular point. Kende classified wars of liberation (or, conversely, wars to maintain colonial power) as anti-regime wars. We decided to treat these cases as a separate type altogether:

- *Decolonialization wars* (type D). In contrast to the other categories this type is specific to the post-1945 historical epoch, especially the 1950s and 1960s (omitting the anti-colonial emancipation of North America in the late 18th and Latin America in the early 19th century). For type D we count only those wars where a formal colonial power forms one side as the main opponent (Belgium, France, Great Britain, the Netherlands, Portugal, and Spain). Despite its occasional quasi-colonial behavior we do not regard the USA as a colonial power; the reasons for her prominent war-like activities lie elsewhere. Wars for the securing or elimination of indirect, *neo*-colonial influence also do not fall under type D. Nor do wars in which the former colonial power intervenes in the former colony, or such wars in which a post-colonial state attempts to annex neighboring territory which previously belonged to the colony but has not yet achieved sovereign statehood (e.g., Morocco/West Sahara, Indonesia/East Timor, or the war between India and Pakistan in Kashmir).

In many cases a war does not fit unambiguously into one or the other category. Every typology that tries to capture "real life" contains some unavoidable haziness. Moreover, and more importantly, it often happens that one and the same war fits several categories simultaneously or consecutively. We classify such wars as "mixed types" and characterize them by the respective combination of types A,B,C or D. If two or more types appear simultaneously in one war, then we are dealing with a "simultaneous mixed-type." If the type changes during a war without any marked interruption of hostilities, and if the subsequent phases can be clearly differentiated, then we note this with the use of a slash between the type (e.g., A/B). In the statistical portrayal and analysis, however, we neglect these successive-mixed types since they occur relatively infrequently. We also leave out of our statistical considerations the rare occurrence when a war begins as a B war and broadens later into an AB-war (in the war register this would be marked either as B/AB or AB). This particular mixed type is always statistically considered to be AB (mixed internal war). A war of the type B/BC/B will always be statistically considered as a BC-war (intra-/international mixed war with B components). If we speak occasionally of *complex* wars or complex types then in most cases we mean that more than two types appear. There are only few such cases so far: the combination ABC occurs only twice (war # 92: Chad; war # 120: Lebanon) and the combination ABCD (War # 62: Angola) occurs only once. Due to this infrequency we generally assign complex wars to the mixed-war category.

Wars with third-party or foreign participation in which at least one foreign power participates (see section 2.2 above) are noted, as in Kende, by adding the numeral "1" after the war type. A "2" following the war type means that there is no such third-party participation.

The following grid summarizes the typologies:

	with foreign participation	**without foreign participation**
Anti-regime wars	A-1	A-2
Other intrastate wars	B-1	B-2
Interstate wars	C-1	C-2
Decolonization wars	D-1	D-2

Additionally we have the mixed types, whereby we actually only encounter the following cases (ignoring differences between simultaneous and successive mixed-types):[13] AB-1, AB-2, AC-1, AC-2, AD-2, BC-1, BC-2, BD-1, BD-2, CD-2, ABC-1, ABCD-1.

For certain questions it is useful to differentiate more narrowly. In these cases we combine all A-, B- and AB- types into *intranational* or *internal* wars, all C-, D- and CD types as *international*, and all AC-, AD-, BC-, BD-, ABC-, and ABCD types as *intra-/international mixed* wars.

2.6 Forms of Conflict

Despite loose scientific usage the term "war" is not synonymous with "conflict." War is a special expression of conflict where one group uses the organized, collective force of arms to advance its goals or interests against opposing goals or interests of another side. Accordingly, our war directory notes the type of conflict out of which war emerges as well as the interests and goals of the respective aggressors. In the "war list" (the tabular "yellow pages" designed for quick reference) we have created general descriptive terms to reasonably encompass the myriad subjects of conflict and goals of war. The relevant statistics regarding "types of conflict" in the analytical portion of the book are based on these classifications. Since these designations consist of preliminary, inductive and generalized classifications, we cannot presently speak of these concepts in a strict scientific sense.

Our systematization of the object of conflict orders and then comparatively evaluates different and specialized disputes according to content. This should help in the development of a more substantive and informative typology of war than is currently possible with the A-B-C-D classifications. This goal has hardly been adequately accomplished through our listing of types of conflict.[14] Further intensive research combined with continued theoretical efforts are required for its perfection. In addition to the inherent difficulty of delineating clear-cut borders (e.g., between system formation, systemic conflict, power conflicts, particular interests, etc.) lies the special problem that conflicts often intersect, overlap and become mutually entangled. Our attempt at systematization is therefore preliminary. We would be grateful to the reader for suggestions and comments.[15]

The objects of conflict for various wars are not necessarily the same as the causes of the wars, though they are similar. Our research is still far from creating a comprehensive typology or even theory of the causes of war, not to speak of an "etiology" of war. We must therefore proceed pragmatically. If the aggressing side states its goals and if their statements are credible then we have recorded them accordingly. If such a statement is absent or is not credible then we derive the goals or the content of the conflict from societal circumstances. It is not possible to exclude errors from this process, despite all efforts at diligence and carefully weighed judgments. It should be noted that our admittedly preliminary breakdown of conflict types contains more ordered information than other war lists.

Below are the designations and explanations of the types of conflicts (conflict groups):

I. Systemic Conflicts

The subject of this form of conflict is the whole societal system, the social order of the country, or a least such an essential part of the social order (e.g., the political-administrative, social and economical, ideological or cultural system) that its transformation entails substantial consequences for the whole society. The following types of conflict belong to this group:

1. *System creation*: the issue here is the fundamental structural transformation (revolution) of the total societal system, or, conversely, its defense.
2. *System reform*: the object of this conflict is not fundamental restructuring of the total societal system but rather the essential improvement of the whole or an important part in the framework of the given social order, or, conversely, the prevention of such reforms from being carried out. For many Third World countries the question of agrarian reform is a characteristic example.
3. *System restoration*: this type of conflict concerns itself with the reversal of already instituted or initiated reforms which aim toward the fundamental reorganization of society (counter-revolution, reaction). Whether the condition to be restored was "conservative" or "progressive" is irrelevant.

II. Power Conflicts

This term is insufficient, yet we have not found a better one. First, "power" is defined in myriad ways in scientific literature, and a universal understanding of this concept does not exist. Second, power struggles arise in conflicts other than the kind discussed here. Power is necessary to realize a revolution, institute reform, gain independence for a part of a state, overthrow colonial rule, or, conversely, to prevent any of the above.

Nonetheless we want to regard as "power conflicts" those conflicts in which a certain government or ruling class is targeted to be replaced and the respective incumbents of power are thus provoked into defending themselves. Admittedly, revolutions and counter-revolutions also overthrow their respective governments but we classify these cases as systemic conflict. A straightforward power conflict occurs only when government personnel or certain elites are exchanged without simultaneously placing the entire system in question. Note, however, that a coup d'état does not in itself constitute a war (see section 2.1).

We also regard as power conflicts those conflicts where relatively clear, limited, and specialized interests are advanced or prevented and which possess little or no relevance for systemic transformation.

Power conflicts encompass the following types of conflict:

1. *Conflicts of governance*: coup d'état; exclusion of an opposition group or party; rivalries between different factions or personalities within a ruling group or class; ideological conflicts concerning the "right" religion or the

"right" strategy for the realization of certain goals, so long as these conflicts do not aim to fundamentally alter the societal system; attempts to force the recognition of a particular ruler's legitimate claim to govern, and so forth.

2. *Diversionary conflicts*: A government or ruling group attempts to save its own position which is threatened by internal political or sociopolitical tensions by portraying a national minority or another state as a threat regardless of the actual capabilities or intentions of that group. The government or ruling group calls for "defense" or even aggressive action against the "enemy" in order to distract from internal tensions and to rally oppositional elements within the population to their side.
3. *Particularistic conflicts*: A particular, usually economic interest is advanced for control over a region with raw materials; the use of certain transportation routes; the control of key positions in the state or administrative bureaucracy through corruption or graft (not always clearly distinguishable from conflicts of governance); blockade or reversal of refugee influxes from neighboring countries when the refugees are perceived, realistically or not, to negatively impact the economic and social situation, etc.
4. *Tribal or ethnopolitical conflicts*: Conflicts in which two or more ethnic or tribal[16] groups or their leaders struggle for priority or certain advantages in the political-administrative system or feud with each other over ideological or cultural prejudices. It is difficult to distinguish clearly between ruling or particularistic interests, especially since tribal conflicts, like religious ones, stem primarily from material and social structural problems. One cannot deny that so-called ethnic or tribal issues play a role in many kinds of conflicts. The question is, however, whether the ethnic component is really the cause of a conflict and not merely a pretext invoked to mask socioeconomic-political divergence, whether historically rooted or a product of a colonial past. Our impression is that ethnic or tribal identity is overwhelmingly abused by elite groups to mobilize support within the population by appealing to prejudices and enemy images, especially when no other effective means for social organization exists (e.g., parties or unions), thereby securing the elite groups' own privileged position. Our previous studies suggest that the frequent designation of conflicts as ethnic or religious relies all too often on superficial judgments. Unlike Moser (1983), for example, and more in line with Eikenberg (1987), Waldman (1989) and Tetzlaf (1991) we tend to be parsimonious in our classification of conflicts as ethnic or tribal.[17] We only proceed thus when these aspects have, in our opinion, an extraordinary significance (e.g., for the conflict between Tutsis and Hutus in Rwanda and Burundi). Even the few conflicts remaining in this category could probably be classified differently upon closer examination. Conceivably this could lead to the eventual removal of the category of "tribal and ethnic conflicts."
5. *Military-strategic conflicts*: Actions or goals which have a recognizable and limited military purpose: preemptive actions against threatening measures taken by an opponent (e.g., against the concentration of troops); con-

trol over a strategically important territory (such as the high ground of the Siachen Glacier between Pakistan and India); a security strategy in order to avoid being drawn into a conflict as a third party (such as the successful 1970-71 attack in Jordan against the PLO by the Jordanian monarchy); attacks against a government involved in a war with a third party in order to occupy its troops to relieve the third party (e.g., the 1979 attack against Vietnam by the People's Republic of China while Vietnam was in the middle of occupying and controlling neighboring Cambodia); and similar military-strategic conflicts, although such conflicts are naturally never divorced from political ambitions.

III. Territorial Conflicts

State governance is contractual rule over a delimited territory which is accorded external reciprocal recognition and respect by other states. This concept of the territorial or national "state" is, on the one hand, a receding model as the world undergoes a process of greater integration and socialization (see Diner 1985). On the other hand, in most parts of the world territorial rule continues to play a significant role for the foreseeable future, since temporally unequal stages and rates of social development can only be expressed territorially, that is, through *space-time*.

Borders are nonetheless often disputed, for example between the People's Republic of China and Taiwan, the People's Republic of China and India, between Somalia and Ethiopia (Ogaden), Iraq and Kuwait, or between Ecuador and Peru. Correspondingly, groups lay claim to territory which lies in a neighboring state, or a certain local population (often a disadvantaged national minority) strives for greater self-sufficiency or even secession from their respective state.

All conflicts in which territorial demarcation plays a primary role are included in this group, and are broken down into the following types:

1. *Expansion*: All goals and measures which aim to expand a territory through conquest and annexation of an area or target areas belonging to another society, regardless whether the latter is a formal state or not or is still in the process of state formation.
2. *Territorial restoration*: The restoration of an earlier situation, usually the unification of separated state territories which formerly belonged together.
3. *Territorial revision*: Disputes about state borders, due to historical ambiguity in the process of state formation, a treaty, or the physical roughness of the border area. The dispute could also the result from arbitrary borders created by former colonial powers or by coerced border "treaties," etc.

Conflicts which involve territorial revision through the annexation of a state or part of a state (e.g., Kashmir) belong here as well.

4. *Autonomy/Secession*: A population (or several population groups, though usually a national minority) in one part of a state endeavors to gain greater self-sufficiency (internal autonomy), political-administrative goals (self-administration), economic goals (receiving a greater share of resources or profit), social goals (prevention of racist or similar forms of discrimination) and/or cultural goals (preservation of local traditions or religious independence) in opposition to the central ruling power. Secessionist or separatist intentions go beyond this: the area in question seeks to abandon its current state to either form its own sovereign state or to annex itself to a neighbor. A significant difference between "autonomy/secession" conflicts and the previous types of territorial conflict is that the initiative for territorial revisions comes not from the central ruling power but from the ruled in their respective part of the state, regardless of how politically manipulative such an initiative may in fact be. In this sense such conflicts are often analytically difficult to distinguish from "power conflicts."

IV. Foreign Rule

This term is used to describe two types of conflict:

1. *Colonial rule*: Conflicts between a colonial power and those who wish to overthrow their direct rule in favor of constituting their own sovereign state (decolonization, national liberation). Note, however, that many resistance movements in the Third World call themselves "national liberation fronts" or something similar even when their country is no longer a colony. What is meant by this is the liberation from rule within their own country which is perceived as illegitimate, or liberation from foreign, usually capital-rich, exploitative influences. Such conflicts do not belong to the "colonial rule" type of conflict.
2. *Influence/Hegemony*: Conflicts stemming from the introduction, maintenance or expansion of *indirect* rule by one state or a group of states over another state or the states of an entire region. Often this entails an alliance of interests with the indigenous elite who either profit or are promised profit from hegemonial rule.

In many wars two or more of the fourteen types described above occur simultaneously or consecutively. Correspondingly, a systemic or power conflict can be the subject of both an interstate and an anti-regime war. For this reason the breakdown of wars according to types of conflict does not correspond directly with the formal categorization of war types. Certain patterns, however, can be reasonably expected: "Systemic conflicts" are generally A-wars, "power conflicts" tend to be A-

and C-wars, "territorial conflicts" fall into B- and C-war categories, and "conflicts about foreign powers" fall above all into the D-war category, although they can also appear in A- and C-wars.

2.7 War Termination and War Results

We have also recorded how the wars through December 31, 1992 have ended:[18] by the military victory of one side over the other, by an agreement between the warring parties, whether via the mediation of a third party or without such help, or simply by the cessation of hostilities. When the wars are terminated through mediation we note to whom the honor is due.

Every war in the register is followed by remarks concerning its outcome. This information, however, is not systematically ordered and was not devised according to unitary criteria, since we have not yet found an acceptable method of doing so.[19]

Notes

1. With the descriptive term "national" we adopt the loose meaning with which US social science, particularly international relations, refers to the terms *nations, national*, etc. in light of the social-scientific modernization process of the 50s to the 70s. In no sense do we imply with this usage a national homogeneity and integration, a special social or cultural peculiarity of the respective society, or any nationalist aspects.

2. Cf. the theory of state in Thomas Hobbes' *Leviathan*. Elias (1980, volume 2) uses the prototypical case of France to portray how the state monopoly of violence gradually emerged in Europe and to illustrate the respective roles of the ruling interests and social processes involved in the transformation.

3. It would be nonetheless absurd to maintain that private violence disappears where the state appears, bringing with it one hundred percent security. The current situations in Algeria or Colombia are testimony to this.

4. Our definatory basis in this sense optimistic. For a related model of "civilization" see Senghaas (1994:17-49). The opposite (an anarchical) position is to be found in Krippendorff (1985); following his (in our view ahistorical) path through history he sees the fundamental causes of war precisely in the existence of the state and the military.

5. We can live with the oft-heard reproach that we are Eurocentric. This accusation contains a curiously narrow view of history, an arrogant view of humanity, and it immunizes one against political criticism. We also reject the critique aimed at comparative war studies, which claims that we ignore the specific cultural difference of the societies concerned. Regarding the first objection: it is historical fact, though one may lament it, that the Europeans conquered the rest of the world and foisted the law of development upon them, be it through war, colonialism, imperialism or free trade. For this reason it is justified to analytically (!) view European-style societal development as the standard for global development. Regarding the second objection: this reminds us fatally of the all-to-often invoked saying that one cannot compare apples and oranges. But one can: both are fruit, indeed both grow on trees and have seeds. Indeed it is the initial comparison which makes the unique differences visible.

6. An exile government does not count here.

7. They classify a violent confrontation *between states* as war only when at least a *total* of 1,000 *soldiers* have been killed. If an internationally recognized state fights against a society without state sovereignty (e.g., a colony), then they call this "extra-systemic war" and raise the number of dead to at least 1,000 military casualties *per year* and *only on the side of the so-called state. Intrastate* warfare only counts as a civil war when at least 1,000 people *per year* are killed, regardless of whether they are military personnel or civilians only of the state in question. All military confrontations below these thresholds are classified by Small and Singer as *military confrontations* or *serious disputes*, and are regarded at best as occurrences of a prewar phase.

8. For an example of such a distortion we turn once more to Small and Singer (1982:233): had they fixed the number of casualties at which point a military confrontation becomes a war at 100 instead of 1,000 then the frequency ratio of intra-state to inter-state wars in their time-frame of 1816-1980 would not be approximately 1:1, but rather 2:1!

9. Citha D. Maass (1989:149-157) had welcome comments concerning our classifications using the example of the 1971 war in Bangladesh. He recommends that both characteristics be used (153): the "aggressor" should refer to that war party which formally began the hostilities, and "initiator" should refer to the side which provoked the outbreak of war through prior political posturing or through military mobilization. It would greatly improve our war register if we could incorporate this suggestion. For the moment we unfortunately lack the personnel resources to differentiate all wars since 1945 in this manner.

10. Previously in Gantzel/Meyer-Stamer we differentiated solely between North and Black Africa.

11. The term "Southern Africa" refers to the region itself, in contrast to "South Africa," which is the colloquial term for the Republic of South Africa (RSA, formerly the South African Union).

12. However, we allocate the wars between the People's Republic of China against India and Vietnam to the South and Southeast region.

13. Cf. the Help List ordered by type in the Register.

14. This is perhaps the reason why, despite the widespread reception and use of the war data in our earlier book (Gantzel/Meyer-Stamer 1986), the breakdown according to types of conflicts remained unnoticed.

15. An attempt to group the conflicts somewhat differently and classify them more concretely can be found in Gantzel (1994:145f., 148-150). Unfortunately this attempt came about after the publication deadline for this book and could not be used here. An additional issue that needs to be considered is whether environmental conflicts present their own class of conflict. Since 1991 the Environment and Conflict Project Team (ENCOP) has dealt with the problem of the environment as a cause of violent conflicts. Funded by the Swiss Peace Foundation, the team consists of scientists in Bern, Zürich and Hamburg, plus colleagues in Bangladesh, Nigeria and Sudan. Out of the 35 to 40 cases of environmental conflict selected for comparative evaluation by this project team 21 cases so far have reached the stage of war. For further information please contact in Hamburg: Volker Böge c/o Research Center for War, Armament and Development, Institute for Political Science, University of Hamburg, Allende-Platz 1, 20146 Hamburg, telephone (40) 4123-6194, fax (40) 4123-2460.

16. The term *tribe* was introduced by the English and has a specific colonial function (division of the conquered). Despite this ideological content we retain the concept (indeed, there are many words which one would otherwise have to eliminate from the language because they are historically burdened) and we do not impute any discriminatory meaning to our usage.

17. On methodological, conceptual and ideological objections to the concept of "ethnic war" or "ethnic conflict" see, among others, Schlichte (1994), Singer (1994:9f., 16) and Siegelberg (1994b).

18. In a far more differentiated way than in our 1986 book.

19. The disputed and accordingly questionable classification from our earlier book is not repeated here.

3

Reference List of Wars, 1945-1992

In this chapter we compile a tabular overview of the war list based on the definitions and inclusion criteria of the preceding chapters and on the information contained in the war register. Presenting the data in this tabular form makes it more adaptable to the reader's specific purposes. The following description of the column headings are meant to facilitate a quick orientation. Only those abbreviations that appear in the list are repeated here.

The list is ordered according to the regions described in the preceding chapter. It is made up of sub-lists corresponding to the regions Central America/Caribbean, South America, Europe, North Africa, West Africa, Central Africa, East Africa, Southern Africa, Near and Middle East, South and Southeast Asia (including Oceana) and East Asia.

- Column 1*:* "Nr." = Number of the war.
  The war numbers are merely for identification. With few exceptions they are identical with those in Gantzel/Meyer-Stamer (1986). For practical reasons, however, they no longer follow the chronological sequence of the wars.
- Column 2*:* "Name/Country" = Country or countries, and/or the name of the war, when applicable.
  The country names follow colloquial usage. In cases of internal wars the local region may also be noted in parentheses, e.g., Spain (Basques) or Iraq (Kurds). For interstate wars the aggressing country is named first, with the name of the aggressed country following the forward slash. Certain wars have come to be known by their own names, e.g., "the Mau-Mau Rebellion," "the Soccer War," "Chimurenga," "Shifta War," or "(Second) Gulf War." When appropriate these "nicknames" will also be noted. Only those

wars which meet our definition are included here; undecided cases are left out.

- Column 3: "Time frame"
  The first war since the Second World War began on December 3, 1944 (the First Greek Civil War). In order to avoid statistical distortions we consider the war as beginning in January 1945. With this minor yet justifiable manipulation the timeframe of the list encompasses 48 years from January 1, 1945 to December 31, 1992.
  Insofar as the exact starting and ending dates are known for a war, we present them according to day, month and year. If only the month and the year are known, the month is separated from the year by a forward slash. If it is not possible to determine the month, then only the year is given.

- Column 4: "War Type"

A =Anti-regime War	}	
B = Other Internal War	}	-1 = with foreign participation
C = Interstate War	}	-2 = without foreign participation
D= Decolonialization War	}	

  AB-1, BC-2, ABC-2 etc. = Mixed Types
- Column 5: If it is not possible to unambiguously determine the aggressor, then the warring parties are presented simply as "Side A" and "Side B."
- Column 6: "Object of Conflict" or, respectively, the goal of the attacking side.

This designation follows the classification schema in section 2.6 of the preceding chapter.

Central America and the Caribbean

Nr.	Country	Time Period	War Type	Warring Parties	Object of Conflict	Outcome
014	Costa Rica	12.3.1948 - 19.4.1948	A-2	Aggressor (Side A): Legión del Caribe (Caribbean Legion) Aggressed (Side B): Costa Rica	System reform (social democratic power-grab for the purpose of economic structural reform, elimination of the communist worker's movement and the corrupt liberal elite; breaking elite control over agrarian exports).	Third-Party Mediation (Diplomatic Corps)
016	Costa Rica (Coup attempt)	10.12.1948 - 24.12.1948	A-2	Aggressor (Side A): Exiled Costaricans Aggressed (Side B): Costa Rica	System reform (overthrow of the social democratic yet quite repressive "revolutionary Junta") and power conflict (Herrschaftskonflikt) (securing of the Somoza regime in Nicaragua by destroying the anti- Somaza Exile Brigade for the "Liberation of Nicaragua" in Costa Rica).	Third-Party Mediation (OAS)
031	Guatemala (Operation Success)	18.6.1954 - 1.7.1954	A-1	Aggressor (Side A): Exiled Guatemalans and Mercenaries under US command Aggressed (Side B): Guatemala Intervention favoring side A: USA	System reform (prevention of socio-political reform and Guatemalan sovereignty).	Military victory Side A
037	Costa Rica (2nd Coup attempt)	11.1.1955 - 21.1.1955	A-1	Aggressor (Side A): Exiled Costaricans Aggressed (Side B): Costa Rica Intervention favoring side A: Nicaragua	System reform (overthrow of the elected social-democratic government of Costa Rica) and power conflict (Herrschaftskonflikt) (Securing of the Somoza Regime in Nicaragua).	Military victory Side B
045	Cuban Revolution	12/1956 - 1/1959	A-2	Aggressor (Side A): Ejército Rebelde (Movement of July 26) Aggressed (Side B): Cuba	System transformation (overthrow of the Batista dictatorship).	Military victory Side A

Central America and the Caribbean

Nr.	Name/Country	Time Period	War Type	Warring Parties	Object of Conflict	Outcome
058	Cuba (Bay of Pigs invasion)	17.4.1961 - 20.4.1961	A-2	Aggressor (Side A): Exiled Cubans and mercenaries from other countries Aggressed (Side B): Cuba	System restoration (overthrow of Castro's socialist government) and hegemonic conflict (securing U.S. influence in the Caribbean, preventing the expansion of leftist revolutionary movements in Latin America).	Military victory Side B
056	Guatemala	1962 - 1968	A-2	Aggressor (Side A): Fuerzas Armadas Rebeldes (FAR) Aggressed (Side B): Guatemala	System reform (overthrow of the military dictatorship and re-establishment of democracy).	Military victory Side B
083	Dominican Republic	25.4.1965 - 3.9.1965	A-1	Aggressor (Side A): Conservative Military Aggressed (Side B): Dominican Republic Intervention favoring side A: USA Intervention to end the war: Honduras / El Salvador / Nicaragua / Costa Rica / Brazil / Paraguay	Power conflict (Herrschaftskonflikt) (Successor of the Cabral Junta).	Military victory Side A
100	El Salvador / Honduras ("Soccer War")	14.7.1969 - 18.7.1969	C-2	Aggressor (Side A): El Salvador Aggressed (Side B): Honduras	Diversionary conflict (from internal tensions set against a background of socioeconomic crises in both countries).	Third-Party Mediation (OAS)
129	Nicaraguan Revolution	10/1977 - 19.7.1979	A-1	Aggressor (Side A): Frente Sandinista de Liberación Nacional (FSLN) Aggressed (Side B): Nicaragua Intervention favoring side B: Honduras	System reform and transformation (overthrow of the Somoza dictatorship and democratization / revolutionary Sandinista movement for socialist transformation).	Military victory Side A

Central America and the Caribbean

Nr.	Name/Country	Time Period	War Type	Warring Parties	Object of Conflict	Outcome
146	Guatemala	since 1980	A-2	Aggressor (Side A): Fuerzas Armadas Rebeldes (FAR) / Organización del Pueblo en Armas (ORPA) / Ejército Guerrillero de los Pobres (EGP) / Partido Guatemalteco del Trabajo (PGT) Aggressed (Side B): Guatemala	System reform (expansion of social and political participation).	Ongoing
140	El Salvador	10.1.1981 - 1.2.1992	A-2	Aggressor (Side A): Frente Farabundo Martí para la Liberacíon Nacional (FMLN) Aggressed (Side B): El Salvador	System reform (democratization of the society).	Third-Party Mediation (UN)
151	Nicaragua (Contra War)	4/1981 - 1990	A-1	Aggressor (Side A): Contras Aggressed (Side B): Nicaragua Intervention favoring side A: Honduras	System restoration (overthrow of the leftist Sandinista revolutionary government, counter-revolution) and autonomy (for the Miskito Indians).	Third-Party Mediation (regional state(s))
156	Grenada-Invasion (Operation Urgent Fury)	24.10.1983 - 1.11.1983	C-2	Aggressor (Side A): USA Aggressed (Side B): Grenada	hegemonic conflict (securing of US hegemony in the Caribbean; overthrow of the marxist-oriented government of Grenada and deterrent demonstration of US superiority against leftist governments and movements in Latin America, especially Nicaragua).	Military victory Side A
170	USA / Panama	20.12.1989 - 24.12.1989	C-2	Aggressor (Side A): USA Aggressed (Side B): Panama	hegemonic conflict (securing of US hegemony in the Panama Canal).	Military victory Side A

South America

	Name/Country	Time Period	War Type	Warring Parties	Object of Conflict	Outcome
008	Bolivian Revolution	1946 - 4/1952	A-2	Aggressor (Side A): Movimiento Nacionalista Revolucionario (MNR) Aggressed (Side B): Bolivia	System transformation (social-revolutionary system transformation against traditional oligarchy and US dominance in the economic sector).	Military victory Side A
011	Paraguay	3/1947 - 12.8.1947	A-2	Side A: Paraguay Side B: Coalition of governmental opponents	System reform (overthrow of the authoritarian government by a heterogenous alliance consisting of Febreristen, Liberals and Communists).	Military victory Side A
019	Colombia (Violencia)	21.5.1949 - 10.5.1957	A-2	Side A: Colombia Side B: Liberal guerrilla groups / Communist guerilla and Campesino guerillas	System reform (first phase (1949-1953): liberal and communist guerrillas against increasing militarization by the conservative government; second phase (1953-1957): communist guerrilla for agrarian reform and redistribution of property as well as better work and lease conditions, above all in the coffee sector).	Fighting stopped
053	Venezuela	2/1963 - 1967	A-2	Aggressor (Side A): Fuerzas Armadas de Liberación Nacional (FALN) Aggressed (Side B): Venezuela	System reform (social reform transformation of the formally democratic system).	Military victory Side B
060	Colombia (FARC)	since 27.5.1964	A-2	Aggressor (Side A): Colombia Aggressed (Side B): Fuerzas Armadas Revolucionarias de Colombia (FARC) Side B: Ejército Popular de Liberación (EPL)	System reform (agrarian reform, since 1982 also political reforms).	Ongoing
079	Colombia (ELN)	since 7.1.1965	A-2	Aggressor (Side A): Ejército Popular de Liberación (EPL) / Ejército de Liberación Nacional (ELN) Aggressed (Side B): Colombia	System transformation (following the model of the Cuban Revolution).	Ongoing

South America

Nr.	Name/Country	Time Period	War Type	Warring Parties	Object of Conflict	Outcome
086	Peru (MIR and ELN guerilla revolts of 1965)	6/1965 - 1/1966	A-2	Aggressor (Side A): Movimiento de Izquierda Revolucionaria (MIR) / Ejército de Liberación Nacional (ELN) Aggressed (Side B): Peru	System transformation (following the model of the Cuban Revolution).	Military victory Side B
104	Argentina	1968 - 1977	A-2	Aggressor (Side A): Montoneros / Ejército Revolucionario del Pueblo (ERP) Aggressed (Side B): Argentina	System reform (overthrow of the military government) and system transformation (Marxist guerrillas); also power conflict (Herrschaftskonflikt) (left-peronist guerrillas in favor of a new coup by Peron).	Military victory Side B
137	Colombia (M-19)	1/1974 - 8.3.1990	A-2	Side A: Colombia Side B: Movimiento 19 de abril (M-19)	System reform (especially expansion of political and economic participation).	Agreement without mediation
147	Peru (Sendero Luminoso)	since 17.5.1980	A-2	Aggressor (Side A): Sendero Luminoso (Shining Path) Aggressed (Side B): Peru	System transformation (construction of a state following Maoist theory).	Ongoing
149	Peru / Ecuador	28.1.1981 - 2.2.1981	C-2	Aggressor (Side A): Peru Aggressed (Side B): Ecuador	Territorial revision (border conflict) and particularist conflict (control of resources).	Third-Party Mediation (OAS)
154	Falkland/ Malvinas War (Argentina / Great Britain)	25.4.1982 - 15.6.1982	C-2	Aggressor (Side A): Argentina Aggressed (Side B): Great Britain	Territorial revision (regarding the Falkland-/Malvinas islands—ruled by Great Britain and claimed by Argentina) and diversionary conflict (from internal crises).	Military victory Side B

South America

Nr.	Name/Country	Time Period	War Type	Warring Parties	Object of Conflict	Outcome
163	Surinam (Jungle War)	21.6.1986 - 7.6.1989	B-2	Aggressor (Side A): Jungle Command of the Maroons/(Surinamese National Liberation Army (SNLA)) Aggressed (Side B): Surinam	Autonomy (for the Maroons) and System reform (overthrow of the military government).	Agreement without mediation
204	Peru (MRTA)	since 11/1987	A-2	Aggressor (Side A): Movimiento Revolucionario Tupac Amaru (MRTA) Aggressed (Side B): Peru	System reform (nationalization and democratization).	Ongoing

Europe

Nr.	Name/Country	Time Period	War Type	Warring Parties	Object of Conflict	Outcome
001	Greek Civil War (1stPhase)	3.12.1944 - 12.2.1945	A-1	Side A: Ethnikon Apeleftherotikon Metopou / Ethnikos Laikos Apeleftherotikos Stratos (EAM/ELAS) Side B: Greece Intervention favoring side B: Great Britain	System transformation (socialist revolution).	Third-Party Mediation (third-party states)
004	Spain 1945-1950	1945 - 1950	A-2	Aggressor (Side A): Remains of the Republican Army Aggressed (Side B): Spain	System transformation (socialist revolution).	Military victory Side B
006	Greek Civil War (2nd Phase)	6/1946 - 9.10.1949	A-1	Aggressor (Side A): Dimokratikos Stratos Ellados (DSE) Aggressed (Side B): Greece Intervention favoring side B: Great Britain	System transformation (socialist revolution).	Military victory Side B

Europe

Nr.	Name/Country	Time Period	War Type	Warring Parties	Object of Conflict	Outcome
038	Cyprus I	1.4.1955 - 24.2.1959	DB-2	Aggressor (Side A): Ethniki Organosis Kiprion Agoniston (EOKA; National Union of Cypriot Fighters) Aggressed (Side B): Great Britain / Vulkan /Türk Mukavement Teskilati (TMT)	Decolonization (independence from Great Britain) and Territorial revision (Greek Cypriots for annexation to Greece, Turks for annexation to Turkey).	Third-Party Mediation (neighboring state[s])
043	Hungarian Revolt	24.10.1956 - 10.11.1956	A-1	Aggressor (Side A): Heterogeneous opposition and parts of the Hungarian Army Aggressed (Side B): Hungary Intervention favoring side B: USSR	System reform (Anti-Stalinism, political and economic liberalization, in part also reactionary counter-revolution; dissatisfaction with the low standard of living) and hegemonic conflict (elimination vs. maintenance of Soviet dominance).	Military victory Side B
073	Cyprus II	21.12.1963 - 11.8.1964	B-2	Side A: Greece / Cyprus / Ethniki Organosis Kiprion Agoniston (EOKA; National Union of Cypriot Fighters) Side B: Türk Mukavement Teskilati (TMT) / Turkey	Territorial revision (attempt by the Greek majority to complete the annexation to Greece) and secession (attempt by the Turkish minority to divide the island).	Third-Party Mediation (U.N.)
106	Spain (Basque Country)	7.6.1968 - 1979	AB-2 / B-2	Aggressor (Side A): Euskadi ta Askatasuna (ETA; Basque Country and Freedom) Aggressed (Side B): Spain	Secession (construction of independent Basque state).	Fighting below the level of war
105	Northern Ireland	since 14.8.1969	B-2	Side A: Great Britain / Ulster Defence Association (UDA)/Ulster Freedom Fighters (UFF) / Ulster Volunteer Force (UVF)/Protestant Action Force (PAF) Side B: Irish Republican Army/Provisionals (IRA) / Irish National Liberation Army (INLA)/Irish Republican Socialist Party (IRSP)	System reform (favoring the poorer, politically and socially discriminated against Catholic population) and secession (united Ireland with socialist orientation).	Ongoing

Europe

Nr.	Name/Country	Time Period	War Type	Warring Parties	Object of Conflict	Outcome
116	Cyprus III	20.7.1974 - 16.8.1974	BC-1	Aggressor (Side A): Turkey Aggressed (Side B): Cyprus Side B: Ethniki Organosis Kiprion Agoniston-B (EOKA-B; National Union of Cypriot Fighters) Intervention favoring side B: Greece	Secession (one part of Cyprus favoring the creation of the Turkish-Cypriot minority as independent state against Greek-Cypriot dominance) and territorial revision (prevention of annexation to Greece planned by the Greek military government and Greek-Cypriot nationalists).	Third-Party Mediation (U.N.)
169	Romania	17.12.1989 - 28.12.1989	A-2	Aggressor (Side A): Opposition factions of the Romanian army Aggressed (Side B): Romania	System reform (overthrow of the Ceaucescu regime).	Military victory Side A
200	Yugoslavia (Slovenia, Croatia, Bosnia)	since 26.6.1991	BC-2	Aggressor (Side A): Yugoslavia Aggressed (Side B): Slovenia / Croatia / Bosnia Side A: Serbian militias and Chechniks Side B: Croatian militias in Bosnia-Hercegovina	Secession (non-recognition of the secession and sovereignty of Slovenia, Croatia, and Bosnia by the Serbian leadership) and expansion / territorial revision (varying territorial claims).	Ongoing
205	Moldavia (Dneistr Republic)	3/1992 - 8/1992	B-1	Side A: Moldavia Side B: Armed forces of the Dneistr Republic and Cossacks Often Changing: "Third Power" Intervention to end the war: Russia	Secession (of the Russian-language minority).	Third-Party Mediation (neighboring state[s])

North Africa

Nr.	Name/Country	Time Period	War Type	Warring Parties	Object of Conflict	Outcome
030	Morocco	7.12.1952 - 2.3.1956	D-2	Aggressor (Side A): Istiqlal Aggressed (Side B): France first B, then A: Berber tribes	Decolonization (independence from France).	Agreement without mediation
035	Algeria	1.11.1954 - 19.3.1962	D-2	Aggressor (Side A): Front de Libération Nationale (FLN) / Mouvement National Algérien (MNA) Aggressed (Side B): France Side B: Organisation de l'Armée Secrète (OAS)	Decolonization (independence from France).	Agreement without mediation
040	Sudan (1st Sudan Civil War)	18.8.1955 - 2/1972	B-2	Aggressor (Side A): Anya-Nya Aggressed (Side B): Sudan	Autonomy/secession (of the disadvantaged, Christian and black African southern Sudan from the Arabic and Muslim dominated central government).	Third-Party Mediation (various)
047	Spanish Morocco (Ifni)	10/1957 - 4/1958	D-2	Aggressor (Side A): Ejército de Liberación (EL) Aggressed (Side B): Spain	Decolonization (independence from Spain).	Third-Party Mediation (former colonial power)
057	Tunisia (Bizerta Crisis)	19.7.1961 - 22.7.1961	D-2	Aggressor (Side A): Tunisia Aggressed (Side B): France	Decolonization (independence from France).	Third-Party Mediation (U.N.)
070	Morocco / Algeria ("Tindouf War")	10/1963 - 2/1964	C-2	Aggressor (Side A): Morocco Aggressed (Side B): Algeria	Territorial revision (border conflict) and particularist conflict (control of resources through annexion of the West Algerian Sahara by Morocco).	Third-Party Mediation (regional state[s])

North Africa

Nr.	Name/Country	Time Period	War Type	Warring Parties	Object of Conflict	Outcome
123	Western Sahara	18.11.1975 - 6.9.1991	D-2 / B-1	Aggressor (Side A): Frente Popular de Liberación de Seguia el Hamra y Rio de Oro (POLISARIO) Aggressed (Side B): Spain / Side B: Morocco / Mauritania Intervention favoring side A: Algeria	Decolonization (independence from Spain), followed by territorial expansion (annexion of the former Spanish colony by Morocco and Mauritania as well as against Moroccan claims to the whole of West Sahara).	Third-Party Mediation (U.N.)
158	Sudan (Southern Sudan Conflict, 2nd Sudan Civil War)	since 9/1983	BA-2	Aggressor (Side A): Sudanese People's Liberation Movement / Army (SPLM bzw. SPLA; since August 1991 SPLA-Garang Faction) Aggressed (Side B): Sudan Side B: Sudanese People's Liberation Army (SPLA-Macarfraktion)	Secession / autonomy (against discrimination by Southern Sudan and the Islamification policies of the central government) and particularist conflict (control of resources).	Ongoing
206	Algeria	since 1/1992	A-2	Side A: Algeria Side B: Mouvement islamique armé (MIA) / Groupe Islamique Armé (GIA)	Power conflict (Herrschaftskonflikt) (prevention of the Islamic opposition from taking over the government and changing the constitution).	Ongoing

West Africa

Nr.	Name/Country	Time Period	War Type	Warring Parties	Object of Conflict	Outcome
077	Guinea-Bissau (War of Independence)	1/1963 - 25.4.1974	D-2	Aggressor (Side A): Partido Africano da Independência da Guiné e Cabo Verde (PAIGC) / Frente da tutta para la Independencia Nacional da Guiné (FLING) Aggressed (Side B): Portugal	Decolonization (independence from Portugal).	Fighting stopped
184	Nigeria (Tiv-revolt 1964)	8/1964 - 12/1964	AB-2	Aggressor (Side A): Armed Tiv Militias Aggressed (Side B): Nigeria	Power conflict (Herrschaftskonflikt) (revolt of the Tiv population against Haussa-Fulani dominance and the Northern People's Congress in northern Nigeria).	Fighting stopped

West Africa

Nr.	Name/Country	Time Period	War Type	Warring Parties	Object of Conflict	Outcome
097	Nigeria (Biafra War)	6.7.1967 - 15.1.1970	B-2	Aggressor (Side A): Nigeria Aggressed (Side B): Biafra	Secession (the Christian, Igbo-dominated, oil-rich province of Biafra against the central government, controlled mostly by Muslim northern Nigerians).	Military victory Side A
185	Nigeria (Maitatsine Revolt in Kano/North Nigeria)	18.12.1980 - 29.12.1980	B-2	Aggressor (Side A): Maitatsine Movement Aggressed (Side B): Nigeria	System reform (against povery and Western materialism).	Military victory Side B
180	Mali / Burkina Faso (border war)	21.12.1985 - 31.12.1985	C-2	Aggressor (Side A): Mali Aggressed (Side B): Burkina Faso	Territorial revision (border conflict).	Third-Party Mediation (various)
168	Liberia, Sierra Leone	since 24.12.1989	A-1	Aggressor (Side A): National Patriotic Front of Liberia (NPFL) / Independent National Patriotic Front of Liberia (INPFL) / United Liberation Movement for Democracy (ULIMO) Aggressed (Side B): Liberia / Sierra Leone Intervention to end the war: Gambia / Mali / Senegal / Guinea / Sierra Leone / Ghana / Nigeria	System reform and power conflict (Herrschaftskonflikt) (overthrow of the Doe regime, followed by a power-struggle between the different organizations).	Ongoing
191	Senegal (Basse-Casa-mance Conflict)	since 4/1990	B-2	Aggressor (Side A): Mouvement des Forces Démocratiques de la Casamance (MFDC) Aggressed (Side B): Senegal	Autonomy (der Basse-Casamance) and particularist conflict (control of resources).	Ongoing

West Africa

Nr.	Name/Country	Time Period	War Type	Warring Parties	Object of Conflict	Outcome
197	Tuareg (Mali, Niger)	since 7.5.1990	B-2	Aggressor (Side A): Mouvement Populaire de l'Azawad (MPA) / Front Islamique Arabe de l'Azawad (FIAA) / Front Populaire pour la Libération de l'Azawad (FPLA) / Alliance Révolutionnaire pour la Libération de l'Azawad (ARLA) Aggressed (Side B): Mali / Niger	Autonomy (for the Tuareg).	Ongoing

Central Africa

Nr.	Name/Country	Time Period	War Type	Warring Parties	Object of Conflict	Outcome
039	Cameroon (De-colonization)	5/1955 - 1963	DA-2	Aggressor (Side A): Union des Populations du Cameroun (UPC) Aggressed (Side B): France / Cameroon	Decolonization (independence from France) and system transformation (social revolution).	Military victory Side B
055	Zaïre (Secession and Anti-regime war)	11.7.1960 - 14.1.1963	AB-1	Aggressor (Side A): Confédération des Associations Tribales de Katanga (CONAKAT) / Mouvement National Congolais/Lumumba (MNC/L) / Mouvement National Congolais/ Kalondji (MNC/K), Kasai Secessionists Aggressed (Side B): Zaïre Intervention favoring side A: Belgium Intervention to end the war: Ethiopia / Ghana / India / Ireland / Italy / Canada / Morocco / Sweden / Sudan / Tunisia /	Secession (of the Katanga (Shaba) and Kasai provinces, which are rich in raw materials) and power conflict (Herrschaftskonflikt) (revolt by supporters of the disempowered head of government Lumumba against the central government).	Military victory Side B

Central Africa

Nr.	Name/Country	Time Period	War Type	Warring Parties	Object of Conflict	Outcome
072	Rwanda (Tutsi Invasion via Burundi, Uganda and Zaïre)	20.12.1963 - 1966	A-2	Aggressor (Side A): Inyenzi Groups of the Rwandese Tutsi Aggressed (Side B): Rwanda	Power conflict (Herrschaftskonflikt) (re-establishment of the Tutsi domination (interrupted in 1961) over the Hutu majority).	Military victory Side B
078	Zaïre (Anti-regime revolts)	1/1964 - 24.9.1966	A-1	Aggressor (Side A): Mulelists / Conseil National de Libération (CNL) and his troops, the Armée Populaire de Libération (APL) / Katangian soldiers and mercenaries under Ferdinand Tshimpala Aggressed (Side B): Zaïre Intervention favoring side B: USA / Belgium	Power conflict (Herrschaftskonflikt) (revolt of different groups with unclear social and political goals against the central government).	Military victory Side B
092	Chad (Civil War)	since 6/1966	ABC-2	Front de Libération Nationale (FROLINAT) (until its disbanding in 1977) fought against troops of the Chad government. Since 1979 the Forces Armées Tchadiennes (FAT) of President Malloum (recruited in the south) fought against the Forces Armées du North (FAN) lead by Prime Minister Habré, supported by the Forces Armées Populaires (FAP) of later Habré-opponent Oueddei, who together with the transition government built the Gouvernement d'Unité National Transitoire (GUNT). Since 1980 Oueddeis' FAP fought with support of Libyan troops against Habrés' FAN (President since 1983), supported by French troops. There were also ongoing fights between Chad and Libyan troops over the Aouzou strips. Since 1989 the Mouvement pour la Salvation Nationale Tchadienne (MOSANAT), later the Mouvement Patriotique du Salut (MPS) under Déby fought against the troops of Präsidenten Habré. After Habrés' fall in 1990 fighting emerged between supporters of Habrés and troops loyal to President Déby.	System reform (revolt against the disadvantaged position of the black African south), power conflict (Herrschaftskonflikt) (power struggle between different organizations) and at times territorial revision/particularist conflict (war with Libya over the Aouzou strips).	Ongoing
096	Zaïre (Mercenary Rebellion)	5.7.1967 - 6.11.1967	A-2	Aggressor (Side A): Mercenaries and elements of the former "Gendarmerie Katangaise" Aggressed (Side B): Zaïre	Power conflict (Herrschaftskonflikt) (re-installing of Moise Tschombé).	Military victory Side B

Central Africa

Nr.	Name/Country	Time Period	War Type	Warring Parties	Object of Conflict	Outcome
113	Burundi	29.4.1972 - 6/1973	A-2	Aggressor (Side A): Hutu guerillas Aggressed (Side B): Burundi	Power conflict (Herrschaftskonflikt) (Cementing of the Tutsi minority's dominance over the Hutu majority).	Military victory Side B
128	Zaïre (Shaba I, "80-Day War")	8.3.1977 - 26.5.1977	A-1	Aggressor (Side A): Front de Libération Nationale Congolais (FLNC) Aggressed (Side B): Zaïre Intervention favoring side B: Morocco	System reform (overthrow of the Mobutu regime).	Fighting stopped
133	Zaïre (Shaba II)	12.5.1978 - 30.5.1978	A-1	Aggressor (Side A): Front de Libération Nationale Congolais (FLNC) Aggressed (Side B): Zaïre Intervention favoring side B: Belgium / France	System reform (overthrow of the Mobutu regime).	Military victory Side B
189	Rwanda (Rebel invasion from Uganda)	since 1.10.1990	AB-1	Aggressor (Side A): Front Patriotique Rwandais (FPR) Aggressed (Side B): Rwanda Intervention favoring side B: Zaïre	Power conflict (Herrschaftskonflikt) (overthrow of the Habyarimana regime, Hutu-Tutsi Antagonismus and return of the refugees from Uganda).	Ongoing

East Africa

Nr.	Name/Country	Time Period	War Type	Warring Parties	Object of Conflict	Outcome
012	Madagascar (De-colonialization War)	29.3.1947 - 1948	D-2	Aggressor (Side A): PANAMA / JINA / Mouvement de la Rénovation Malgache (MDRM) Aggressed (Side B): France	Decolonization (independence from France).	Military victory Side B
029	Kenya (Mau-Mau Uprising)	10/1952 - 11/1956	D-2	Aggressor (Side A): Mau-Mau resistance fighters Aggressed (Side B): Great Britain	Decolonization (independence from Great Britain).	Military victory Side B

East Africa

Nr.	Name/Country	Time Period	War Type	Warring Parties	Object of Conflict	Outcome
063	Ethiopia (Eritrea Conflict)	1962 - 5/1991	B-2	Aggressor (Side A): Eritrean Peoples Liberation Front (EPLF) / Eritrean Liberation Front-Popular Liberation Forces (ELF-PLF) / Eritrean Liberation Front (ELF) Aggressed (Side B): Ethiopia	Secession (reversal of Eritrea's annexion by Ethiopia).	Military victory Side A
071	Ogaden Revolt (Ethiopia 1963/64)	8/1963 - 12/1964	BC-2	Aggressor (Side A): Ogaden-Befreiungsfront Aggressed (Side B): Ethiopia Side A: Somalia	Autonomy/secession (of Ethiopean Ogaden, populated by Somalis)/territorial revision (annexation to Somalia).	Military victory Side B
075	Kenya (Shifta War)	11/1963 - 10/1967	B-2	Aggressor (Side A): Shifta-Guerilla (Kenya) Aggressed (Side B): Kenya	Secession (from the Somali-populated region in northern Kenya with the goal of annexation to Somalia).	Third-Party Mediation (OAU)
112	Uganda / Tanzania (I)	8/1971 - 7.10.1972	AC-2	Aggressor (Side A): Uganda Aggressed (Side B): Tanzania Side B: Uganda People's Army	Power conflict (Herrschaftskonflikt) (re-installation of the overthrown Obote civilian government).	Third-Party Mediation (OAU)
122	Ethiopia (Tigray Conflict)	1975 - 5/1991	B-2 / BA-2	Aggressor (Side A): Ethiopian Peoples Democratic Movement (EPDM) / Ethiopian Peoples Revolutionary Party (EPRP) / Ethiopian Democratic Union (EDU) / Tigray Peoples Liberation Front (TPLF) / Tigray Liberation Front (TLF) Aggressed (Side B): Ethiopia	Autonomy (of the Tigrays), later system reform (democratization, decentralization).	Military victory Side A
127	2nd Ogaden War (Ethiopia)	1975 - 1984	BC-1	Aggressor (Side A): Somalia / Western Somali Liberation Front (WSLF) Aggressed (Side B): Ethiopia Intervention favoring side B: Cuba	Secession (of the Somali-populated Ethiopian Ogaden).	Third-Party Mediation (other third-party states)

East Africa

Nr.	Name/Country	Time Period	War Type	Warring Parties	Object of Conflict	Outcome
131	Ethiopia (Oromo Conflict)	since 1976	AB-2	Aggressor (Side A): Oromo-Liberation-Front (OLF) / Islamic Front for the Liberation of Oromiya (IFLO) Aggressed (Side B): Ethiopia	Secession, since 1990 autonomy (for the Oromo).	Ongoing
134	Uganda / Tanzania (II)	10/1978 - 2.6.1979	AC-1	Aggressor (Side A): Uganda Aggressed (Side B): Tanzania Side B: Uganda National Liberation Army (UNLA) Intervention favoring side A: Libya	Power conflict (Herrschaftskonflikt) (overthrow of the Amin regime).	Military victory Side B
152	Uganda	2/1981 - 5/1992	A-2	Phase 1 (until 1986): National Resistance Movement/Army (NRM/A) and Uganda Freedom Movement (UFM) against Uganda Phase 2: Uganda Freedom Movement (UFM) / Holy Spirit Movement/Batallion / Uganda People's Democratic Movement/Army (UPDM/A) / Uganda People's Front/Army (UPF/A) plus many smaller groups against Uganda	System reform (overthrow of the Obote government, then overthrow of the Okello government), since 1986 power conflict (Herrschaftskonflikt) (overthrow of the Museveni government).	Military victory Side B
166	Somalia (Civil War)	since 5/1988	A-2 / AB-2	Aggressor (Side A): Somali National Movement (SNM; Isaaq Clan) / Somali Salvation Democratic Front (SSDF; Darod/Majertain Clan) / Somali Patriotic Movement (SPM; Ogadeni/Darod Clan) / United Somali Congress (USC; Hawiye Clan), faction under Mohammed Farah Hassan "Aidid" (subclan Abgal) / United Somali Congress (USC; Hawiye Clan), faction under Somali Interim President Ali Mahdi Mohammed (subclan Habr Gedir) / Somali Democratic Movement (SDM; Rahanweyn Clan) / Somali Democratic Alliance (SDA; Gadabursi Clan) / Somali National Alliance (SNA) / Southern Somali National Movement (SSNM; Dir Clans in southern Somalia), Somali National Front (SNG, Darod-Marehan) and further organizations Aggressed (Side B): Somalia	System reform (overthrow of the Barré regime and democratization), since 1991 power conflict (Herrschaftskonflikt) (between different clans) and secession (North Somalia).	Ongoing

East Africa

Nr.	Name/Country	Time Period	War Type	Warring Parties	Object of Conflict	Outcome
190	Djibuti	since 10/1991	AB-2	Aggressor (Side A): Front pour la Restauration de l'Unité et de la Démocratie (FRUD) Aggressed (Side B): Djibuti	System reform (elimination of political and socio-economic disadvantages of the Afar people).	Ongoing

Southern Africa

Nr.	Name/Country	Time Period	War Type	Warring Parties	Object of Conflict	Outcome
062	Angola	since 4.2.1961	D-1 / ABC-1 / A-2	Phase 1 (bis 1975): Movimento Popular de Libertaçao de Angola (MPLA) / União Nacional pela Independência Total de Angola (UNITA) / Frente Nacional de Libertação de Angola (FNLA) against Portugal Phase 2: União Nacional pela Independência Total de Angola (UNITA) / Frente Nacional de Libertação de Angola (FNLA) / Frente de Libertação do Enclave de Cabinda (FLEC) against Angola Intervention favoring side MPLA/Angola: Cuba Intervention favoring side FNLA/UNITA: Zaïre / South Africa	Decolonization (independence from Portugal), since 1975/76 System transformation (MPLA) / system restoration (UNITA and FNLA); since October 1992 above all power conflict (Herrschaftskonflikt) (following UNITA's electoral defeat); also secession (FLEC for the independence of Cabinda).	Ongoing
081	Mosambique (War of Independence)	25.9.1964 - 7.9.1974	D-1	Side A: Frente de Libertação de Moçambique (FRELIMO) Side B: Portugal Intervention favoring side B: Rhodesia / South Africa	Decolonization (independence from Portugal).	Agreement without mediation

Southern Africa

Nr.	Name/Country	Time Period	War Type	Warring Parties	Object of Conflict	Outcome
091	Rhodesia ("Chimu-renga II")	28.4.1966 - 21.12.1979	A-1	Aggressor (Side A): Zimbabwe African National Union (ZANU) / Zimbabwe African People's Union (ZAPU) / Patriotic Front (PF) Aggressed (Side B): Rhodesia Intervention favoring side A: Mozambique Intervention favoring side B: South Africa	System transformation (against the white settler regime, which unilateraly declared its independence from Great Britain in 1965).	Third-Party Mediation (various)
093	Namibia (Liberation and War of Independence)	26.8.1966 - 22.12.1988	B-2	Aggressor (Side A): South West Africa People's Organisation (SWAPO) Aggressed (Side B): South Africa	Secession (Namibian independence and the end of the illegal occupation of former Southwest Africa by the Republic of South Africa).	Third-Party Mediation (U.N.)
119	Mozambique (Destabilization campaign)	since 1975	AC-2 / A-1	Aggressor (Side A): Resistência Nacional Moçambiquana (RENAMO) / Rhodesia Aggressed (Side B): Mozambique Intervention favoring side A: South Africa Intervention favoring side B: Nigeria / Tanzania / Zimbabwe / Malawi	Power conflict (Herrschaftskonflikt) (securing rule by the white minority regimes of Rhodesia and South Africa) and system restoration (overthrow of the socialist government).	Ongoing
125	South Africa (Anti-Apartheid War)	since 1976	A-2 / AB-2	Aggressor (Side A): African National Congress (ANC) Aggressed (Side B): South Africa also: Inkatha (Zulu Movement)	System transformation (elimination of the apartheids system) and power conflict (Herrschaftskonflikt) (Inkatha vs. ANC).	Ongoing
181	Zambia (Mushala Revolt)	14.1.1976 - 20.11.1982	A-2	Aggressor (Side A): Mushala Army Aggressed (Side B): Zambia	System reform (elimination of one-party rule and socioeconomic disadvantages in Zambia's northwest province).	Military victory Side B

Southern Africa

Nr.	Name/Country	Time Period	War Type	Warring Parties	Object of Conflict	Outcome
160	Zimbabwe (Matabele Conflict)	1/1983 - 5/1988	A-2	Aggressor (Side A): "Dissidents" Aggressed (Side B): Zimbabwe	Power conflict (Herrschaftskonflikt) (*inter alia* between ZANU and ZAPU and between Shona and Ndebele).	Agreement without mediation

Near and Middle East

Nr.	Name/Country	Time Period	War Type	Warring Parties	Object of Conflict	Outcome
179	Iraq (Barzanis)	8/1945 - 10/1945	B-1	Aggressor (Side A): Barzanis Aggressed (Side B): Iraq Intervention favoring side B: Great Britain	Autonomy (for the Kurds).	Military victory Side B
005	Mahabad (Iran)	14.3.1947 - 13.4.1947	B-2	Aggressor (Side A): Iran Aggressed (Side B): Barzanis	Secession (for the Kurds).	Military victory Side A
017	North Yemen 1948	2/1948 - 14.3.1948	A-2	Aggressor (Side A): Yeminite tribal fighters and oppositional factions of the army under the leadership of Crown Prince Ahmad Aggressed (Side B): North Yemen	Power conflict (Herrschaftskonflikt) (Occupation of the Imamats).	Military victory Side A
018	1st Arab-Israeli War	15.5.1948 - 7.1.1949	C-2	Aggressor (Side A): Egypt / Syria / Lebanon / Jordan / Iraq / Saudi Arabia Aggressed (Side B): Israel	Territorial restoration (destruction of the Israeli state).	Third-Party Mediation (U.N.)
044	Protectorate Aden / North Yemen	1955 - 1958	C-2	Side A: Great Britain Side B: Yemen	Territorial revision (disputed border).	Fighting stopped

Near and Middle East

Nr.	Name/Country	Time Period	War Type	Warring Parties	Object of Conflict	Outcome
042	2nd Arab-Israeli War (Suez War)	29.10.1956 - 6.11.1956	C-2	Aggressor (Side A): Great Britain / France / Israel Aggressed (Side B): Egypt	Israel: Military-strategic conflict (to break the sea blockade (Gulf of Aqaba, Suez Canl)) and territorial expansion; Great Britain and France: hegemonic conflict (securing of neo-colonial influence, regaining the Suez Canal which was nationalized by Egypt).	Third-Party Mediation (U.N.)
048	Muskat and Oman (Imam-Sultan Con-flikt)	7/1957 - 1959	B-1	Aggressor (Side A): Imam from Oman Aggressed (Side B): Muskat and Oman Intervention favoring side B: Great Britain	Territorial restoration (elimination of the Sultan of Muskrat's control over Oman) and particularist conflict (securing of British oil interests).	Military victory Side B
049	Lebanese Civil War 1958	9.5.1958 - 14.10.1958	AB-2	Side A: Lebanon Side B: United National Front (UNF)	Securing of governmental rule (power struggle between traditional leaders) and System reform (constitutional reform).	Third-Party Mediation (U.N.)
061	1st and 2nd Iraqi--Kurdish War	16.9.1961 - 10.2.1964	B-1	Aggressor (Side A): Iraq Aggressed (Side B: Barzanis and allied Stämme / Kurdish Democratic Party-Iraq (KDP-Iraq) Intervention favoring side A: Syria	Autonomy (for the Kurds).	Agreement without mediation
068	Yemenite Civil War	10/1962 - 1969	A-1	Aggressor (Side A): Republican troops Aggressed (Side B): Yemen Intervention favoring side A: Egypt	System transformation (Imamats vs. Republic).	Third-Party Mediation (Arabic League)

Near and Middle East

Nr.	Name/Country	Time Period	War Type	Warring Parties	Object of Conflict	Outcome
087	Protectorate Aden (South Yemen War of Independence)	14.10.1963 - 30.11.1967	D-2	Aggressor (Side A): National Liberation Front (NLF) / Front for the Liberation of South Yemen (FLOSY) Aggressed (Side B): Great Britain	Decolonization (independence from Great Britain).	Agreement without mediation
085	3rd Iraqi-Kurdish War	4.4.1965 - 29.6.1966	B-2	Aggressor (Side A): Iraq Aggressed (Side B): Kurdisch Democratic Party - Iraq (KDP-Iraq)	Autonomy (for the Kurds).	Agreement without mediation
088	Oman Revolt in Dhofar	9.6.1965 - 11/1975	AB-1 / A-1	Aggressor (Side A): Popular Front for the Liberation of Oman (PFLO) Aggressed (Side B): (Muskat and) Oman Intervention favoring side B: Great Britain / Iran / Jordan	First secession (Dhofar province) and system reform (overthrow of the Sultans), then system transformation (Omans and the other Gulf states by means of social revolution); hegemonic conflict (Great Britain: securing influence before withdrawal from Gulf region; Iran: hegemonic interests and securing of the straits of Hormus).	Military victory Side B
095	3rd Arab-Israel War ("Six Day War")	5.6.1967 - 10.6.1967	C-2	Aggressor (Side A): Israel Aggressed (Side B): Egypt / Syria / Jordan	Territorial revision (securing the border, later expansion by Israel).	Third-Party Mediation (UN)

Near and Middle East

Nr.	Name/Country	Time Period	War Type	Warring Parties	Object of Conflict	Outcome
165	Israel, Palästina	since 1968	B-2	Aggressor (Side A): Palestinian Liberation Organization (PLO) with its key organizations al-Fatah, People's Front for the Liberation of Palestine (PFLP), Democratic Front for the Liberation of Palestine (DFLP), Palestine Liberation Front (PLF), PFLP-General Command (PFLP-GC), and since the second half of the 1980s Hamas and Jihad Islam) Aggressed (Side B): Israel	Territorial revision (founding of the Palestinian state).	Ongoing
098	People's Republic of Yemen	6/1968 - 11/1968	A-2	Aggressor (Side A): Front for the Liberation of South Yemen (FLOSY) / South Arabian League (SAL) Aggressed (Side B): DPR Yemen	System restoration (overthrow of the socialist government).	Military victory Side B
102	4th Iraqi-Kurdish war	3.1.1969 - 11.3.1970	B-2	Aggressor (Side A): Iraq / Talaban Faction of the Kurdish Demokratic Party - Iraq Aggressed (Side B): Kurdish Democratic Party - Iraq (KDP-Iraq)	Autonomy (for the Kurds).	Agree-ment without mediation
103	Egypt / Israel	6.3.1969 - 7.8.1970	C-1	Side A: Egypt Side B: Israel Intervention favoring side A: USSR	Territorial restoration (regaining the Sinai peninsula).	Third-Party Mediation (U.N.)
101	DPR Yemen / Saudi Arabia	26.11.1969 - 12/1969	C-2	Side A: DPR Yemen Side B: Saudi Arabia	Territorial revision (disputed border).	Third-Party Mediation (Arab League)

Near and Middle East

Nr.	Name/Country	Time Period	War Type	Warring Parties	Object of Conflict	Outcome
108	Jordan (Black September)	16.9.1970 - 19.7.1971	AB-1	Aggressor (Side A): Jordan Aggressed (Side B): Palestine Liberation Organization (PLO) Intervention favoring side B: Syria	System reform (overthrow of the Monarchy) and Military-strategic conflict (prevention of guerilla attacks on Israel).	Military victory Side A
114	AR Yemen / DPR Yemen	26.9.1972 - 13.10.1972	C-2	Side A: AR Yemen Side B: DPR Yemen	System reform (reciprocal support of the respective opposition forces) and territorial revision (border conflict).	Third-Party Mediation (Arab League)
115	4th Arab-Israeli war (October War)	6.10.1973 - 26.10.1973	C-2	Aggressor (Side A): Egypt / Syria / Iraq / Palestine Liberation Organization (PLO) Aggressed (Side B): Israel	Territorial restoration (regaining the occupied territories from Israel).	Third-Party Mediation (UN)
117	5th Iraqi-Kurdish war	14.3.1974 - 3/1975	B-2	Side A: Iraq Side B: Kurdish Democratic Party - Iraq (KDP-Iraq)	Autonomy (for the Kurds).	Military victory Side A
120	Lebanon War	since 13.4.1975	ABC-2	Side A: Forces Libanaises / South Lebanese Army (SLA) / Israel Side B: Lebanon / National Movement / Palestine Liberation Organization (PLO) often switching sides: Syria	System transformation (elimination of the proportional representation system, socialist transformation, Islamification), power conflict (Herrschaftskonflikt) (*inter alia.* within the Maronite and the Shi'ite camps), particularist conflict (*inter alia.* securing of Maronite dominance) as well as territorial expansion (Israel) and hegemonic conflict (securing of Syrian influence).	Ongoing

Near and Middle East

Nr.	Name/Country	Time Period	War Type	Warring Parties	Object of Conflict	Outcome
126	6th Iraqi-Kurdish War	since 5/1976	BA-3	Aggressor (Side A): Kurdish Democratic Party (KDP-Iraq) / Patriotic Union of Kurdistan (PUK) / SocialistParty (Iraqi)-Kurdistans (PSK) / Communist Party of Iraq Aggressed (Side B): Iraq Intervention favoring side A: Iran Intervention favoring side B: Turkey	Autonomy (for the Kurds) and system reform (democratization).	Ongoing
135	AR Yemen	1978 - 1982	A-1	Aggressor (Side A): National Democratic Front (NDF) Aggressed (Side B): AR Yemen Intervention favoring side A: DPR Yemen	System reform (creating a strong central state, reducing the influence of tribes and of Saudi Arabia).	Third-Party Mediation (Arab League)
141	Afghanistan	since 10/1978	A-1	Aggressor (Side A): Conservative and traditional-oriented resistance movement / Muslim fundemental resistance groups Aggressed (Side B): Afghanistan Intervention favoring side B: USSR	System restoration (overthrow of Communist rule, maintenance of traditional production relations and Islam); hegemonic conflict (securing of Soviet influence).	Ongoing
144	Iran (Kurdistan)	7/1979 - 1988	B-2	Side A: Iran Side B: Kurdish Democratic Party-Iran (KDP-Iran) / Komalah	Autonomy (for the Kurds).	Military victory Side A
148	Iraq / Iran (1st Gulf War)	22.9.1980 - 20.8.1988	C-2	Aggressor (Side A): Iraq Aggressed (Side B): Iran	Territorial revision (border at Schatt-el-Arab), Territorial restoration (return of the Gulf islands occupied in 1971 by Iran), territorial expansion (Iraqi annexion of an Iranian oil province), hegemonic conflict (dominance in the Gulf), power conflict (Herrschaftskonflikt) (against Iran's support of the Shi'ite revolution).	Third-Party Mediation (U.N.)

Near and Middle East

Nr.	Name/Country	Time Period	War Type	Warring Parties	Object of Conflict	Outcome
153	Syria 1982	2.2.1982 - 2/1982	A-2	Aggressor (Side A): al-Jabha al-Islamiyya (Islamic Front) Aggressed (Side B): Syria	System transformation (Islamification) and particularist conflict (maintenance of Hamas' economic structure).	Military victory Side B
159	Turkey (Kurds)	since 15.8.1984	B-2	Aggressor (Side A): Partiya Karkeren Kurdistan (PKK; Kurdish Workers Party) Aggressed (Side B): Turkey	Secession (of the Kurds).	Ongoing
162	South Yemen 1986	13.1.1986 - 29.1.1986	A-2	Side A: DPR Yemen Side B: Left wing of the Yemenite Socialist Party (JSP)	Power conflict (Herrschaftskonflikt) (power struggle between wings of the JSP).	Military victory Side B
174	USSR (Azerbaijan, Armenia); Nagorno-Karabach	since 20.1.1990	B-2	Phase 1: Aggressor (Side A): USSR Aggressed (Side B): People's Front of Azerbaijan (also: National People's Front) / Armenian National Army (ANA) Phase 2: Side A: Azerbaijan Side B: Armenian Militias of Nagorno-Karabach	Secession (national independence of Azerbaijan and Armenia from Moscow - only the first phase); territorial revision (accession of the enclave Nagorno-Karabach (in Azerbaijani territory) to Armenia).	Ongoing
182	Iraq / Kuwait	2.8.1990 - 4.8.1990	C-2	Aggressor (Side A): Iraq Aggressed (Side B): Kuwait	Territorial expansion (annexation of Kuwait) and particularist conflict (control of the entire Rumaila oil fields and raising of oil prices).	Military victory Side A
202	Georgia (South Ossetia)	12/1990 - 7/1992	B-2	Aggressor (Side A): Georgia Aggressed (Side B): South Ossetian Nationalists	Secession (of South Ossetia).	Third-Party Mediation (neighboring state(s))

Near and Middle East

Nr.	Name/Country	Time Period	War Type	Warring Parties	Object of Conflict	Outcome
194	Anti-Iraq Coalition / Iraq (2nd Gulf War)	17.1.1991 - 27.2.1991	C-2	Aggressor (Side A): USA / Great Britain / France / Italy / Egypt / Syria / Saudi Arabia / Kuwait and other states Aggressed (Side B): Iraq	Territorial revision (liberation of Kuwait) and hegemonic conflict (weakening of Iraq).	Military victory Side A
196	Iraq (Shi'ites)	since 2.3.1991	A-2	Aggressor (Side A): Iraqi Shi'ites Aggressed (Side B): Iraq	System transformation (Islamification).	Ongoing
201	Georgia (Anti-Regime War)	since 9/1991	A-2	Aggressor (Side A): Oppositional National Movement Aggressed (Side B): Georgia	Power conflict (Herrschaftskonflikt) (power-struggle).	Ongoing
208	Tadjikistan	since 8/1992	A-2	Side A: Tadjikistan Side B: Coalition of the Western-oriented Democratic Party of Tadjikistan, the nationalist movements Rastokhez and Lali Badakhshan and the Islamic Renaissance Party	System transformation (elimination of the socialist system).	Ongoing
207	Georgia (Abkhazia)	since 14.8.1992	B-2	Aggressor (Side A): Georgia Aggressed (Side B): Abkhazian militias with support from armed volunteers from the Confederation of Caucus' Mountain People and different autonomous Republics of the Russian Federation	Secession (of Abkhazia).	Ongoing
209	North Ossetian district of Prigorodny	since 30.10.1992	B-2	Aggressor (Side A): Armed Ingush forces Aggressed (Side B): Autonomous Republic of North Ossetia	Territorial revision (annexation of the Prigorodny district to Ingushia).	Ongoing

South and Southeast Asia

Nr.	Name/Country	Time Period	War Type	Warring Parties	Object of Conflict	Outcome
003	Indonesia	29.9.1945 - 27.7.1949	D-2	Aggressor (Side A): Indonesian Liberation Movement Aggressed (Side B): Great Britain / Netherlands	Decolonization (independence from Netherlands).	Third-Party Mediation (UN)
010	Philippines (Luzon Island)	1946 - 1954	A-2	Aggressor (Side A): Hubko ng Bayan sa Hupon (also: Hukbalahap; HUK; Anti-Japanese People's Army) Aggressed (Side B): Philippines	System reform (Land reform and political participation).	Military victory Side B
009	Indochina (1st Indochina War)	9.3.1946 - 1.6.1954	D-2	Aggressor (Side A): France Aggressed (Side B): Khmer Vietminh (später "Khmer Rouge") / Vietminh Side B: Pathet Lao	Decolonization (independence from France) and system transformation (socialist revolution).	Third-Party Mediation (international conference)
013	Pakistan / India (1st Kashmir War)	22.10.1947 - 1.1.1949	A-2 / AC-1	Aggressor (Side A): Pashtun Aggressed (Side B): Kashmir Side A: troops Azad-Kashmirs (Azad-Kashmiri-Forces) Intervention favoring side A: Pakistan Intervention favoring side B: India	Territorial expansion (annexation of Kashmir to India vs. Kashmir independence).	Third-Party Mediation (U.N.)
020	Malayan Federation ("Malayan-Emergency")	1948 - 2/1960	DA-2	Aggressor (Side A): Malayan Communist Party (MCP) Aggressed (Side B): Great Britain / Malayan Federation	Decolonization (independence from Great Britain) and system transformation (socialist revolution, sectoral reforms and elimination of discrimination against the Chinese minority).	Military victory Side B

South and Southeast Asia

Nr.	Name/Country	Time Period	War Type	Warring Parties	Object of Conflict	Outcome
021	Myanmar (Burma)	since 1948	AB-2	Aggressor (Side A): Burmese Communist Party (BCP) / Groups of ethnic minorities under the umbrella of Democratic Alliance of Burma (DAB) Aggressed (Side B): Burma often switching sides: Warlords	System transformation (communist party for socialist revolution), autonomy or secession (ethnic minorities) and particularist conflict ("Warlords" pursuing their own economic goals).	Ongoing
015	India / Hyderabad	13.9.1948 - 17.9.1948	C-2	Aggressor (Side A): India Aggressed (Side B): Hyderabad	Territorial expansion (annexation of Hyderabad to the Indian Union).	Military victory Side A
023	Indonesia (South Moluk-ken)	7/1950 - 12/1950	B-2	Aggressor (Side A): Indonesia Aggressed (Side B): Republic of South Molukken (RMS)	Secession (of South Molukken from Indonesia).	Military victory Side A
024	Nepal 1950	11.11.1950 - 1/1951	A-2	Aggressor (Side A): Nepalese Congress Party Aggressed (Side B): Nepal	System reform (overthrow of the feudal systems and construction of a parlamentary democracy).	Military victory Side B
026	India (Nagas)	1954 - 8/1964	B-2	Aggressor (Side A): "Federal Army" of the "Federal Republic of Nagas" Aggressed (Side B): India	Secession (formation of an independent Naga state, including settlements in Burma).	Third-Party Mediation (church)
041	Vietnam (2nd Indochina War)	1957 - 29.4.1975	A-1/AC-1	Side A: Front National de Libération (FNL) Side B: South Vietnam Intervention favoring side A: People's Republic of China / North Vietnam Intervention favoring side B: USA / South Korea / Philippines / Australia / New Zealand	System transformation (communist revolution in South Vietnam), territorial restoration (union with communist North Vietnam) and hegemonic conflict (prevention of North Vietnamese support of South Vietnam's revolutionary war by the USA and allies, spread of hostilities to neighboring states).	Military victory Side A

South and Southeast Asia

Nr.	Name/Country	Time Period	War Type	Warring Parties	Object of Conflict	Outcome
052	Indonesia (Civil War)	3/1958 - 7/1961	AB-1	Aggressor (Side A): Heterogeneous Rebel Army Aggressed (Side B): Indonesia Intervention favoring side A: USA	Autonomy (against Java's dominance) and system reform (against parliamentary change in a "steered democracy" and for decentralization of power).	Military victory Side B
034	Laos (1st Laotian Civil War)	1.7.1959 - 16.6.1961	A-1	Phase 1 (until 8.8.1960): Pathet Lao and North Vietnamese troops against Laotian government troops Phase 2: Pathet Lao, North Vietnamese troops and neutrals under Kong Le against remaining Laotian government troops (lead by conservatives), Thai troops and Meo-Guerilla (also called Hmong)	System transformation (socialist revolution).	Third-Party Mediation
065	Nepal 1962	1/1962 - 2/1962	A-2	Aggressor (Side A): Nepalese Congress Party Aggressed (Side B): Nepal	System reform (overthrow of the Monarchy and social reforms).	Military victory Side B
064	Indonesia (West Papua)	14.1.1962 - 18.8.1962	C-2	Aggressor (Side A): Indonesia Aggressed (Side B): Netherlands	Territorial expansion (Indonesian goal: annexation; Dutch goal: independence for West Papua).	Third-Party Mediation
066	PRC / India	20.10.1962- 20.11.1962	C-2	Aggressor (Side A): People's Republic of China Aggressed (Side B): India	Territorial revision (conflict in northeastern (Aksai Chin) and northwesten (McMahon-Linie) India about competing territorial claims).	Military victory Side A
067	Brunei	12/1962 - 12/1962	A-1	Aggressor (Side A): North Kalimantan National Army (NKNA) Aggressed (Side B): Brunei Intervention favoring side B: Great Britain	Secession (creation of an independent "North Borneo" with neighboring territories instead of annexation on the planned Malay Federation, which was Britain's desired outcome) and system reform (overthrow of the anti-communist Sultan Monarchy).	Military victory Side B

South and Southeast Asia

Nr.	Name/Country	Time Period	War Type	Warring Parties	Object of Conflict	Outcome
076	Laos (2nd Laotian Civil War)	31.3.1963 - 4/1973	A-1	Aggressor (Side A): Pathet Lao Aggressed (Side B): Laos / Neutrals under Kong Le / Meo-Guerilla (also called Hmong) Intervention favoring side A: North Vietnam Intervention favoring side B: USA / Thailand / South Vietnam	Military-strategic conflict (escalated to theater of war in the Vietnam conflict as the U.S. attacked supply lines leading through Laos from North to South Vietnam) and power conflict (Herrschaftskonflikt) (between Pathet-Lao and the Neutrals).	Agreement without mediation
074	Indonesia / Malaysia ("Confron-tation")	12.9.1963 - 11.8.1966	C-1	Aggressor (Side A): Indonesia / Heterogeneous Rebels Aggressed (Side B): Malaysia Intervention favoring side B: Great Britain / Australia / New Zealand	Hegemonial conflict (securing of Indonesian regional dominance), territorial restoration (defeat of Malaysia) and diversionary conflict (from domestic and economic problems in Indonesia).	Agreement without mediation
178	Vietnam (FULRO)	20.9.1964 - 10/1992	B-2	Side A: Force unifiée pour la libération des races opprimées (FULRO) Side B: North Vietnam	Autonomy (the Dega fought against the advancement of the Vietnamese in the highland region).	Fighting stopped
130	Indonesia (West Irian and West-Papua)	since 1965	B-2	Aggressor (Side A): Organisasi Papua Merdeka (OPM; Organisation for a Free Papua) Aggressed (Side B): Indonesia	Secession (from Indonesia; founding of the independent state of West Papau).	Ongoing
082	India / Pakistan (Rann-from-Kutch)	9.4.1965 - 1.7.1965	C-2	Side A: India Side B: Pakistan	Territorial revision (border conflict).	Third-Party Mediation

Nr.	Name/Country	Time Period	War Type	Warring Parties	Object of Conflict	Outcome
084	Pakistan / India (2nd Kashmir War)	5.8.1965 - 23.9.1965	BC-2	Aggressor (Side A): Mujahids (muslim guerillas) Side A: Pakistan Aggressed (Side B): India	Territorial expansion (Pakistani attempt to " liberate" the Indian part of Kashmir) and hegemonic conflict (Indian attempt to gain status after their defeat in the Sino-Indian War).	Third-Party Mediation
089	Thailand	7.8.1965 - 1980	A-2	Side A: Thailand Side B: Communist Party of Thailand (CPT)	System transformation (communist revolution).	Unknown
090	India (Mizos)	1.3.1966 - 31.7.1980	B-2	Aggressor (Side A): Mizo National Front (MNF) / Buddhist Chakmas Aggressed (Side B): India	Secession (of the Christian Mizos from India, including settlements in Burma).	Agreement without mediation
099	Cambodia	17.1.1968 - 17.4.1975	AC-1	Side A: Khmer Rouge / Front Uni National du Kampuchea (FUNK; National Unity Front of Kampuchea) Side B: Cambodia Intervention favoring side A: North Vietnam Intervention favoring side B: USA / South Vietnam	System transformation (communist revolution, since 1970 strong nationalist components) and military-strategic conflict (participation in the Vietnam War as of 1970; fighting against communist bases and ending of Vietminh supply routes in Cambodia by Cambodian government, South Vietnamese and U.S. forces).	Military victory Side A
172	India (Nagas)	1969 - 6.11.1975	B-2	Aggressor (Side A): "Federal Army" of the "Federal Republic of Nagas" Aggressed (Side B): India	Secession (of Nagas).	Agreement without mediation

South and Southeast Asia

Nr.	Name/Country	Time Period	War Type	Warring Parties	Object of Conflict	Outcome
109	Philippines (NPA)	since 1970	A-2	Aggressor (Side A): New People's Army (NPA) Aggressed (Side B): Philippines	System reform (overthrow of the government and instigation of a national, democratic revolution as prerequisite for far-reaching land reform and as step toward a socialist society).	Ongoing
199	Philippines (Mindanao)	since 1970	B-2	Aggressor (Side A): Moro National Liberation Front (MNLF) / Moro Islamic Liberation Front (MILF) Aggressed (Side B): Philippines	Secession (of Mindanao and Sulu islands).	Ongoing
111	Bangladesh War	25.3.1971 - 17.12.1971	B-2 / B-1 / BC-2	Aggressor (Side A): Pakistan Aggressed (Side B): Various guerilla movements; also some police forces and paramilitary units Intervention favoring side B: India	Secession (of East Pakistan) and hegemonic conflict (India/Pakistan)	Military victory Side B
110	Ceylon (JVP revolt)	5.4.1971 - 5.6.1971	A-1	Aggressor (Side A): Janatha Vimukthi Peramuna (JVP) Aggressed (Side B): Ceylon Intervention favoring side B: India	System transformation (socialist revolution following the Chinese model).	Military victory Side B
177	Bangladesh (Chittagong Hill Tracts)	since 1973	B-2	Aggressor (Side A): Shanti Bahini (SB; Peace Army) Aggressed (Side B): Bangladesh	Autonomy (of the Chittagong Hill Tracts, against settelments of farmers from the lowlands).	Ongoing
171	Pakistan (Belutchistan)	3/1973 - 7/1977	B-2	Aggressor (Side A): Balutchistan People's Liberation Front (BPLF) Aggressed (Side B): Pakistan	Autonomy (of Belutchistan) and system reform (political and social reforms).	Military victory Side B

South and Southeast Asia

Nr.	Name/Country	Time Period	War Type	Warring Parties	Object of Conflict	Outcome
183	Laos (1st Meo Guerrilla War [Hmong])	1975 - 1979	B-1	Side A: Laos Side B: Meo Guerrillas Intervention favoring side A: Vietnam	System reform (against political persecution).	Military victory Side A
121	Indonesia (East Timor)	since 11.8.1975	B-2	Aggressor (Side A): União Democratica Timorese (UDT) / Indonesia Aggressed (Side B): Frente Revolucionaria de Timor Leste Independiente (FRETELIN)	Phase 1: System transformation (socialist orientation after withdrawal of the Portugese colonists); Phase 2: Territorial expansion (incorporation into Indonesia); Phase 3: Secession (from Indonesia) and system transformation (introduction of a socialist society).	Ongoing
132	Cambodia (3rd Indochina War)	since 12/1975	C-2 / A-1	Phase 1 (until 16.1.1979): Aggressor (Side A): Cambodia Aggressed (Side B): Vietnam Side B: Front d'Union Nationale du Kampuchea pour le Salut National (FUNKSN) Phase 2: Side A: Khmer Rouge / Armée Nationaliste Sihanou-kienne (ANS) / Kambodian People's National Liberation Front (KPLNF) Side B: Cambodia Intervention favoring side B: Vietnam	From 1975-1979 hegemonic conflict (hegemonic interests of Vietnams vs. Cambodian independence); since 1979 system restoration (Khmer Rouge fighting to regain power) and power conflict (Herrschaftskonflikt) (power-struggle between various organizations).	Ongoing
142	PRC / Vietnam ("Education Campaign")	17.2.1979 - 1988	C-2	Aggressor (Side A): People's Republic of China Aggressed (Side B): Vietnam	Military-strategic conflict (support for Democratic Kampuchea, which was attacked by Vietnam and allied with China).	Fighting stopped

South and Southeast Asia

Nr.	Name/Country	Time Period	War Type	Warring Parties	Object of Conflict	Outcome
155	India (Punjab)	since 7/1982	B-2	Aggressor (Side A): Khalistan Liberation Force (KLF) / All India Sikh Student's Federation (AISSF) / Khalistan Commando Force (KCF) Aggressed (Side B): India Side B: Shiv Sena (Shivas Army)	Secession (of the majority Sikh-populated Punjab from India).	Ongoing
157	Sri Lanka ("Tamil Conflict")	since 7/1983	B-2 / AB-1 / B-1	Aggressor (Side A): Liberation Tigers of Tamil Eelam (LTTE; also "Tamil Tigers") / Eelam Revolutionary Organisation of Students (EROS) and further small Tamil guerrilla movements; Janatha Vimukhti Peramuna (JVP; People's Liberation Front) Aggressed (Side B): Sri Lanka Side B: Citizen's Volunteer Force (CVF) / Tamil National Army (TNA) / Diverse Death Squads. erst A, dann B: Eelam People's Revolutionary Liberation Front (EPRLF) / People's Liberation Organization of Tamil Eelam (PLOTE) Intervention favoring side B: India	Secession (of the Tamils) and occasional military-strategic conflict (prevention of an Indo-Sri Lankan peace treaty) as well as power conflict (Herrschaftskonflikt) (power struggle among Tamil organisations, of which some switched between government and Singhalese sides).	Ongoing
161	Pakistan / India (Siach glacier)	4/1984 - 1989	C-2	Aggressor (Side A): Pakistan Aggressed (Side B): India	Territorial revision (historical conflict from the previous two Kashmir wars, militarily limited and with regional power-political components).	Fighting below the level of war
187	Pakistan (Sind province)	since 11/1986	B-2	Aggressor: Jiye Sind Mahaz (Jeeya-Sind-Bewegung) Aggressed (Side B): Pakistan Side A: Muhajir Qaumi Movement (MQM) / Pashtunic Drug and Weapon Mafia and their armed militias	Autonomy and secession (of moderate and radical Sindhi, respectively) and power conflict (Herrschaftskonflikt) (restoration of the Muhajir's old power position).	Ongoing

Nr.	Name/Country	Time Period	War Type	Warring Parties	Object of Conflict	Outcome
164	Thailand / Laos	3.11.1987 - 19.2.1988	C-2	Aggressor (Side A): Thailand Aggressed (Side B): Laos	Territorial revision (border conflict).	Agreement without mediation
167	Papua New Guinea (Bougainville)	since 2/1989	B-2	Aggressor (Side A): Bougainville Revolutionary Army (BRA) Aggressed (Side B): Papua New Guinea	Autonomy (of Bougainville), Territorial revision (annexation to the Salomon Islands) and particularist conflict (against environmental destruction causeed by tin mining; control of resources).	Ongoing
203	India (Assam)	1990 - 1992	B-2	Aggressor (Side A): United Liberation Front of Assam (ULFA) Aggressed (Side B): India	Secession (Assams).	Fighting below the level of war
188	India (Kashmir)	since 1/1990	B-2	Aggressor (Side A): Jammu and Kashmir Liberation Front (JKLF) / Hizbul-Mujaheddin; many other organizations Aggressed (Side B): India	Secession (of Kashmir), currently territorial revision (annexation with the Pakistani part of Kashmir).	Ongoing
192	Laos (2nd Meo Guerilla War (Hmong)	1/1990 - 7/1992	B-1	Side A: Laos Side B: Lao United Front (LUF) / Ethnic Liberation Organization of Laos (ELOL) Intervention favoring side A: Vietnam	System reform (against communist, government friendly to Vietnam and against Vietnamese occupation; later against participation of high- and lowland Laotians).	Fighting stopped
198	Indonesia (Aceh)	since 5/1990	B-2	Aggressor (Side A): Movement for the Disruption of Public Order (GPK) Aggressed (Side B): Indonesia	Secession (of the fundamentalist muslim Aceh against Java's colonialist-like behavior).	Ongoing

East Asia

Nr.	Name/Country	Time Period	War Type	Warring Parties	Object of Conflict	Outcome
007	China (Chinese Civil War)	15.3.1946 - 21.4.1950	A-2	Side A: China Side B: Guerillas of the Chinese Communist Party	System transformation (socialist revolution).	Military victory Side B
025	Korean War	25.6.1950 - 27.7.1953	C-1	Side A: North Korea Side B: USA / Canada / Colombia / Great Britain / Netherlands / Belgium / Luxemburg / France / Greece / Ethiopia / South Africa / Turkey / South Korea / Thailand / Philippines / Australia / New Zealand Intervention favoring side A: People's Republic of China	Territorial restoration (unification) and system transformation (of the respective halves of the divided country); hegemonic conflict (USA for containment of communism in East Asia).	Agreement without mediation
022	PRC / Tibet	7.10.1950 - 22.10.1950	C-2	Aggressor (Side A): People's Republic of China Aggressed (Side B): Tibet Side B: Khambas	Territorial expansion (incorporation of Tibet in the PRC).	Military victory Side A
033	Tibet / PRC	8/1954 - 22.3.1959	AB-2	Side A: Tibet Side B: People's Republic of China	System transformation (attempt to maintain traditional Tibetan society and resistance against Chinese socio-economic reforms, forced settlement policies and increasing military occupation) and secession (following escalation of resistance for separation from China).	Military victory Side B
032	PRC / Taiwan (1st Quemoy Crisis)	3.9.1954 - 25.9.1954	C-1	Side A: People's Republic of China Side B: Taiwan Intervention favoring side B: USA	Territorial expansion (incorporation of the island groups into the PRC).	Fighting stopped

East Asia

Nr.	Name/Country	Time Period	War Type	Warring Parties	Object of Conflict	Outcome
036	PRC / Taiwan (Da Chen- and Nan Ji Island Groups)	18.1.1955 - 22.1.1955	C-2	Side A: People's Republic of China Side B: Taiwan	Territorial expansion (incorporation of those island groups directly adjacent to the mainland into the PRC).	Fighting stopped
050	PRC / Taiwan (2nd Quemoy Crisis)	23.8.1958 - 25.10.1958	C-2	Side A: People's Republic of China Side B: Taiwan	Territorial expansion (incorporation of those island groups directly adjacent to the mainland into the PRC) and hegemonic conflict (Test of U.S. and USSR hegemonic behavior).	Fighting stopped

4

Evaluating War Since 1945

4.1. Methodological Remarks

We have recorded a total of 184 wars occurring between 1945 and 1992 following our definitions in chapter 2. This chapter presents an annotated overview of the trends, tendencies and important characteristics of the wars during this period. The period after 1945 is an epoch unto itself, for reasons that we will soon make apparent. All of the statistics in this chapter can be reconstructed, checked, and if necessary, modified from the accompanying war list and/or war register and the lists in the appendix.

The portrayal of statistical material occurs in two forms: tables and figures (diagrams, pie charts, etc.). The statistical analyses in this book are not exhaustive, for we have restricted ourselves to what we considered most meaningful. With few exceptions we refrain from listing the finer points of statistical analysis (e.g., means and distributions) and investigations of statistical relations (e.g., associations and correlations). As a rule we are satisfied with percentages in order to convey an idea of scale or for comparative purposes. It should be noted, however, that the smaller the number of cases ("N"), the less meaningful the percentage. Very small "N" percentages are nonsensical: e.g., 1+1=2 or 2+2=4 are both equivalent to 50%+50%=100%. In figures with different "Ns," unfortunately, it is not always possible to avoid this situation. Furthermore, we avoid calculating tenths of a percent when faced with small

"Ns" or uncertain data and only give the full percentage because rounding percentages also do not always add up to 100. The symbol "R" (or also R-%) means "row percentage" and the symbol "C" (or also C-%) means "column percentage."

For those readers who are relatively unfamiliar with statistics and quantitative research we offer this comment: from a political or ethical viewpoint the number of wars are certainly too high—indeed, even one war would be too much. For statistical methods, however, which work best with mass events, the number of wars is not high enough to adequately analyze the data, create the necessary subdivisions or pursue greatly differentiated experiments. Yet the 184 wars between 1945 and 1992 are sufficient to allow a host of satisfactory insights into their structure and development. Of course with social issues there are always single instances where statistical generalizations or predictions do not apply. Occasional exceptions, however, do not automatically refute statistically supported statements, though a high number of exceptions can call a statistic into question.

Table 1: Wars Begun, Terminated, and Waged Per Year

Year	Wars Begun	Wars Ended	Wars Waged
1945	4	2	4
1946	5	0	7
1947	4	2	11
1948	7	5	16
1949	1	4	12
1950	4	4	12
1951	0	1	8
1952	2	1	9
1953	0	1	8
1954	5	4	12
1955	6	2	14
1956	3	4	15
1957	3	1	14
1958	3	4	16
1959	1	4	13
1960	1	1	10

Year	Wars Begun	Wars Ended	Wars Waged
1961	4	4	13
1962	7	5	16
1963	10	2	21
1964	5	6	24
1965	9	3	27
1966	4	5	28
1967	3	5	26
1968	5	2	26
1969	6	3	30
1970	3	3	30
1971	3	3	30
1972	2	3	29
1973	3	3	29
1974	3	3	29
1975	8	5	34
1976	4	0	33
1977	2	3	35
1978	4	1	36
1979	2	5	37
1980	4	3	36
1981	4	1	37
1982	3	4	39
1983	4	1	39
1984	2	1	40
1985	1	1	40
1986	3	1	42
1987	2	0	43
1988	1	6	44
1989	4	4	42
1990	10	3	48
1991	5	4	50
1992	5	7	51
Total	184	140	1265
Mean	3.8	2.9	26.4

4.2 The Steady Increase of the Global Burden of War

Since 1945 (more precisely, since December 3, 1944) until the end of 1992, 184 wars were begun, an average of 3.8 wars in each of the 48 years (see table 1).[1] 140 of those had also been ended during that time, so that the New Year 1993 dawned upon 44 *theaters* of war, as they are cynically called by the military and the media.[2] The register contains a list of those wars still active beyond December 31, 1992. Some wars drag on for decades: the longest war or, more precisely, the complicated web of wars in Burma (Myanmar) began in 1948.

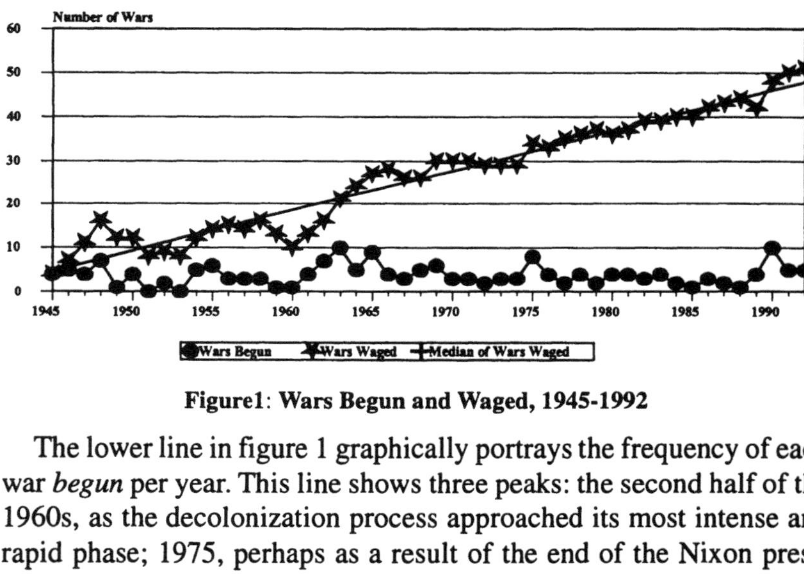

Figure1: Wars Begun and Waged, 1945-1992

The lower line in figure 1 graphically portrays the frequency of each war *begun* per year. This line shows three peaks: the second half of the 1960s, as the decolonization process approached its most intense and rapid phase; 1975, perhaps as a result of the end of the Nixon presidency and the general malaise and introspection in the U.S. following the disastrous end of the Vietnam War; and 1990, which can be attributed to the disintegration of the Soviet empire and the Soviet Union itself. This speculation notwithstanding it is important to note that over time we can point to no regularity, no cycle or even trend regarding the outbreak of war. This statement accords with previous research results by other authors.

If one looks, however, at the upper line in figure 1 a terrible picture presents itself. This line shows how many wars are waged per year. The term "waged wars" includes the wars begun before the given year, in addition to new wars begun in that year. It is therefore irrelevant whether these "older" wars end in the given year, or how long they persist past it. This line shows an *almost steady rise in the worldwide susceptibility to war.* On average in each year almost one more war (0.95) is waged than in the previous year. Again on average more wars are begun in every year than are terminated, and the new wars last longer. In 1992 more wars were waged than at any time since the end of the Second World War.

Until approximately the mid-1960s this upper line clearly shows war outbreaks in a wave-like form expressive of two world-historical, partially overlapping processes: first is the uncertainty which the Cold War, with its nuclear arms race and deterrence, brought to the behavior of international actors (with all its consequences for the orientation of the developing world); second is the decolonialization process.

One could also interpret the wave-like outbreak of wars as swinging close to the sides of a relatively straight line: the decolonization process is by and large finished as a historical event, and the Cold War quickly settled into routine patterns, at least after the Cuban Missile Crisis of 1962, and evolved into an even more stable system with the dawn of détente. This interpretation would also mean, however, that the East-West conflict was only of minimal importance for the causes of wars since 1945. None of this changes the fact that in 1990 and the following years 1991-92 a rise in wars is clearly noticeable. This rise is attributable to the demise of the Soviet Union, and its very disintegration shows that the socialist reorganization of society since 1917 was far from permanent. The socialist modernization process only partially substituted traditional social conditions. As the Soviet Empire shattered ignominiously, nationalism and separatism once again arose in a haunting repetition of what had long since worked through the European core in the nineteenth and early twentieth centuries.

The term "proxy-war" is often used in politics and the media (and in parts of the scientific community as well) to subsume all wars since 1945. This term serves to hide the real foundations of these wars. It is a formulaic phrase with its origins in ignorance, ideology, or pure propaganda. This expression creates the false impression that the warring parties in the Third World—the governments and any opposition groups—were mere puppets of the superpowers or their respective blocs.[3] The worst thing about the usage of this term is that it distorted interpretation of the structural causes of real conflicts. It was symptomatic of this problem that neither the USA in Cuba, Vietnam, Lebanon or Grenada, nor the Soviet Union in Hungary, Czechoslovakia or Afghanistan, really comprehended what was being played out in these places. This ignorance alone cost the lives of untold thousands, not to mention other short- and long-term consequences of the respective (usually disingenuous) national interests or sociopolitical ideologies. This was the "Rule of Folly," as Barbara Tuchman (1984) formulated it. Or perhaps Karl W. Deutsch hits the nail even more squarely on the head when he wrote in 1973 that those who have power never learn. To put it succinctly: power makes one dumb!

The seemingly steady worldwide increase of wars indicates that *war seems to be becoming more and more the norm* (though the majority of countries in the world remain free of war[4]). Yet another indicator can

Table 2: Yearly Account of Countries' War Participation by Month

Year	Months	Year	Months	Year	Months
1945	19	1961	246	1977	497
1946	74	1962	263	1978	530
1947	111	1963	228	1979	577
1948	206	1964	356	1980	507
1949	131	1965	428	1981	557
1950	232	1966	395	1982	564
1951	313	1967	363	1983	583
1952	308	1968	365	1984	613
1953	229	1969	427	1985	621
1954	110	1970	450	1986	634
1955	161	1971	475	1987	669
1956	178	1972	453	1988	661
1957	183	1973	394	1989	565
1958	212	1974	382	1990	625
1959	145	1975	402	1991	726
1960	189	1976	487	1992	773
				Total	**18617**

Legend: Each number indicates the total number of months in which all countries participated in war in the given year. The number of sub-state warring parties (e.g., Guerrillas) is not reflected here.

control for the development of wars: a yearly account of countries' war participation by month. All the months in which all countries participated in a war are added together for each year. The number of months is accordingly increased if a country was involved in several wars in the same year. For internal wars not every individual actor, such as guerrillas or feuding military factions are considered, but only the warring parties (including any intervening third parties). This indicator shows the frequency of the wars waged each year, the number of countries participating in them and the duration of their participation.[5]

Figure 2: Yearly Account of Countries' War Participation by Month

Altogether the participating countries waged 18,617 months of war during the time frame of our study (table 2), or approximately 387.9 months per year. The corresponding graph (figure 2) shows somewhat greater fluctuation, but overall it shows a rather steady increase as does the frequency of wars waged per year (figure 1).

The fact remains that *the burden of war has more or less increased continuously since 1945, that is for almost a full half century.* When we compare the averages of the first and the last five years of the study we see that the number of wars waged per year has increased almost five-fold and the frequency of the yearly account of countries' participation by month has increased by more than six times. We have not been able to establish any significant influence of the East-West conflict and the decolonization process on this trend. Apparently this trend is the result of long-term and deeply-engraved processes.

4.3 War in the Peripheries, Peace in the Center: The Regional Distribution of the Wars

In this section we move from the temporal to the spatial: to the war zones. How is the frequency of wars distributed among the different regions of the world?[6] Where do more wars take place, and where fewer? How has this distribution developed over time?

	Wars	Percent
North America	0	0
Europe	12	7
Latin America and Caribbean	29	16
Near and Middle East	41	22
Africa	48	26
Asia, Oceania, Australia	54	29
Total	**184**	**100**

If one uses the continents as primary categories then the frequency of the 184 wars from 1945 to 1992 are distributed in the following order:

This overview shows a significant rank order related, perhaps, to the average developmental level of the respective continent.

Because the continents differ greatly in their political and socioeconomic aspects they must be broken down further. As we explained in section 2.3 this sub-division is particularly difficult and our solution is not yet fully satisfactory.

In table 3, and even more dramatically in figure 3, we notice that North America shows no war at all (and hence is not even included in these figures) and that there were only twelve wars in Europe. All of the remaining 172 wars (93 percent) occurred—or are still occurring—in the regions known popularly as the "South," the "Third World," the "periphery," or more optimistically as "developing countries" or "transition-societies." From this geographical area the industrialized and war-free islands of Japan, Australia and New Zealand must be excluded.

The economic, political and military centers of the world have thus, since the Second World War, been either free of war or experienced very few wars. Closer inspection of the twelve European wars shows

Table 3: Regional War Frequency

	Wars	Percent
Central America/Caribbean	15	8
South America	14	8
Europe	12	7
North Africa	9	5
West Africa	8	4
Central Africa	10	5
East Africa	13	7
Southern Africa	8	4
Near and Middle East	41	22
South and Southeast Asia	47	26
East Asia	7	4
Total	**184**	**100**

Note: For South and Southeast Asia 33 of the 47 wars took place in India.

Figure 3: Number of Wars by Region

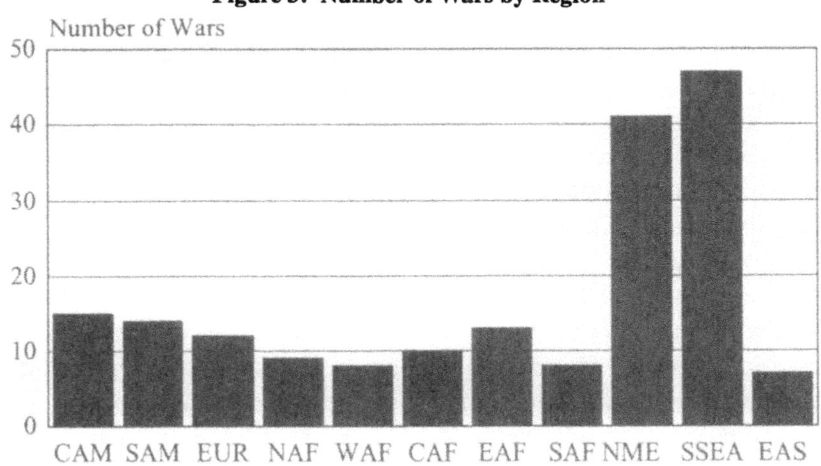

that these wars occurred without exception on the periphery or margins of Europe, that is, in regions plagued by socioeconomic problems (Greece; Cyprus; Spain, including Basque country; Northern Ireland; Hungary; Romania; Moldavia; Yugoslavia).[7] From this we can derive the rough hypothesis that lasting peace in a region is more likely the more its social form is middle class and capitalist and the greater its economic development, including the correlates of rule of law, democracy, pluralism and social welfare.

The East Asian region (China [including Tibet], Taiwan, Japan, the two Koreas, Mongolia and the Asiatic portion of the former USSR/ Russian Federation), however, comes away with even fewer wars than Europe, only seven in all. This is the case although only Japan counts as a mature capitalist industrial power (subject to a few qualifications) and South Korea and Taiwan can be considered threshold capitalist countries, while socialist China, despite its early capitalist "special economic zones" in its Eastern zones, remains a giant developing country. Furthermore both figures show that several African sub-regions (with the largest number of poor countries in the world) have less wars than Europe, and that the Latin/South America regions and East Africa record only an insignificantly greater number of wars. To be viable, the hypothesis about the relative peace of the industrialized countries requires the consideration of these additional aspects. To these belong the observation that the Central America/Caribbean region shows the same susceptibility to war as South America. Both these regions share similar political and social structures, yet the former consists of small countries (Mexico aside) while the latter consists overwhelmingly of large countries.

Both figures, however, clearly show a high susceptibility to war in two distinct regions: the Near and Middle East (NME, 41 wars) and South/Southeast Asia (SSEA, 47 wars). Outside of Africa these are in fact the most explosive flash points in the world. These region also present the greatest problems for analytical demarcation. It is not possible to meaningfully subdivide the NME region despite its political, socioeconomic and ethnographic heterogeneity. The complexity of the conflict structure, where conflict in one region spills quickly over into another, speaks against any subdivision. Yet paradoxically, perhaps, this region is less intensely and less directly imbued with European colonial structures than, for example, Africa, the Indian subcontinent or Indochina.

Even greater problems emerge from the heterogeneous SSEA region which stretches from Pakistan to New Zealand. Despite a host of interconnections we can subdivide this region into four parts: the Indian subcontinent (from Pakistan to Bangladesh and Bhutan) with 20 wars (11 percent); Indochina (from Myanmar (Burma) to Vietnam) with 13 wars (7 percent); the rest of Southeast Asia (Malaysia, Indonesia, Brunei, Papua New Guinea, the Philippines and Oceania) with 14 wars (8 percent); and the industrialized countries of Australia and New Zealand with no wars. Viewed in this disaggregated way rather than as one unit, with the exception of the Indian subcontinent the SSEA region does not differ markedly in its war-proneness from most other regions.

Despite internal regional differences and the imperfections of our subdivisions we are comfortable stating that *the bourgeois centers of the world are by and large pacified. War occurs in the comparatively backward regions*, if one takes as a point of measurement for backwardness the fully unfurled capitalist mode of re-production with its attendant general welfare and respective social and political achievements—and what other measure could there be? Furthermore, when one compares the (sub-)regions, even South and Southeast Asia, the Near and Middle East still shows extreme susceptibility to war, while the other regions are not so strongly differentiated: the numbers of wars span from 7 (in East Asia) to 20 (in the Indian subcontinent) with a mean of 12 wars. No regionally specific patterns about the distribution of war frequency occur at this level of aggregation.

Figure 4
Wars in Each Region in Five-Year Periods

Central America/Caribbean

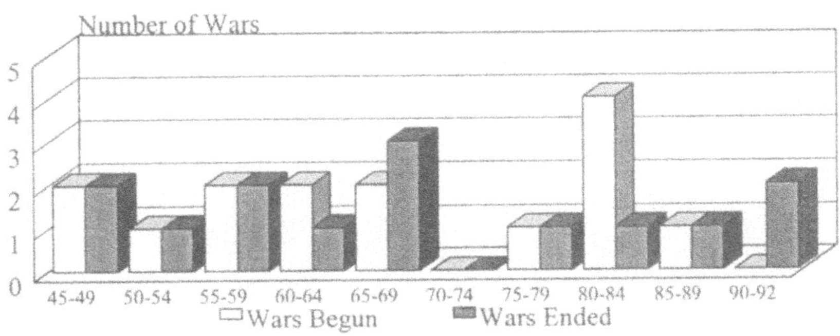

South America

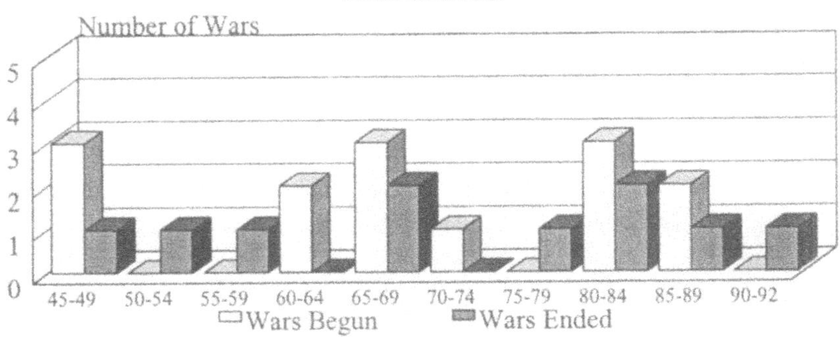

Figure 4
Wars in Each Region in Five-Year Periods, cont'd.

Europe

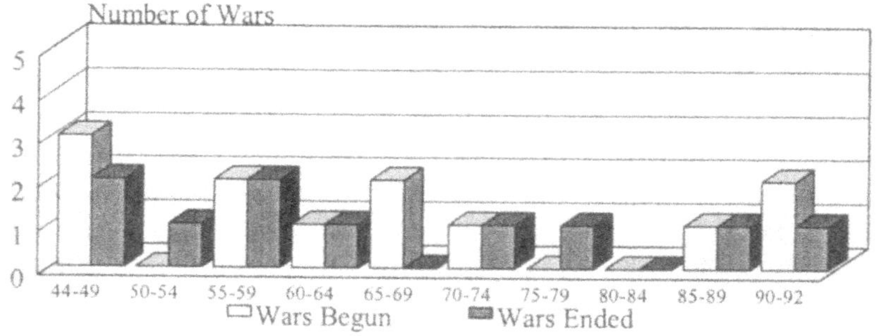

North Africa

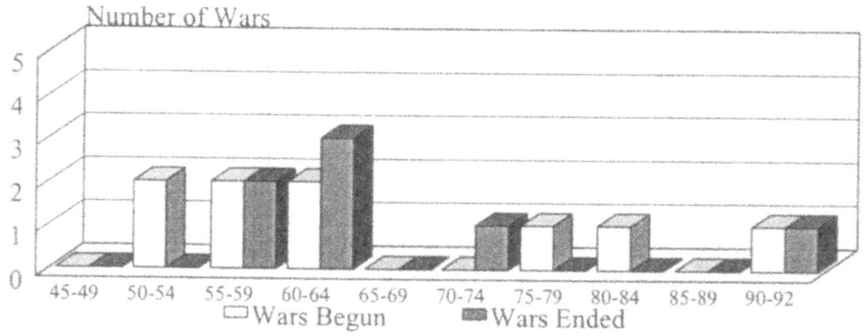

West Africa

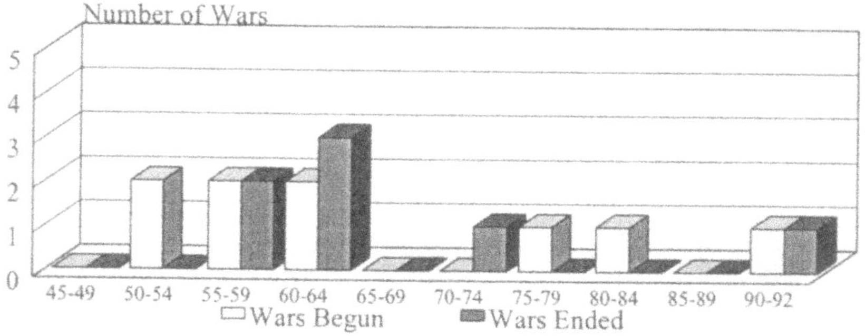

Figure 4
Wars in Each Region in Five-Year Periods, cont'd.

Central Africa

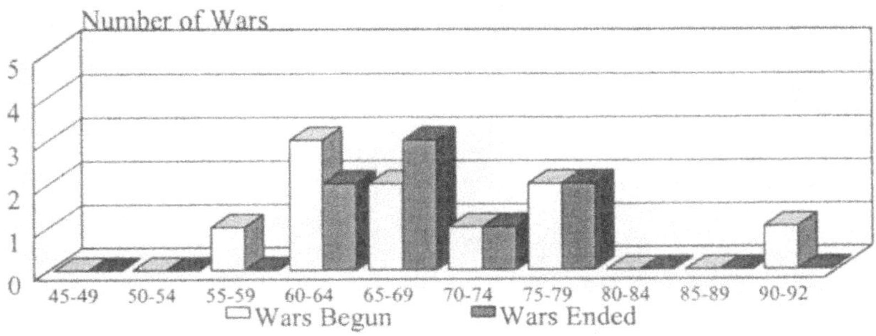

East Asia

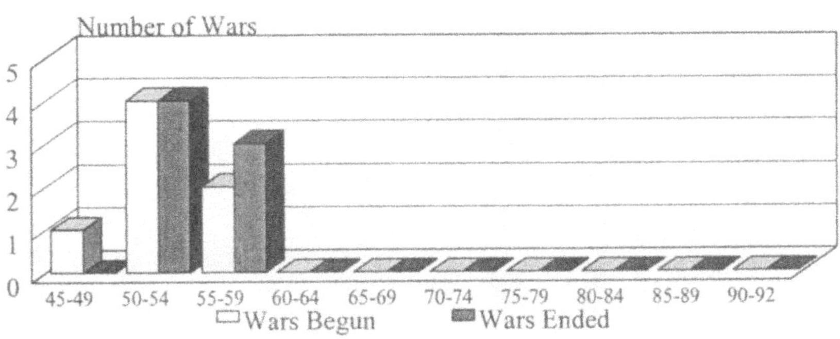

Southern Africa

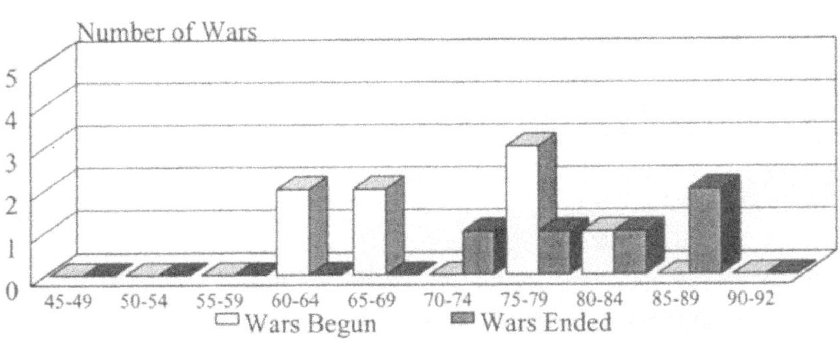

Figure 4
Wars in Each Region in Five-Year Periods, cont'd.

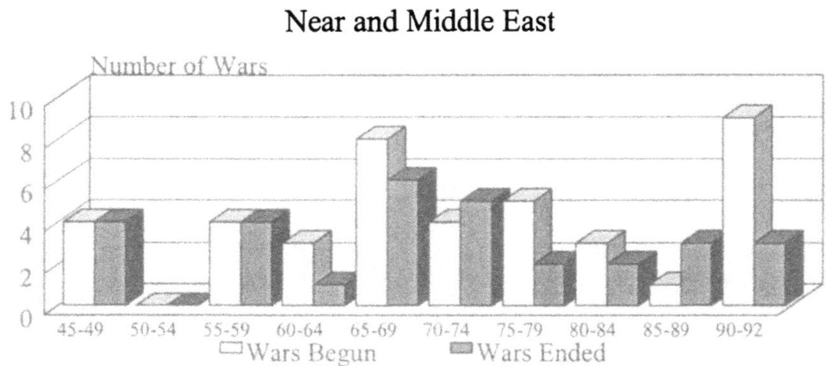

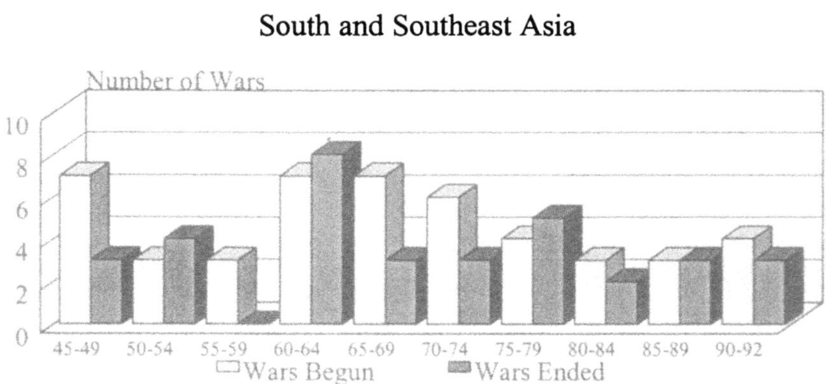

Let us now consider the trends of regional war frequency over time. Figure 4 shows how many wars were begun in each five-year period.[8] Even a cursory glance at the graphs suggests that the regional tendencies are, in part, quite different.

The frequency of wars in Central America/Caribbean and South America is almost identical (15 and 14 wars, respectively). Even the temporal distribution is very similar, although no single war occurred in SAM in the 1950s. This similarity becomes even clearer when one adds the number of wars begun per period in both regions (sometimes differences hide more than they illuminate):

	45-49	50-54	55-59	60-64	65-69	70-74	75-79	80-84	85-89	90-92
CAM	2	1	2	2	2	-	1	4	1	-
SAM	3	-	-	2	3	1	-	3	2	-
Total	**5**	**1**	**2**	**4**	**5**	**1**	**1**	**7**	**3**	**-**

The periods when Latin America was most prone to wars were the late 1940s, the 1960s, and the 1980s (above all the first half). The causes behind this cycle must still be researched. It also remains unclear whether one can speak of a pacification trend in the 1970s when fewer wars were instigated (12 as opposed to 17 in the previous period).

The case of Europe requires more detail. We see here a similar decisive point in the number of wars soon after the downward trend in Latin America: nine of the twelve wars broke out before 1974. Four of these nine wars arose in connection with the Cold War (Greece 1944/45 and 1946-49, Spain 1945-50, Hungary 1956), three concerned themselves at first with anti-colonialism, but became increasingly nationalist regarding Cyprus (1955-59, 1963-64, 1974), and the two longest wars in postwar Europe broke out at the end of the 1960s with the separatist struggle in Basque country (1968-79) and the civil war in Northern Ireland (since 1969). Yet beginning with the middle of the 1970s an almost fifteen-year-long period followed in which no new war broke out. In this sense we can speak of a clear tendency toward pacification. Only since the end of 1989 have three new wars broken out, all of which are connected more or less with the fall of the Soviet bloc: the short but intense war in Romania in 1989/90 against the Ceaucescu dictatorship, the separatist war between nationalities in Moldavia in 1992, and above all the terrible war in the ruins of Yugoslavia.[9] It is also uncertain how the neighboring nationalist conflicts in the southern Balkans and northeastern and eastern Mediterranean will develop (Serbs/Kosovo Albanians; Albania/Greece; Greece/Macedonia; Turkey/Greece; the division of Cyprus; and possibly even Turkey/Bulgaria).[10]

Still we should not overlook the progress toward peace in this period as well: Northern Ireland seems to be moving gradually toward a peaceful settlement; the occasional violent acts in Basque country have not qualified as war for some time; the wars in Romania and Moldavia were over quickly; and the Yugoslavian tragedy will eventually produce a new state order in the region which over time will most likely return to some sort of federation in order to compete in the world economy. In sum, when the aftershocks of the collapse of the USSR and the Eastern Bloc subside Europe will most likely continue its historical process of pacification.

A quick glance at the diagrams depicting the African regions shows notable differences. Wars in connection with the decolonization process occur earlier in North Africa (NAF) and East Africa (EAF) than in the other African regions. But while a pacification trend is discernible for NAF we cannot assume the same for EAF. In West Africa (WAF), in Central Africa (CAF) and in Southern Africa (SAF) the periods of war began mostly in the 1960s. But while wars increased significantly in West Africa in the 1980s after an approximately ten-year pause, in SAF and CAF the frequency of wars declined over the same period. Thus we have a very heterogeneous situation. Perhaps once again the necessary regional differences hide the structural process behind these events. For this reason we summarize the frequency of the outbreak of wars in Africa:

	45-49	50-54	55-59	60-64	65-69	70-74	75-79	80-84	85-89	90-92
NAF	-	2	2	2	-	-	1	1	-	1
WAF	-	-	-	2	1	-	-	1	2	2
CAF	-	-	1	3	2	1	2	-	-	1
EAF	1	1	-	3	-	1	4	1	1	1
SAF	-	-	-	2	2	-	3	1	-	-
Total	**1**	**3**	**3**	**12**	**5**	**2**	**10**	**4**	**3**	**5**

Here we discern a sudden increase in anti-regime, independence and secession wars during the first half of the 1960s (12 wars). In the second half of the 1960s the frequency of war diminished, but remained nonetheless relatively high (5 wars). The first noteworthy reduction in the number of wars occurred in the first half of the 1970s. This was, however, only the quiet before the storm, for the second half of the 1970s shows once again a high number of new wars, especially anti-regime wars (10 wars). After this the frequency sinks again, but not to such a low point as in the preceding period—now the frequency of new wars lies somewhere in the middle. The situation in Africa remains very unstable.

Similarities appear when one compares the diagrams for the Near and Middle East (NME) and for South/Southeast Asia (SSEA). After the apex of war susceptibility in the 1960s a reduction in the number of wars is apparent for both regions. Only at the start of the 1990s is a new rise in wars observable, less in SSEA and much more so in NME, attributable to conflicts in the south of the former USSR.

The figures for East Asia are especially notable. Since the beginning of the 1960s no new war has broken out there. The previous wars belonged almost entirely to China: the Chinese Civil War, the war for the conquest and control of Tibet, the battles over islands with Taiwan (then called Nationalist China). China also participated as an intervening power in the Korean War. We attribute two further instances—the Chinese incursion into India in 1962 and the "educational campaign" against Vietnam in 1977-78—to the SSEA region. The pacific order in East Asia is nonetheless precarious. It depends completely on the stability of the People's Republic of China, which itself is increasingly challenged by the inevitable march of the capitalist mode of reproduction and the corresponding democratic movement within China. The political inflexibility of China's leaders is a greater threat to peace than North Korea's threatening nuclear gestures.

Figure 5 indicates the regional burden of war by showcasing wars waged per year. We have already determined that wars are increasing overall when all regions are considered. This rise in wars is apparently attributable to the following regions: Latin America, especially South America (Columbia and Peru), East Africa and Southern Africa (though there has been a decrease in recent years); and above all South and

Figure 5: Wars Waged Yearly by Region

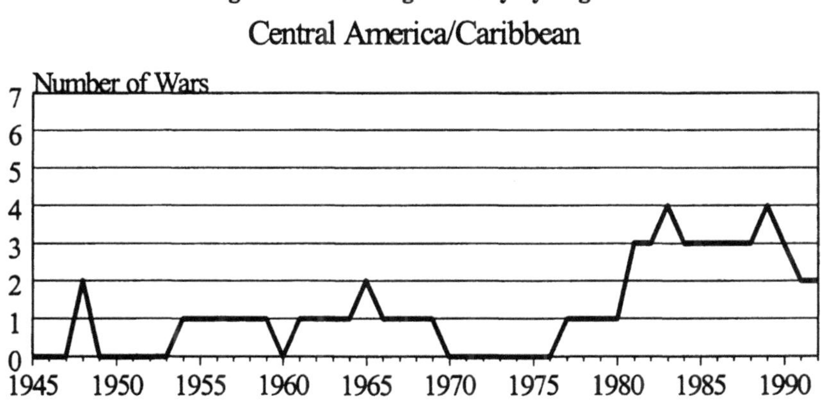

Figure 5
Wars Waged Yearly by Region (cont'd.)

South America

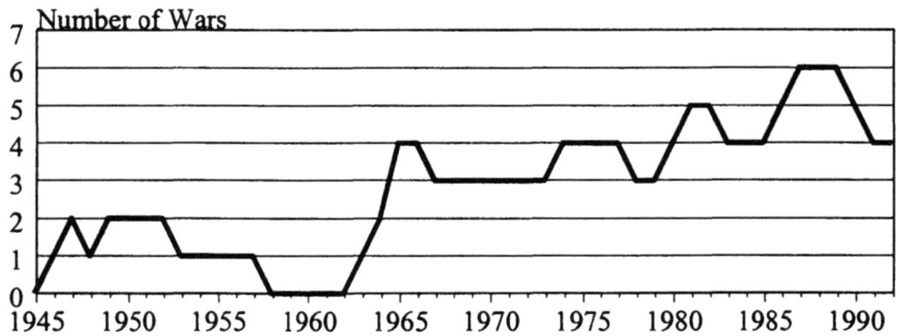

Europe

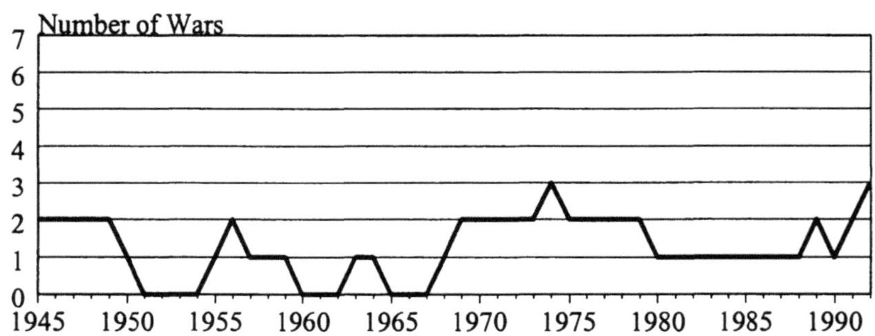

Africa

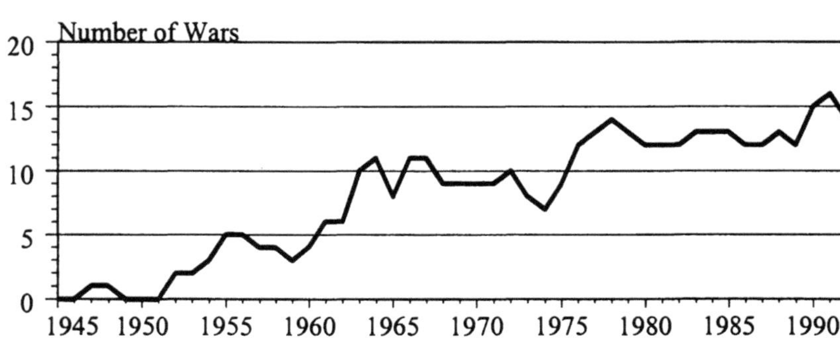

Figure 5
Wars Waged Yearly by Region (cont'd.)

North Africa

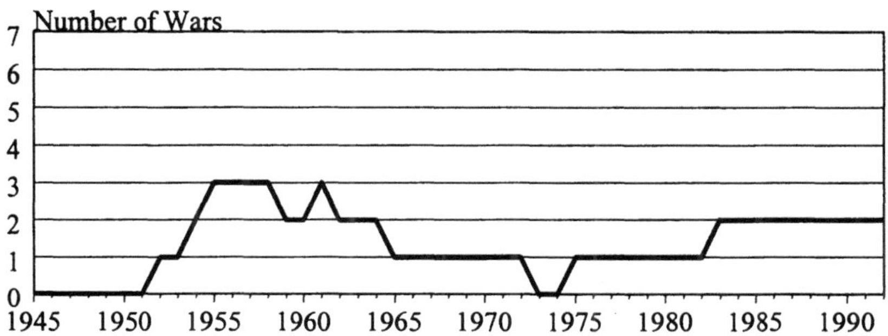

West Africa

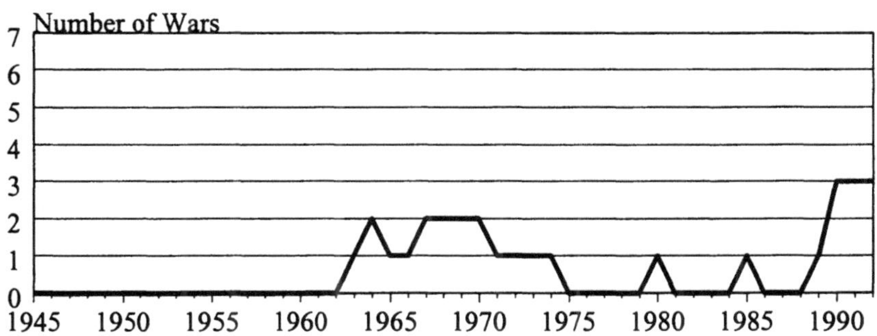

Central Africa

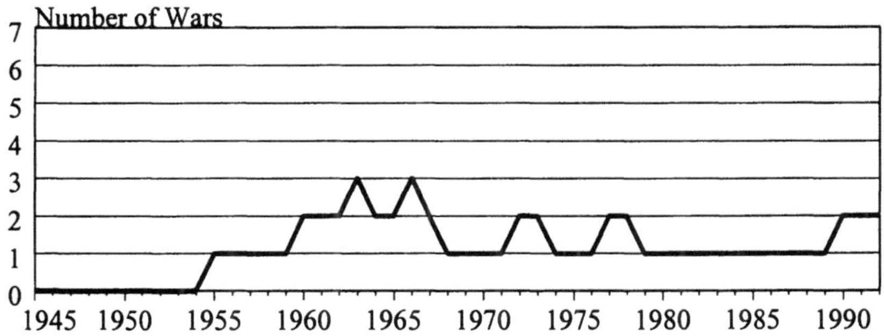

Figure 5
Wars Waged Yearly by Region (cont'd.)

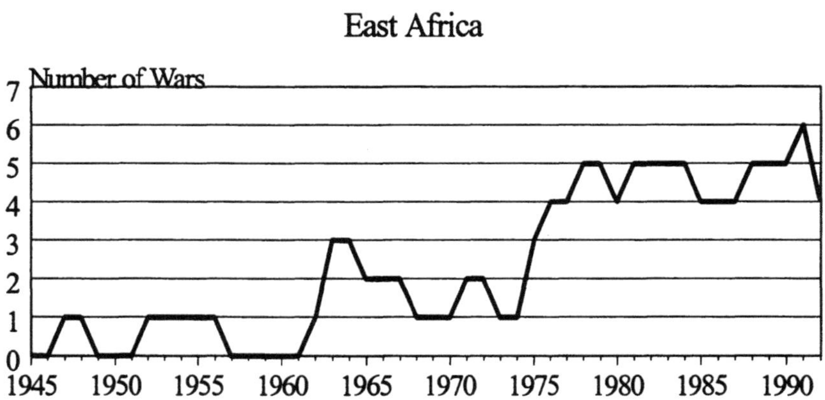

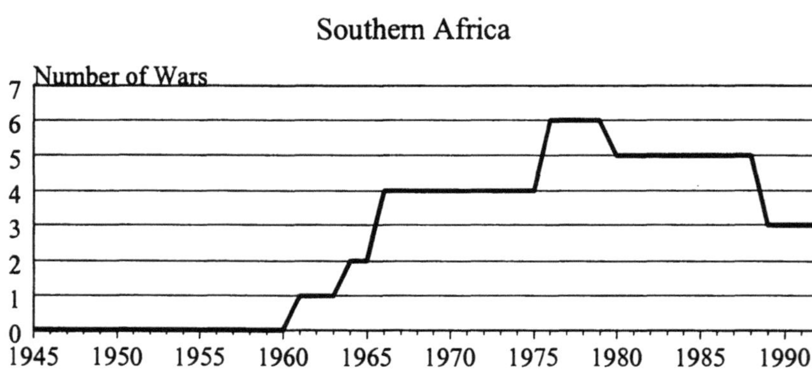

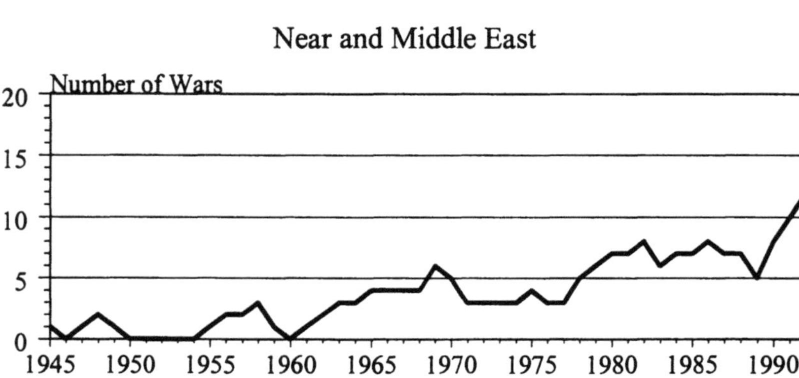

Figure 5
Wars Waged Yearly by Region (cont'd.)

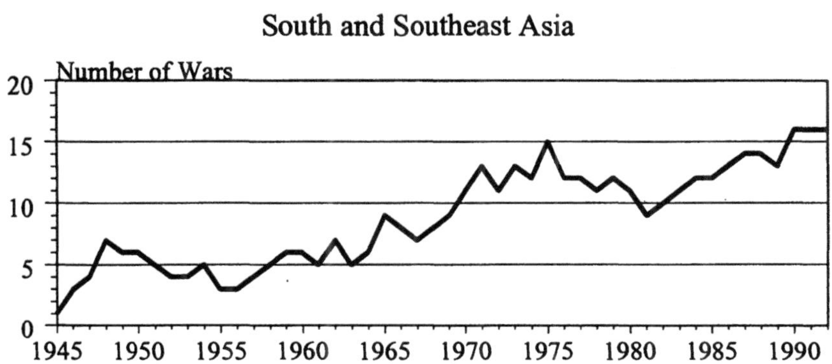

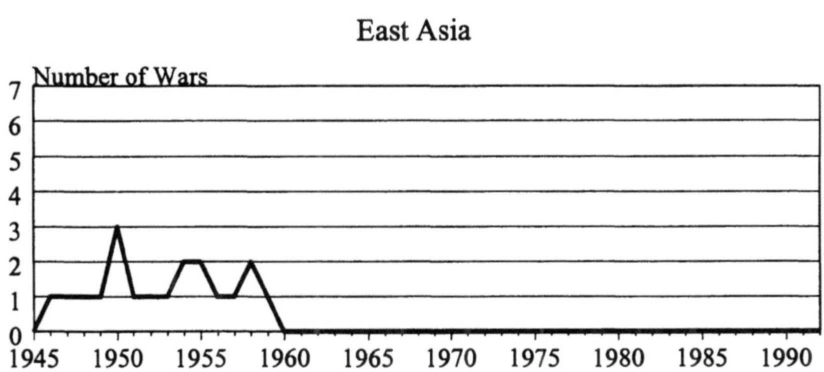

Southeast Asia and the Near and Middle East, where both regions show a significant and steady rise starting from an already high frequency of wars.

For the rest of the regions—Europe, West and Central Africa and East Asia—the level of war frequency is relatively low and fluctuating, which implies shorter wars. There is an overall increase for Africa when viewed as a whole, but it is not steady. A gradual increase in yearly-waged wars was followed by more of a plateau in the early 1960s, with waves of new wars in the late 1970s and early 1990s.

To summarize, *with the exception of East Asia* (where apparently special conditions apply which require more research) the statistics show *no fundamentally divergent distributions or trends regarding the war-proneness and burden of war in the peripheries* from which one could conclude that regional characteristics alone can explain the causes of these wars. There are differences in magnitude, but even these differences are dispelled when one further differentiates the sub-regions. This does not deny the existence or importance of regional characteristics, but suggests that such conclusions cannot be directly inferred with the methods used here.[11] (Though we cannot exclude the possibility that the spread of capitalism has advanced so far in the peripheries that regional differences are leveling out.[12]) Most importantly we observe that *wars have shifted from the center of capitalist society (the "core" of Western Europe, North America, Australia and New Zealand) completely to the periphery*. This shift occurs after the excessively violent state and civil wars which plagued the center in the nineteenth century and the first half of the twentieth. This shift marks a veritable "qualitative leap" in world history.

4.4 The Wagers of War: The Third World at War with Itself

The subtitle of the previous section, "war in the peripheries, peace in the center," is a play on the expression which originated at the time of the French Revolution: "peace in the slums! war in the palaces!" Our portrayal of events, however, leads to a reverse formula: war in the slums! peace in the palaces! This characterization is both correct and incorrect, for reality is more complicated. There have doubtless been anti-colonial or anti-imperialist wars of the "slums" against the colonial powers' "palaces." But afterwards the "native" successors, often the military, tended to move quickly into the abandoned palaces while the slum dwellers remained in their humble abodes. Occasionally these denizens attempted to storm the palaces of their new rulers. In such cases the former "palace inhabitants," the colonial powers, would once again intervene. To gain greater insight into the wagers of war themselves we turn now to the actors or, in the sober language of our databank, "the war participants."[13]

Table 4 lists all countries involved in wars in rank order according to the frequency with which they were drawn into war, whether for offensive, defensive, or intervening reasons. In contrast to our conclusions at the end of the last section this table shows that, when measured by the frequency of participation in war, several Western industrial countries join Third World states at the top of the list. Along with India, Iraq, the People's Republic of China and Syria we find Great Britain (18 participations), the USA (12 participations) and France (11 participations). All together these Western industrial states account for 11 percent of all participation in wars. This merits closer investigation. Is internal peace compensated for by increased external aggression, as anthropological hypotheses and psychoanalytic analogies would often have us believe? Are the civilizational achievements of industrial societies merely facades behind which pent-up political violence threatens to explode?

Table 4: Frequency of Participation in War, 1945-1992 (including armed U.N. participation)

Rank	State	Nr.	Percent
1.	Great Britain	18	4.8
2.	India	16	4.3
3.	United States of America	12	3.2
	Iraq	12	3.2
4.	France	11	2.9
5.	People's Republic of China	10	2.7
6.	Syria	8	2.1
	Subtotal 7 Countries	87	23.3
7.	Egypt	7	1.9
	Ethiopia	7	1.9
	Indonesia	7	1.9
	Israel	7	1.9
	Pakistan	7	1.9
	South Africa	7	1.9
	Zaïre	7	1.9
	Subtotal 14 Countries	136	36.4
8.	AR Yemen	5	1.3
	DPR Yemen	5	1.3
	Greece	5	1.3
	Iran	5	1.3
	Columbia	5	1.3
	Laos	5	1.3

	Nigeria	5	1.3
	North Vietnam	5	1.3
	Philippines	5	1.3
	Thailand	5	1.3
	Turkey	5	1.3
	USSR/RF	5	1.3
	Subtotal 26 Countries	196	52.4
9.	Belgium	4	1.1
	Costa Rica	4	1.1
	Honduras	4	1.1
	Jordan	4	1.1
	Cuba	4	1.1
	Morocco	4	1.1
	Nicaragua	4	1.1
	Peru	4	1.1
	Spain	4	1.1
	Vietnam	4	1.1
	Subtotal 36 Countries	236	63.1
10.	Algeria	3	0.8
	Australia	3	0.8
	El Salvador	3	0.8
	Georgia	3	0.8
	Guatemala	3	0.8
	Lebanon	3	0.8
	Mali	3	0.8
	New Zealand	3	0.8
	Netherlands	3	0.8
	Portugal	3	0.8
	Rhodesia	3	0.8
	Saudi Arabia	3	0.8
	Somalia	3	0.8
	Sudan	3	0.8
	South Vietnam	3	0.8
	Taiwan	3	0.8
	Tanzania	3	0.8
	Uganda	3	0.8
	Subtotal 54 Countries	290	77.5

Rank	State	Nr.	Percent
11.	Argentina	2	0.5
	Ceylon/Sri Lanka	2	0.5
	Ghana	2	0.5
	Italy	2	0.5
	Cambodia	2	0.5
	Canada	2	0.5
	Kuwait	2	0.5
	Libya	2	0.5
	Mozambique	2	0.5
	Nepal	2	0.5
	(Muskat and) Oman	2	0.5
	Paraguay	2	0.5
	Rwanda	2	0.5
	Senegal	2	0.5
	South Korea	2	0.5
	Tibet	2	0.5
	Tunisia	2	0.5
	Zimbabwe	2	0.5
	Cyprus	2	0.5
	Subtotal 73 Countries	328	87.7
12.	Afghanistan	1	0.3
	Angola	1	0.3
	Azerbaijan	1	0.3
	Bangladesh	1	0.3
	Burma	1	0.3
	Bolivia	1	0.3
	Bosnia	1	0.3
	Brazil	1	0.3
	Brunei	1	0.3
	Burkina Faso	1	0.3
	Burundi	1	0.3
	Dominican Republic	1	0.3
	Djibuti	1	0.3
	Ecuador	1	0.3
	Gambia	1	0.3
	Grenada	1	0.3
	Guinea	1	0.3
	Hyderabad	1	0.3

Rank	State	Nr.	Percent
12. cont'd	Ireland	1	0.3
	Yugoslavia	1	0.3
	Cameroon	1	0.3
	Kashmir	1	0.3
	Kenya	1	0.3
	Croatia	1	0.3
	Liberia	1	0.3
	Luxembourg	1	0.3
	Malay Federation	1	0.3
	Malawi	1	0.3
	Malaysia	1	0.3
	Mauritania	1	0.3
	Moldavia	1	0.3
	Niger	1	0.3
	North Korea	1	0.3
	North Ossetia	1	0.3
	Panama	1	0.3
	Papua-New Guinea	1	0.3
	Romania	1	0.3
	Zambia	1	0.3
	Sweden	1	0.3
	Sierra Leone	1	0.3
	Slovenia	1	0.3
	Surinam	1	0.3
	Tadjikistan	1	0.3
	Chad	1	0.3
	Hungary	1	0.3
	Venezuela	1	0.3
	Grand Total 119 Countries	374	100.0

Without delving more deeply into theoretical considerations (see Siegelberg 1994) the hypotheses at the end of the last paragraph contradict several empirical facts. First, there are important industrial societies which are either missing entirely from the rank order or rank at the bottom of the list, and then most only because of participation in U.N. actions (e.g., Japan, Canada, Scandinavia, Austria, Switzerland, Italy and the two former German states). Among these countries we find some of the most fearsome warmongers of the first half of the twentieth century. Second, Great Britain and France were the largest former colonial powers. At the beginning of the decolonization process they reacted primarily with violence against the dissolution and self-determination of their colonies.[14] Even after decolonization they intervened preferentially in the internal struggles of their former colonies for reasons of direct interest, as a hegemonial power, or because they were asked to by the warring parties. This behavior follows a historically vanishing model. The old colonial maps, however, have yet to disappear fully from the minds of the "motherlands'" governmental, bureaucratic and military structures. The increasing number of U.N. or U.N.-sanctioned interventions in recent years gives hope that the traditional colonialist approach is disintegrating in favor of a "world-order" solution.[15]

The high degree of direct U.S. participation in wars can be explained in two ways: first, the U.S. has always claimed control over the so-called Western Hemisphere, meaning above all Latin America. Here, and in several pacific territories (the Philippines, Hawaii) the U.S. has acted identical to a colonial power. Central America and the Caribbean have especially been considered the "backyard" of the United States. Four of the United States' twelve[16] war participations fall in this region.[17] The majority of the time, however, U.S. interventions can be explained in the context of the U.S. hegemonial role as leader and protector of the capitalist world, a task which fell decisively to the U.S. at the end of World War II. This required containing and confronting socialist alternative models on the one hand[18] while supporting the decolonization process on the other and working to bind emerging states into the U.S.-dominated "one-world" free-market model (even if against the interests of their allied colonial powers). The role of the U.S. as protector of a capitalist world order continues after the end of the East-West conflict.[19]

Yet military power has its limits even for the mightiest nation. These limits came to light during the Vietnam War. We briefly sketch here some more recent examples:

- The small Caribbean island of Grenada (91,000 inhabitants) allegedly threatened the security of the free world by building an airstrip with Cuban assistance. Despite a rather chaotic performance by the American military they managed to invade and successfully "liberate" the country from its supposedly communist rulers (1983). There were approximately 100 dead.
- The U.S. presence in the many-faceted Lebanese war was practically eliminated in one fell swoop by a devastating terrorist attack on a marine bar-

racks that killed 241 soldiers (1983). A helpless military response followed with U.S. warships firing rounds into the mountains for a few days. One cannot even consider this response as intervention.

- In northern Angola, soldiers from Cuba defended installations owned by U.S. oil concerns against attacks by the right-leaning rebel movement UNITA, which was supported by the U.S. and South Africa (1975-1991).
- The U.S. navy provoked the militant though militarily weak country of Libya after its leader Quaddafi was accused of inciting and aiding terrorism. The U.S. air force then bombed the cities of Tripoli and Bengasi with British logistical help (which France refused to provide) (1986). The result was dead civilians, including many children.
- The attempt to liberate U.S. hostages held by the "Islamic Revolution" in Iran ended in total disaster in the desert (1980).
- The U.S. supported the Contra rebels in Nicaragua for almost ten years (1981-1990), in part illegally, in their common quest to destabilize the Sandinista government. The Contras emerged from the defeated Somoza dictatorship, yet it was the Sandinista's socialism which allegedly threatened to undermine the freedom of the Western hemisphere, and the U.S. tried to force them to compromise with the liberals and the conservative oligarchy. 60,000 people were killed.
- It took five days for an invading army to capture and bring before a U.S. court the dictatorial president of Panama, General Manuel Noriega (1989). Although originally trained by the U.S. and for a while in their pay, he was suspected of participation in drug smuggling and money laundering. Furthermore, he did not play the official game regarding U.S. interests in the Panama Canal. The result of the invasion was between 600 and 7,000 dead, 15,000 homeless, and billions of dollars of damage.
- At the head of a coalition of 28 states[20] the U.S. directed a massive, television-perfect campaign to drive Iraq out of Kuwait which it had occupied in violation of international law six months earlier (1991). Although the West had more or less openly supported and supplied Iraq with arms in its war against Iran, it was successful in turning its power against a former customer and rapidly decimated Iraq's military and economic potential. In Iraq there were between 85,000 and 250,000 soldiers and between 40,000 and 180,000 civilians killed, not counting those killed as a result of the embargo following the war (100,000 people, including 30,000 children, according to Iraqi figures). On the allied side there were 343 casualties not including 4,000 to 7,000 dead Kuwaitis and 13 Israelis. Yet despite all this the Iraqi leader Saddam Hussein was not ousted, the Shiites in the south continue to be repressed, the Kurds in the north continue to be endangered, the democratization of Kuwait did not occur, Islamic fundamentalism in Afghanistan was indirectly strengthened, and the militarization of the region was intensified. The U.S. was not even able to finance such an undertaking itself and required the contributions of other states.[21]

In short, the time of great and independent gunboat diplomacy by the industrial powers limps toward its end, just as colonialism and imperialism and its corresponding military campaigns have long disappeared. The cultivation and securing of a capitalist world order can no longer be carried out by nation-states, rather its organization occurs more and more at the global or regional level.[22] Even a militantly nationalist Russia, China and the larger threshold countries must recognize this trend. The heightened importance of human rights, U.N. reform and the rising number of interventions by U.N. or regional organizations even in internal conflicts all point to a world-historical logic. The failure of such interventions in ex-Yugoslavia, Somalia or Rwanda does not contradict this trend. If social learning processes within states are notoriously slow, even in industrialized states, then why should they be any quicker at the global level?

Let us summarize the results so far: *Peace within Western industrialized countries after 1945 contrasts with a relatively high degree of warlike behavior in the Third World. A closer look shows, however, that specific historical circumstances play a significant role (decolonization, U.S. hegemony, the failed Soviet experiment). As the influence of these circumstances dwindles we see a multitude of indicators which herald the commencement of a new international order.*

Table 4 contains a further bit of important information. The leading group of seven countries accounts for close to a quarter (23 percent) of all war participation since 1945. Together with another seven countries who participated seven times each in wars and twelve countries who participated in only five wars each (22 percent of the 119 countries who were engaged in at least one war between 1945 and 1992) these twenty-six countries make up more than 50 percent of all 374 cases of participation in wars. Fifty-four countries produce over three- quarters of participation in war (78 percent). 46 countries (39 percent of the 119) have only participated in one war each or were only affected by one war. Furthermore, there are still approximately 70 countries which enjoy peace within and beyond their borders. *We can thus observe a high concentration of war susceptibility among a relatively small number of countries*. A quick glance at table 4 also shows that *the rank order of the countries does not follow any particular pattern*, aside from the industrialized Great Powers. The list places large countries next to small ones, the relatively rich next to the extremely poor, moderately democratic governments next to authoritarian or dictatorial ones, capitalist-oriented economies next to socialist-oriented ones, private capitalism next to state capitalism, politically stable next to politically unstable lands, old next to new states, and all this without any regional clumping. It cannot be determined conclusively here whether susceptibility to war admits to law-like behavior. Against common perception and even some scientific literature there is no simple correlation between power potential or social and political structures on the one hand and the susceptibility of a country to war on the other.[23]

We should not overlook, however, that 106 (90 percent) of the countries which participate in war belong to the developing or threshold countries, accounting for 82 percent of all participation in war.[24] The Third World does seem more war-prone than other regions and is, so to speak, at war with itself, while the industrialized societies are a stronghold of peace (see Senghaas 1984:274). Using the data of the rather simple war list from Kende, Gantzel and Fabig (1982) Senghaas tried to show that this conception is false, or at the very least must be considerably relativized (see Senghaas' testimony before the *Bundestag* Committee for Economic Cooperation on February 22, 1984). It is undeniable that the industrial states supported the militarization of the Third World through weapons exports and military aid, among other activities, and thereby also contributed indirectly to the hindering of economic development and democratization.[25] Yet four arguments can be raised which weaken the significance of this truism. First, it is equally true that indigenous arms production is growing in developing countries, even if initially as a subsidiary or partner of an international firm.[26] Secondly, Third World countries, above all China and Brazil, are increasingly exporting weapons. Thirdly, the global arsenal is so ample and the arms market so highly developed and uncontrollable that anybody wishing to wage war faces no problem purchasing weapons. A causal connection between armament and war is anyhow questionable, or at least remains unproved, even for the Third World (see Ferdowsi 1990). Brzoska (1990) only showed a weak parallel between the frequency of wars and large-scale weapons sales. Many weapons transfers go to states already engaged in war. We recognize that the indirect social and economic effects of armament can kill without a single shot being fired by redirecting resources away from sociopolitical goals. Nonetheless we are concerned here with "actual" war: organized, politically motivated killing and destruction caused by weapons rather than by destructive intentions or potential alone.

Table 5: Third World Wars in Ten-Year Periods

Absolute Frequencies

Third World Wars by type of War participation	**Periods in which wars began**					**Total**
	1945-54	**1955-64**	**1965-74**	**1975-84**	**1985-92**	
Wars with participation of at least one highly developed industrialized state						
pure and mixed decolonialization wars	7	7	0	1	0	15
other Wars	4	11	5	4	3	27
Total	11	18	5	5	3	42
Wars with foreign participation only of other Third World states	1	4	4	11	3	23
Wars without foreign participation	17	18	29	21	22	107
Third World Wars Total	**29**	**40**	**38**	**37**	**28**	**172**

Row Percentages

Third World Wars	**1945-54**	**1955-64**	**1965-74**	**1975-84**	**1985-92**	**1945-92**
With participation of industrialized states	26%	43%	12%	12%	7%	100%
With foreign participation only of Third World states	4%	17%	17%	48%	13%	99%
Without any foreign participation	16%	17%	27%	20%	21%	101%
Third World Wars Total	**17%**	**23%**	**22%**	**22%**	**16%**	**100%**

Column Percentages

Third World Wars	**1945-54**	**1955-64**	**1965-74**	**1975-84**	**1985-92**	**1945-92**
With participation of industrialized states	38%	45%	13%	14%	11%	24%
With foreign participation only of Third World states	3%	10%	11%	30%	11%	13%
Without any foreign participation	59%	45%	76%	57%	79%	62%
Third World Wars Total	**100%**	**100%**	**100%**	**101%**	**101%**	**99%**

Table 5 records wars in the Third World with attention to the participation of industrialized states.[27] The term "Third World" is a very general term which has undergone many mutations in its meaning. Today the "Third World" refers mostly to the developing countries. In other words, the socioeconomic dimension has replaced the dimension of non-alignment so central to the term during the East-West conflict. The criteria by which a country is judged to be developing or industrialized is admittedly controversial and disparate. We cannot delve deeper into these theoretical and empirical-methodological problems at this point. We have simplified the issue in table 5 by counting as a Third-World war any war which did not take place in Europe. These amount to 172 (93 percent) out of the 184 wars between 1945 and 1992. Despite such generalizations we can see from the given percentages in the lower two tables that the share of wars in the Third World in which at least one industrialized state participated has decreased sharply, especially following the crest of the decolonization process in the mid-1960s. The absolute frequencies in the upper table also show that the diminishing decolonization process does not explain this decrease alone, but it is also connected with the declining participation of industrialized states in other kinds of Third World wars. The share of those wars in which a Third World state intervened has only occasionally increased. The share of Third-World wars without any third-party participation shows no observable trend, though it lies notably above the level of wars with foreign participation. So against Senghaas' contention we maintain that *the Third World is more and more "at war with itself."* The trend observed in Gantzel/Meyer-Stamer (1986:72ff.) through 1984 continues almost ten years later.

This is no coincidence. The Third World is unfortunately emancipating itself from the First World by imitation and not by adopting a leading progressive role. This is morally regrettable, and such "imitation" should be contained where possible. This form of "emancipation" indicates that the Third World is following the same bloody path which the industrialized societies seem to have put behind them. This fact hardly justifies arrogance or ignorance about wars in the Third World, but serves to remind us that the history of capitalism and democracy were always accompanied by violence. Hopefully the Third World will not require the centuries it took "us" to develop, for now reality requires only a slight modification of the French Revolutionary slogan with which this chapter began: *war in the slums, i.e., the Third World, peace in the palaces, i.e., the industrialized world.*

4.5 Types of War: A Historical Shift in the Struggle over Statehood and Power?

When "war" or "war and peace" are mentioned, most people, conditioned by the historical conservative fixation on the state, spontaneously imagine fighting between two states or groups of states.[28] These "borders in the mind" (Gantzel 1987:79/82) have resulted in two problematic consequences for research: first, international war is treated as its own special form of conflict—its own genus requiring specialized categories, theories and methods for research and not admitting of comparison with other forms of organized, collective violence.[29] Secondly, this understanding corresponds with an almost exclusive concentration of empirical research on states and on war between states, while scant attention is paid to research on "internal" wars, civil war and revolution.[30] As explained in chapter 2 we concern ourselves with *all* forms of war as long as they satisfy our criteria.

Figure 6a portrays our war data according to the types of war set out in section 2.5. This figure reveals the extent to which today's research ignores reality and contributes to political errors made by the U.N., regional organizations and individual governments regarding internal conflicts. Two-thirds (66.3 percent) of all wars since 1945 have been "internal" and only a close quarter (22.8 percent) were international wars, including "pure" decolonization wars. A further 11 percent of the 184 wars concern cases where an internal and international war overlapped or were interconnected. The wars since the Second World War are thus dominated by "internal" wars.

The beginning of this period with the Greek Civil War is a symbolic prologue for our epoch. Does the high occurrence of internal wars since World War II represent a historical peculiarity? Unfortunately we cannot answer this question fully, since our register does not go back further than 1945. To take a stab at the question we turn to data from other authors. We use here the war list from Small and Singer (1982)[31] because of its methodological and operational exactitude, even though we have substantial misgivings about their inclusion criteria and definitions. By compiling the number of wars which broke out in each of the periods below we get the following ratio of international to internal wars: [32]

	Frequency of international to internal wars	corresponds to the ratio
1816-1855	24 : 14	1 : 0,6
1856-1899	38 : 25	1 : 0,7
1900-1943[33]	26 : 23	1 : 0,9
1944-1980	30 : 44	1 : 1,5

Figure 6a
Frequency of Types of War, 1945-1992

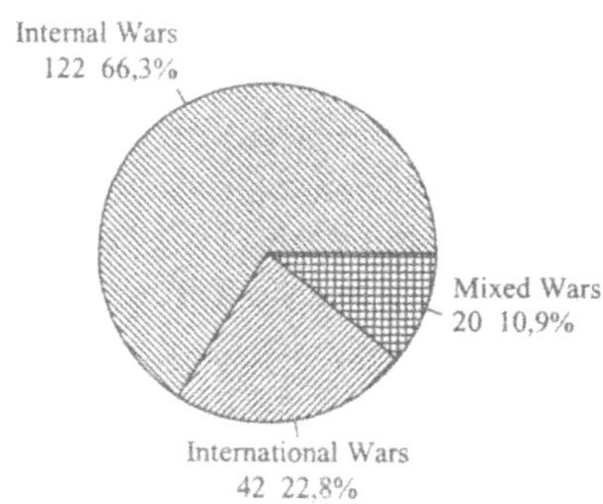

Figure 6b
Frequency of Types of War, 1945-1992

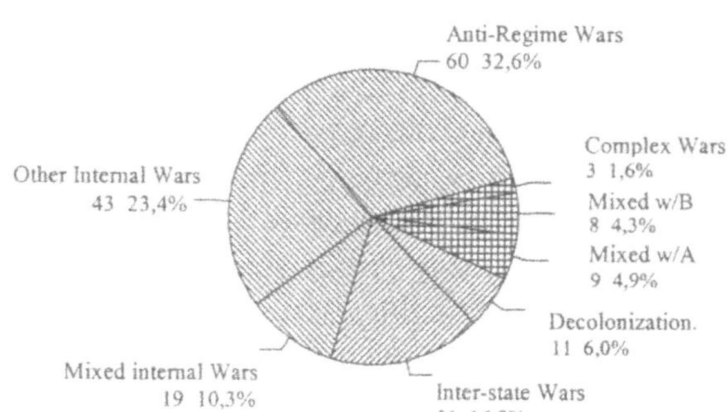

Following the Second World War the number of internal wars becomes dominant for the first time since the Napoleonic wars. Compared with the period of 1816-1855 the ratio is exactly reversed. The change begins gradually in the nineteenth century. The rate of change quickly accelerates and the jump from 0.9 to 1.5 from the first to the second half of the twentieth century is greater than ever before. We are dealing with a qualitatively historical change.

We turn next to Figure 6b, in which the three kinds of war are further broken down by type. There is a noticeably high share of anti-regime wars. This type A is the most frequent type and constitutes almost half of all internal wars. This type also plays a role in mixed internal wars (AB wars), in intra-/international mixed wars with A-components and in the three complex wars. Altogether 49.5 percent of all wars since 1945 concerned themselves in whole or in part with the attributes of anti-regime war. We can infer from this that the most significant influence on wars since 1945 comes from struggles for the shape of society and the state and for power within the state. We understand "state" here in the Weberian sense of the monopoly of physical force in a given territory.[34] However, precisely this type of state legitimacy is lacking in most, if not all, of the Third World countries where wars are waged. The principles, rules and institutional forms which allow peaceful systemic change and the sharing of power are lacking. While this is not the place for further rumination on this point, one thing is certain: the European conquest of the world through colonialism, imperialism and capitalism necessarily destroyed traditional societies without preparing the soil for modern forms of civil society. Furthermore, the lack of a developed bourgeoisie means that power is often sought for personal gain and by violent means (we see this especially concerning access to natural resources). While not forgetting the bloody centuries needed for the consolidation of power in the West (see Elias 1980, vol.2) our data confirms the hypothesis that the majority of internal wars since 1945 express the Third World's struggle to play "catch-up" with the West by unfortunate imitation.

Peaceful state formation and power sharing rests on the existence of social consensus. Precisely this consensus is missing in many Third World countries for a variety of reasons. Some lie in the failed attempts at national integration and their consequences: weak economic structures, crass income inequality and the arbitrary privileging of particular groups (see "other internal wars" which account for almost 24 percent of the total in figure 6b[35]). The lists of war by type and by subject of conflict (see appendix) show that the vast majority of pure type B wars involve ethnic or religious groups engaged in struggles for greater autonomy, secession, or annexation. Such ambitions played a major role in 38 of the 43 pure type B wars. Even among the mixed types of war we find eight similar cases and another eight among mixed wars with B components. The aggressors are not always victims of discrimination but also privileged groups who fight to maintain their privileged positions (e.g., in Biafra or Croatia) or who fight to regain lost privilege (e.g., Sikhs, Basques) due to economic or political change. Even in wars which seem exclusively about the distribution of resources ethnicity is usually manipulated to mobilize the population by one or both sides. It is for this reason that we remain skeptical about classifying a war as "ethnic." Wars are caused not by ethnic differences *per se* but by the interests of the rulers or would-be rulers who twist these differences for their own purposes. Though the source of B-type wars needs further analysis, a substantial number of internal wars result from incomplete or failed social integration.

In many cases, both A- and B-type wars are indirect consequences of former colonial influence (an observation which should not be interpreted as an alibi for postcolonial rule!). When one considers, however, that after 1945 an enormous amount of the globe was still under colonial rule,[36] then the relative peacefulness of the decolonization process is quite impressive.[37] Only eleven wars were pure decolonization wars, beginning in 1945 with the struggle of the Indonesian liberation movement against Great Britain and Holland and ending in 1976 with Spain's withdrawal from the Western Sahara. After this we find four mixed types with D components[38] as well as the war in Angola (Nr. 62, D/ABC/A), which began in 1961 as a liberation war against the Portuguese, grew into an ABC conflict in 1975 with internal power-struggles between the MPLA and UNITA (and occasionally other groups such as the FNLA and FLEC) and the struggle against the South African invasion (with Cuban and Zairian intervention), and finally transformed in 1992 into a pure anti-regime war pitting UNITA against the MPLA government. The reasons why decolonization was relatively peaceful and rapid (measured by the frequency of wars—see figure 7) lie in the weakened position of Western colonial powers after World War II, pressure from the United States, and the capacity of capitalist social struc-

Figure 7
Decolonization Wars / Anti-Colonial Liberation Wars

Colonial-power	Colonized country	Type	194. 56789	195... 0123456789	196 0123456789	197 0123456789	198 0123456789	199 012
France	Indochina	D	xxxx	xxxxx				
	Madagascar	D	xxx					
	Morocco	D		xxxxx				
	Algeria	D		xxxxxx	xxx			
	Cameroon	DA		xxxxx	x			
	Tunisia	D			x			
Netherlands	Indonesia	D	xxx					
Great Britain	Indonesia	D	xx					
	Malaya	DA		xx	xxxxxxxxxx	x		
	Kenya	D			xxxxx			
	Cyprus	DB			xxxxx			
	Aden, Yemen	D				xxxxx		
Spain	Span. Morocco (Ifni)	D		xx				
	West Sahara	D/B				xx		
Portugal	Angola	D/ABC/A			xxxxxxxxxx	xxxxxx		
	Guinea-Bissau	D			xxxxxx	xxxxx		
	Mozam-bique	D			xxxxx	xxxxx		

tures to adapt and learn. Competition between economies replaced power struggles (see Siegelberg 1994:126). Only Portugal adapted slowly because their own structural economic problems bound them to a colonial mindset vis à vis their African colonies. The epoch of direct, violent colonialism and imperialism has drawn to an end and with it the Third World completed an historically important, if preliminary, step toward emancipation and state formation.

Barely 17 percent of the wars belong to the "classical" type of inter-state war. On the basis of our data, which shows 31 pure C-type wars and 16 mixed wars with C components, we can nonetheless make the following statements (see the list ordered by war type in the appendix):

- Six wars (Nr. 18, 42, 95, 103, 115) emerged out of the Palestine conflict, that is, the establishment of the Israeli state in its Arab environment and the effect upon its neighboring countries (Nr. 120);
- Six C-type wars were consequences of the Chinese Revolution, its claims to territorial restitution,[39] and the connection with the East-West conflict, including the rivalry between China and the USSR (Nr. 22, 32, 36, 50, 66,

142); the Korean War (Nr. 25), the Vietnam War (Nr. 41 and the Cambodian wars (Nr. 99, 132) belong more or less in this context;
- Seven wars dealt with border conflicts, and internal political tensions played a role in some of them (Nr. 70, 71, 100, 112, 127, 134, 149, 164, 180);
- Five cases belong to the Indian-Pakistani rivalry, whose roots lie in both hegemony-seeking and decolonization wars (Nr. 13, 82, 84, 111, 161);
- Three wars share a (semi)colonialist background (Nr. 15, 47, 92, 154);
- Three wars take place in the Yemenite sub-region (Nr. 44, 101, 114);
- Three wars resulted from Iraqi hegemonic ambitions (Nr. 148, 182, 194);
- Two wars can be attributed to the territorial expansion of Indonesia (Nr. 64, 74, compare also wars Nr. 23, 130, 121, 198);
- Two wars were affected by South Africa's obsolete attempts to destabilize its environment in order to uphold apartheid (Nr. 62, 119);
- Two wars with C-components are the result of the collapse of state unity or failed social integration (Nr. 116, 200);
- Two wars are attributable to the U.S. desire to control its "backyard," which also entails a high degree of internal political calculations (Nr. 156, 170).

We cannot, however, observe any overall trends from these groupings.

In table 6 and figure 8, the wars waged in each year are categorized by the three types of war: "internal war" (types A, B and AB), "international war" (type C, D and CD) and the "intra-/international mixed wars," called here simply "mixed type." These figures show the temporal development of the distribution portrayed in figure 6 above.[40] These data convincingly support our finding that wars are becoming more internal. The nearly constant increase in wars since the Second World War is wholly a result of the increase of internal wars. This trend reproduces in microcosm what Small and Singer's data show for the nineteenth and first half of the twentieth century. It should come as no surprise that mixed types of wars are also on the rise, although at a significantly slower rate. The number of international wars remains on average rather low. Accordingly, security policy's almost exclusive focus on *international* threats and conflict prevention does little to address the real problems at hand.

Table 6
Types of War by Wars Waged Per Year

Year	internal	international	Mixed types
1945	3	1	0
1946	5	2	0
1947	7	3	1
1948	9	5	2
1949	7	3	2
1950	8	3	1

Table 6 cont.

1951	5	2	1
1952	4	4	1
1953	3	4	1
1954	6	5	1
1955	6	5	3
1956	7	5	3
1957	7	3	4
1958	8	4	4
1959	8	1	4
1960	6	1	3
1961	8	2	3
1962	10	3	3
1963	13	4	4
1964	16	5	3
1965	19	5	3
1966	21	4	3
1967	19	4	3
1968	20	2	4
1969	21	5	4
1970	23	3	4
1971	22	2	6
1972	21	3	5
1973	22	3	4
1974	22	2	5
1975	25	0	9
1976	26	0	7
1977	28	0	7
1978	28	0	8
1979	28	1	8
1980	27	2	7
1981	27	3	7
1982	29	3	7
1983	29	3	7
1984	30	3	7
1985	30	4	6
1986	33	3	6
1987	33	4	6
1988	34	4	6
1989	34	2	6
1990	41	1	6
1991	42	1	7
1992	45	0	6
Total	**122**	**42**	**20**

The vast majority of internal wars occurred in Central and South America, Europe, Black Africa, the Near and Middle East and South/ Southeast Asia (table 7). International wars occurred more frequently than internal wars in only two regions: North Africa and East Asia. The number of wars and the differences between the two types in these cases is, however, relatively small, so that in table 8 only the large regions are considered.[41]

In Central and South America anti-regime wars are equally dominant. The main points of conflict are social reform (often accompanied by much revolutionary and anti-imperialist rhetoric), democratization and contested political power. All these types of conflict concern the inner structure of the society, since the state formation process has been successfully completed. Territorial disagreements are not burning issues, and only in one case did a short and fairly mild C-type war erupt

Figure 8
Frequency of the Forms of War Per Year

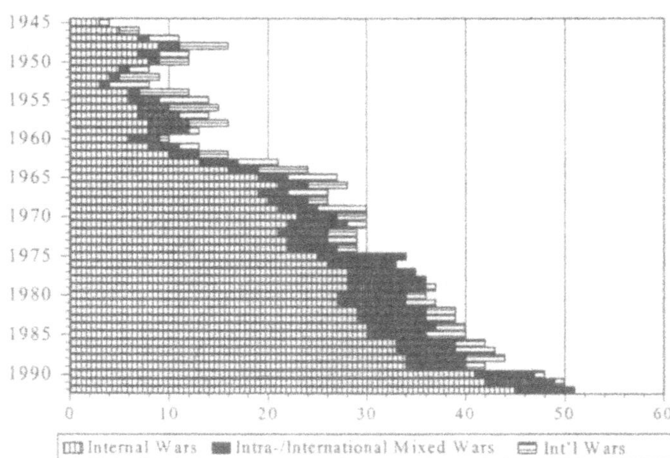

over the drawing of borders (between Peru and Ecuador). Even the short but fierce war between El Salvador and Honduras (the "Soccer War") had essentially domestic politics at its root. The remaining two C-type wars were U.S. invasions (Grenada and Panama) which had nothing to do with border issues. Even in Europe the wars were internal in nature. If we count mixed internal wars (AB-types) and mixed wars with B components together with other internal wars (B-types) we see that autonomy and secession played a greater role than anti-regime sentiment. What anti-regime sentiment there was existed in the context of the East-West conflict or its aftermath (Greece (twice), Spain, Hungary and Romania). Wars in Africa are above all internal wars, whereby A- and B-type wars are relatively equal in frequency. Anti-regime wars in Africa mostly concern shifts in power, since nondemocratic transitions often entail violent solutions. Decolonization wars also occurred disproportionately in Africa. The Near and Middle East and Asia are very similar in the frequency of war types. Internal wars dominate both regions, but B-type wars are more common than A-types, despite several socialist upheavals and numerous rivalries. Interstate conflicts play a far greater role than in any other region.

Table 7
Frequency of War Types, 1945-1992, by Sub-Region

Regions

War types	CAM	SAM	EUR	NAF	WAF	CAF	EAF	SAF	NME	SSEA	EA	Total
Anti-regime	12	11	5	1	1	6	1	3	10	9	1	60
other internal	-	1	3	1	4	-	2	1	14	17	-	43
mixed (AB)	-	-	1	1	1	2	4	1	4	4	1	19
internal Wars	12	12	9	3	6	8	7	5	28	30	2	122
Interstate	3	2	-	1	1	-	-	-	11	8	5	31
De-colonization	-	-	-	4	1	-	2	1	1	2	-	11
International Wars	3	2	-	5	2	-	2	1	12	10	5	42
AC	-	-	-	-	-	-	2	1	-	4	-	7
AD	-	-	-	-	-	1	-	-	-	1	-	2
With A components	-	-	-	-	-	1	2	1	-	5	-	9
BC	-	-	2	-	-	-	2	-	-	2	-	6
BD	-	-	1	1	-	-	-	-	-	-	-	2
With B components	-	-	3	1	-	-	2	-	-	2	-	8
Complex wars	-	-	-	-	-	1	-	1	1	-	-	3
Mixed wars	-	-	3	1	-	2	4	2	1	7	-	20
Total	**15**	**14**	**12**	**9**	**8**	**10**	**13**	**8**	**41**	**47**	**7**	**184**

Table 8: Frequency of War Types by General Region

	General Regions										Total	
	Latin America		Europe		Africa		Near & Mid- East		Asia			
War types	Nr.	%	Nr.	%	Nr.	%	Nr.	%	Nr.	%	Nr.	%
Anti-regime	23	79	5	42	12	25	10	24	10	19	60	33
Other internal	1	3	3	25	8	17	14	34	17	31	43	23
Mixed	-	-	1	8	9	19	4	10	5	9	19	10
Internal Wars	24	83	9	75	29	60	28	68	32	59	122	66
Interstate	5	17	-	-	2	4	11	27	13	24	31	17
Decolonization	-	-	-	-	8	17	1	2	2	4	11	6
Int'l Wars	5	17	-	-	10	21	12	29	15	28	42	23
With A components	-	-	-	-	4	8	-	-	5	9	9	5
With B components	-	-	3	25	3	6	-	-	2	4	8	4
Complex wars	-	-	-	-	2	4	1	2	-	-	3	2
Mixed wars	-	-	3	25	9	19	1	2	7	13	20	11
Total	29	100	12	100	48	100	41	99	54	100	184	100

Critics have oft pointed out that the frequency of wars must "naturally" increase given an increase in the number of independent states following decolonization. In response let us look at table 9 and figure 9. To begin with it is difficult to settle on a definition of "state." We tie statehood primarily to recognition by the United Nations in the context of international law, although we remain fully aware of the shortcomings of such a narrow definition. Admission into the U.N. results less from "objective" factors than from interests of Security Council members and occurred in clusters, first in 1945, then 1955/56, 1960-64, 1975/76 and 1991/92 (see appendix). To avoid statistical errors this definition also neglects small states with 500,000 to one million inhabitants and the many small states with populations under 500,000 which clamored for recognition between 1964 and 1991. Most of these states consist of small islands with minimal war-fighting capability.

The figures do not show any significant correlation between the number of new states and the number of wars. This is true not only for international and mixed wars, but also for internal wars.[42] The same applies to the false assumption that the total growth of states counts rather than growth in each individual period. This concern is addressed in table 9 with the cumulative number of states worldwide. This response may not conclusively counter the original assertion, but it strongly suggests that the enormous number of internal wars cannot be explained by statehood in any conventional sense. We must turn, rather, to societal problems and the interests and perspectives of the actors within the affected countries for a more adequate explanation. *These wars were not fought for international recognition, but result from internal conflicts within the respective countries.*

Table 9: Number of States Recognized by the U.N. in Five-Year Periods and the Frequency of Wars Begun in the Same Periods

Period	Number of States		Number of Wars Begun		
	Period	cumul-ative	internal Wars	intra-/ international mixed Wars	international Wars
1945-1949	59[b]	59	14	2	5
1950-1954	1	60	5	-	6
1955-1959	23	83	8	3	5
1960-1964	31[c]	114	17	2	8
1965-1969	6	120	19	3	5
1970-1974	7	127	9	3	2
1975-1979	7[d]	134	13	6	1
1980-1984	2	136	12	-	5
1985-1992	16[e]	152	25	1	5

a. Recognition means acceptance as a member of the United Nations. Three non-U.N. members are counted here: Switzerland, Taiwan and the Arabic Republic of Sahara (Western Sahara). Small and mini-states with less than one million inhabitants (a total of 45) are not included here with the exception of those which were involved in wars (Brunei, Djibuti, Gambia, Grenada, Suriname, Western Sahara and Cyprus. See the list in the appendix.

b. Including Switzerland and Taiwan.

c. Without Zanzibar.

d. Including Western Sahara.

e. Minus the GDR and South Yemen.

This begs the question, however, of how much influence the societal system, especially the political system, exerts on war susceptibility. In peace research terms this is the old question about the relationship between democracy and peace, which spurred Immanuel Kant's famous thoughts on "Eternal Peace" in 1795. This debate has been recently rekindled in the field of international relations. There are no firm results, but see Risse-Kappen (1994) who succinctly summarizes the debate so far.[43] Despite the multitude of definitions, variables, methods and time frames the empirical research from many different studies allows the following observation (adapted from Risse-Kappen 1994:159):

Figure 9
Number of States Recognized by the U.N. and Wars in Five-Year Periods

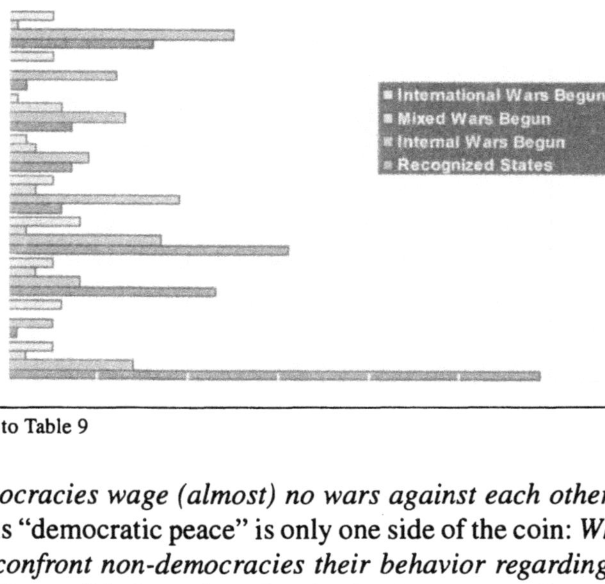

See notes to Table 9

Democracies wage (almost) no wars against each other. Nonetheless, this "democratic peace" is only one side of the coin: *When democracies confront non-democracies their behavior regarding conflict is barely discernible from that of authoritarian systems.* Democracies are drawn into war as often as non-democratic systems.

While this finding is empirically undebatable, the theoretical ability to explain it remains wanting (Nielebock 1993, Risse-Kappen 1994). The democratic peace theory is tautological when it comes to the problem of internal wars, at least in the theoretical sense that "true" democracy is characterized by its ability to peacefully regulate societal conflicts. In reality, however, the concrete prerequisites and conditions for democracy (not to mention the definition of the term itself) play a substantial role. While most internal wars do take place in non-democratic states (see table 4), even the "mother of democracy," Great Britain, cannot make peace in Northern Ireland, and local wars break out constantly in India, to name only two instances of wars affecting democratically constituted states. Unfortunately the relation of democracy to internal peace has not been well addressed by existing research, though new studies of democratization in developing countries may bring new insights. The crux of the matter is that internal peace is not solely a function of the political form of government but depends on considerations that affect every level of society. Democracy is the result of particular economic, social, legal, administrative and cultural developments and a political learning process. It cannot be reproduced cookie-cutter style. For this reason the world will continue to confront many internal wars in the Third World. Seen in this light the debate about "humanitarian intervention" assumes a growing importance.

4.6 Third-Party Intervention–A Losing Proposition?

Let us now turn to third-party participation—"intervention" for short—which requires an active role in the fighting.[44] We only consider

Table 10
Types of War with and without Intervention, 1945-1992

	Absolute Frequencies		Row-%	
	with Interv.	**without Interv.**	**with Interv.**	**without Interv.**
Intranational Wars				
Pure anti-regime Wars	18	40	31	69
Pure other internal Wars	5	37	12	88
Mixed internal Wars	6	12	33	67
Total internal Wars	29	89	25	75
International Wars				
Interstate Wars	3	26	10	90
Decolonization wars	1	10	9	91
Total international Wars	4	36	10	90
Intra-/international mixed Wars				
With A components	6	3	67	33
With B components	4	4	50	50
Complex Wars	1	2	33	67
Total Mixed wars	11	9	55	45
Wars Total	44	134	25	75

interventions in support of one or the other side, and not those which claim to "police" the situation and treat all sides equally. These latter cases include interventions to force a cease-fire,[45] to end a war, or interventions by the U.N. or under its supervision.[46]From the row totals in table 10 we see that wars with one or more interventions are not nearly as frequent as the public might believe from the selective reporting of the mass media. Only one-quarter (44 wars, or 25 percent) of the 178 wars include foreign participation. We find almost exactly the same relation (25 percent) regarding foreign participation in internal wars (see also figure 10). Among these the anti-regime type of war results most often in intervention, even in the mixed AB variation. This is not surprising, since anti-regime wars usually concern issues of the system's ultimate orientation toward one existing camp or coalition. Other internal wars are fought primarily for secession or autonomy and are thus generally of less interest to third parties. The picture changes, however, when we consider intra-/international mixed wars. Their number is relatively small overall, but in 55 percent of their cases we find intervention, especially when the war also contains anti-regime components. *The vast majority of all wars are fought without the intervention of a third party. When intervention occurs, it is within the context of support for one or both of the main warring parties.*

Figure 10
War Types with and without Intervention, 1945-1992

Absolute Frequencies

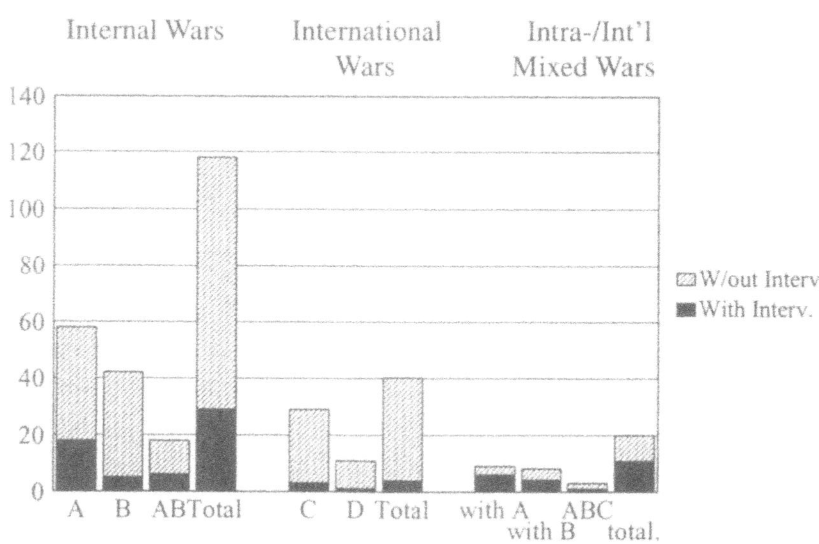

Percent Distribution per War Type

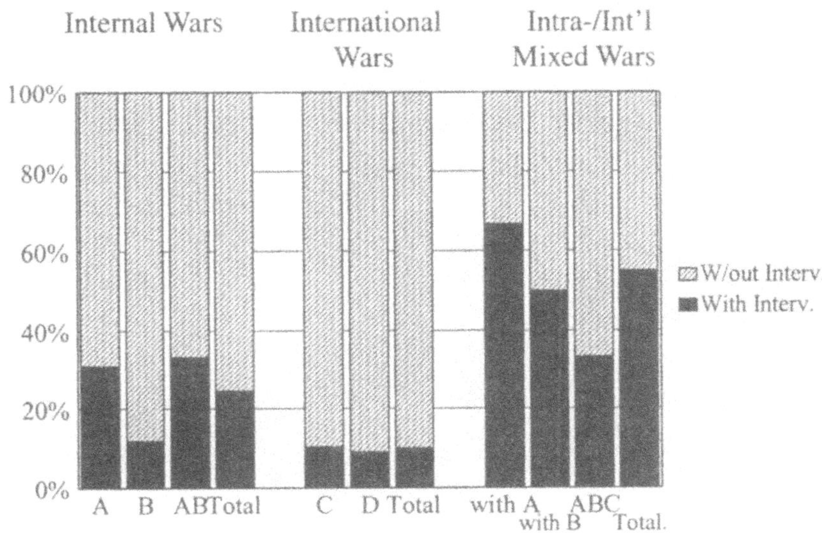

Table 11
Development of War Frequency with and without Intervention, 1945-1992, in Five-Year Periods following the Outbreak of War
Part A: All Wars

Period of	With Intervention		Without Intervention		Total	
War Outbreak	Nr.	%	Nr.	%	Nr.	%
1945-49	4	19	17	81	21	100
1950-54	2	20	8	80	10	100
1955-59	6	38	10	63	16	101
1960-64	8	31	18	69	26	100
1965-69	4	15	22	85	26	100
1970-74	4	29	10	71	14	100
1975-79	12	60	8	40	20	100
1980-84	2	12	15	88	17	100
1985-89	-	-	10	100	10	100
1990-92	2	11	16	89	18	100
Total	44	25	134	76	178	100
Average for each period	4,4	23,4	13,4	76,6	17,8	

Part B: Breakdown by Form of War

Wars	Intranational Wars			International Wars			Intra-/international mixed Wars		
starting between	with Interv.	with-out Interv.	Total	with Interv.	without Interv.	Total	With Interv.	without Interv.	Total
1945-49	3	11	14	-	5	5	1	1	2
1950-54	1	4	5	1	4	5	-	-	-
1955-59	5	3	8	-	5	5	1	2	3
1960-64	5	11	16	2	6	8	1	1	2
1965-69	2	16	18	1	4	5	1	2	3
1970-74	2	7	9	-	2	2	2	1	3
1975-79	7	6	13	-	1	1	5	1	6
1980-84	2	10	12	-	5	5	-	-	-
1985-89	-	7	7	-	3	3	-	-	-
1990-92	2	14	16	-	1	1	-	1	1
Total	29	89	122	4	36	40	11	9	20

Do changes occur over time? We can ascertain from the percentages in table 11, part A, that the share of wars with intervention fluctuates between 15 percent and 38 percent until the mid-1970s. The amount of third-party participation then rises sharply in the second half of the 1970s to over half of all wars begun in this period.[47] Starting in the early 1980s, however, the following period lies far below the average with rates of interventions fluctuating between 0 percent and 12 percent, even well before the end of the East-West Conflict. Perhaps wars with third-party intervention will become rarer as states recognize its small payoffs and difficult justification. We should be cautious with such prognostication, however, since the last time period covers only three years instead of five, although those three years have the most wars of any of the last three periods.

Parts B and C of table 11 and figures 11 and 12 further differentiate our previous statements about war-proneness and the forms and types of war. The share of internal wars with foreign participation fluctuates over time without hinting at any clear trend. This concerns above all anti-regime wars (type A-1). Other internal wars with intervention (type B-1) occur more sporadically. Mixed internal wars with intervention occur once in each period since the 1950s except for 1985-89 (type AB-10). The intervention of third parties in international wars has decreased since the 1970s, a noteworthy fact when we look back at interstate wars

Part C: Breakdown by War Type

Wars	Pure Anti-Regime Wars			Pure Other Internal Wars			mixed Internal Wars			Interstate Wars		
starting between	with Interv.	with-out Interv.	Total	with Interv.	without Interv.	Total	With Interv.	without Interv.	Total	With Interv.	without Interv.	Total
1945-49	2	9	11	1	1	2	-	1	1	-	2	2
1950-54	1	1	2	-	2	2	-	1	1	1	1	2
1955-59	3	1	4	1	1	2	1	1	2	-	4	4
1960-64	4	6	10	1	4	5	-	1	1	1	3	4
1965-69	1	6	7	-	9	9	1	1	2	1	4	5
1970-74	1	3	4	-	4	4	1	-	1	-	2	2
1975-79	5	1	6	1	1	2	1	4	5	-	1	1
1980-84	1	6	7	-	3	3	1	1	2	-	5	5
1985-89	-	3	3	-	3	3	-	1	1	-	3	3
1990-92	-	4	4	1	9	10	1	1	2	-	1	1
Total	18	40	58	5	37	42	6	12	18	3	26	29

Part D:

	Decolonization Wars			Intra-/international mixed Wars with A components			Intra-/international mixed Wars with B components			Complex Wars		
Wars starting between	**with Interv.**	**with-out Interv.**	**Total**	**with Interv.**	**without Interv.**	**Total**	**With Interv.**	**without Interv.**	**Total**	**With Interv.**	**without Interv.**	**Total**
1945-49	-	3	3	1	1	2	-	-	-	-	-	-
1950-54	-	3	3	-	-	-	-	-	-	-	-	-
1955-59	-	1	1	1	1	2	-	1	1	-	-	-
1960-64	1	3	4	-	-	-	-	1	1	1	-	1
1965-69	-	-	-	1	-	1	-	1	1	-	1	1
1970-74	-	-	-	-	1	1	2	-	2	-	-	-
1975-79	-	-	-	3	-	3	2	-	2	-	1	1
1980-84	-	-	-	-	-	-	-	-	-	-	-	-
1985-89	-	-	-	-	-	-	-	-	-	-	-	-
1990-92	-	-	-	-	-	-	-	1	1	-	-	-
Total	**1**	**10**	**11**	**6**	**3**	**9**	**4**	**4**	**8**	**1**	**2**	**3**

(type C-1), since no pure decolonization wars began after the mid-1960s and rarely (only once, in fact) involved direct third-party intervention. The decrease since the 1980s in wars with foreign participation lies in the absence of new intra-/international mixed wars with anti-regime components. This is significant because precisely these kinds of wars are prone to third party intervention, even if they are not the most frequent type of war.

Several differences become apparent when comparing the relative distribution of wars with and without intervention by region (table 12). Wars in Central America, Europe, Central Africa, Southern Africa and South/Southeast Asia are more likely to include third-party participation. In the Near and Middle East the ratio is about the same, while in all other regions the share of wars without intervention is higher. All the wars in South America were waged without direct foreign participation.

Table 12
Regional Distribution of Wars with and without Third-Party Participation

		Wars	
Region	**With Interv. %**		**Without Interv. %**
Central America/Caribbean	9		8
South America	-		10
LATIN AMERICA	9		18
EUROPE	9		5
North Africa	2		6
West Africa	0		5
Central Africa	9		4
East Africa	5		8
Southern Africa	9		3
AFRICA	25		26
NEAR/MIDDLE EAST	23		22
South/Southeast Asia including Oceana	32		24
East Asia	2		4
ASIA	34		28
Total	10	0	99

Figure 11
Forms of War with and without Third-Party Participation in Five-Year Periods following the Beginning of the War

1. Intranational Wars

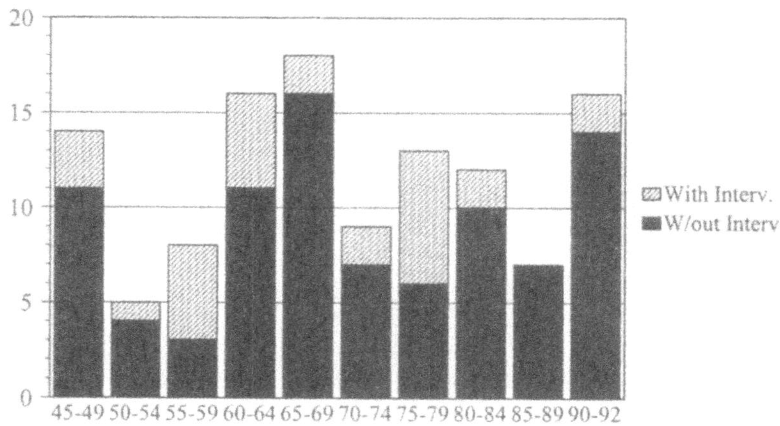

2. International Wars

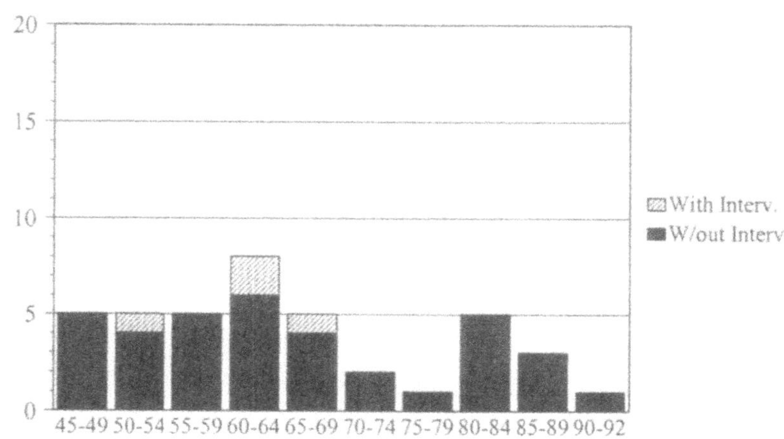

3. Intra-/International Mixed Wars

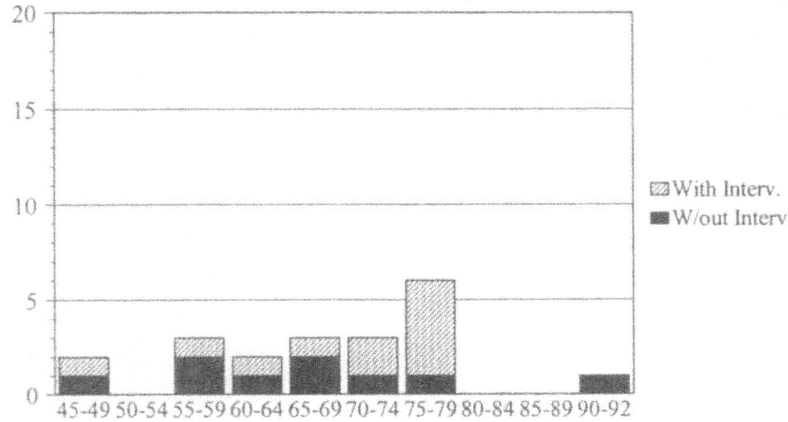

Figure 12
Frequency of Types of War with and without Intervention in Five-Year Periods following the Beginning of the War

1. Pure Anti-Regime Wars

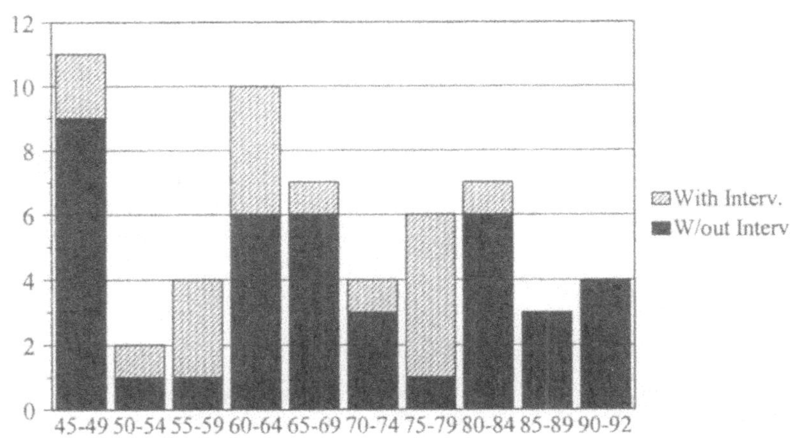

Figure 12 (cont'd)
Frequency of Types of War with and without Intervention in Five-Year Periods following the Beginning of the War

2. Pure Internal Wars (Others)

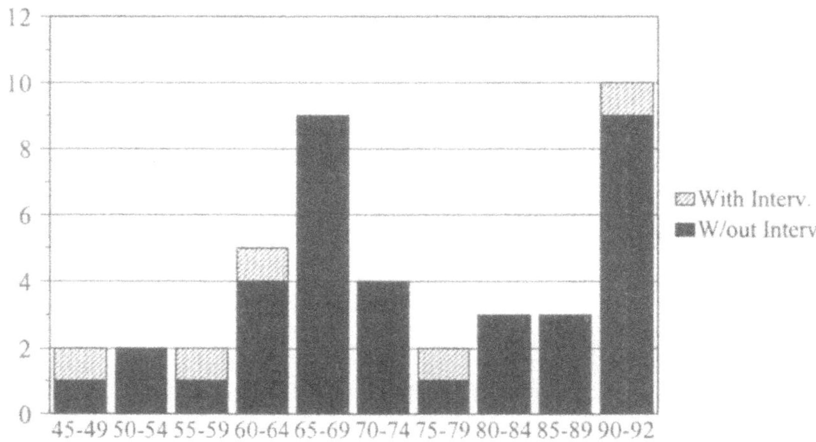

3. Mixed Internal Wars

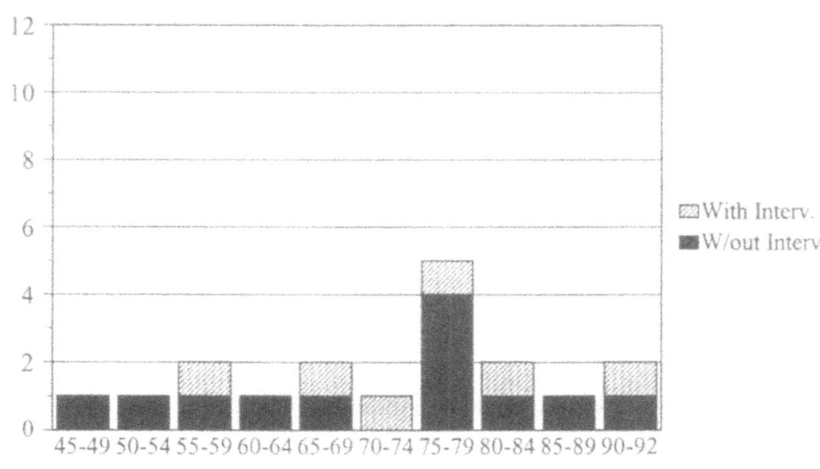

Figure 12 (cont'd)
Frequency of Types of War with and without Intervention in Five-Year Periods following the Beginning of the War

4. Interstate Wars

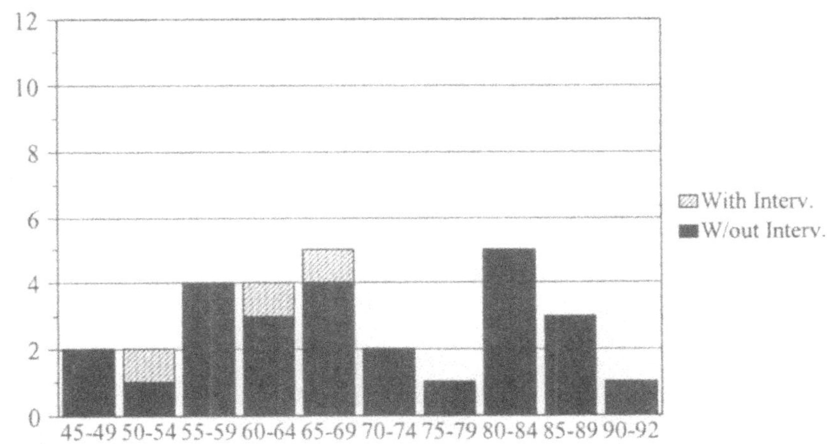

5. Decolonization Wars

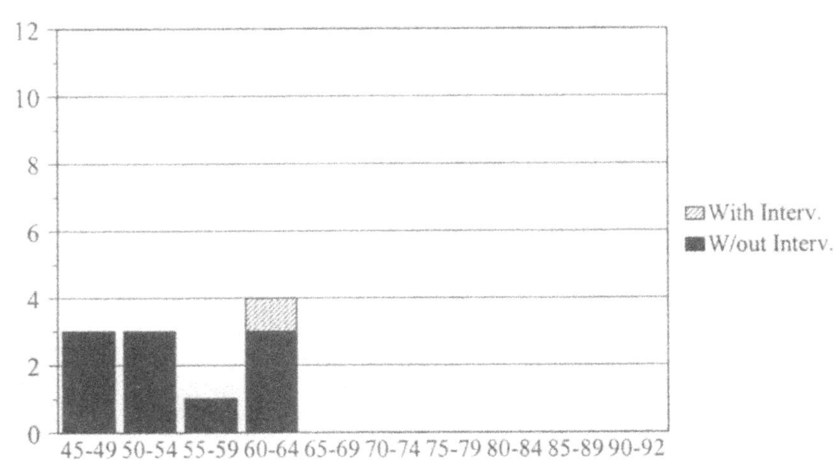

Figure 12 (cont'd)
Frequency of Types of War with and without Intervention in Five-Year Periods following the Beginning of the War

6. Intra-/International Mixed Wars with A-Components

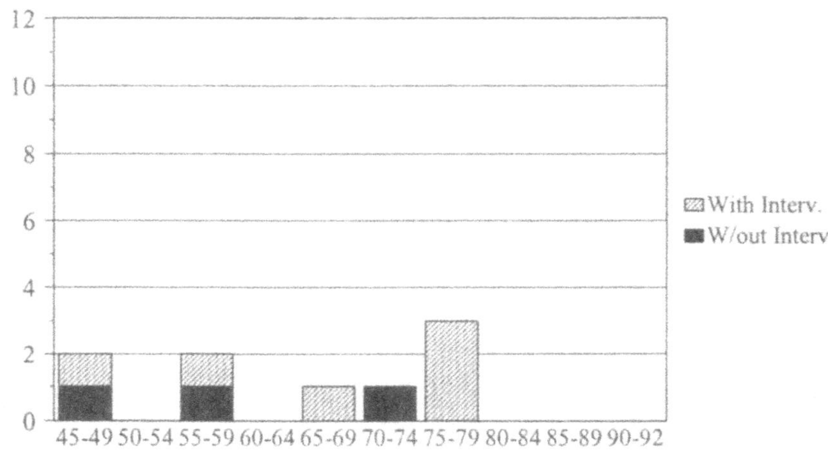

7. Intra-/International Mixed Wars with B-Components

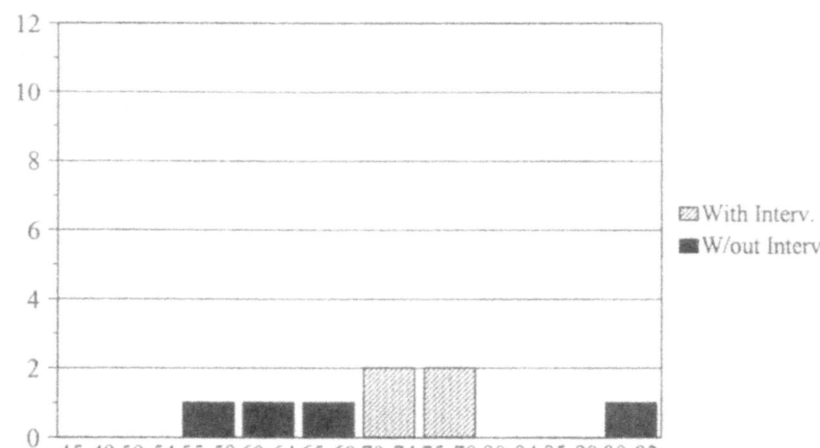

Figure 12 (cont'd)
Frequency of Types of War with and without Intervention in Five-Year Periods following the Beginning of the War

8. Complex Wars

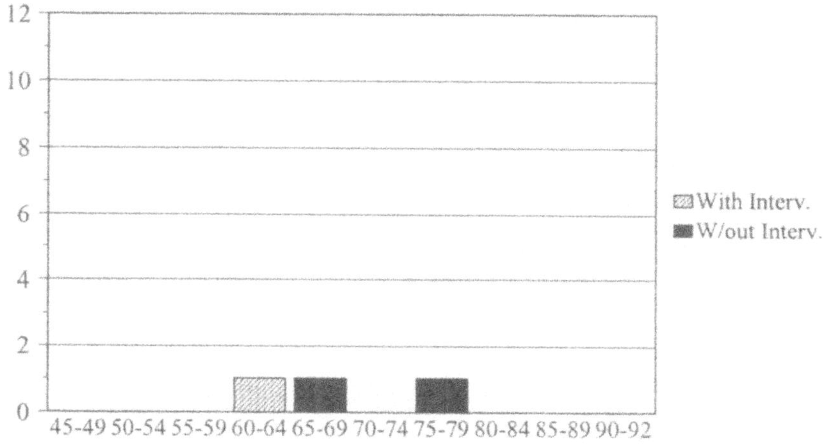

Table 13 reveals which warring party is supported by the respective interventions. Intervention *occurs more often on the behalf of the aggressed than the aggressor, and also more on behalf of the state or government than of opposition forces.* To explain this we must consider the following: In internal and mixed-type wars the opposition (guerrillas, rival forces, separatists, etc.) is usually the aggressor. The intervening forces usually represent other governments or state institutions who are more oriented toward preserving state structures than overthrowing them. In addition the intervening forces usually come from neighboring states who fear regional destabilization. In many cases, of course, ideological or hegemonic interests also play a role.[48]

Is there a connection between third-party participation and the length of war? Comparing the percentages in table 14 shows that shorter wars (up to 6-12 months) are usually those without intervention. Yet most short wars are interstate wars, and intervention is relatively rare in such cases anyway.

Table 13
Warring Parties which Benefit from Intervention in the Wars, 1945-1992*

Form of War	Aggressor **	Aggressed **	State/ Gov't	Opposition
Internal Wars	10	18	22	11
International Wars	-	2	4	-
Intra-/international mixed Wars	4	7	8	7
Total	**14**	**27**	**34**	**18**

* The frequencies are greater than those for wars with foreign participation because in a whole row of cases intervening forces fought on both sides.

** Only those wars are considered here which allow for the clear designation of an aggressor.

Table 14: Length of War with and without Intervention*

	with Intervention		without Intervention	
Months	**Nr.**	**%**	**Nr.**	**%**
up to 6	11	20	43	80
up to 12	3	33	6	67
up to 24	5	29	12	71
up to 60	9	28	23	72
up to 120	7	27	19	73
over 121	9	23	31	77
Total	**44**	**25**	**134**	**75**

* Including wars which, while not yet over, are counted as ending on December 31, 1992. The values are rounded to the nearest full month. In cases where no information about the month was available we used July (the middle of the year) for the start or end of the war.

While a good number of interventions were brief (21) or the war itself proved short (table 15), the majority of interventions lasted a long time: in 16 cases the interventions lasted between one and five years, in another 16 cases they persisted for more than five years. In 13 cases, however, the length of the intervention is unknown.

To end this section we turn once more to the nature of the interventions themselves (table 16). The largest number belong to the U.S. and Great Britain. In our explication of table 4 we speculated why these two countries seem so eager to intervene. Now we see that out of twelve interventions, eight of them were in direct support of one of the warring parties. The last was 1970-73 in the context of the Vietnam War. Great Britain participated 18 times in wars, but "only" seven of those were partisan participations (the last one in Oman in 1975). If we discount the four decolonization wars, however, then Great Britain's ratio of intervention to other forms of war participation is no longer so different from that of the U.S. The picture changes when we look at France: she participated in wars eleven times, but only two of those (since 1966 in Chad and 1978 in Zaire) are interventions. Unlike Great Britain, however, France waged six decolonization wars. France parted with its colonies more arduously than England, and accordingly was more willing to play the role of policeman in her former territories. Four of the USSR/Russian Federation's seven involvements in wars were interventions. With one exception (Egypt/Israel 1969/70) these interventions were closely tied to the hegemonic interests of the Soviet Union: Vietnam, Hungary and Afghanistan. South Africa intervened four times, and all for the destabilization of the "leftist" liberation movements in southern Africa in order to maintain their doomed system of apartheid (Angola, Mozambique, Rhodesia/Zimbabwe, and again Mozambique). North Vietnam's four interventions all occurred in the context of the Vietnam War (South Vietnam, Laos and Cambodia). India's four interventions only affected the neighboring states of Ceylon/Sri Lanka, Bangladesh and Kashmir (which was still independent in 1947).

Table 15: Length of Individual Interventions*

Duration	Nr. of interventions
up to 6 Months	21
7 to 12 Months	4
13 to 24 Months	5
25 to 60 Months	11
61 to 120 Months	12
121 Months and over	4
without or with unreliable figures	13

* Including wars which, while not yet over, are counted as ending on December 31, 1992. The values are rounded to the nearest full month. In cases where no information about the month was available we used July (the middle of the year) for the start or end of the war.

There is nothing systematic about the remaining interventions in table 16. Although often militarily successful they rarely achieved any lasting political solutions, and they occasionally failed militarily before political solutions could even be considered. Partisan interventions no longer seem worthwhile. This is certainly one reason why the respective powers have recently become so reticent to intervene.[49]

Table 16
Rank Order of Third-Party States by the Frequency of Their Interventions (not including interventions solely to end wars or U.N. actions)

Number of Interventions	States	Number of interventions
8	USA	1
7	Great Britain	1
4	India, North Vietnam, Soviet Union/Russia, South Africa	4
3	Belgium, Honduras	2
2	Australia, People's Republic of China, France, Iran, Cuba, New Zealand, Nicaragua, Nigeria, South Vietnam, Syria, Thailand, Vietnam, Zaire	13
1	Egypt, Algeria, Brazil, Costa Rica, El Salvador, Gambia, Ghana, Greece, Guinea, Jordan, Libya, Malawi, Mali, Morocco, Mozambique, Pakistan, Paraguay, Philippines, Rhodesia, Senegal, Sierra Leone, South Yemen, South Korea, Tanzania, Turkey, Zimbabwe	26

Let us summarize: Third-party interventions for partisan reasons occur primarily in anti-regime wars or mixed-type wars with anti-regime components. These types of wars are in the minority and have been decreasing since the 1980s. Interventions seldom happen in interstate wars, and since the 1970s no new war has begun where a third party has intervened. Intervention usually benefits the aggressing party and/or the state or its government. Wars with interventions are generally of medium-length lasting at least one year, and sometimes much longer. Rarely does intervention bring the desired results for all concerned.

Despite a low overall level of intervention if one adds together all of the internal wars with partisan intervention (29), all international wars (40), all intra-/international mixed wars (20), and all wars where intervention occurred to terminate the war or under U.N. auspices (6), then we find that over half of all 184 wars have a strong international component (95=52 percent). It would be an exaggeration to say that war is being internationalized (or transnationalized) in comparable ways to capital and production, migrations, environmental issues, and so forth. Nonetheless there is a certain contradiction between the dominance of internal wars on the one hand and the high international component of wars since 1945 on the other.

4.7 Time and Suffering: Duration and Victims of War

Until now we have used the frequency of and participation in war as the indicator for war susceptibility. Clearly this is not sufficient. To "force the opponent to fulfill one's will (von Clausewitz 1832/34, book 1)" requires tenaciousness (long wars), intensity (a high level of victims), and the mobilization of the whole of society. Von Clausewitz described the three ideal interactions which, according to the logic of war, lead to escalation of violence, and the impediments which prevent the mutual annihilation of both sides.[50] These observations are historically relevant not only for the East-West Conflict but for new nuclear powers from the ranks of the Third World. Without delving deeper into this subject we note that, once begun, wars develop their own often unpredictable dynamics. Nonetheless the duration, number of victims and cost of the war remain important indicators for susceptibility to war.

We turn now to the duration of war, which is relatively quantifiable. The following statistics include wars that were unfinished as of 1992, so for statistical purposes we use December 31, 1992, as the date which those wars are assumed to have ended. Omitting these unfinished wars would be even more distorting than "artificially" ending them on December 31, 1992: 22 of the 44 unfinished wars have persisted more than ten years and five have lasted between five and ten years. The sole effect of our artificial ending of the three new and unfinished wars since July 1992 is an increase in the number of very short wars.

The row for all wars in table 17 reflects a twofold division: wars are either rather short (six months or less) or, more frequently, they are quite long (from two to ten years). The results are disturbing: 40 wars lasted longer than ten years. On average this means 225 months (almost 19 years), which corresponds to 8,998 war-months (almost 750 war-years), and all this within "only" the 48 calendar years (576 calendar months) of our study.[51] *All the wars together add up to 13,305 months or 1,109 years—each war required on average 72 months or six years.* Ideally these numbers should be multiplied by the number of people who actively conducted the wars or whose productive efforts were hindered by them, which unfortunately we cannot do. What would the world look like if all of this time had been used constructively instead of destructively? Yet let us leave moral indignation aside for now and return to the sober presentation of facts.

Table 17: Length of Wars by Month and Type *

War type	Months -1	2-6	7-12	13-24	25-60	61-120	121 and above	Total
Anti-regime	13	9	2	5	9	11	11	60
other internal	2	5	2	5	13	4	12	43
mixed	-	2	1	1	5	2	8	19
INTERNAL WARS								
Number	15	16	5	11	27	17	31	122
Row percent	12	13	4	9	22	14	25	99
inter-state	17	5	2	1	3	3	-	31
Decolonization wars	1	1	-	1	4	3	1	11
INTERNATIONAL WARS								
Number	18	6	2	2	7	6	1	42
Row percent	43	14	5	5	17	14	2	100
intra-/international mixed Wars:								
with A components	-	-	1	2	-	2	4	9
with B components	1	1	1	2	1	1	1	8
Complex wars	-	-	-	-	-	-	3	3
MIXED WARS								
Number	1	1	2	4	1	3	8	20
Row percent	5	5	10	20	5	15	40	100

WARS								
Total Number	34	23	9	17	35	26	40	184
Row percent	18	13	5	9	19	14	22	100
Total Months of war	15	83	81	303	1408	2417	8998	13305
Average Duration per War	0,4	3,6	9,0	17,8	40,2	93,0	225,0	

* Including wars which, while not yet over, are counted as ending on December 31, 1992. The values are rounded to the nearest full month. In cases where no information about the month was available we used July (the middle of the year) for the start or end of the war.

Looking behind the totals we find differences among the types of war: Anti-regime wars (type A) tend to be either very short or quite long, i.e., they are either over after six months or last between two and ten years. The other types and the mixed-types (types B and AB) tend overwhelmingly to last longer (from two to ten years). This indicates that issues of "national" unity or its converse, territorial autonomy or secession are particularly obstinate issues. The majority of interstate wars (type C) end within a month, six months, or one year. Most of these types of war are terminated quickly because of the relatively clear "game rules" of international law, the available diplomatic means (including sanctions), the many international organizations capable of involvement and, last but not least, the need for foreign support, especially regarding military technology. Decolonization wars, on the other hand, tend to last longer. Intra/international wars tend to be drawn out as well, but they have the potential to end within six months to two years.

Table 18
Length of Wars by the Period They Began *

Months	**1945 -49**	**1950 -54**	**1955 -59**	**1960 -64**	**1965 -69**	**1970 -74**	**1975 –79**	**1980 -84**	**1985 -89**	**1990 -92**	**Total**
up to 1	5	3	4	5	3	3	1	4	4	2	34
Clm-%	24	27	25	19	11	21	5	24	36	10	18
2-6	3	2	3	2	5	1	1	1	1	4	23
Clm-%	*14*	*18*	*19*	*7*	*19*	*7*	*5*	*6*	*9*	*20*	*13*
7-12	1	-	-	2	1	3	1	-	-	1	9
Clm-%	*5*	-	-	*7*	*4*	*21*	*5*	-	-	*5*	*5*
13-24	2	-	2	1	3	2	1	-	-	6	17
Clm-%	*10*	-	*13*	*4*	*11*	*14*	*5*	-	-	*30*	*9*

Table 18
Length of Wars by the Period They Began *(cont'd)

25-60	4	4	4	8	1	1	2	-	4	7	35
Clm-%	*19*	*36*	*25*	*30*	*4*	*7*	*10*	-	*36*	*35*	*19*
61-120	4	1	1	4	3	-	4	7	2	XXX	26
Clm-%	*19*	*9*	*6*	*15*	*11*	-	*20*	*41*	*18*	*XXX*	*14*
121 and above	2	1	2	5	11	4	10	5	XXX	XXX	40
Clm-%	*10*	*9*	*13*	*19*	*41*	*29*	*50*	*29*	*XXX*	*XXX*	*22*
Summen	21	11	16	27	27	14	20	17	11	20	184
Clm-%	*100*	*100*	*100*	*100*	*100*	*100*	*100*	*100*	*100*	*100*	*100*

* Including wars which, while not yet over, are counted as ending on December 31, 1992. The values are rounded to the nearest full month. In cases where no information about the month was available we used July (the middle of the year) for the start or end of the war.

In table 18 wars are broken down by the length and period in which they began to assess change over time. The percentages are more important than the total numbers in the last column. The share of very short and short wars (up to six months) varies somewhat, but no clear trend is recognizable even if the share of these wars was somewhat greater until the end of the 1950s and the share of short wars (2-6 months) decreased in the 1970s. We do not analyze the data from the last period 1990-92 for the aforementioned reasons. The wars of medium duration (from six months to two years) occur primarily between the second half of the 1950s and the end of the 1970s. The share of wars with durations up to five years was relatively high until the mid-1960s and decreased notably after that. It remains to be seen whether the increased share of such wars since 1985 suggests a return to the earlier trend. What emerges is that *the share of very long wars increases over time*, especially after the mid-1960s. This corresponds with our previous observations. From table 17 we know that most long wars are usually internal and mixed wars, and from table 9 we know that these types increased after the 1960s and that these are the types of war which account for the steady yearly rise in wars (see figure 8).

Table 19
Regional Distribution of Wars with a Duration of More than Ten Years *

Region	All Wars Nr.	All Wars %	Wars > 10 Years Nr.	Wars > 10 Years %
Central America/Caribbean	15	8	2	5
South America	14	8	4	10
LATIN AMERICA	29	16	6	15
EUROPE	12	7	2	5
North Africa	9	5	2	5
West Africa	8	4	1	3
Central Africa	10	5	1	3
East Africa	13	7	4	10
Southern Africa	8	4	5	13
AFRICA	48	26	13	33
NEAR/MIDDLE EAST	41	22	5	13
South/Southeast Asia including Oceania	47	26	14	35
East Asia	7	4	-	-
ASIA	54	29	14	35
Total	**184**	**100**	**40**	**101**

* Including wars ongoing as of December 31, 1992.

The share of Latin American[52] and European war zones[53] with very long wars corresponds to the share of all wars overall (table 19). Note the comparatively small share of very long wars attributable to the Near and Middle East region.[54] Africa, especially East and Southern Africa, and South/Southeast Asia are disproportionately represented in the category of very long wars, while East Asia has not even one.[55] More research is necessary to explain these differences. Regarding the Near and Middle East we speculate that the Great Powers have more interest in controlling this region because of Israel and oil, which, however, does not necessarily translate into direct military intervention. Additionally, Israel's military-strategic and logistical interests lie in short wars with its neighbors. Even the rich oil-producing countries are so dependent on the global market that they cannot afford long disruptions in production (unless it is a matter of rivalry to weaken the other, as in the Iran-Iraq war 1980-1988). Furthermore, the state formation process is far more advanced and consolidated in the Near and Middle East than in Black Africa. Finally the physical geography and settlement patterns may not make the Middle and Near East as suitable for guerrilla warfare as are Africa or Asia.

The most forceful indicator for the intensity of a war is the number of victims, especially those whose deaths were caused by weapons (battles, massacres, bombardments, mines, etc.) or direct effects of the war (hunger, disease, etc.). We need to preface our remarks before proceeding further: In section 1.2 we already discussed the problems of using such indicators as the basis of a definition. Even greater problems arise in trying to empirically record the number of dead. There are a host of objective (e.g., poor administrative conditions or abilities) and subjective (e.g., tactical or propaganda interests of the warring parties) reasons for this difficulty. For this reason the data on war victims is incomplete and not very reliable, although this does not mean that in every case the numbers are always exaggerated or understated. We are nonetheless amazed at how cavalierly the media and even scientific publications report number of victims.[56] We have tried to gather with the utmost care all available reasonable data about the victims of wars since 1945, including when possible data about the injured, expellees and refugees, war costs, levels of destruction, etc. The preliminary result can be found in the corresponding list following the register.

We only deal here with those killed in war, since the data about other war effects are even more unreliable. Often there are different numbers of victims for one and the same war. For the following statistics we use only the lowest figure or estimation. With few exceptions all of the sources themselves give only approximate figures and no exact "head count." High numbers of victims are usually rounded to the nearest hundred, thousand, ten-thousand or hundred-thousand. We also consider the victims of those wars which are still going on to the extent possible.

Unfortunately, we cannot clearly delineate between fallen soldiers and dead civilians due to lack of information. Most likely the percentage of civilians (noncombatants) killed in relation to military personnel (combatants) has risen steadily since the industrialization of war, beginning with the U.S. Civil War (1861-1865). In a war zone today a soldier is often safer than a civilian.[57] The modern technology of destruction has undermined the other achievement of modernity: the ability to distinguish between combatants (the object of the fighting) and civilians (who are to be protected). The spatial distance between marksman and target has increased considerably along with the weapon's speed and destructive capacity. Many weapons cannot distinguish between threatening enemy military and noncombatant civilians.[58] Certain guerrilla strategies purposely camouflage fighters as civilians, and certain military strategies intentionally attack "soft targets"—the general population—in order to weaken the support for and the fighting spirit of the "hard targets," the military. Distant industrial and agricultural areas are increasingly important for offensive and defensive actions (through bombardment, for example). We are unable, however, to confirm the hypothesis that civilians constitute the greatest number of war victims since the Second World War.[59]

Despite all limitations we do have figures about victims for 135 of the 184 wars. We have no reliable data for 49 wars (27 percent). In order to ascertain how these missing 49 wars affect our analyses we selected four variables to be applied to the war victim data and tested these variables to see if they are similarly distributed for all 184 wars. The results were encouraging:

Percent Distribution of Wars by Types of War[60]

	internal Wars	intra-/international mixed Wars	international Wars
All Wars	66	11	23
Wars with Data for Victims	65	14	21

Percent Distribution of Cases by General Regions

	Latin America	Europe	Africa	Near & Mid-East	Asia
All Wars	16	7	26	22	29
Wars with Data for Victims	12	8	33	19	29

Percent Distribution of Cases by Length of War

Months	-1	2-6	7-12	13-24	25-60	61-120	over 120
All Wars	18	13	5	8	20	14	22
Wars with Data for Victims	16	13	5	8	19	12	27

Percent Distribution of Cases by Periods in Which Wars Began

Years	45-49	50-54	55-59	60-64	65-69	70-74	75-79	80-84	85-89	90-92
All Wars	11	6	9	15	15	8	11	9	6	11
Wars with Data for Victims	11	10	20	9	6	5	19	13	3	4

We see some differences in the regional distribution which might result from the following discrepancy: there are few reliable data about the Kurdish victims in Iraq and Iran, and accordingly the Near and Middle East is probably under-represented. The 1950s are probably somewhat over-represented as a period when wars began. Otherwise, however, the distributions are so similar that we can consider the 135 wars for which we have victim data to be representative for all wars.[61]

Table 20 breaks down intensity of war by form of war.[62] According to these figures in the lower half of the table, where the "real" victim numbers are given, 12.5 million human lives were extinguished by the 135 wars. Projecting these numbers to include all 184 wars we arrive at a number of war dead equal to nearly 17 million since 1945. On average that would mean approximately 354,000 war dead per year. Now there are those who would point out that this number is merely, for example, one-sixth of Hamburg's population, or 0.01 percent of the world's population.[63] The political and ethical dimensions appear in a different light, however, when one imagines a yearly pile of 354,000 corpses, not to mention the incalculable "investment" of spiritual and material care which has gone into each and every one.

The numbers in table 20 show unambiguously what we already suspected: the vast majority of war victims results from internal wars, followed by mixed wars and international wars. When measured by the average per war, however, mixed wars result in the largest numbers of losses.[64] Other factors are at work here as well, such as the internal dynamics created by the start of a war. All this questions Small and Singer's (1982) operational definition of war in which they classify instances with fewer than one thousand casualties merely as military "disputes."

Table 20
War Dead in the 135 Wars, Broken Down by War Type *

Number Killed	Number of Wars: Intra-national Wars	Intra-/ internat. Mixed	Inter-national Wars	Total Nr.	%
up to 1.000	17	2	13	32	24
up to 5.000	20	4	5	29	21
up to 10.000	13	2	1	16	12
up to 50.000	23	4	4	31	23
up to 100.000	5	-	1	6	4
up to 500.000	7	4	3	14	10
up to 1 Mil.	3	2	1	6	4
over 1 Mil.	-	1	-	1	1
Total	88	19	28	135	99
Number of dead**	6.164.000	4.370.000	2.005.000	12.539.000	
Average per War **	70.000	230.000	72.000	93.000	
Minimum	60	360	9		
Maximum	1.000.000	2.000.000	854.197		

* Compiled from the lists in the register. There are no data for 49 wars. In cases of multiple figures we always used the lowest. To the extent possible the data for victims includes dead soldiers and civilians. Ongoing wars are also included when possible.

** Rounded to the nearest thousand in order not to give the impression that these figures are anything more than an estimate.

The information in table 21 is not surprising regarding regional differences. Even when broken down by general category there remains a noticeable difference between continents. The rank order of both the total sum of war dead and the average of those killed per war mirrors the level of state formation: Europe (self-evidently also North America and Australia) and Latin America come first, followed by the Near and Middle East, Asia, and finally, Africa.

Table 21
War Dead in the 135 Wars, Broken Down by Region *

Number Killed	Latin America	Europe	Number of Wars Africa	Near/Middle East	Asia
up to 1.000	5	5	11	4	7
up to 5.000	3	2	8	6	10
up to 10.000	1	-	3	5	7
up to 50.000	4	3	8	6	10
up to 100.000	2	-	3	-	1
up to 500.000	1	1	8	3	1
up to 1 Million	-	-	3	1	2
over 1 Million-	-	-	-	1	
Total	**16**	**11**	**44**	**25**	**39**
Number of dead **	396.000	238.000	5.354.000	1.895.000	4.656.000
Average per War **	25.000	22.000	122.000	76.000	119.000
Minimum	9	60	60	100	100
Maximum	126.000	150.000	950.000	1.000.000	2.000.000

* Compiled from the lists in the register. There are no data for 49 wars. In cases of multiple figures we always used the lowest. To the extent possible the data for victims includes dead soldiers and civilians. Ongoing wars are also included when possible.

** Rounded to the nearest thousand in order not to give the impression that these figures are anything more than an estimate.

Even a cursory glance at the appendix shows no necessary connection between the length of a war and its number of victims. Such a connection would falsely assume that every day of a given war is waged with the same intensity. Of course there are both short and severe or long and less intense wars, just as there are short, bloodless wars and long, extremely bloody ones. While table 22 shows no significant relationship between the number of victims and the duration of war it does indicate that wars with the greatest number of victims tend to be the longest wars.

Table 22: War Dead in the 135 Wars, Absolute Values, Broken Down by War Duration *

Number Killed	Number of Wars with a Duration of ... Months							
	up to 1	2-6	7-12	13-24	25-60	61-120	over 120	Total
up to 1.000	9	7	3	5	6	1	1	32
up to 5.000	7	5	1	2	6	3	5	29
up to 10.000	2	1	-	-	5	4	4	16
up to 50.000	3	4	2	1	3	2	16	31
up to 100.000	-	-	-	2	1	1	2	6
up to 500.000	1	-	1	1	2	5	4	14
up to 1 Mil.	-	-	-	-	3	-	3	6
over 1 Mil.	-	-	-	-	-	-	1	1
Total	**22**	**17**	**7**	**11**	**26**	**16**	**36**	**135**

* Compiled from the lists in the register. There are no data for 49 wars. In cases of multiple figures we always used the lowest. To the extent possible the data for victims includes dead soldiers and civilians.

Table 23
War Dead in the 135 Wars, Percentage Values,
Broken Down by War Duration *

Number Killed	Share of Wars with a Duration of ... Months							
	up to 1	2-6	7-12	13-24	25-60	61-120	over 120	Total
up to 1.000	41	41	43	45	23	6	3	24
up to 5.000	32	29	14	18	23	19	14	21
up to 10.000	9	6	-	-	19	25	11	12
up to 50.000	14	24	29	9	12	13	44	23
up to 100.000	-	-	-	18	4	6	6	4
up to 500.000	5	-	14	9	8	31	11	10
up to 1 Mil.	-	-	-	-	12	-	8	4
over 1 Mil.	-	-	-	-	-	-	3	1
Total %	**101**	**100**	**100**	**99**	**101**	**100**	**100**	**99**

* Compiled from the lists in the register. There are no data for 49 wars. In cases of multiple figures we always used the lowest. To the extent possible the data for victims includes dead soldiers and civilians.

4.8 Does Power Propel War?

A dominant interpretation of politics considers it self-evident that nations strive for power, and that states who possess relatively more power than others are more willing to go to war. The corollary of this interpretation is that only power can fight power. This primitive understanding comes to international politics from feudal times and persists today in highly stylized forms such as "balance of power politics" and "deterrence," which have become not only theories but guidelines for policy. It would be beyond the scope of this work to expose this interpretation to detailed criticism,[65] but we observe that it has more to do with psychology, ideology, and even propaganda than with explanations of reality. It is true enough, yet trivial, to note that power potential is required for fighting a war. It does not follow from this axiom that the most powerful nation is therefore the most war-like. Such an approach contributes little toward understanding the internal wars which have become the most common form of violent conflict. Additionally, despite its beloved and common usage in the social sciences the term "power" itself is rarely clearly defined.

We pose a relatively simple question: is there a connection between the status of a society in the international hierarchy of power and susceptibility to war? The following observations can only provide an introduction, and a full exploration requires more differentiated studies.[64]

When we talk of "status" we are referring to the *relative* position of the warring countries. Which standard should we apply here? Proceeding pragmatically, it is indubitably the case that during the time frame of our study the U.S. and USSR/Russia are great powers ("superpowers") when measured by economic, political and military potential and political influence. For all other countries we limit ourselves to considering the following: relative national economic power, the level of development (starting from the dominant interpretation that more power leads to more war), material resources and productive capacity. It is true that neither social science, nor policymakers, nor international organizations can agree on how to measure development to everyone's theoretically and empirically satisfaction, with the result that many different lists and assessments exist (see Menzel/Senghaas 1985). We have prepared a provisional classification in table 24, whereby the classification occurs retrospectively, i.e., according the status in the most recent time period. Note that we are talking here again about the war participation of individual countries, not war itself.

Table 24
Country Status and the Frequency of Their Participation in War 1945-1992 *

Status Group	Participation in Wars Nr.	Per Country
Great Powers (2): USA, USSR/Russia	17	8,5
Main Industrialized Countries(9): Australia, Belgium, France, Great Britain, Italy, Canada, Luxembourg, Netherlands, Sweden	45	5,0
other relatively highly developed Countries (6): Argentina, Brazil, Costa Rica, Ireland, Spain, New Zealand	15	2,5
Combined	*77*	*4,5*
Threshold Countries(20): Egypt, Algeria, Greece, India, Iraq, Israel, Yugoslavia (i.e. Serbia and Montenegro), Croatia, Kuwait, Portugal, Romania, Saudi Arabia, Slovenia, South Africa, South Korea, Taiwan, Turkey, Hungary, Venezuela, People's Republic of China	91	4,6
Developing Countries (82): all other countries participating in wars	206	2,5
Total Participation in War	**374**	**3,1**

* Including participation in U.N. military operations. The criteria for a country's respective status is its economic output per capita for the most recent period. Military potential and global influence were also considered for the Great Powers.

While it is true that the developed world shows less participation in war than the threshold countries, and these less than the developing countries, the reason lies in the different size of these groups. The number of participations in a war per country shows that the popular conception is not entirely false: the developing and threshold countries are noticeably more prone to war than developed countries. Within the developed countries there is an even more prominent rank order, although the averages tend to level differences, for example, between the two super powers, Great Britain and Italy, Costa Rica and Spain, Egypt and China or Ethiopia and Oman. In this respect our status groupings are still too general. Nonetheless we do find that higher economic levels of development enable states to wage war more frequently than states which are less developed.

Table 25
Country Status and the Development of Their Participation in War 1945-1992 *

Status Group	45-54	55-64	65-74	75-84	85-92	Total
2 Great Powers						
USA	25	33	17	8	17	100
USSR/Russia	-	20	20	20	40	100
Combined	18	29	18	12	24	101
9 main industrialized countries	40	40	7	7	7	101
6 other relatively highly developed countries	27	33	27	13	-	100
Combined	32	36	13	9	9	99
20 threshold countries	22	25	24	13	15	99
82 developing countries	12	24	22	25	18	101
all 119 countries participating in wars	18	27	21	19	16	101
7 former colonial powers **	36	40	9	9	7	101

* Row percent = Year of the start of wars in ten-year periods. Including participation in U.N. military operations. The criteria for a country's respective status is its economic output per capita for the most recent period. Military potential and global influence were also considered for the Great Powers. See table 24 for a breakdown of the status groups.

** Five belong to the main industrialized countries, Spain belongs to the other relatively highly developed countries, and Portugal belongs to the threshold countries.

Table 25 attempts to present the above averages in light of chronological developments. Here we begin to see why the sweeping historical claims made above must be relativized and differentiated. While no clear trend is discernible for the great powers (save perhaps for the decreasing participation of the U.S. in wars), among the other industrialized countries we see a marked reduction in their participation in wars since the 1960s. This results in part from the decrease in decolonization wars, since the most important former colonial powers belong to this

Table 26
Frequency of Forms of Conflict by Region for Wars, 1945-1992

	Regions											
Forms of Conflict	**CAM**	**SAM**	**EUR**	**NAF**	**WAF**	**CAF**	**EAF**	**SAF**	**NME**	**SSEA**	**EA**	**total**
System transformation	2	5	3	1	-	1	-	2	4	13	3	34
System reform	7	5	3	-	-	-	-	1	5	4	-	25
System restoration	4	-	-	-	-	-	-	-	3	3	-	10
SYSTEMIC CONFLICT TOTAL	13	10	6	1	-	1	-	3	12	20	3	69
Power struggle	4	3	-	1	2	9	5	4	7	1	-	36
Diversion conflict	1	1	-	-	-	-	1	-	2	1	-	6
Particularist conflict	-	3	-	3	3	4	1	-	2	4	-	20
Tribal conflict	-	-	-	-	-	1	1	-	1	-	-	3
mil/strategic conflict	-	-	-	-	-	-	-	1	8	3	3	15
POWER CONFLICT TOTAL	5	7	-	4	5	14	8	5	20	9	3	80
Expansion	-	-	-	1	-	-	-	-	5	4	4	14
Territ. Restoration	-	-	1	-	-	-	-	-	10	3	1	15
Territ. Revision	-	2	2	1	1	1	3	-	6	5	-	21
Autonomy/ Secession	1	1	6	3	4	1	6	-	13	19	1	55
TERRITORIAL CONFLICT TOTAL	1	3	9	5	5	2	9	-	34	31	6	105
Colonial rule	-	-	1	4	1	1	2	4	1	4	-	18
Influence/ Hegemony	4	-	1	-	-	-	-	-	10	7	-	22
FOREIGN RULE CONFLICT TOTAL	4	-	2	4	1	1	2	4	11	11	-	40
Total	23	20	17	14	11	18	19	12	77	71	12	294

middle group. Yet even among the threshold countries (of which only Portugal is a former colonial power) there is an observable decrease in war participation. Developing countries, on the other hand, remain constant or even slightly increase their participation in wars. So while the susceptibility of "rich" states for participation in war abates, "poor" states remain involved in wars at the same or higher level. Apparently more is at work here than the balance of power principle alone.

4.9 What is Fought For and Where: The Forms of Conflict

Although we are unable to pinpoint the exact *causes* of war we can ascertain *what* is being fought about, which conflicts lead to wars, and whether there are noteworthy regional differences in these matters. In section 2.6 we presented our classification scheme for conflicts and explained why these classifications should be regarded as general approximations. We also pointed out that in nearly every war several conflicts overlap and intermingle such that the number of types of conflict (294) is much greater than the number of wars (184).

Table 26 contains the absolute frequencies of conflict forms by individual regions. This rather complex table reminds us that the differences between the regions within a continent are significantly small, thus enabling the approach in table 27 which considers only the general regions.

Table 27
Frequency and Relative Share of Forms of Conflict by Region for Wars 1945-1992 *

Forms of Conflict	General Regions										Total	
	Latin America		Europe		Africa		Near/Mid East		Asia			
	Nr.	%	Nr.	%	Nr.	%	Nr.	%	Nr.	%	Nr.	Clm -%
System transformation	7	16	3	18	4	5	4	5	16	19	34	12
System reform	12	28	3	18	1	1	5	6	4	5	25	9
System restoration	4	9	-	-	-	-	3	4	3	4	10	3
SYSTEMIC CONFLICT TOTAL	23	53	6	35	5	7	12	16	23	28	69	24
Row percent	*33*		*9*		*7*		*17*		*33*		*100*	

Power struggle	7	16	-	-	21	28	7	9	1	1	36	12
Diversion	2	5	-	-	1	1	2	3	1	1	6	2
Particularist conflict	3	7	-	-	11	15	2	3	4	5	20	7
Tribal conflict	-	-	-	-	2	3	1	1	-	-	3	1
mil/strategic conflict	-	-	-	-	1	1	8	10	6	7	15	5
POWER CONLFICT												
TOTAL	12	28	-	-	36	49	20	26	12	14	80	27
Row percent	*15*		-		*45*		*25*		*15*		*100*	
Expansion	-	-	-	-	1	1	5	6	8	10	14	5
Territ.Restoration	-	-	1	6	-	-	10	13	4	5	15	5
Territ. Revision	2	5	2	12	6	8	6	8	5	6	21	7
Autonomy/Secession	2	5	6	35	14	19	13	17	20	24	55	19
TERRITORIAL CONFLICT												
TOTAL	4	9	9	53	21	28	34	44	37	45	105	36
Row percent	*4*		*9*		*20*		*32*		*35*		*100*	
Colonial rule	-	-	1	6	12	16	1	1	4	5	18	6
Influence/Hegemony	4	9	1	6	-	-	10	13	7	8	22	7
FOREIGN RULE CONFLICT												
TOTAL	4	9	2	12	12	16	11	14	11	13	40	14
Row percent	*10*		*5*		*30*		*28*		*28*		*101*	
TOTAL	43	99	17	100	74	100	77	100	83	100	294	100
Row percent	*15*		*6*		*25*		*26*		*28*		*100*	

* For legibility the row percentages are italicized.

The totals column for table 27 indicates that the most frequent forms of conflict are territorial, which can be traced primarily to desires for greater autonomy within a state, secession or irredentist annexation. This undermines the assumption that mostly B-type wars result from territorial interests.

"Power conflicts" (*Machtkonflikte*) are the second-largest form, yet this classification belies yet an even deeper conflict: the struggle for state power (*Herrschaftskonflikte*). This corroborates our earlier thesis that, in addition to struggles *for* the state (i.e., for territory), the control of power *within* a state is a driving force behind wars since 1945.

The third-largest form of conflict is revolution: the fight for the very shape of state and society. This form was most prevalent during the peak of the East-West conflict, but has become meaningless since capitalism has emerged as the only remaining path. Amazingly, anti-colonial wars have a subordinate rank despite the epochal historical significance of decolonization.

How are the forms of conflict distributed across the general regions? A comparison of the bottom rows of table 27 and table 3 shows that the frequency of forms of conflict and the number of wars are similarly distributed across regions. Within the forms of conflict, however, we find significant differences: systemic conflicts are concentrated mostly in Latin America and Asia, power struggles are found mostly in Africa, territorial conflicts arise again mostly in Asia, followed by the Near and Middle East, and conflicts involving foreign rule erupt mostly in Africa, though the Near and Middle East and Asia are close behind. Let us look more closely at the column percentages for the general regions.

Wars in Latin America revolve mostly around systemic reforms, not revolutions. The latter follow in second place, together with struggles for state power. The widely touted specter of Communist revolution in Latin America which resounded from the halls of U.S. military academies was either a result of ignorance, ideological blindness or pure propaganda. It was the U.S. which pushed Fidel Castro toward Communism. As Che Guevara's diaries reveal, his campaign was a series of almost comical mishaps. Chile's striking transportation workers and table-thumping middle-class housewives have fallen into silence, and we have already expressed our opinion about the allegedly "revolutionary power-kegs" of Nicaragua and Grenada. Even in the one case where the "Shining Path" does follow a twisted ideology (with urban intellectuals terrorizing mountain peasants) the government is unable to halt atrocities and responds in kind with its military. These examples should suffice to show how even the simple empirical science practiced here can dispel myths and errors.

The scant European wars primarily concerned autonomy or secession (Cyprus, Basque country, Northern Ireland, Yugoslavia, Moldavia). Systemic issues also occurred (in Greece, Spain, Hungary and Romania), but no longer play any role.

Africa is mostly plagued by conflicts about state power, resulting from the fact that state formation is the least consolidated there. Correspondingly, the external state formation process (i.e., the marking of territory) is rich in conflicts, and we see a high share of autonomy and secession conflicts. Both these major forms result from earlier colonial rule. No region experienced colonial penetration as much as Africa, which explains why the majority of struggles against foreign rule occur here. It is therefore all the more astounding that territorial restoration conflicts (irredentism) play no role at all, and conflicts involving territorial revision (such as between Mali and Burkina Faso or between Libya and Chad) play only a subordinate role. One of the few successes of the Organization for African Unity (OAU) and the rulers themselves has been the acceptance of the often arbitrary colonial borders.

In the Near and Middle East territorial questions dominate, followed by autonomy and secession conflicts and territorial restoration (e.g., for the Kurds, Palestinians, between Syria and Lebanon and between Israel and its neighboring states), and further still by hegemonic conflicts (Syria, Iraq, Saudi Arabia, and Turkey). This region is also the only one in which military-strategic conflicts are consequential.

Territorial issues are also most frequent in Asia, where they predominate because of autonomy or secessionist movements (above all in India, Sri Lanka, Indonesia and Laos). The second most common form of conflict is systemic change followed by "expansion." Both these forms were determined by the Vietnam and Indochina wars and conflicts in Indonesia. Indonesia tried like no other post-colonial state to violently realign their boundaries while the world community, here as elsewhere, looked on silently. Centralization was the watchword in Indonesia, perhaps stemming from a pre-modern conception of power. Against this trend arose demands of decentralization (federal systems, secession, etc.) which were the opening bids of the post-colonial era. Many B-type wars result from neglecting such demands.

Diversionary wars and ancient ethnic or tribal conflicts are of lesser importance for the different regions despite the attention they receive in the mass media. *The shape of the state system, control over state power and territorial definition of sovereignty are the essential sources of conflict.*

4.10 How War Ends: Termination and Outcome

Prevention is always better than cure, but once a war is started the most important task is to bring it to an end. What are the different ways in which wars have come to an end?

Table 28
Types of War Termination by War Type *

	Forms of War Termination											
	Military Victory				Cessation of Hostilities		Treaty without Mediation		Third-Party Mediation		Total	
	Aggressor		Aggressed									
War Types	**Nr.**	**%**	**Nr.**	**%**	**Nr.**	**%**	**Nr.**	**%**	**Nr.**	**%**	**Nr.**	**%**
Anti-regime	8	*17*	25	*53*	2	*4*	3	*6*	9	*19*	47	*99*
other internal	7	*27*	4	*15*	2	*8*	6	*23*	7	*27*	26	*100*
Mixed	2	*25*	4	*50*	1	*13*	-	-	1	*13*	8	*100*
INTERNAL WARS	17	*21*	33	*41*	5	*6*	9	*11*	17	*21*	81	*100*
Inter-state	7	*23*	1	*3*	5	*17*	3	*10*	14	*47*	30	*100*
Decolonization	-	-	2	*18*	1	*9*	4	*36*	4	*36*	11	*99*
INTERNATIONAL WARS	7	*17*	3	*7*	6	*15*	7	*17*	18	*44*	41	*100*
Intra-/international Mixed form												
with A components	2	*29*	3	*43*	-	-	-	-	2	*29*	7	*101*
with B components	-	-	2	*29*	-	-	-	-	5	*71*	7	*100*
MIXED WARS	2	*14*	5	*36*	-	-	-	-	7	*50*	14	*100*
Total	**26**	***19***	**41**	***30***	**11**	***8***	**16**	***12***	**42**	***31***	**136**	***100***

* Not included are four cases in which the war was terminated according to our definition yet fighting still continued below the threshold of war. The mixed form ABC does not appear here because the three complex wars of this type (Angola, Lebanon, Chad) did not end by December 31, 1992.

Table 28 shows types of war termination by type of war. The bottom row illustrates a fundamental fact: *Barely one-fifth (19 percent) of all wars were settled by the aggressor's eventual military success.* Far more often the aggressed turned the military tables (30 percent),[67] and in 8 percent of the cases the wars simply ended when fighting halted, either from exhaustion or common sense. Somewhat more often (12 percent) the wars ended through a treaty (either a cease-fire, formal peace treaty or substantial compromise). The most frequent end to wars—and this came as a surprise—*occurs through the mediation of a third party.* Unfortunately we have no corresponding data for the period before 1945 so we cannot establish whether this augurs an historical shift.

Table 28 indicates that for internal wars, especially anti-regime wars, the chance is significantly higher that the aggressed side wins against the aggressor. Note, however, that in "other internal wars" (B-type wars) the chances for victory do lie in favor of the aggressor, but precisely these types of war are most often (23 percent or 27 percent) settled through treaties (with or without the mediation of a third party).

Mediation is successful more in international wars than for any other type. This corresponds with the relatively short duration of international (above all interstate) wars. Moreover, no outcome for a decolonization war has ever been decided by the aggressing side, which usually means the liberation movement.[68] Although military struggles for liberation are of tremendous symbolic importance, in the end decolonization was ushered in by competition between the industrialized countries, economic rationality and colonialism and imperialism's general loss of political legitimacy.

The figures for mixed wars are too small to yield any far-reaching conclusions. It seems that third-party mediation is equally successful, especially in mixed wars with B-type components. Yet mixed wars necessarily involve a third party in the cessation of hostilities. Endless fighting or forging an agreement seem to be the only alternatives. This corresponds to our earlier observation that mixed wars last disproportionately long.

Table 29: Temporal Development of Forms of War Termination, 1945-1992 *

	Forms of War Termination											
	Military Victory				Cessation Of Hostilities		Treaty without Mediation		Third-Party Mediation		Total	
	Aggressor		Aggressed									
Periods of War Termination	Nr.	%	Nr.	%	Nr.	%	Nr.	%	Nr.	%	Nr.	%
1945 - 1954	8	*33*	7	*29*	1	*4*	1	*4*	7	*29*	24	*100*
1955 - 1964	2	*6*	13	*39*	5	*15*	3	*9*	10	*30*	33	*100*
1965 - 1974	3	*9*	10	*30*	1	*3*	9	*18*	13	*39*	33	*99*
1975 - 1984	6	*27*	9	*41*	1	*5*	2	*9*	4	*18*	22	*100*
1985 - 1992	7	*29*	2	*8*	3	*13*	4	*17*	8	*33*	24	*100*
Total	**26**	***19***	**41**	***30***	**11**	***8***	**16**	***12***	**42**	***31***	**136**	***100***

* Not included are four cases in which the war was terminated according to our definition yet fighting still continued below the threshold of war.

We determine next whether there is any temporal variation in the termination of war. According to table 29 this is not the case; at least we discern no trends. The share of cases in which the aggressed side prevailed militarily remains constant between 30 and 40 percent (the last period 1985-1992 is atypical). Between 1955-74 the aggressor seldom won. During this period the decolonization wars unfolded, none of which were won militarily by the aggressors. Furthermore the share of wars which simply stopped between 1955-64 was disproportionately high. War termination through mediation fluctuated between 18 and almost 40 percent. *Mediation by a third party has apparently become an institution of global governance.*

The share of the two main forms of war termination (victory of the aggressed side or mediation by a third party) are distributed similarly across regions, independent of the number of wars per region (see table 30). The share of "defensive victories" is about 30 percent—the runaway case of Central Africa can be explained by the recurring attempts at secession and revolts against the Mobutu-regime in Zaire. Aggressors seldom win in Africa, as opposed to Latin America or the Near and Middle East, the only two regions where aggressors prevail more often than the aggressed, if only barely. The Near and Middle East is also the region where a conspicuously high percentage of wars were ended through mediation (40 percent of 30 wars). Mediation is similarly effective in Europe, where aggressors have less chance of prevailing. The mistakes regarding the war in Bosnia have to be seen in historical context. The conditions for mediation in North Africa have been even better: five of the seven wars were thus ended, and the remaining two ended through direct negotiations between the warring parties. In none of the cases was there a decisive military result. The ongoing war in Algeria between the Islamic Salvation Front and the government will most likely end in negotiations. In East Asia, whose wars occurred in the 1940s and 1950s, mediation played no role due, in part, to the special position and politics of the People's Republic of China and the vagaries of the East-West conflict. Mediation is also noticeably weak in South America, in contrast to Central America and the Caribbean.

Table 30
Forms of War Termination by Region 1945-1992 *

Region	Military Victory: Aggressor Nr.	%	Military Victory: Aggressed Nr.	%	Cessation Of Hostilities Nr.	%	Treaty without Mediation Nr.	%	Third-Party Mediation Nr.	%	Total Nr.	%
Central America and Caribbean	6	*43*	3	*21*	-	-	-	-	5	*36*	14	*100*
South America	2	*20*	4	*40*	1	*10*	2	*20*	1	*10*	10	*100*
LATIN AMERICA	8	*33*	7	*29*	1	*4*	2	*8*	6	*25*	24	*99*
EUROPE	1	*11*	3	*33*	-	-	-	-	5	*56*	9	*100*
North Africa	-	-	-	-	-	-	2	*29*	5	*71*	7	*100*
West Africa	1	*20*	1	*20*	2	*40*	-	-	1	*20*	5	*100*
Central Africa	-	-	7	*88*	1	*12*	-	-	-	-	8	*100*
East Africa	2	*20*	5	*50*	-	-	-	-	3	*30*	10	*100*
Southern Africa	-	-	1	*20*	-	-	2	*40*	2	*40*	5	*100*
AFRICA	3	*9*	14	*40*	3	*9*	4	*11*	11	*31*	35	*100*
NEAR/ MIDDLE EAST	7	*23*	6	*20*	1	*3*	4	*13*	12	*40*	30	*100*
South/Southeast Asia and Oceania	6	*19*	9	*29*	3	*10*	5	*16*	8	*26*	31	*100*
East Asia	1	*14*	2	*29*	3	*43*	1	*14*	-	-	7	*100*
ASIA	7	*18*	11	*29*	6	*16*	6	*16*	8	*21*	38	*100*
Total	26	*19*	41	*30*	11	*8*	16	*12*	42	*31*	136	*100*

* Not included are four cases in which the war was terminated according to our definition yet fighting still continued below the threshold of war.

Regional differences persist, as we have seen, but mediation is a more successful form of war termination than it might appear.[69] Figure 13 indicates the mediators. The United Nations ranks first, with 40 percent of all successful mediations. In deeds the U.N. appears better than its much-maligned reputation. Their 17 cases constitute only 12 percent of all terminated wars, but how high was the success rate for the League of Nations? The U.N. is considerably more effective than the three regional organizations listed, despite the expectation that regional organizations exert more influence because of their intimate knowledge of the situation and actors. The remaining mediators are rather

Figure 13
Wars Ended by Mediation, 1956-1992, and the Mediator(s)

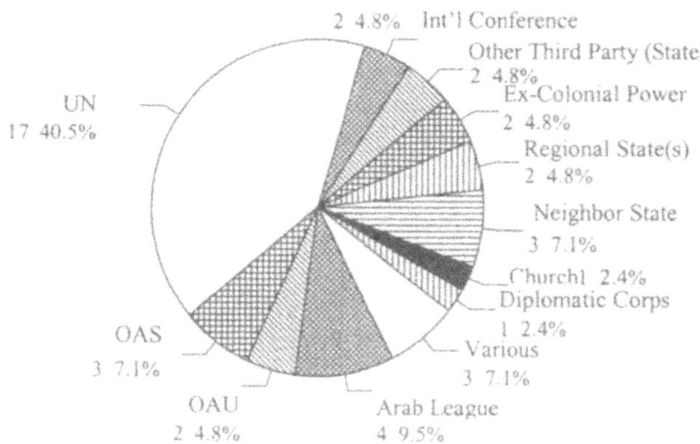

arbitrarily lumped together. This highlights the importance of formally institutionalizing mediation and the need for training qualified personnel.

Table 31
Wars, 1945-1992, with Armed Intervention by U.N. or U.N.-Authorized Forces

War Nr.	War	Duration	Type	Troops' Country of Origin
25	Korean War	1950-53	C	Ethiopia, Australia, Belgium, France, Greece, Great Britain, Canada, Columbia, Luxemburg, New Zealand, Netherlands, Philippines, South Africa, Thailand, Turkey, USA
55	Zaïre (Congo)	1960-63	AB	Ethiopia, Ghana, India, Ireland, Italy, Canada, Marocco, Sweden, Sudan, Tunesia
194	Anti-Iraq -Coalition	1991	C	Egypt, France, Great Britain, Italy, Kuwait, Saudi Arabia, Syria, USA *

* Includes only troops actually involved in fighting and not, e.g., German aircraft stationed in Turkey.

Table 32
Wars, 1945-1992, with Intervention by Third States or Regional Organizations to Prevent Further Hostilities

War Nr.	War	Duration	Type	Troops' Country of Origin
83	Dominican Republic	1965	A	OAS (Brazil, Costa Rica, El Salvador, Honduras, Nicaragua, Paraguay)
157	Sri Lanka (Tamil)	since 1983	B/ AB/B	India (from 1987-1990)
168	Liberia	since 1989	A	ECOWAS (Gambia, Ghana, Guinea, Mali, Nigeria, Senegal, Sierra Leone* (since 1990)
205	Moldavia (due to Dniester Republic)	1992	B	Russian Federation

* itself attacked since March 1991.

In addition to mediation and peace-keeping functions the U.N. (through the Security Council) relies on a variety of methods to pursue peace—from sanctions to peace enforcement. Up to now the U.N. has thrice participated or authorized others to participate in direct confrontation. These cases are listed in table 31. These interventions were not particularly successful: the Korean War resulted in a hard-won restoration of the status quo ante bellum, achieving neither peace nor a political solution to the division of the country. U.N. intervention in the Congo was catastrophic and Zaire turned into one of the most repressive dictatorships in Africa, characterized by recurring internal rebellions. The expulsion of Iraq from Kuwait was successful, and Iraq was forced to submit to inspections and dismantling of its weapons of mass destruction. Yet thousands lost their lives in the process (see the disputed numbers in the register), the Iraqi regime remains firmly ensconced in Baghdad, the border issues between Iraq and Kuwait have not been solved, the Kurds in the north live in a precarious protective zone and the Shiites in the south are still oppressed. The military victory solved few political problems. The U.N. action in Somalia (which escalated to fighting only after December 31, 1992) also eventually failed. These failures do not in themselves justify criticism of the role of the U.N. and its right to impose sanctions or to intervene, but they do indicate that the future cannot proceed like the past. We should not overlook, however, the U.N.'s success in mediating or assisting in mediating the end of 17 wars.

Table 33
Outcome of Internal Wars, 1945-1992, by Type of War *

	War Outcome							
	Favors the Government		Favors the Opposition		Undecided or Status quo ante		Total	
Region	**Nr.**	**%**	**Nr.**	**%**	**Nr.**	**%**	**Nr.**	**%**
Anti-regime Wars (Type A)	39	*65*	13	*22*	8	*13*	60	*100*
Other internal Wars (Type B)	26	*60*	5	*12*	12	*28*	43	*100*
Mixed internal Wars (Type AB)	9	*41*	3	*14*	10	*45*	22	*100*
Internal Wars Total	**74**	***59***	**21**	***17***	**30**	***24***	**125**	***100***

* Including ongoing wars. Ongoing wars and those wars since terminated by mediation or cessation of hostilities are treated as if they ended by December 31, 1992. The complex wars (type ABC) are included in mixed internal wars and all three belong to the category of "undecided or status quo ante."

Five further cases in which individual states or groups of states intervened to end a war apart from the U.N. can be found in table 32. The success rate here is also poor. The OAS intervention in the Dominican Republic did end the war (during which U.S. troops supported the conservative aggressor against the government), but by supporting the conservative rebels the intervention hardly contributed to the solution of political and social conflicts. The occasional interventions by India in Sri Lanka against the separatist Tamil Tigers had no effect. The West African troops in ECOWAS managed to control Monrovia and the surrounding area in their Liberia intervention, but otherwise the situation remained a stalemate. In these instances one cannot speak of war termination, not to mention peace creation. The Russian intervention in Moldavia did end the war and prevent its separation, but the old-style communists of the "Dniester Republic" achieved most of their goals. So far national and regional intervention is a less than perfect recipe for conflict resolution.

Table 34
Outcome of Internal Wars, 1945-1992, by Region *

Region	War Outcome Favors the Government Nr.	%	Favors the Opposition Nr.	%	Undecided or Status quo ante Nr.	%	Total Nr.	%
Central America and Caribbean	6	*50*	6	*50*	-	-	12	*100*
South America	10	*83*	1	*8*	1	*8*	12	*99*
LATIN AMERICA	16	*67*	7	*29*	1	*4*	24	*100*
EUROPE	5	*56*	2	*22*	2	*22*	9	*100*
North Africa	1	*33*	-	-	2	*66*	3	*100*
West Africa	3	*50*	-	-	3	*50*	6	*100*
Central Africa	7	*78*	1	*11*	1	*11*	9	*100*
East Africa	2	*29*	2	*29*	3	*43*	7	*100*
Southern Africa	2	*33*	2	*33*	2	*33*	6	*99*
AFRICA	15	*48*	5	*16*	11	*35*	31	*99*
NEAR/MIDDLE EAST	16	*55*	5	*17*	8	*28*	29	*100*
South/Southeast Asia and Oceania	21	*70*	1	*3*	8	*27*	30	*100*
East Asia	1	*50*	1	*50*	-	-	2	*100*
ASIA	22	*69*	2	*6*	8	*25*	32	*100*
Total	**74**	***59***	**21**	***17***	**30**	***24***	**125**	***100***

* Including ongoing wars. Ongoing wars and those wars since terminated by mediation or cessastion of hostilities are treated as if they ended by December 31, 1992. The complex wars (type ABC) are included in mixed internal wars and all three belong to the category of "undecided or status quo ante."

In our 1986 book we attempted to attach plus, minus or zero values to the political outcome of wars measured by the aggressor's war goals, followed by statistical analysis. We have decided not to repeat this because it was far too vague. The challenge remains, however, to develop both a unitary scheme for data collection and a coding process which enables quantification and comparison without sacrificing validity and differentiation. Here we limit ourselves to classifying the outcome of war as ending either favorably for the government/rulers or for the opposition. We recognize, of course, that "opposition" is not always identical with the aggressed side, and that a war can end with no discernible gain for either side (restoration of the status quo ante bellum). The question is which side most closely achieved their stated goal. The term opposition does not, however, imply that once successful their behavior is necessarily more exemplary than that of their former enemies.

These figures apply only to internal wars for two reasons: first, internal wars constitute the vast majority of all wars. Second, the outcome of decolonization wars are well known: even when the oppositional liberation movement never prevailed militarily, the end result was always formal independence—in other words an unfavorable outcome for the "government" (here the colonial power). Out of 31 interstate wars only three were resolved in favor of the aggressed state and eight in favor of the aggressor, irrespective of the specific form of war termination. There is a lesson, however, in that *the outcome of over two-thirds (20) of the 31 interstate wars remained politically undecided or merely restored the* status quo ante bellum. This includes all interstate wars for which the aggressor could not be clearly determined. It is particularly difficult to justify the human and material cost of such apparently pointless wars.

For internal wars, whose outcomes are tabulated by war type in table 33, the results must be discouraging for all oppositional elements. The rulers maintained the upper hand in the majority of all internal wars, but especially in anti-regime wars. Mixed wars produced mixed results: they usually ended undecided or merely restored the status quo. *The opposition has no significant statistical chance of success in any of the three war types.* Apparently Friederich Engels' warning to European workers that the time of the barricades is over now resonates globally. Third World opposition groups who possess little chance for democratic participation or interest articulation must develop strategies more effective and less destructive than violence and war.

Table 34 shows the results of the prevailing side by region. These figures substantiate our general observations regarding all regions. Latin America is one exception, where practically no wars were left undecided, as is South America, where the opposition apparently possessed not even the slimmest chance of success. Central Africa was also similar (cf. the Zaire War). Yet despite this we cannot conclude that regional influences significantly affect the outcome of an internal war.

Our data corroborates the hypothesis that since 1945 *war is increasingly ineffective as a political instrument.* This remains the case even if the growing number of wars in the Third World are part of an unavoidable historical process of catch-up vis à vis the formerly violent European states.

Notes

1. Between 1992 and December 31, 1994 three new wars have occurred: in Mexico (January 1994), in Yemen (April 1994) and in Chechnya. The war in Rwanda which so brutally entered the public consciousness in 1994 had in reality its start in 1990 (see the War Register Nr. 189).
2. See the comments about the term "war zone" in the glossary of important terms.
3. Gantzel/Meyer-Stamer (1986:93 ff.) have long since called attention to this exaggerated category of classification.
4. We explicate below the ways in which the observable increase in the number of countries participating in a war effects our findings.
5. Unfortunately we are not able to present similarly the development of the yearly intensity of war as measured by the amount of war dead.
6. For details on the breakdown of the regions see section 2.3 or the legend for the war list (chapter 3).
7. In order to avoid misunderstandings: We do not want to injure anyone's patriotic pride by our allocation of these countries to the European periphery. We proceed only according to objective economic data and do not attach any cultural value to these statements.
8. The last period (1990-1992) encompasses only three years.
9. Up to now there have been actually three wars in Yugoslavia which overlap in part but which we must separate analytically: the Serb-Slovenian war, the Serb-Croatian war, and the war between Serbs, Muslims and Croats in Bosnia-Herzegovina.
10. We wager to predict that any of these conflicts which involve Greece will not escalate into war because, despite the confrontational rhetoric of Greek domestic politics, Greece is simply too politically and economically integrated into European institutions and too vulnerable to sanctions.
11. See the comparative studies on South America (Chile, Columbia) from Kurtenbach (1991) and on Central America by Niebling (1992). For a further example of the mixture of socialist and traditional societies in the southern Caucasus as a legacy of the Soviet Union see Dorenwendt (1993) on Uzbekistan. We also hope to know more about how regional characteristics influence the causes of war after the completion of an ongoing project supported by the German Research Society.
12. In the sixties and early seventies one spoke in Red China and Vietnam of "global villages."
13. Whenever we mention the name of a state we always mean, of course, the respective government or possessor or power. We wish to remind the reader here of the problem of levels of analysis which until now have operated

primarily at the systemic level in our discussion of war. Now we are switching to the agent level, although we will return to the system level. Such changes in the level of analysis requires a (methodo)logical and even dialectical tolerance on the part of the reader, for they are unavoidable in our scientific quest to understand the structures behind historical processes.

14. The same goes for Belgium, the Netherlands, Spain and Portugal, although to a lesser degree. Portugal's resistance in their former colonies of Mozambique and Angola was particularly stubborn as a consequence of the economic backwardness maintained until 1975 by Portugal's fascistic-authoritarian government. The roots of this backwardness, however, can already be found in the effects of the specific form of "capitalist" trade expansion in the fifteenth and sixteenth centuries and in Portugal's dependence on England since the seventeenth century.
15. In this respect it is not so important that the U.N. missions do not meet yet with great success. The fact that the oft-used and terrible term "humanitarian intervention" obscures more than it illuminates is also unimportant here.
16. Since the U.S. involvement in Somalia began in 1993 we exclude it from our present considerations.
17. Guatemala in 1954, Dominican Republic in 1965, Grenada in 1983 and Panama in 1989.
18. The seven U.S. interventions in the context of the East-West conflict all take place—with the exception of U.S. participation in the anti-regime war in the Congo 1964-66—in South/Southeast Asia: Korea 1950-53, Quemoy Crisis of 1954, Vietnam 1957-1975, Indonesia 1958-61, Laos 1963-73, and Cambodia 1968-75. In determining the significance of U.S. hegemony as a background factor for intervention it is irrelevant whether the leadership position of the United States was acquired voluntarily or not. It is also of no concern here whether the U.S. perception of communist threat was realistic or not, whether the interventions were efficient or whether the decisions in each case were correct or counterproductive. Finally, for this context it is also unimportant whether the U.S. administrations followed particularistic national interests or the logic of domestic politics.
19. At this point a brief remark about the five war participations of the USSR/Russian Federation is appropriate. For four of these the Soviet Union was responsible, of which only one is directly related to the East-West conflict (Hungary 1956; the invasion of Czechoslovakia in 1968 does not count as war following our definition). The remaining case, in which the Russian Republic intervened, belongs to the disintegration-process of the former Soviet Union, or to put it somewhat differently, to the new ordering of nations emerging from the old multinational state (Moldavia 1992; the Russian Federation intervened militarily in Tadjikistan only after 1992, and in North Ossetia the Russian government had only sent troops in 1992). We do not wish to excuse or gloss over the fatal politics of the Soviet

Union, but in hindsight we can determine that the Soviet Union was at its core a defensive power despite all its threatening gestures.

20. Of which only eight participated directly in the fighting: mainly the USA, then Great Britain, Italy, France, Egypt, Syria, Saudi Arabia and Kuwait.
21. These subsidies actually generated an excess for the U.S., which was naturally not refunded.
22. A future "world-state" will probably emerge slowly and as a result of many small, almost unremarkable processes. It will bear little resemblance to the historical and present concept of state, and lack those elements of Hobbes' Leviathan which are so fearsome. While it won't be a Hobbesian state, neither will it be a version of the Kantian world federation of republics, for Kant's image remains constrained within the confines of the nation-state. Most likely we will encounter a decentralized network of international organizations and regimes which can implement those state functions which nation-states no longer efficiently fulfill. Even newer international structures will emerge to accommodate the high degree of necessary coordination.
23. Whether socioeconomic status and the political system have direct influence on a state's susceptibility to war is a question we will turn to later.
24. The industrialized societies in table 4 are Great Britain, the U.S., France, the USSR/Russia, Belgium, Australia, New Zealand, the Netherlands, Canada, Italy, Luxembourg and Sweden.
25. For further information on the development of the international arms trade, especially heavy weaponry (airplanes, tanks, artillery, warships, etc.) see the data in the SIPRI yearbooks; for the last few years see the graphic portrayal in Global Trends 93/94:145-173. Information about small weapons trade (automatic rifles and machine guns, grenade launchers, mines, jeeps and trucks, etc.) is unsurprisingly sparse and less reliable. These types of weapons are of much greater importance in the types of wars fought in the Third World, and we consider this point later in the text. For the best data regarding small weapons see Ezell (1988) and, for current information, the regularly appearing "Small Arms World Report (1990ff.)."
26. See the data in Anthony (1993) and Brzoska (1994).
27. We should note here that methodologically we are moving away from the agent-level analysis (i.e., from the individual participants) and back to a system-level analysis where war is the object of our observation rather than the actors themselves.
28. We can only mention here that such a division between "inside" and "outside" has had fatal consequences for the public understanding of foreign and security policy.
29. See, for example, Small and Singer (1982), and more recently Holsti (1991).
30. Little has changed on this point, despite an attempt to deal with this lack by Zimmermann (1981).
31. The following calculation is from Gantzel (1988a:45). It is based on the

data from Small and Singer (1982:82-99, 223-232). A different historical basis for the break down of the time period 1816-1980 would naturally led to other results, however the general development of the ratios would not change significantly. If one shortens the time period then we would naturally encounter more fluctuations. We are interested here, however, more in the historical current than in the eddies and swirls (to borrow an image from Richardson 1972:174).

32. Using the indicator "frequency of wars *waged* per year" leads to almost the same results. The data for this are to be found in Gantzel (1987:44-45).
33. The break 1943/44 instead of 1944/45 is used here because the first war following the Second World War broke out in December 1944.
34. Compare section 2.1 above. Certainly this only describes the internal core of the ideal state. This concept does not explain which objective functions the monopoly of power serves or in whose subjective interests.
35. We have already emphasized in section 2.5 that we are not entirely satisfied with this classification.
36. See the chronology of decolonization in the appendix.
37. Our study, however, does not include the overthrow of Japanese colonial rule during the Second World War.
38. Nr. 20, Malayan Alliance (AD); Nr. 39, Cameroon (AD); Nr. 38, Cyprus (BD); Nr. 123, Western Sahara (BD).
39. The legitimacy of these claims is unimportant here.
40. See also table 1 and figure 1.
41. In this figure we also left out the detailed breakdown of mixed wars from table 7, which indicates how they are distributed and which C- and D-type components are present. The strong mean variation gives us no statistically useful values for the subcategories of mixed wars, which is why we neglect further detailed analysis.
42. We can see from the mean distribution of the numbers that even an exact correlation with a time lag would not show any such connection which could support the assumption of a causal relation between the number of states and the frequency of war.
43. See also his extensive bibliographical references on this topic. See also Bächler (1988), Dixon (1993), Kittel (1993) and especially Nielebock (1993).
44. Let us take this opportunity to remind the reader that the number following the war type has the following meaning: "-1" means with intervention and "-2" means without.
45. War Nr. 83 (Dominican Republic), 168 (Liberia) and 205 (Moldavia).
46. War Nr. 55 (Zaire); the wars Nr. 25 (Korea) and 194 (Gulf War of 1991) are excluded in this analysis.
47. It remains unclear why there was so much apparent willingness to intervene between 1975-79. With two exceptions the eleven wars in question belong either to pure or mixed A-types, which in itself is not surprising. Cuba participated in the second Ogade War (Somalia/Ethiopia), the USSR

began its fatal intervention in Afghanistan, and Belgium and France aided the government of Zaire in the second Shaba war; in all other cases a neighboring state became involved. From these events we cannot infer any particular reason for the relatively high rate of intervention during this period.

48. It would be a worthwhile to more closely analyze the comparative motivations and results of intervention.
49. For the results of wars with interventions see the register.
50. The potential for mutual annihilation certainly exists in our nuclear age! So far, however, practical reason has had the last word, even if by dialectical turns. Reagan and Gorbachev seemed to recognize during their summit in Reykjavik that the search for usable nuclear weapons was hopeless aside from their utility as bargaining chips. We suspect that this logic cannot also escape India and Pakistan, Israel and Iraq, and even China and North Korea.
51. And over half of these wars were not yet really over by December 31, 1992!
52. Including three Colombian wars plus Guatemala, El Salvador and Peru.
53. Basque Country and Northern Ireland.
54. The five cases are the rebellion in Dhofar, the Palestine Intifada, Lebanon, the sixth Iraq-Kurdish war and the war in Afghanistan.
55. The longest war in East Asia was the Chinese defeat of the Tibetan resistance from 1954 to 1959 (56 months). The second longest was the Chinese Civil War (50 months). Far behind that follows the Korean War (37 months).
56. For example, numbers of victims were reported daily by a main press agency for the Kurdish conflict in Turkey, but seldom did the same for the war in Bosnia, as if that war produced no victims.
57. And soldiers are apparently also more important, since when distinct figures are indeed available, then usually only the soldiers are accounted for in detail, while the civilian casualties are only roughly estimated.
58. Worse than the meanest weapons used in Vietnam (napalm, pressure and shrapnel bombs, defoliation sprays, etc.) and even worse than high-speed high-tech weaponry are landmines (both anti-personnel and anti-tank mines), which are especially hard to detect when manufactured from plastic. They are inexpensive and millions have already been buried in war zones across the world, especially in Angola and Afghanistan. The mines do not distinguish who treads on them, whether soldier or tank or child or tractor. Additionally they are not removed when the wars end, so they remain in place. Even after fifty years certain parts of North Africa remain contaminated by German mines from the Second World War.
59. For many years the following relation between civilian and military casualties has made its rounds through the media and scientific literature: in the First World War "only" 5 percent of war casualties were civilians, in the Second World War this number rose to 48%, then to 84 percent in the Korean War and 95 percent in the Vietnam War. We cite these figures

from "Ami-antimilitarismus information," Volume 10, Issue 3, March 1980, p. I-69. The original source is unfortunately not given, as is the case in many scientific peace-research books where this figure appears. To the best of our knowledge these figures first appeared in a flier called "Army without a Mission" from the "Conscientious Objectors International" in the 1970s, issued by the late Helmut Michael Vogel, the former chairman of the German Peace Society/United Conscientious Objectors and vice president of the International Peace Office in Geneva. Our efforts at discovering the methodology and the numbers behind these figures have been unsuccessful. Perhaps they were just vague estimations cited primarily for agitation. See William Eckhardt (1992:254) for an attempt to construct a quantitative overview of these issues for the last five hundred years and the first ninety years of the twentieth century.

60. Row percentages.
61. This test is naturally only an indication, not proof, that the 135 wars are indeed representative. Absolute proof is not possible in this situation.
62. Despite the probability that our victim numbers are representative we decided not to break them down further by the individual type of war. Fine statistical figures are not possible given the relative ambiguity and unreliability of the data on victims.
63. Based on an estimated world population of 3.5 billion at the mid-point of our study's time frame.
64. We dispense with more exact statistical analyses here because of the precarious condition of the data. The maximum and minimum values in table 18 are a general starting point.
65. See citations in Gantzel (1982).
66. For example, the power distance between states must be more clearly determined, especially under consideration of regional contexts and the relations with neighboring states (whereby "neighborhood" for a highly integrated economic nation with a highly mobile military such as the U.S. or Great Britain must be defined differently than for Chad and Libya or Ecuador and Peru). Furthermore, changes in status over time must also be taken into account. The status, for example, of Thailand or Taiwan (Formosa) is no longer the same as it was in the 1950s, and that of the Russian Federation is certainly quite different than the Soviet Union in the 1970s. Beyond this a comparison of states which had no wars must be possible. We cannot undertake even these "simple" differentiations here. Gaps in the data as well as extremely limited personnel and time resources are among the main reasons for this, but above all each differentiation unleashes a host of unsolved problems regarding theoretical conceptualization and operationalization.
67. Even Von Clausewitz already saw that defense was the stronger form of war.
68. One exception here is the first Indochina war which France began in 1946 in order to restore its colonial empire which lay in ruins after both Japa-

nese occupation during the Second World War and colonial declarations of independence following the war.

69. Of course we must also look more closely at the number of cases where mediation failed.

5

Afterword: Warfare in 1995

Dietrich Jung, Klaus Schlichte, and Jens Siegelberg[1]

Fifty years after the end of World War II warfare continues unabated across most of the world. In 1995, wars were waged in 32 countries. Iraq, Colombia and the Philippines each saw two wars occurring simultaneously in their lands. In addition, there were fifteen armed conflicts whose intensity fell only narrowly below the threshold of war.

Africa bore the brunt of the 35 total wars with twelve wars, followed by Asia with eight, the Near and Middle East with seven, Latin America with six, and Europe with two. The total number of wars, however, has decreased compared with the previous years. The apex of wars since the Second World War was reached in 1992 with its 52 wars; the following three years show a steady decline. Twelve wars were removed from the war list in 1994/95 alone.

There are two reasons, however, why this positive development should be evaluated with cautious optimism. First, wars since 1945 have tended to fluctuate in number, so that the trend continues to show an overall increase. Second, the end of a war is not synonymous with the end of massive, politically motivated violence. The war in South Africa ended in 1994, yet that very year 17,000 people fell victim to violence, more than the total casualties of that sixteen-year-long war.

Furthermore, many conflicts which have fallen below the threshold of war exhibit no major structural changes in their respective situations. Many of these conflicts flame up again as war, as happened in Senegal and Pakistan during 1995. Four wars began in 1995, starting with the war in Uganda and the border war between Peru and Ecuador.

Figure 14
Wars Begun, Ended and Waged, 1945-1995

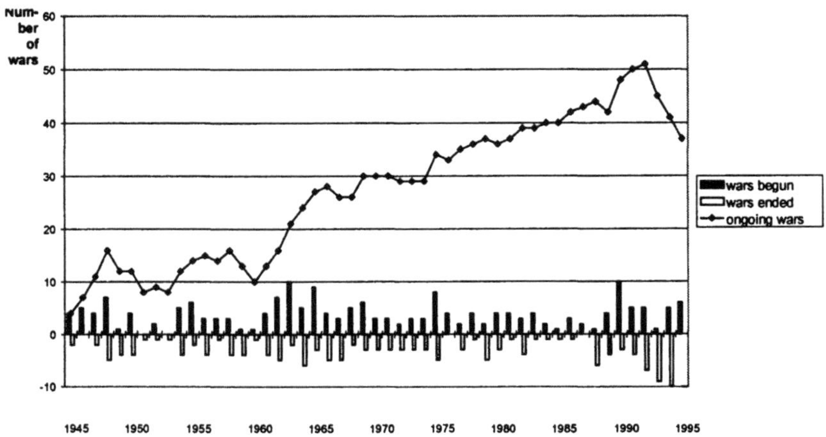

5.1 Main Trends in Warfare after 1945

Three primary tendencies characterize warfare since 1945, including the wars of the 1990s. These tendencies affect the number of wars as well as their typological classification and regional distribution.

The total number of wars has steadily increased since 1945. In the 1950s there were an average of twelve wars per year. In the 1960s and 1970s this average reached 22 and 32, respectively, and by the 1980s an average of 40 wars were fought yearly. The 1990s will exceed the average of the 1980s. The extremely high number of wars between 1990-92 alone, mostly related to the demise of the Soviet Union, insures that the average for the first half of the 1990s lies significantly higher than that of the 1980s.

In contrast to the interstate wars of the preceding one and a half centuries, internal wars have come to dominate the form of war in the second half of the twentieth century. Interstate wars have shrunk dramatically to only 17 percent of all wars between 1945 and mid-1995. The cause of wars can thus be connected to the global modernization process which is playing itself out through internal wars within the transitional societies of the Second and Third Worlds. The dominant types of war are, accordingly, anti-regime, autonomy, and secessionist wars.

The regional distribution of war is also unambiguous: between 1945 and 1992 Asia was the most affected region with 54 wars, followed by Africa with 49, the Near and Middle East with 43, Latin America with 31, and Europe with 13. One can see clearly that there have been relatively few wars within and between the developed capitalist states during the last fifty years. Over ninety percent of the 194 wars between 1945 and mid-1995 occurred in the Third World. Additionally, the former Eastern Bloc states have emerged as a new center of crisis in world politics.

Twelve conflicts have escalated into war in this latter region, which had been nearly devoid of war until 1989. This change signifies the only really new development, for aside from the spate of wars to be found in the former Eastern Bloc and remains of the Soviet Union there have been no fundamentally new elements in warfare during the 1990s.

5.2 Causes of War Since 1945

The persistence of war in the 1990s demonstrates that warfare since the Second World War was by no means solely a byproduct of the bipolar conflict and its so-called proxy wars. The end of the East-West Conflict did not lead to the termination of numerous ongoing wars. To the contrary, instead of an increase in peaceful democratic states we find an escalation of the conflict potential latent in capitalist modernization. The former Eastern Bloc provides an example of this: before capitalism can create a civil society with the rule of law, democracy, human rights, high living standards and the peaceful resolution of social conflicts it first leads to tremendous upheavals in the social order, to conflicts and to wars. It is the rare exception when a society can successfully implement modernization without authoritarian or violent means. This observation remains valid for both historical Europe, the current Third World and the transitional societies in the former Soviet sphere of influence. The same sequence will doubtless repeat itself in the future collapse of China's sphere of influence.

The third great state-building process is now taking place on the territory of the former Soviet Union, similar to the often bloody state-building processes of decolonization from the 1940s to the 1960s, and the European experience of the eighteenth and nineteenth centuries. Following the territorial consolidation of these new states we will be confronted with a form of conflict recognizable from the Third World experience after its formal independence, namely, internal societal conflicts emerging from the pursuit of belated state consolidation.

A structural characteristic specific to transitional societies is discernible in all of these wars: the institutions of the modern state are insufficient to settle existing social conflicts. The borders of state and society are apparently incongruous. At the bottom of this gap between political and social integration lies a chronic deficit of legitimacy for Third World rulers. In Second and Third World countries the state fails to function as the locus for the crystallization of collective identity, but rather as the space for the pursuit of particular interests by actors who have wrested room for maneuver either during the colonial era or from the structure of international law and the state system. As a result, family, tribal or religious institutions become the focus of personal identity and political loyalties rather than the state.

When this occurs, the modern state—with its claim to a monopoly on the legitimate use of force—comes to exist solely as a territorial and legal shell awaiting the consolidation of its societal content. This consolidation, however, is anything but a straightforward linear development. The legitimacy of the existing political order is invariably and consistently challenged by regional, ethnic or religious groups, often by violent means. In many cases lengthy civil wars result destroying what internal consolidation may have been achieved.

5.3 The Fall of States and the Diffusion of Violence

Warfare in the 1990s indicates additional trends which pose cause for concern. The end of most wars in the 1990s did little to end politically motivated violence. Eleven of the nineteen wars technically terminated between 1992 and 1994 continued in the form of armed conflicts under the threshold of war. In three further cases—Pakistan, Uganda, and Senegal—ongoing societal conflicts escalated anew into war.

Only the greatest instances of political violence caused by the disintegration process in transitional societies are recorded as wars. Worldwide we observe a process whereby violence is becoming increasingly diffused. Examples include the activities of death squads in Latin

America and the semi-criminal activities accompanying the struggle for power in countries such as Pakistan or Sierra Leone.

A significant rise in violence is not always the result of a long-suffering war. It is true that former theaters of war provide easy access to weapons and leave behind a culture of violence. Yet beyond this disadvantage many societies show a tendency to dissolve into stateless spaces where emerging political structures compete in the gray zone of the global economy. This disintegration is spurred by the worldwide trade in drugs and weapons, as well as the exploitation of natural resources, which together form an economic basis which undermines attempts at state consolidation.

An increase in global interdependence is thus not advantageous for transition societies, since they find themselves unable to control the consequences. The "end of territoriality" (Badie 1995) has had devastating consequences for the state-building process: states fall apart before they achieve internal consolidation, because the development of national economies is no longer possible.

5.4 A Political and Social Dilemma

The expectations on these fragile political structures is nonetheless increasing from all sides. In the world of international politics the distribution of resources such as credits and security guarantees are bound more and more to the rapid development of democracy. Even the largest donor countries make their development aid dependent on the realization of democratic rights and formal institutions for free elections. A functioning democratic order rests, however, not in the form but in the content of modern statehood. Democratic political structures rest ultimately on the state's legitimate claim to the monopoly of force and the existence of an expert civil service. The division between political office and the state budget must be complete. This assumes that citizens and civil servants alike are not responsible to familial or religious instances but follow formal, legal rules which regulate social relations on the basis of formally equal contractual partners.

This is, however, rarely the case in transitional societies, a fact which leads to a paradox: the implementation of a state monopoly of force cannot keep pace with a demand for rapid democratization. Democratization requires a high level of social integration, which relies on the successful consolidation of the state, meaning in large part a state monopoly of the use of force. Exactly this monopoly, however, must first be established as a pre-condition for democracy. The desired result of the social transformation–democracy–is turned into a requirement for its realization.

The desire for democratization is not only a product of foreign interests, but also arises within the transitional societies. Here the reality of capitalist transformation runs up against the promises of its eventual rewards. Failed promises of wealth, better education and increased political participation become a gauge for dissatisfaction regarding worsening living conditions. Religious fundamentalism and national chauvinism both express nothing other than the growing contradiction between unmet expectations and lived experience.

For most Second and Third World states the crisis of the capitalist world system has led to a heightened opposition between political expectations and economic possibilities. Development becomes blocked, and tendencies toward integration and disintegration fail to balance each other. The end of the state socialist model has also removed the attractiveness of social revolutionary goals. Socialist utopia has been replaced by religious or nationalist communities and their troubled pasts.

The noble call for democratization also opens the path for fundamentalist groups to achieve state power. As a result, the much sought after stabilizing effect of democratization threatens to achieve exactly the opposite. The consequences can include internal conflicts and wars fought over the nature of the state and society.

The same observation applies for many states in regard to regional and irredentist movements. If two regionally separated population groups struggle over state power, the result can often be a movement for autonomy or secession. The existence of multiple language communities or different religions within the territory of a single state are common reasons for secession or autonomy movements. Economic motives, however, have played an equally important role in the development and intensity of such conflicts. These include control over resources with export potential and anger at the neglect of certain regions in development policies.

The increase in so-called ethnic or religious conflicts in the 1990s says much less about changes in the causes of war and more about the perception of the wars' observers. The end of the East-West conflict has allegedly ushered in a new era where the "clash of cultures" figures as the new point of orientation and interpretation. This approach, however, is as wrong as the previously dominant interpretation of wars as "proxy wars" between the superpowers.

In reality little has changed. The very actors who sought international support by flying the banner of socialism or liberalism now indulge in the rhetoric of national self-determination and cultural identity to drum up solidarity at home and abroad. This is no epochal shift. The broad spectrum of international warfare is characterized by the goals of those social groups who seek international recognition as bearers of state power and their claims to present the sole "national" path to well-being, security and self-determination. At the same time these groups expect exclusive access not only to their own national resources but to the financial resources available from international creditors and development organizations. Yet sovereign statehood hardly ever stands alone as a solution to social problems and conflicts. An ethnically or culturally legitimated consolidation of power is doomed to failure as soon as the myth of homogeneity is challenged by the inevitable process of social differentiation. In violent conflicts, an actors' recourse to ethnicity, culture or religion as a base for solidarity is nothing more than a model for social organization and mobilization, even when they portray their respective cause as anchored in ancient rights or as the basis for a new society.

Despite the rhetoric, ethnic tension is not the cause but the result of social conflicts. It is the result of a process which destroys all those social mechanisms which allow people to live together peacefully. The destruction of these social mechanisms, rules and institutions in favor of a simple dualism based on ethnic categories is the real cause of the conflicts. The diffuse and overly broad term "ethnic conflict" works to hide precisely those complex processes which social scientific analysis must be concerned with.

The use of ethnic and religious symbols in warfare is by no means the sole province of the 1990s: it is a typical element accompanying all of modernity. A renewed reliance on cultural symbols always arises when the transition from traditional to modern modes of living are blocked, when the old crumbles and the new has yet to find its form. The process of social transformation remains at the core of any return to cultural symbols in violent conflict.

The majority of the Third World, now in its third decade since independence, and the regions of the former Soviet Union since 1989 face a similar situation: traditional social structures are being wholly destroyed, yet integration into new structures is not occurring. Caught in this uncertain moment, the return to ancient symbols and the creation of new traditions appear as survival strategies, indispensable for the mobilization and organization of groups.

The structural causes of war have therefore remained the same through 1995. The fundamental transformation masked by the euphemism "development" has caused the successive displacement of traditional structures in the Third World. This crisis-prone process makes societal consensus nearly impossible, even at the most basic level of state integration. Precisely this consensus, however, is an essential cornerstone for a functioning political system.

5.5 On Ending Wars and Regulating Peace

Each war is doubtless unique in its details, which vary from region to region and case to case. It is nonetheless possible to observe common problems which arise as obstacles to instituting peace.

There is no patent prescription for ending war and restoring peaceful relations. Both the reasons for beginning and the methods used to end wars differ substantially in each case, as do their effects. Whereas international pressure on Rwanda to democratize more quickly led to an undesired escalation of the war there, the termination of the war in Mozambique and the subsequent peaceful consolidation of state power would have been unthinkable without international support.

Appearances aside, individual governments, and within them individual personalities, play a more substantial role in mediation than large international organizations. Regional organizations are only rarely the most appropriate institutional setting for ameliorating conflicts. This holds true even for the United Nations, whose duties fall primarily in the area of implementing peace accords. The U.N.'s engagement reflects the power politics of its member states. For this reason it is unlikely that much will change in the U.N.'s current practice of expending tremendous energy on selected conflicts while all but ignoring others.

The internal dynamic of wars also presents fundamental challenges to their termination and the peaceful settlement of conflict. The disintegration process in the Third World is accelerated through war. The disruptive effects of the capitalist-induced transformation process and noncohesive rebel movements encourage the division of warring parties into competing factions, greatly reducing the possibility of ending the war. The development of wars in Afghanistan or in Southern Sudan show that supposedly united fronts collapse as soon as the constellation of the conflict changes. Rebel movements fall into fratricidal factions who turn on each other.

Frequently the very efforts to peacefully regulate the conflict can cause the disintegration of united fronts. This was the case in the rebellion in Mali and for the guerillas in Peru and Guatemala. These developments again show how the label of "ethnic group" is ultimately insufficient to mask differing interests. If political leadership cannot fulfill the material expectations of its fighters, then new rival groups form to pursue the original promises on their own.

For this very reason the leaders of many militant groups adopt policies of patronage not unlike the warlords of old. Social support services offered by the Hindu nationalist "Bharatiya Janata Party" in India, the construction of streets and hospitals by the drug kingpin Khun Sa in Burma or by the Islamic Hizb-Allah in Lebanon serve to build ties stronger than the existing loose connection of a common language or religious beliefs. In war zones such programs are only possible by wringing financial means out of the gray zone of the global market. For this very reason the trade in drugs, weapons, and natural resources has formed the spine of the economy in war-torn regions. Once established, this economy makes the peaceful regulation of conflict even more difficult, for its existence becomes dependent on a continued state of war. Drug production in Colombia, Somalia and Afghanistan thrives in the shadow of war, as does the rapid depletion of Liberia's and Burma's rain forests or the smuggling of diamonds out of Sierra Leone and Angola. Because these economic activities only flourish under conditions of war, they institutionalize interests which seek to prolong the conflict.

Military intervention alone is not an adequate means to stop a war; it requires the political will to pull together divergent interests and ideas present in a war's own dynamic. This political requirement cannot be forced with military means. Military intervention also faces a dilemma of the proper relation of its coercive means to its peace-seeking ends. If peace is restored by military might, then the interveners turn into warring parties, as happened with the United Nations in Somalia and with

France in Rwanda. Yet if intervention policy is restricted to humanitarian assistance, then the foreign forces risk becoming mere pawns of the warring parties.

Agreements between warring parties are less concerned with a final settlement of the issues and more with a temporary halt in the fighting. Peace talks all too often prove futile, and treaties are prone to being broken. This has been the experience of the United Nations, which has now undertaken peace missions across the globe after years of being tethered by the Cold War. It is partially for this reason that the U.N. remains unsuccessful as an effective instrument for securing peace. Yugoslavia, Rwanda and Somalia are examples of its failure.

There is no magic formula to end violence. Any realistic concept for addressing the sources of violence, however, must understand that the framework for global development needs to emerge above all from the Western power centers. While the political classes in these states are not directly responsible for the level of warfare in the world, they must face up to their failure to intervene when help is an imperative. Is it a crime of omission or commission when, for example, the German government allowed Germany in 1994 to join the ranks of the highest weapons exporters in the world and weakened the export prohibitions on dual-use products exported to those very states which appear in this book's register at a time when the German government sought to participate as peacekeepers in the war in former Yugoslavia?

Only when the Western centers create a politics of global responsibility and social reform will new spaces open to improve the development perspectives of the former Second and Third Worlds and will movement start down the difficult path to peace. The chances today for such a politics are slight, but one point remains clear: military intervention is no solution. Conflict regulation can only take place before battle lines are drawn. As necessary as the efforts to end war undoubtedly are, they remain a Sisyphean task as long as the conditions for war and violence grow faster than the means to counter them.

There is little reason to hope for substantive change in the near future. Not even increasing pressure on the Third World for democratization can, by itself, help the consolidation of political order. Just as the mere formal recognition of a state says nothing about its internal form of rule, so elections by themselves say nothing about actual democratization. The countries of the Third World and the former Soviet Union confront a gigantic challenge in creating democracy, a market economy and consolidating state power all within a very short period of time. Their ability to contain the social conflicts inherent in such an undertaking, however, remains hindered by global recession, indebtedness and the structural weakness of their own economies. As long as these conditions persist, a high level of war-like conflicts will continue to accompany the process of transformation.

Note

1. A version of this chapter originally appeared in Klaus Schlichte, ed., *"Das Kriegsgeschehen 1995"* in *Interdependenz* (Bonn: Materialien und Studien der Stiftung Entwicklung und Frieden, Nr. 20, 1996) pp. 11-19.

The War Register, 1945-1992

The Americas

Central America and the Caribbean

14 Costa Rica
12.3.1948 - 19.4.1948; A-2[1]
War ended by: Third-party mediation (diplomatic corps)

Warring parties:
Aggressor (Side A): Legión del Caribe[2]
Aggressed (Side B): Costa Rica

Object and goals of the conflict

In February of 1948 an opposition coalition of conservatives and social democrats won the presidential election, apparently by manipulating the electorate, whereupon the Costa Rican Congress annulled the outcome and set the stage for war. Tensions had risen steadily since 1942, when, besieged by economic crisis, the governing Republican party sought alliances to solidify the country's liberal agrarian export model. On the one hand they supported traditional accumulation structures with classical economic measures, while on the other they sought to strengthen their political legitimacy through social reforms by alliances with both the politically strong reform Communists and (until 1945-46) the reform-oriented Catholic Church. Through these contradictory alliances the government managed to alienate both the originally supportive capital sector (agriculture, trade, and banking) and the newer social democratic movement. Despite deep differences between social democrats and traditional capitalists both found in anticommunism a sufficient basis for common opposition to the government. The ideological cleft worsened as the international climate changed from antifascism to anticommunism in the wake of the Second World War. The social democratic troops, led by José Figueres, fought to overthrow the Picado government and eliminate Communist influence and implement social democratic concepts of development. The social democrats gained the upper hand in their alliance with the conservatives, not least because their broad social networks allowed for superior military organization.

Outcome of the war

The Picado government fell to the opposition, and an 18-month ruling junta was installed with Figueres as leader. The small government army proved more or less incompetent when fighting broke out, and the militias which supported the government quickly surrendered.[3] A peace treaty negotiated by diplomats was signed in the Mexican embassy on April 19, 1948.

The new regime persecuted former government supporters and communists, outlawing both the Republican and Communist parties and destroying the independent union movement. The repression compromised the social democratic junta's ability to expand its social basis, despite actually implementing social reform. They rapidly lost legitimacy and soon abdicated in favor of a conservative-liberal alliance. This alliance pursued measures to increase productivity and nationalize key industries (banks, utilities). At this time the regular army was disbanded and replaced by a national guard responsible for internal security (see War Nr. 16).

Notes

1. On April 17, 1948, Nicaraguan troop occupied the city of Villa Quesadas in northern Costa Rica. The Nicaraguan government justified their action by claiming national security interests: They wanted to protect their borders from a feared invasion by exiled Nicaraguans fighting on the side of the Costa Rican government.
2. The name stems from the participation of international brigades in the social democratic troops, consisting mostly of emigrants from other Central American and Caribbean countries.
3. The Communist militias decided to surrender when they heard that, in

addition to Nicaraguan intervention, the U.S. planned to invade Costa Rica. This would have drastically changed the balance of internal forces against the Communists, and they hence unwillingly entered negotiations whose outcome impacted them negatively. It is not known if the U.S. really intended to intervene.

Sources

Niebling 1984b; Schifter 1979 (UN)

---

16 Costa Rica (coup attempt)
10.12.1948 - 24.12.1948; A-2
War ended by: Third-party mediation (OAS)

Warring parties:

Aggressor (Side A): Exiled Costa Ricans[1]
Aggressed (Side B): Costa Rica[2]

Object and goals of the conflict

Following the civil war of March and April 1948 (see War Nr. 14) many supporters of the ousted Republican party fled to Nicaragua. They had lost their social basis in the defeat, especially in the civil-service sector, and their political power evaporated when the Republican party was outlawed. With the crucial support of Nicaragua's dictator, Somoza, the exiles invaded Costa Rica with the hope of overthrowing the now-weakened social democratic junta. Somoza hoped that the invasion would allow him to destroy the rebellious "Army for the National Liberation of Nicaragua" founded by Nicaraguan Legionnaires and based in Costa Rica. [3] These exiled Nicaraguans had financial and other support from Figueres (in accordance with the "Pacto del Caribe" [4]), and they represented a real threat to Somoza's dictatorship which, faced with a heterogeneous opposition within Nicaragua, found itself vulnerable to destabilization. Unlike the "Calderonistas," Somoza's first goal of the invasion was to disband the Legión del Caribe. Furthermore the mere existence of a social-democratic governing model in the country next door, especially in a country where the Nicaraguan opposition was concentrated, was itself a threat to Somoza's dictatorship.

Outcome of the war

The aggressors were successfully repulsed after Costa Rica invoked the Inter-American Alliance Treaty (TIAR) which they had signed on

December 3, 1948 and after the OAS stepped in to mediate. The political status quo ante bellum was restored, and the Legión del Caribe was disbanded. The structural conflict, however, between Costa Rica's social democratic government and Somoza's military dictatorship remained, and caused continuing attempts at reciprocal destabilization. By 1955 the tension escalated again to war (see War Nr. 37).

Notes

1. Exiled Costa Ricans were supported by the Somoza regime in Nicaragua; supporters of the former Costa Rican Republican Presidents Calderón Guardia and Picado.
2. Volunteer militia of Figueres' social democratic government.
3. The Legión del Caribe had split into two groups: one sought to first overthrow the Trujillo dictatorship in the Dominican Republic, while the others (mostly Nicaraguans) sought the overthrow of the Somoza dictatorship.
4. The "Pacto del Caribe" came together in 1947 from a politically heterogeneous alliance of opposition groups from different Central American and Caribbean countries. Its goal was the military overthrow of regional dictatorships.

Sources

Niebling 1985; Hoivik/Aas 1981 (UN)

---

31 Guatemala (Operation Success)

18.6.1954 - 1.7.1954; A-1

War ended by: military victory for side A

Warring parties:

Aggressor (Side A): Exiled Guatemalens and mercenaries under U.S. Command

Aggressed (Side B): Guatemala[1]

Intervention favoring side A: United States of America[2]

Object and goals of the conflict

The exiled Guatemalens sought to overthrow the democratically elected, social reform minded and nationally oriented Arbenz government. Jacobo Arbenz (in office since 1950) continued the structural reform politics of his predecesor, Arévalo, who came to power after the "revolution" of 1944. He was supported by a wide range of party alliance, including the Communist party. The cornerstone of his policies consisted of land reform and the introduction of a capitalist industrialization process geared toward the domestic market. Although he pursued capitalist reforms the U.S. interpreted his democratic activities within the context of the Cold War threat of Communism, especially his active involvement of unions and peasants and his policy of sovereignty. One major point of conflict was the expropriation of fallow lands belonging to the United Fruit Company as part of agrarian reform. The U.S. supported the opposition, consisting of the local oligarchy and rightwing military, against the Arbenz government and had the CIA organize the invasion of 1954.

Outcome of the war

After the Arbenz government fell, its reforms were reversed by a military government representing oligarchical interests. Worker and peasant organizations were destroyed, forced labor was reintroduced, and supporters of Arbenz were persecuted and murdered. They instituted an agrarian and industrial development program which focused on exports and on the interests of multinational companies, policies which caused increasing and widespread misery. Social and political conflicts remained unsolved. Since the fall of the Arbenz government civil war has been the norm in Guatemala (see Wars Nr. 56 and 146), with only brief interruptions (usually following a wave of repression).

The negative experience of the Arbenz government, especially regarding the untrustworthiness of the military and the defenselessness of organized people's movements, consequently influenced debates, policies and strategies of later social-revolutionary movements in Latin America concerning the means of getting out from under U.S. hegemonic claims. Lessons for social-revolutionary survival tactics came to include disbanding and total replacing (to the extent possible) of the existing military and arming of the general population.

Notes

1. Only a portion of the governmental troops remained loyal to Arbenz.
2. U.S. Air Force pilots.

Sources

Pearce 1981; Schlesinger/Kinzer 1986; Black 1984 (UN)

37 Costa Rica (second coup attempt)
11.1.1955 - 21.1.1955; A-1
War ended by: military victory for side B

Warring parties:
Aggressor (Side A): Exiled Costa Ricans[1]
Aggressed (Side B): Costa Rica[2]
Intervention favoring side A: Nicaragua

Object and goals of the conflict

Once again, with material and logistical support of the Somoza dictatorship, exiled Costa Ricans living in Nicaragua attempted to overthrow the social democratic government of José Figueres (elected President in 1953).[3] The Somoza regime continued to see itself threatened by the democratic model in Costa Rica and by Figueres' active support for the Nicaraguan opposition. The social democratic Costa Rican government, for its part, capitalized on the enemy image of Somoza's dictatorship to help solidify alienated domestic support: the worker movement rejected the social democrats because of their repression against communist-led unions, and the conservatives opposed their reform policies.

Outcome of the war

The invasion was defeated after the U.S. delivered fighter jets on short notice to Costa Rica. The OAS installed a demilitarized buffer zone between the two states in accordance with the TIAR (Inter-American Alliance Treaty). The U.S. pushed a peace plan through the OAS (which it dominates) which revitalized a Friendship Treaty from February 1949 (after the December 1948 war—see War Nr. 16) stipulating that both countries must prevent "terrorist activities" against their neighbor and control weapons transfers. The main U.S. interest was a rapid restoration of the status quo in Central America, which was achieved through the OAS peace efforts. While further attempts at mutual destabilization were not wholly prevented by this, they did occur at a reduced level. Tensions reached a high level once more only after the popular rebellion in Nicaragua in 1978/79. The 1960s and first half of the 1970s had even seen partial cooperation between Costa Rican and Nicaraguan police, for example when Costa Rican police arrested and deported FMLN guerrillas.

Notes

1. Exiled Costa Ricans were supported by the Nicaraguan Somoza regime; supporters of the previous Costa Rican Republican Presidents Calderón and Picado attacked the capital of Costa Rica by air from bases in Nicaraguan territory.
2. Volunteer units of the Costa Rican government that received figher jets from the U.S. on the basis of the TIAR. The regular Costa Rican army had been disbanded in 1948 and replaced with a national guard for domestic security. Costa Rica's external defense was essentially delegated to the Inter-American security system with their signing of the TIAR on December 3, 1948.
3. The social democratic junta was unable to hold on to power for long after the civil war of 1948 (see War Nr. 14). It was 1953 before a democratic government was elected.

Sources

AdG 4970F; Hoivik/Aas 1981; Niebling 1984b; Niebling 1985; Schifter 1979; Wheelock 1980 (UN)

---

45 Cuban Revolution
12/1956 - 1/1959; A-2
War ended by: military victory for side A

Warring parties:

Aggressor (Side A): Ejército Rebelde (Movement of July 26)
Aggressed (Side B): Cuba

Object and goals of the conflict

The social and political crisis of the corrupt, U.S.-supported Batista regime forms the background for this conflict. Almost a decade after the overthrow of the Machado dictatorship social reform groups in the 1940s won a degree of influence (for example the legitimation of opposition parties) but were unable to change the ruling structure itself. Political pressure from the opposition was increasingly met with repression. The failure of reform politics culminated in 1952 in a coup by Batista (Fulgencio Batista y Zaldívar) who prevented presidential elections and ushered in a new military dictatorship.

Against this background members of the mainly petite-bourgeoise intelligencia decided to pursue social change through military means. Their leader was Fidel Castro, who himself had planned to run as a candidate for the "Partido del Pueblo 'Ortodoxo'" in June of 1952. The first military action was an unsuccessful attack on the Moncada military base on July 26, 1953. While it failed militarily the ensuing public trial brought the rebels political success. Tensions between the government and opposition continued to rise, and in December 1956 rebels landed in eastern Cuba to begin the armed struggle to oust Batista. By 1958 this rebellion assumed the character of a massive revolt.

Outcome of the war

On New Year's Eve 1958, Batista, isolated politically and under pressure from the advancing guerrilla troops, fled. A general strike arose when Batista's supporters tried to impose a military junta in his place, preventing this from happening as it did in 1933 when Machado fled. The rebel army entered Santiago de Cuba on January 2, 1959, and six days later occupied Havana.

Following Batista's overthrow the U.S. military mission was thrown out of Cuba, the army was disbanded, and state administration was reorganized as part of greater fundamental structural reforms. The conflict between the former ruling class and Castro's social-revolutionary movement escalated to war again in 1961 (see War Nr. 58). The subsequent orientation of the Cuban regime toward socialism must also be seen in the light of fierce U.S. opposition to the new Cuban government: Even though Castro clearly wanted social reform, his sociopolitical orientation at the time of his rise to power remained open.[1]

Note

1. Castro's goals for structural reform included nationalization of electricity and telephone concerns which were U.S. property, a sharing of 30 percent of industrial profits with their workers and 55 percent of profits from the sugar industry, and finally land reform which benefited leasees and small farmers. The philosophy and ideology of the Movement of July 26 were strongly influenced by the egalitarian and national social thought of José Martís.

Sources

AdG 7485ff; Pearce 1981 (UN)

---

58 Cuba (Bay of Pigs)

17.4.1961 - 20.4.1961; A-2

War ended by: military victory for side B

Warring parties:

Aggressor (Side A): Exiled Cubans and mercenaries from other countries[1]

Aggressed (Side B): Cuba

Object and goals of the conflict

The May 1959 decision to implement agrarian reforms began a direct and open political conflict between the United States and the new Cuban government. The U.S.-owned sugar and cattle industries in Cuba were most affected by a new rule which limited the amount of land available for private use to 30 *caballerias* (about 402 hectares). The U.S. demanded immediate cash compensation for the expropriated property, which the Cuban government refused. As a result the U.S. aided the arming and training of Batista supporters who had emigrated to the U.S., imposed a sugar and export embargo (in the summer and fall of 1960, respectively), and isolated Cuba politically within the OAS, which the U.S. controlled.[1] On April 17, 1961, trained by the U.S. and led by the CIA , exiled Cubans and mercenaries invaded Cuba at the Bay of Pigs.

Outcome of the war

The invasion failed and only served to consolidate the revolutionary government. Cuba set out on its own path to socialist development despite new economic and military dependence on the Soviet Union. The U.S. continued attempts to destabilize the regime, yet forswore further direct military intervention after the Cuban missile crisis of 1962 threatened to escalate into direct military confrontation with the Soviet Union. Despite a brief period of detente under U.S. President Carter, the U.S. continued to try and destabilize the Castro regime (through sabotage, assassination attempts, economic embargos, etc.), reaching even new levels of intensity under President Reagan.

The consolidation of the Cuban revolution had far-reaching consequences for all of Latin America. Guerrilla movements saw in Castro's rebel army a model. Reformers saw the possibility of transforming society with the significant support of the people even under conditions of underdevelopment.

Note

1. The first bombardments began in the fall of 1959. In March 1960, a French

ship loaded with Belgian weapons exploded in Havana Harbor. A counterrevolutionary group formed in the Escambray mountains who focused on sabotage and isolated attacks.

Sources

AdG 4038; Hippler 1984; Pearce 1981 (UN)

---

56 Guatemala

1962 - 1968[1]; A-2

War ended by: military victory for side B

Warring parties:

Aggressor (Side A): Fuerzas Armadas Rebeldes (FAR)

Aggressed (Side B): Guatemala

Object and goals of the conflict

After the fall of the democratically elected Arbenz regime (see War Nr. 31) social and political conflict was held in check by repression. But a failed revolt by democratic and nationalist-oriented officers in November 1960 led to the founding of a guerrilla army. The revolt opposed General Ydigoras Fuente's decision to make Guatemala the base of operations for the planned Bay of Pigs invasion. The following February 1962 saw a call for a national uprising. Various guerrilla groups, especially from the eastern rural regions, joined together in December 1962 as the FAR. Their common goal was the overthrow of the military government, the restoration of Guatemalen sovereignty and the implementation of social and political reforms a lá Arbenz. Due to the ideological divergences among these groups (Trotskyists, Communists, left socialists) they usually operated independently of one another.

Outcome of the war

The disparate character of the rebels contributed to their defeat at the hands of a "Counter-Insurgency Offensive" of the Guatemalen military (directed by the U.S.) and paramilitary death squads which operated against both rebels and civilians. The rebel's military actions were largely unconnected to the political and social struggles of the urban opposition despite feelings of commonality. Furthermore they had little organized contact with the Indian population, who constituted the majority in the regions where the rebels operated.

The defeat of the FAR led to a revision of guerrilla strategies. At the same time their defeat strengthened military rule, which expanded in the 1970s to include the ruling economic factions. The war furthered the implementation of an economic model of accumulation based on exploitation which structurally excluded social reform.

Note

1. It is difficult to pinpoint the end of the war. The Counter-Insurgency Offensive of 1966-68 under Colonel Arana Osorio essentially wiped out the guerrillas by the end of the sixties. Yet isolated guerrilla activity continued until 1972, mostly in the cities. The remaining guerrillas turned toward cultivating a political base with other social and political forces in the 1970s. One of the four new guerrilla organizations, Ejército Guerrillero de los Pobres (EGP), began again in 1975 with military activity, leading by the end of the 1970s to a new war (see War Nr. 146).

Sources

Black 1984; Boris/Hiedl/Sieglin 1984; Schlesinger/Kinzer 1986; Pearce 1981 (UN)

---

83 Dominican Republic
25.4.1965 - 3.9.1965; A-1[1]
War ended by: military victory for side A[2]

Warring parties:

Aggressor (Side A): Conservative military elements
Aggressed (Side B): Dominican Republic[3]
Intervention favoring side A: United States of America[4] (28.4.1965 - 3.9.1965)
Intervention to end the war: Brazil / Costa Rica / El Salvador / Honduras / Nicaragua / Paraguay (all: 29.5.1965 - 3.9.1965).

Object and goals of the conflict

The bloodless coup of April 24, 1965 against the Cabral Junta left a power vacuum in which traditional power centers (including the U.S.) and new reform forces failed to decide upon who should form the new government. The opposition sought to implement reforms against a background of crushing poverty and unemployment. They consisted mostly of young officers with the support of the capital and the army loyal to social democract Juan Bosch, who had been overthrown in 1963 by the military. The conservative military elite felt threatened by the prospect of reform and formed a new junta which they tried to impose on April 25, 1965. When this attempt failed the U.S. intervened on April 28 in the hopes of establishing a government favorable to their interests.

OUTCOME OF THE WAR

While the U.S. was successful in maintaining its hegemony in the Caribbean, their invasion intensified rather than solved both the misery affecting most of the population and the deeply rooted social conflicts. Occasional military actions below the level of war continued sporadically. 2,000 to 3,000 persons were killed in the war.

Notes

1. Even though the OAS troops from Latin America followed their peace mission they did not always remain neutral, and often favored the conservative side outright. This had much to do with U.S. leadership in the OAS and the dictatorial regimes of many of the Latin American countries who contributed troops to the OAS.
2. It was not the mediation attempts which led to the end of the war but rather the military defeat of the opposition, whose position became untenable after the introduction of U.S. troops.
3. The Dominican Republic was represented by those forces who overthrew the Cabral Junta. These were the young officers who enjoyed broad support among the mobilized population. Those citizens who took part in the fighting were primarily supporters of the "social democratic" Partido Revolucionario Dominicano and other leftist organizations.
4. The U.S. formally advocated no particular party, but de facto they supported the conservative military junta. At the end of May 1965, U.S. troops formally became part of the OAS troops.

SOURCES

AdG 11875-11884, 12008f.; Atkins 1981; Barnet 1968; Bosch 1969; Carey 1967; Chang-Rodriguez 1969; Draper 1968; Draper 1971; Ferguson 1973; Goff/Locker 1969; Kumar 1987; Kurzman 1965; Lowenthal 1969; Lowenthal 1972; Moreno 1970; Niedergang 1966; Slater 1970; Szulc 1965; Wiarda 1980 (TR)

---

100 El Salvador / Honduras ("Soccer War")
14.7.1969 - 18.7.1969; C-2
War ended by: Third-party mediation (OAS)

WARRING PARTIES:

Aggressor (Side A): El Salvador
Aggressed (Side B): Honduras

OBJECT AND GOALS OF THE CONFLICT

El Salvador was more highly developed than its neighbors and profited most from the integration of a common Central American market. It also ran more quickly against the limits of economic growth in a system which lacked consumers and dumped excess industrial products. The military and agrobusiness cartel in El Salvador tried to stave off economic crisis by expanding their markets into Honduras while also encouraging migration to neighboring lands, especially Honduras. Honduras responded with protectionist policies, non-renewal of a migration treaty between the two lands and the forced repatriation of Salvadorans. The latter action was also connected with Honduran domestic politics, for El Salvadoran settlers were "squatting" on land which the Honduran government was trying to bring under its control through land reform. Nationalist sword waving in both countries also served to distract the population from domestic problems. The Salvadoran government was worried that repatriation of the migrants would close an important "vent" for "undesired" parts of their population, and they railed against the treatment of the "squatters" to mobilize anti-Honduran sentiments at home. The ensuing war would close the border and keep the migrants abroad. The nationalist card united even otherwise antagonistic groups, such as union movements, together with the government. The external cause of the war was riots and unrest accompanying two soccer matches between the two countries.

OUTCOME OF THE WAR

The Salvadoran forces made rapid inroads into Honduras, and were only stopped when the population fought alongside the poorly equipped Honduran troops. A cease-fire was eventually instituted under pressure and with mediation from the OAS and the U.N. General Secretary.

The status quo ante bellum was restored, and in 1970 Honduras quit the Common Central American Market MCCA). The nationalist movements were largely demobilized in both countries, though in El Salvador tensions rose internally through the 1970s, culminating in internal strife (see War Nr. 140).

Approximately 3,000 people were killed in the war.

Sources

AdG 14780, 14854; Cable 1971; Camacho u.a. 1979; Carías/Slutzky 1971; CEPAL 1980; CIDAMO 1980; Jung 1980a; Meza 1980; Niebling 1984a; Niebling 1991; Richter 1976; Rowles 1980 (UN)

---

129 Nicaraguan Revolution
10/1977[1] - 19.7.1979; A-1
War ended by: military victory for side A

Warring parties:

Aggressor (Side A): Frente Sandinista de Liberación Nacional (FSLN)
Aggressed (Side B): Nicaragua
Intervention favoring side B: Honduras (3/1978)

Object and goals of the conflict

Three goals united the opposition in Nicaragua: the overthrow of the Somoza dictatorship, established in 1937; the struggle for national self-determination; and the realization of civil rights. The revolutionary protagonists came from the heterogeneous Sandinista movement. The "bourgeoise" opposition suffered from a lack of legitimacy and never achieved the status of a movement. This was because so many of their members sought to share power with Somoza and remained caught in the mind-set of oligarchy. The U.S. government supported this opposition, though their solutions remained inadequate. The Sandinistas won some room for maneuver during the Carter Administration, which however, was reluctant to commit the U.S. to direct intervention in the aftermath of the Vietnam debacle. [2]

Outcome of the war

At first the OAS was successful in mediating an end to the war. Somoza had fled, and the intention was for the interim President Urcoyo to hand over power on July 17, 1979 to a "Governing Junta for National Reconstruction" founded by the FSLN in exile. The junta consisted of three FSLN members, the founder of the liberal MDN party Robelo, and Violeta Camorro, widow of the slain opposition leader. The national guard was also to be "cleansed." When Urcoyo proclaimed his intention of remaining in office until 1981 this solution failed. He called upon the Somozan National Guard to fight on, and the cease-fire evaporated. But with Somoza gone and the U.S. unwilling to intervene directly, the National Guard offered little resistance and FSLN troops triumphantly took Managua on July 19, 1979.

The National Guard was disbanded and replaced by the Sandinista army. The Sandinista government undertook a wide array of reforms, including expropriation of Somozan property, literacy campaigns, social and political mobilization, nationalization of banking and foreign trade, elections in 1984 and eventually a peaceful democratic change in power in 1990. Conflict between the varied groups involved in the war continued, however, in the years after Somozas fall (see War Nr. 151).

Notes

1. The war began in the sense of a nationwide conflict with the FSLN's October Offensive in 1977. Although formed in 1961, the FSLN was active primarily at the guerrilla level until 1974, when it staged a spectacular hostage-taking. The Somoza regime quickly forced them underground again until 1977.
2. The U.S. did try to convince the OAS on June 21, 1979 to send an international force to Nicaragua,. This proposal was almost unanimously rejected, however, for Somoza's dictatorship was virtually isolated in Latin America.. Costa Rica, and Mexico had already broken off diplomatic relations with Nicaragua.
3. The FSLN accepted this because a short-term military victory over the National Guard seemed unrealistic.

Sources

Black 1981; Gabriel 1987; Harris/Vilas 1985; IES 1982; Invernizzi/Pisani/Ceberio 1986; Jung 1980b; López et al. 1980; Millet 1979; Niebling 1990; Rediske 1984; Vilas 1984 (UN)

---

146 Guatemala
since 1980; A-2

Warring parties:

Aggressor (Side A): Fuerzas Armadas Rebeldes (FAR)[1] / Organización del Pueblo en Armas (ORPA)[2] / Ejército Guerrillero de los Pobres (EGP)[3] / Partido Guatemalteco del Trabajo (PGT)
Aggressed (Side B): Guatemala

Object and goals of the conflict

Ever since Arbenz's reform government was ousted in 1954 (see War

Nr. 31) there arose armed attempts to overthrow the governing military and agrobusiness oligarchy (see War Nr. 56). Extreme inequalities in land distribution and waves of state repression fed the fuel behind such revolts. A major change since the 1960s is the increased participation of the Indian majority. Ethnic and socioeconomic conflicts are further fanned by a lock on control of politics and economics by the Ladino ruling class (descendents of mixed whites and Indians).

Outcome of the war

During the 1980s, state repression began to assume the level of genocide. While effective in holding the guerrillas in check, the government was nonetheless unable to militarily eliminate them. A formally democratic government elected in 1986 did little to change the degree of repression or civil unrest, and the military retained the prerogative in matters of "internal security." In August of 1992 talks were begun again between the URNG and the government.[4]

Over thirty years of intensifying repression had cost approximately 100,000 lives. 40,000 people disappeared during this time, and 250,000 Guatemalens fled to Mexico. Around one million persons were forcibly resettled in so-called "strategic villages" in the context of the military's "Counter-Insurgency Actions."

Notes

1. Some of the guerrillas from the previous war (Nr. 56) continued to fight under their old name.
2. Stemming from parts of the the FAR.
3. Emerging from the remains of the old guerrillas.
4. The four guerrilla movements came together in 1982 as Unidad Revolucionaria Nacional Guatemalteca (URNG).

Sources

Aguilera 1987; Arias 1985; Barry/Preusch 1986; Black 1984; Boris/Hiedl/Sieglin 1984; Brintall 1979; Burgos 1984; CDHG 1989; Davis 1983; Gabriel 1987; Handy 1984; Handy 1986; Informationsstelle Guatemala 1982; Lindig/Münzel 1985; Menchú 1993; Painter 1987; Sandner 1985; Schlesinger/Kinzer 1986; Torres-Rivas 1980; URNG 1982; Veliz 1980 (SK/AAB)

140 El Salvador
10.1.1981 - 1.2.1992; A-2[1]
War ended by: Third-party mediation (U.N.)

WARRING PARTIES:
Aggressor (Side A): Frente Farabundo Martí para la Liberacíon Nacional (FMLN)[2]
Aggressed (Side B): El Salvador

OBJECT AND GOALS OF THE CONFLICT

Politics and society in El Salvador have been dominated by the traditional power sources of the military and agrobusiness, excluding the majority of the population from participation. The socioeconomic crises since the beginning of the 1970s intensified these tendencies and led to growing impoverishment. When the people's movement tried to protest peacefully for social reform and democratic elections the government responded with harsh repressive measures, driving their leaders underground. Many of these consequently joined guerrilla organizations which arose in the early 1970s.[3] Since the mid-1970s the social-reform mass organizations began to create "defense troops" to counter the systematic state repression. One can only speak, however, of a systematic armed defense force under FMLN leadership after the "general offensive" of January 10, 1981, which also served as the beginning of the civil war.

The peasant resistance was significantly anchored in rural Christian communities, which provided a basis for legitimation and organization.

OUTCOME OF THE WAR

Despite several large offensives the FMLN was unable to achieve military victory. Continuous U.S. assistance for Salvadoran forces erased gains made by the FMLN, which led the FMLN to turn to negotiations in order to reach a political solution. A peace treaty was signed in Mexico on January 16, 1992 after drawn-out and difficult negotiations by U.N. General Secretary de Cuellar and the stationing of U.N. observers (UNOSAL) in June, 1991. A cease-fire took effect on February 1, 1992, and has remained in effect since then, if sometimes hesitantly. At the end of 1992 the FMLN lay down their arms and became a political party with the intention of taking part in the next elections. The special police units are to be disbanded at the end of 1993, when the army is also to be reduced by half. Specialized programs are envisioned to integrate former FMLN fighters and demobilized soldiers.

The war cost at least 75,000 deaths, most of which were due to the brutal actions of the security forces and death squads.

Notes

1. The central government was occasionally supported by the Honduran army and U.S. reconnaissance flights.
2. Organization formed from the guerrilla groups FPL, ERP, PCS, RN and PRTC (since 1980).
3. At first these were limited to the kidnapping of members of wealthy families, who were exchanged for communiqués carried by the media or the payment of ransom money used to buy weapons.

Sources

Argueta 1982; Bacevich/Hallums/White 1988; Baumgärtner/Gabriel/Gottwald 1989; Cabarrús 1983; Dunkerley 1986; Frenz/Greinacher/Junk u.a. 1982; Gabriel 1987; Heckhorn 1983; Martín-Baró 1987a; Martín-Baró 1987b; MIPODOK 1988; Niebling/Richter 1983; Niebling 1984a; Niebling 1988; Niebling 1991; Vergara et al. 1987 (UN)

---

151 Nicaragua (Contra War)
4/1981[1] - 1990; A-1
War ended by: Third-party mediation (regional state[s])

Warring parties:

Aggressor (Side A): Contras[2]
Aggressed (Side B): Nicaragua
Intervention favoring side A: Honduras (1981 - ??)

Object and goals of the conflict

Social reforms planned by the Sandinista government after their successful overthrow of the Somoza dictatorship sparked the conflict. Together with personal designs on power, the Contra's main goal was the restoration of the old dictatorial order in Nicaragua. The new Reagan Administration in the U.S. adopted a new Nicaragua policy which played a significant role in allowing the internal conflict to escalate to the level of war. Lacking support among the Nicaraguan population, the Contras were dependent for their survival on enormous financial and military support from the United States.

The Sandinista policy of "national integration," by which they intended to create a unitary nation-state out of the different indigenous ethnic groups, led to conflict between the Spanish-speaking Mestizos of the Pacific side and the Indian minority on the Atlantic coast, above all with the Miskitos.

Outcome of the war

The Sandinistas lost a presidential and parliamentary election on February 25, 1990 to the United Nicaraguan Opposition (UN), an electoral alliance containing returned Contras, traditional left and right parties and a host of other groups. In the lead-up to the election a mechanism was worked out by Central American presidents to disband the Contras by December 8, 1989. Some Contras refused to participate in the demobilization.

The war cost approximately 60,000 lives. Together with a U.S. economic blockade, the war drove Nicaragua into economic ruin and living standards sunk to levels of the 1950s. Responsibility for this has also to be shared with incompetent Sandinista economic and bureaucratic policies.

Notes

1. The beginning and end of the war cannot be precisely determined. Former National Guard members organized themselves into counter-revolutionary rebel groups in Honduras already in 1980 (see War Nr. 129).
2. Consisting of exiled Nicaraguan troops supplied by the U.S. (dominated by former National Guard members of the Somoza regime) and Miskito Indian organizations (MISURA, MISURASATA) who belonged to the culturally and politically unintegrated Atlantic coast. Most Contras operated from bases in Honduras, partially with support of Honduran troops. A small anti-Sandinista opposition group operated out of Costa Rica between 1982 and 1986.

Sources

bfai 1991; Chamorro 1987; CENIDH 1990; CENIDH 1991; Dietrich 1988; Ehrke 1986; Feinberg 1982; Fulbright 1982; Gabriel 1987; Hamilton et al. 1988; Harris/Vilas 1985; Invernizzi/Pisani/Ceberio 1986; Krumwiede 1989; Kurtenbach 1987; Meschkat et al. 1987; Niebling 1990; Schulz 1985; Vilas 1984 (UN)

156 Grenada-Invasion (Operation Urgent Fury)
24.10.1983 - 1.11.1983; C-2[1]
War ended by: military victory for side A

WARRING PARTIES:
Aggressor (Side A): United States of America[2]
Aggressed (Side B): Grenada

OBJECT AND GOALS OF THE CONFLICT

The U.S. intervened to restore the pre-revolutionary balance of power under Eric Gairy, who was overthrown in a bloodless coup while he was travelling in the U.S. The coup against Gairy was led by the New Jewel Movement (NJM) whose leadership Gairy had instructed to be liquidated while he was away. The ensuing "Revolutionary People's Government (PRG) led by Prime Minister Maurice Bishop immediately began to institute political and social reforms. The NJM-led government and its leader, Bishop, enjoyed widespread popularity along with the concept of "revolution." The NJM sought to satisfy basic needs, include the population in political debates and take concrete steps for reform, such as literacy-campaigns and mass mobilization through social programs.

The Reagan Administration had already signified its intention since its arrival in Washington to eliminate Grenada as a precedent-setting example in Latin America. Boycotts, isolation and pressure tactics were employed. During NATO manuevers ("Ocean Venture 81) a practice invasion of the island took place, which was referred to as "our enemy in the East Caribbean." In March of 1993, Bishop unveiled information that the CIA was preparing the overthrow of the revolutionary government and was planning an invasion.

OUTCOME OF THE WAR

The island was occupied by U.S. forces and a provisional government under Scoon was installed, which was to remain in office until elections were called. The U.S. government committed 15 million dollars for "democratic renewal." Further steps included the disbanding of mass organizations, arrests, the firing of 3,500 civil servants, a halt to nearly all social programs and development projects with the exception of the airport expansion, which continued under U.S. direction. The Reagan Administration used the invasion to gain legitimacy in the Congress and with the public, which was especially sought in the aftermath of a terrorist bombing that killed over 200 U.S. troops in Lebanon immediately preceding the invasion of Grenada.

Approximately 100 persons died in the fighting.

Notes

1. The Cuban construction workers who were caught in the middle of the fighting are not to be seen as a warring party, since Castro ordered them only to defend their own lodgings. Grenada's call for military help from Cuba was met by Cuba refusing to participate in Grenada's "internal affairs." 25 Cubans were killed in the attack and 59 wounded.
2. The „Caribbean Peace Keeping Force" stationed police forces in Grenada after the invasion, lending legitimation to the U.S.. The participating countries were: Dominica, St. Lucia, St. Vincent, Antigua, St. Kitts and Nevis, Jamaica und Barbados.

Sources

Hippler 1984; LAB 1984; Agee/Fuchs/Ernst 1984; O'Shaugnessy 1984; Caribbean-Informationszentrum 1984 (UN)

---

170 USA / Panama

20.12.1989 - 24.12.1989; C-2

War ended by: military victory for side A

WARRING PARTIES:

Aggressor (Side A): United States of America

Aggressed (Side B): Panama

OBJECT AND GOALS OF THE CONFLICT

A three-year-long feud between the U.S. government and the Panamanian ruler Manuel Antonio Noriega culminated in a five-day war. Panama's historical dependence on the U.S. was the deeper context, since the interests of the U.S. in Panama since the founding of that state have been inseparable from U.S. strategic interests in the Panama Canal. The main goal of the Reagan and Bush Administrations in Panama had been to foil or at least delay the 1999 return of the Panama Canal as called for by the treaties of 1977. Furthermore, the U.S. had quarrels with Panama concerning the drug trade, especially money laundering.

Resistance was also growing within Panama to Noriega, who as the Commander-in-Chief of the National Guard exerted disproportionate influence on the government and thwarted any movement toward democratization. When the President of Panama tried to remove Noriega at the end of February 1988, he found himself disempowered by the National Assembly. The internal conflict reached its pinnacle in a failed coup on October 3, 1989, after which Noriega's forces were "cleansed." The failed coup, together with sustained criticism of the Bush Administration both from the Panamanian opposition and the U.S. led to the decision to invade on December 20, 1989.

Outcome of the war

The new U.S.-friendly government under Guillermo Endara was sworn in on a U.S. military base the night of the invasion. While the bilateral disputes were left unsolved (regarding the Canal), Noriega was captured and flown to Florida, where he was convicted of selling drugs. A power vacuum ensued when Noriega's military was dissolved, and looting and rioting escalated dramatically. In response, a 13,000-person-strong "Fuerza Pública" was formed, which consisted for the most part of the former military.

Between 600 and 7,000 people died in the five-day war, around 15,000 were left homeless, and the material costs were estimaged at 2.2 billion U.S. dollars, including damages from U.S. sanctions which preceded the invasion.

Sources

Kempe 1990; LaFeber 1989; Sandner 1985; Schubert 1978; Weeks/Gunson 1991 (SK)

South America

8 Bolivian Revolution
1946 - 4/1952; A-2
War ended by: military victory for side A

Warring parties:

Aggressor (Side A): Movimiento Nacionalista Revolucionario (MNR)[1]
Aggressed (Side B): Bolivia

Object and goals of the conflict

A broad social-revolutionary movement joined forces in Bolivia against the agrarian and mining oligarchy, supported by the military, and against foreign capital. The upwardly mobile, nationalist, and urban middle-class intellectuals and the well-organized mine-worker's unions allied themselves with the Campesinos. As militant worker factions within the movement gained influence battles erupted in 1946 with the Bolivian military. From this year on bloody strikes and revolts intensified (especially in 1947, August/September 1949, and May 1950) until the revolution broke out into the open in 1952. The revolutionary goal was the transformation of the society along anti-imperialist and anti-oligarchical lines.

Outcome of the war

The revolutionary movement was victorious and proceeded to introduce radical land reform, nationalization of the tin mines, civil rights for the Indian majority and dissolution of the military. An armed militia numbering 60,000 in 1956 was formed out of Mineros and Campesinos.

Note

1. The MNR can be seen as the political expression of a populist alliance between mine workers, Campesinos, the urban middle class and intellectuals.

Sources

Klein 1969; Malloy 1970; Nohlen 1982 (JS/VL)

---

11 Paraguay
3/1947 - 12.8.1947; A-2
War ended by: military victory for side A

Warring parties:

Aggressor (Side A): Paraguay
Aggressed (Side B): Coalition of Governmental Opposition[1]

Object and goals of the conflict

The PRF (Partido Revolucionario Febrerista) had built a governing coalition with the Colorado Party, which then excluded the PRF from further participation. This exclusion led to the PRF's alliance with the opposition Liberal and Communist parties. The government began a wave of arrests among the opposition, and in March 1947 the Febreristas responded by occupying the police headquarters in Asunción. Part of the military went over to the side of the opposition and tried to help overthrow the government, which received considerable support from the U.S. and Argentina.

Outcome of the war

On August 12, 1947, the Colorados defeated the opposition by capturing their headquarters in the city of Concepción. After their victory the Colorado Party and the military consolidated their war-time alliance. This opened the way for General Alfredo Stroessner's rise to power in 1954. Stroessner had commanded governmental troops in against the opposition rebellion, and he continued the coalition between the military and the Colorado Party.

The war cost approximately 28,000 lives and led to an exodus of some 300,000 people, one fourth of Bolivia's total population.

Note

1. Alliance of the Liberal Party, Partido Revolucionario Febrerista (PRF), Partido Comunista and some military factions.

Sources

amnesty international 1971; Area Handbook for Paraguay July 1972; Gunson/Thompson /Chamberlain 1989; LAB 1980; Miranda 1990; EP 1991 (TF)

---

19 Colombia (Violencia)
21.5.1949 - 10.5.1957; A-2
War ended by: termination of hostilities

Warring parties:

Aggressor (Side A): Colombia
Aggressed (Side B): Liberal Guerrillas (21.5.1949 - 1953) / Communist Guerrillas and Campesino Guerrillas.

Object and goals of the conflict

The first phase (1949-53) revolved around the resistance of liberal and Communist guerrilla organizations against the conservative government of Laureano Gómez. Gómez was trying to transform Colombia into a corporative state following the example of Spain and Portugal. The fundamental social conflict concerned rural power relations, and the centers of hostilities were the coffee-growing regions.

The question of land reform assumed a more central role in the second phase (1953-57), after most liberal guerrilla groups accepted an offer of amnesty from the government, while the Communist and Campesino guerrillas continued to fight for changes in the structure of land ownership.

Outcome of the war

After 1953, the Liberals and most of the Conservative Party worked together, most notably in their common support for a coup against General Rojas Pinilla, whose intransigence they feared would result in war spreading to the cities. In 1957, both parties merged into the National Front. Through a proportional system all posts were to be divided between the party at the local, regional and national level. The rules were to remain until 1974, and aside from a brief disturbance in 1986/87 continue in place today. In practice this system led to the exclusion of all other political forces and to the renewal of clientelism and the inclusion of most of the population in the traditional party system through expansion of the state bureaucracy.

The war took around 200,000 lives, and approximately two million people were expelled. Property relations in the coffee-growing regions were restructured to the benefit of the large land holders.

Sources

Berquist 1986; Gilhodes 1986; Guzman/Fals Borda/Umaña Luna 1962; Kalmanovitz 1988; Krumwiede 1980; LeGrand 1986; Oquist 1978; Palacios 1979; Pécaut 1987; Pearce 1990 (SK)

---

53 Venezuela
2/1963 - 1967[1]; A-2
War ended by: military victory for side B

Warring parties:

Aggressor (Side A): Fuerzas Armadas de Liberación Nacional (FALN)
Aggressed (Side B): Venezuela

Object and goals of the conflict

When the Peruvian dictator Peréz Jiménez fell from power in 1958, the opposition was optimistic about reforming the political and economic system. Rómulo Betancourt (Acción Democrática, AD) returned from exile to be elected with broad popular support. But Betancourt maintained much of the former dictatorship: strong dependence on the U.S., unequal income and land distribution and suppression of the opposition, especially the Venezuelan Communist Party (PCV) and the Movimiento de Izquierda Revolucionaria (MIR), and a rebellious branch of the AD. When these groups gained the support of progressive factions in the military they decided to fight for national liberation by forming the armed FALN on February 20, 1963.

Outcome of the war

The guerrillas suffered military and political defeats, and the government attempted to reintegrate their supporters by granting an amnesty to many prisoners. In 1970, numerous former guerrillas left the PCV and formed their own socialist party called the Movimiento al Socialismo (MAS) which was able, though at a considerable distance, to establish itself as a third political force. In the 1970s, Venezuela's new-found oil wealth allowed for partial integration of the population, largely through expansion of the state sector.

Note

1. Until 1969 there were also sporadic actions by smaller groups.

Sources

Ellner 1988; Ewell 1984; Gott 1973; Werz 1983 (SK)

---

60 Colombia (FARC)

since 27.5.1964; A-2

Warring parties:

Aggressor (Side A): Colombia

Aggressed (Side B): Fuerzas Armadas Revolucionarias de Colombia (FARC)

Also Side B: Ejército Popular de Liberación (EPL)[1] (12/1967 - 12/1990)

Object and goals of the conflict

1966 saw the emergence of FARC (after 1982 the FARC-Ejército Popular (FARC-EP)) from an organization called Bloque Sur. Bloque Sur had its roots in farmer's self-defense groups during the Violencia (see War Nr. 19), when violent conflicts intensified in the Colombian agricultural sector. The goals of the guerrillas until 1982 had primarily been the protection of the rural population AND the implementation of comprehensive agricultural reform. Since 1982 the FARC fought for more general changes in the political structure, especially concerning opportunities for participation in political and economic decisions.

The different guerrilla groups in Colombia occasionally fought amongst themselves, with ideological and strategic disagreements often based in the different international role models each group emulated. The ELN Guerrillas oriented themselves more toward Cuba (see War Nr. 79), while the EPL split from the Communist Party during the Sino-Soviet conflict and became a Maoist faction. Since about 1967 the EPL followed strategies similar to the FARC, although without any close cooperation. An attempt was made in 1987 to reverse the fragmentation by consolidating all Colombian guerrilla groups in the Coordinadora Nacional Simon Bolívar, but to little effect.

Outcome of the war

In light of the revolts, repression increased against all opposition movements, especially the unions and the Unión Patriótica, a party which emerged in 1985 from FARC factions. In 1990 the EPL signed a peace treaty with the government and laid down its weapons, and proceeded to win two seats in the parliamentary assembly at the end of that year. A dissident minority faction of the EPL, however, continues to fight. Since 1991 the ELN and FARC began negotiations with the government, but so far without results.

Note

1. Between 1967 and 1990 the strategy focused on liberating territory (see War Nr. 79).

Sources

Alape 1985; Arenas 1985; Behar 1985; CLADDE-RIAL 1987; Gott 1970; Medina 1986; Meschkat 1980; Molano 1989; Molano/Reyes 1978; Oquist 1978; Pearce 1990; Pizarro Leongómez 1986; Pizarro 1989; Reyes/Bejarano 1988; Sánchez/Peñaranda 1986 (SK)

---

79 Colombia (ELN)
since 7.1.1965; A-2

Warring parties:

Aggressor (Side A): Ejército Popular de Liberación (EPL)[1] (1965-1967) / Ejército de Liberación Nacional (ELN)

Aggressed (Side B): Colombia

Object and goals of the conflict

In response to urbanization and other socioeconomic structural changes there arose a social movement in Colombia's cities, drawing on the Cuban Revolution for inspiration and seeking to open and change the existing system. There were also rural guerrilla groups who wanted to form the core of any people's rebellion. Many representatives of this "founding generation," as the rebels called themselves, had either fought in the Violencia (see War Nr. 19) or experienced it as children or youths. Here, too, the individual guerrilla groups fought amonst themselves, bringing competing ideological and strategic perspectives to bear (see War Nr. 60).

Outcome of the war

The main demand of the ELN since the 1980s has been the nationalization of natural resources, in particular the oil fields. They are most active in those regions where multinational corporations are located.

During Belisario Betancur's Presidency (1982-86) the ELN was the sole guerrilla group not to take part in any negotiations. They agreed to start a dialog in 1991 with the government, but so far without results. Also in 1991 they led a large offensive against the country's physical infrastructure. This action, however, along with ecological damage caused by sabotage in the oil industry, earned massive criticism from the majority of the population.

Note

1. For the time period after 1967 see War Nr. 60.

Sources

Alape 1985; Behar 1985; Berry 1980; Dix 1980; Gott 1970; Leal Buitrago 1984; Levine/Wilde 1977; Pearce 1990; Pécaut 1988; Pizarro Leongómez 1986; Sánchez/Peñaranda 1986; Torres 1972 (SK)

86 Peru (Guerrilla Rebellion of the MIR and ELN, 1965)
6/1965 - 1/1966; A-2
War ended by: military victory for side B

Warring parties:
Aggressor (Side A): Movimiento de Izquierda Revolucionaria (MIR)[1] / Ejército de Liberación Nacional (ELN)[2] (25.9.1965 - 12/1965)
Aggressed (Side B): Peru

Object and goals of the conflict

The Andean highlands were a stronghold of feudal land and labor conditions, and these feudal structures formed a volatile mix when the region enountered the country's more modern economic sector. In the early 1960s a peasant movement led by Hugo Blanco in the Cuzco Province successfully occupied and redistributed land. With an eye toward Blanco's success, intellectuals in the coastal cities sought to reproduce the Cuban Revolution of 1959 in Peru. Yet the guerrilla armies MIR and ELN quickly encountered problems unanticipated in the Cuban model: they knew neither the customs nor the language of the Indians who lived in the highlands, and they failed to understand that their goals of socialist system reform would not be enthusiastically received by the peasants, who were primarily concerned about improving their own economic situation. During the months of fighting the guerrillas were hardly able to mobilize the rural population, and were quickly decimated by the Peruvian army.

Outcome of the war

The war led to no concrete results. Perhaps one can interpret as an indirect consequece the land reform implemented by the military government after its rise to power in 1968.

Notes

1. They consisted of former members of the APRA (which had become a conservative/social-democratic party). MIR started a three-front war in the Piura, Junín and Cuzco provinces. All three met with terrible defeat, with the Junín front holding out the longest until January 1966.
2. The ELN consisted of former members of the Peruvian Communist Party. In Ayacucho Province they garnered the most substantial popular support of the war, but the ELN was not able to defend its supporters against the

military and their popular base was soon liquidated. The rest of the ELN was squashed by December 1965.

Sources

Allemann 1974; Béjar Rivera 1970; Gott 1970; McClintock 1989; Wickham-Crowley 1989 (VL)

---

104 Argentina
1968[1] - 1977; A-2
War ended by: military victory for side B

Warring parties:

Aggressor (Side A): Montoneros[2] / Ejército Revolucionario del Pueblo (ERP)[3] (1970 - 1977)
Aggressed (Side B): Argentina

Object and goals of the conflict

Two guerrilla camps arose during Onganía's military rule (1966-1973): the Marxist ERP and the left-wing Peronist Montoneros. Both sought the overthrow of the government, but while the Marxists wanted to defeat the capitalist system the Montoneros sought the return of Juan Domingo Peron from exile. In 1973, Peron did regain power, and while the Montoneros hoped to share in the spoils the ERP continued its anti-capitalist struggle. Yet the Montoneros were let down by Peron, who failed to sypathize with their radical political ideas, and with the help of the military and right-wing Peronist groups began to repress the left in a campaign which his wife and successor, Isabel Peron, continued after his death in 1974. The ERP and Montoneros continued fighting until 1977, though their spirit and resistance was wrung out of them by the increasingly repressive policies of the military government that followed Isabel Peron's death in 1976.

Outcome of the war

The military government of 1976 created paramilitary terrorist units who were complicit with state ideology. The government fanned the flames of a Terror which expanded to touch almost every segment of the population. Tens of thousands were murdered or kidnapped ("disappeared") by the military or the right-wing terror organizations. The remaining guerrillas lost support in this climate. The military dictatorship remained in power until 1983.

Notes

1. It is not possible to precisely date the beginning and end of the war. The Guerrillas emerged between 1968 and 1970, and they were crushed in the years leading up to 1977. However, the terror instigated by the military government and the right-wing paramilitary units continued long after that.
2. The Montoneros emerged out of a failed military coup which resulted in the Peronist movement being banned. They fall into the category of a radical-military faction of the leftist Peronists. While at the beginning of the war they were strongly nationalist oriented, by the end they had increasingly adopted the radical Marxist demands and positions of the ERP.
3. A Marxist guerrilla organization formed in 1970 as the armed appendage of the Partido Revolucionario de los Trabajaderos (PRT).

Sources

Boris/Hiedl 1978; ERP 1977; Fronius 1977; Giussani 1984; Münster 1977; Ollier 1983; Waldmann 1978 (AMo)

---

137 Colombia (M-19)
1/1974 - 8.3.1990; A-2
War ended by: agreement without mediation

Warring parties:

Aggressor (Side A): Colombia
Aggressed (Side B): Movimiento 19 de abril (M-19)

Object and goals of the conflict

M-19 emerged in opposition to the collusion by the Conservative and Liberal parties to divide administrative and governmental position between them, and thereby exclude all other political forces from meaningful participation (see War Nr. 19). The M-19's founders came from a splinter group of the FARC (see War Nr. 60) and from supporters of the Alianza Nacional Popular (ANAPO), a populist party which had campaigned for General Rojas Pinilla in the 1970 elections. The election results were contested, and it was electoral fraud, real or imagined, that was the catalyst for M-19.

M-19 engaged above all in spectacular acts, such as the theft of Simon Bolívar's sword. In 1980, M-19 garnered worldwide attention when it occupied an embassy in Bogotá and took numerous diplomats hostage. Yet the group neglected building a stable basis, even though they enjoyed a degree of popularity among the marginalized urban population. Their goals were the reform and democratization of the existing system.

Outcome of the war

Between 1978 and 1982 the government's repression of any opposition was institutionalized in a "security statute." After President Betancur's failed peace process in the mid-1980s the M-19 was the first group which opened a dialog with the Barco government in 1989. These talks resulted in an amnesty for M-19's members, its demilitarization and societal re-integration.

Not three days after they had down their weapons the former guerrillas campaigned as a political party in the communal elections of March 11, 1990. In the following elections they established themselves as a third political force. In the December 1990 parliamentary elections they even gained a relative majority, though electoral participation had reached the extremely low level of 20 percent. A new constitution offers the possibility of a partial opening of the political system, but it is too early to tell whether this will lead to fundamental change.

Sources

Alape 1985, Behar 1985; Berry 1980; Gallón Giraldo 1979; Gallón Giraldo 1989; Lara 1986; Leal Buitrago 1984; Pearce 1990; Pécaut 1988; Pizarro 1985; Ramírez Tobon 1989; Ramírez/Restrepo 1989 (SK)

---

147 Peru (Sendero Luminoso)
since 17.5.1980; A-2[1]

Warring parties:

Aggressor (Side A): Sendero Luminoso (Shining Path)
Aggressed (Side B): Peru

Object and goals of the conflict

After colonial independence, Peru's society cemented itself along regional and ethnic lines. The small Spanish-speaking white elite were unwilling to recognize the Indian population as a legitimate ethnic group. On the contrary, they tried to force the Indians to abandon their identity. The Indian highlands experienced little development and was neglected by the economically advanced, and white, coast. In these Indian highland regions the Maoist Shining Path sought and found support.

The Shining Path's ideology drew from Mao and the Peruvian Marxist José Mariátegui, who saw the Indian highlands as the cradle for revolutionary change. The Shining Path seeks to utterly destroy the existing social order by any means necessary as a prerequisite to creating a new socialist order.

The war between the government and the Shining Path intensified toward the end of the 1980s, as right-wing death squads, often in cooperation with the military, expanded their operations beyond the guerrillas to include all opposition (see War Nr. 204).

Outcome of the war

Neither side is currently in a position to prevail in the conflict. Much of the country stands under military control, and the brutal nature of the Peruvian military in the Indian highlands has caused waves of migration to the cities in the coastal region, especially Lima. This helps explain the Shining Path's inroads into urban areas. President Fujimori, although democratically elected in 1990, gathered for himself dictatorial powers when he dissolved parliament in April 1992. This led to increased human rights violations, and in September 1992 the military successfully captured the leader and other prominent members of the Shining Path.

More than 26,000 people have fallen victim to the fighting between the government and the guerrillas. Furthermore, the economic and social conditions in parts of the Andes have been largely destroyed, and the entire country has experienced material damages through sabotage, strikes, and militarization.

Note

1. The U.S. Drug Enforcement Agency (DEA) began campaigns against the coca-growing regions of the Huallaga valley in the 1980s, which also happened to be the main territory of the Shining Path. One might assume that the anti-drug actions were also aimed against the guerrillas.

Sources

Barolo 1985; Favre 1984; Harding 1987; Hertoghe/Labrousse 1990; LAB 1985; LAB 1992; McClintock 1984; McClintock 1989; Oertzen 1992a; Palmer 1985; Waldmann 1989 (VL)

149 Peru / Ecuador
28.1.1981 - 2.2.1981; C-2
War ended by: third-party mediation (OAS)

Warring parties:
Aggressor (Side A): Peru
Aggressed (Side B): Ecuador

Object and goals of the conflict

The conflict along the 200,000 square kilometer border between Ecuador and Peru stems from a map drawn in 1829/30 between Gran Colombia and Peru and never accepted by the latter. Although Ecuador gave some land to Peru in 1890 the matter was never conclusively settled. During the war of 1941, Peru forced the annexation of two-thirds of Ecuadorian territory. The new border was determined by the Rio Protocol of 1942 and mediated by Argentina, Brazil, Chile and the U.S. But in 1960, Ecuador recanted its agreement to the protocol, stating that it was the result of a violence. Peru called for honoring the protocol, but in 1981 the war began with a military attack by Peru, who claimed to be reacting to the Ecuadorian occupation of unmanned Peruvian border posts.

Outcome of the war

The war ended with a cease-fire administered by the countries who mediated the treaty of 1942. The 1942 border remained the official demarcation. Ecuador achieved the promise that future negotiations will be handled by the OAS and not only the states present in 1942. At the same time, Peru achieved a new base for further negotiatons without having to renegotiate the 1942 protocol. In 1984 there was a new border incident, which indicates that the conflict is not yet fully over. Nine soldiers were killed in the fighting.

Sources

International Herald Tribune 11.2.1981; Times 5.2.1981; Eckardtstein 1984; Martz 1983; Martz 1985 (VL)

---

154 Falkland/Malvinas War (Argentina / Great Britain)
25.4.1982 - 15.6.1982; C-2

War ended by: military victory for side B

WARRING PARTIES:
Aggressor (Side A): Argentina
Aggressed (Side B): Great Britain

OBJECT AND GOALS OF THE CONFLICT

In 1833, Great Britain occupied this relatively forsaken island group, which Argentina claimed for its state. Sixteen years of negotiations before the war produced no results. Argentina's claim is grounded in international law, and the islands continue to be considered a British colony. Further motivations for Argentina's claim are strategic, geopolitical (legitimation for Argentina's antarctic claims) and the desire for raw materials. The decision to invade was primarily motivated by domestic politics. The British decision to respond to the invasion with a war was also linked to domestic politics: the loss of the islands would also have spelled the end of Thatcher's conservative government. Great Britain's goal was the restoration of the status quo ante.

OUTCOME OF THE WAR

Restoration of the status quo ante, however, was an increased British military presence on the islands. There were approximately 1,000 dead and billions of dollars in material costs for both countries.

SOURCES

Gambia 1987; Goldblat/Millán 1983; Haffa 1987; Lebow 1983 (JH)

---

163 Suriname (Jungle War)
21.6.1986 - 7.6.1989; B-2
War ended by: agreement without mediation

WARRING PARTIES:
Aggressor (Side A): Maroon Jungle Commando (Surinamese National Liberation Army (SNLA))
Aggressed (Side B): Suriname

OBJECT AND GOALS OF THE CONFLICT

Despite the only 400,000 inhabitants of Suriname, its ethnic heterogeneity resulted in ethnic and social divisions. European, Chinese and

Indian immigrants live primarily in the main coastal colonization zone, and in the rainforest jungle live approximately 50,000 Maroons, who are the descendants of escaped black slaves. Even during the time of Dutch colonization the Maroons commanded recognition of their autonomy.

The catalyst for war was the pursuit by the military of bank robbers who had earned a Robin Hood image by the Maroons by distributing their booty among the poorest. The military was also targeting smuggling, which was a traditional source of Maroon livelihood. The goal of the Maroons was the maintenance of their independence and own lifestyle. The military used existing resentment of indigenous Indians who also lived in the jungle to involve them in military confrontation with the Maroons.

Outcome of the war

Negotiations to end the war concluded on June 7, 1989, and the resulting peace treaty contained many concessions to the Maroons, including plans for regional police units and full withdrawal of the military from Maroon areas. The Indians, who had fought with the military against the Maroons, and the military itself were unhappy with the treaty, and a few independent Indian groups began armed action to undermine the treaty. This lead to further negotiations involving the Indians, which defused the situation. A lasting solution to the conflict, however, does not seem to be at hand, despite a new peace treaty signed in August of 1992.

The war cost about 500 lives, and destroyed in the process the traditional lifestyle of the Maroons. Over 10,000 of them fled to French Guiana, and the majority of those remained even after the end of the war due to the domestic political situation. The weak Suriname economy broke down completely after sabotage acts, particularly against the Bauxite and Aluminum sector.

Sources

Chin/Buddingh 1987; Military Balance 1987-88 and 1988-89; Neumann 1967; Polimé/ Thoden van Velzen 1988; Schmitz 1982; Singelenberg 1982; Thoden van Velzen 1988; Thomas-Hope 1987; AdG 17.5.87, 30.7.87, 30.9.87, 4.8.88, 23.9.88; ICR 2/1987; *International Herald Tribune* 31.1.86, 3.2.87; *Le Monde* 20.11.86; *Le Monde Diplomatique*

März 87; *Latin America Weekly Report* 30.6.88, 4.8.88; NRC Handelsblad 9.6.89 - 1.11.89; NZZ 13.3.82, 25.1.83, 28.1.87; *New York Times* 10.9.86, 17.12.86, 5.9.87; SZ 28.11.87, 31.1.89, 27.12.90, 2.1.91; taz 1.11.89, 4.4.90, 21.6.90, 29.5.91; *Financial Times* 12.9.86, 11.11.86, 17.2.87, 30.4.87, 20.5.88, 27.12.90, 12.2.91, 24.5.91; *The Guardian* 13.9.86, 12.11.86; *Die Zeit* 26.12.86 (JPF)

---

204 Peru (MRTA)
since 11/1987; A-2

Warring parties:
Aggressor (Side A): Movimiento Revolucionario Tupac Amaru (MRTA)
Aggressed (Side B): Peru

Object and goals of the conflict

Unlike the more well-known Shining Path, which strives for the complete elimination of the existing social order, the MRTA mainly seeks to increase the opportunities for social and political participation among the population. The first activities were armed propaganda actions in Lima, where the MRTA robbed food transports and distributed the goods among the poor. The main demands of the MRTA focus on nationalizing foreign firms, raising wages, subsidizing basic food stuffs, and the release of political prisoners. The MRTA has bases in three rural regions north and east of Lima. They achieved tremendous publicity with the occupation of the city of Juanjui in front of television cameras, and much later they achieved even greater notoriety by occupying the Japanese embassy in December of 1996. In certain regions the MRTA and Shining Path (see War Nr. 147) have come to blows over control of the territory.

Outcome of the war

The increased campaign against all opposition guerrillas under President Alberto Fujimori has resulted in many successes against the MRTA, including the arrest or killing of most of their leaders, though their spectacular hostage-taking in 1996 showed that he had not been fully successful.

Sources

El País 8.11.1987; FT 27.11.1987; Informe Latinoamericano 19.11.1987,

26.11.1987, 7.1.1988, 2.3.1989; Instituto de Defensa Legal 1989; Noticias Aliadas 24.12.1987 (SK)

Europe

1 Greek Civil War (Phase 1)
3.12.1944 - 12.2.1945; A-1
War ended by: third-party mediation (other third-party states)

WARRING PARTIES:
Aggressor (Side A): Ethnikon Apeleftherotikon Metopou / Ethnikos Laikos Apeleftherotikos Stratos (EAM/ELAS)[1]

Aggressed (Side B): Greece[2]
Intervention favoring side B: Great Britain

OBJECT AND GOALS OF THE CONFLICT

In the years under the German occupation EAM/ELAS and EDES formed as the largest resistance organizations. After the Germans withdrew both organizations fought against each other in a struggle for the future political structure of Greece.

The EAM/ELAS sought a new political system with socialist orientation and wanted to prevent any restoration of the constitutional monarchy, especially the so-called monarcho-fascism of George II and General Metaxas. The EDES fought for a liberal system. Great Britain had troops stationed in Greece towards the end of the Second World War, primarily around Athens. In the liberated regions of Greece a resistance government arose, but when Royalist Greek brigades refused to be disarmed against the demands of the EAM/ELAS, ELAS themselves pulled out of the general demobilization and left the government on December 1, 1944. There are competing interpretations about the origins and implications of the war:

1. As EAM/ELAS (dominated by the KKE) determined that neither the remaining members of the resistance government nor the British were likely to give up their hold on power, they consciously chose the path of confrontation with the hope that the Soviet Union, whose military was by now in Bulgaria, would come to their aid.

2. The British provoked the confrontation because they feared that the EAM/ELAS would prevail as the strongest political and military power in Greece. This would be a problem, not least because the secret Moscow Treaty of April 9, 1944, between Churchill and Stalin assigned Greece to the British sphere of influence.

Outcome of the war

The warring parties negotiated the Treaty of Varzika on Febraury 12, 1945, which included the disarming and dissolution of the fighting units, amnesty, elections and a plebiscite on the return of the king. Yet a substantial portion of ELAS fighters withdrew to Yugoslavia rather than disband, and many arms and munitions were hidden within Greece. During the war the EAM/ELAS had kidnapped approximately 15,000 citizens to use as insurance against mass arrests by the government. After the war EAM members faced acts of revenge, the EAM/ELAS was politically discredited, and the formal amnesty for political acts ceased in practice, though it existed on paper.

Notes

1. EAM was founded about five months after Greece was occupied by the Axis powers. It was a coalition of socialist groups, but over time the Greek communist party (KKE) came to dominate. Nonethless the EAM was insistent about presenting the alliance character of the organization. ELAS formed in Macedonia and Salonika in April of 1942 and became the military arm of the EAM. Although supported by Great Britain in the fight against the Germans, soon after the Nazi's left they found themselves in conflict with the EDES (Ethnikos Diokratikos Ellinikos Stratosin) Konflikt. Prior to December 3, 1944, EAM/ELAS was the strongest power in Greece, with support particularly in north and central Greece.
2. EDES fought on the government's side as a republican liberation army. Also on the governmental side fought the "Organisation CHI" and the Athenian city policy, as well Royalist Greek troops which returned to Greece in 1944 from Italy.

Sources

Carey/Carey 1968; Esche 1982; Kosiris 1980 (PT)

---

4 Spain 1945-1950

1945 - 1950[1]; A-2

War ended by: military victory for side B

Warring parties:
Aggressor (Side A): Remains of the Repubican Army[2]
Aggressed (Side B): Spain

Object and goals of the conflict

After the Republican Army's defeat in the Spanish Civil War (1936-39) most all opposition forces went into exile. As allied victory drew near in the Second World War plans for a guerrilla army based in France arose with the goal of overthrowing Spanish dictator Fransisco Franco's regime. The driving force behind these plans were officers from the Republican Army who had also participated in the French Resistance and now offered their military expertise to the Spanish Communist Party (PCE), also in exile in France. Along with the PCE various other socialist forces also joined forces with the guerrillas, especially anarchist groups. The plan was for a combination of guerrilla actions, mass strikes and continued international isolation of the Franco Regime to force the restoration of the Republican government which had fled to exile in Mexico.

The prelude to the war was an invasion of 4,000 guerrillas in October 1944 in the Spanish Pyrenees mountain valley of Aran. Though Franco's General Moscardó declared the invasion repulsed, the guerrillas did manage to gain a foothold in Spanish territory. Their Spanish strongholds were Asturia, Andalusia, Extremadura, Castillia and in the Levant. The anarchist guerrillas were particularly active in Catalonia, above all in Barcelona.

Outcome of the war

The PCE gave up its guerrilla tactics in 1948 to concentrate on a strategy of mass strikes. The autonomous guerrilla groups, however, continued their sporadic attacks into the early 1950s. As their support among the population dwindled, however, and as Franco's repression intensified, these groups were also forced to abandon their struggle.

Notes

1. There is no precise date for the end of this war. Official Spanish documents consider 1952 as the year in which the most serious guerrilla groups were completely destroyed (Cossías 1956).

2. There were three different guerrilla organizations which cooperated with the communists and socialists, while the anarchists acted separately (Biescas/Tuñón de Lara 1980:205).

Sources

Aguado Sánchez 1975; Aguado Sánchez 1976; Biescas/Tuñón de Lara 1980; Cossías 1956; Gros 1970; Limia Pérez 1957; Pons Prades 1977; Sorel 1970 (UB)

---

6 Greek Civil War (Phase 2)
6/1946 - 9.10.1949[1]; A-1[2]
War ended by: military victory for side B

Warring parties:

Aggressor (Side A): Dimokratikos Stratos Ellados (DSE)[3]
Aggressed (Side B): Greece
Intervention favoring side B: Great Britain (6/1946 - 3/1947)

Object and goals of the conflict

The dispersed ELAS members, suffering from their defeat in the first Greek civil war (see War Nr. 1), united in 1945 under the leadership of the Greek Communist Party (KKE) and reappeared in December 1946 as the DSE. They sought the democratization of Greece against conservative and anti-socialist tendencies. Their proclaimed goals were free parliamentary elections and the withdrawal of all foreign troops. They began to fight with guerrilla actions against military and police posts in the border areas and against supply and transportation infrastructure, while in Athens they sought a political solution with the government (this was known as the KKE's "double strategy"). The DSE grew quickly and soon controlled fortified territories in the border areas near Albania and Yugoslavia. These areas were declared "liberated Greece" and put under the control of a guerrilla government (the PDK) which never received international recognition. The DSE eventually went over to the strategies of offensives and battles, yet their attempts to conquer large cities failed.

U.N. intervention was requested, but repeatedly foundered on Soviet vetos. Great Britain had given up its support of the Greek government for economic reasons.[4] The Greek government turned to the U.S. after the proclamation of the Truman Doctrine on March 12, 1947, and succeeded in securing extensive financial and military assistance, including military advisors. The Greek army, rapidly rebuilt under the leadership of a Greek-American general staff (formed in November 1947), used these conditions to their advantage and eventual military success.

Outcome of the war

The DSE was utterly defeated, and the KKE leadership fled via Albania and Hungary to Moscow. They had been sorely disadvantaged by the break between Stalin and Tito, which led to Yugoslavia's refusal to supply logistical help to the KKE. The Greek right remained in power until 1963 (partially through electoral manipulation), and tens of thousands of leftists fled into exile or were sent to re-education camps. For years afterwards the KKE had no more influence in Greece.

Twenty-eight thousand children were reported kidnapped during the war, and numbers of casualties span a wide gamut, from between 44,000 (AdG) to 158,000 (Kosiris 1980) for both Greek civil wars. The numbers of refugees are also disputed, running between 80,000 (Esche 1982:342) and 703,000 (AdG). After three wars (the Second World War, War Nr. 1 and Nr. 6) Greece was a devastated country.

Notes

1. The exiled KKE (in Albania) decided to end the fighting on October 9, 1949, though individual groups continued sporadic attacks.
2. While Great Britain actively intervened, U.S. troops participated solely in an advising role.
3. The NOF (the liberation movement for the Macedonians in Greece) came out for a united Macedonia.
4. The withdrawal of British troops was announced in March,1947.

Sources

AdG 30.5.1949; Carey/Carey 1968; Esche 1982; Kosiris 1980(PT)

---

38 Cyprus I
1.4.1955 - 24.2.1959; DB-2
War ended by: third-party mediation (neighboring state[s])

Warring parties:

Aggressor (Side A): Ethniki Organosis Kiprion Agoniston (EOKA; Nationale Vereinigung zypriotischer Kämpfer)[1]
Aggressed (Side B): Great Britain / Vulkan/Türk Mukavement Teskilati (TMT)[2] (2/1957 - 24.2.1959)

Object and goals of the conflict

The EOKA guerrillas fought against the British colonial power for Cyprian independence and eventual annexation to Greece, since approximately 80 percent of the population were Greek-Cypriots. They fought primarily by acts of sabotage and actions against collaborators. The British used mostly Turkish-Cypriot police units against the EOKA (made up of volunteers, including members of the terrorist organization "Vulcan"). Great Britain also pursued divide and conquer tactics to maintain their power. The Turkish Cypriots supported the British, for they feared annexation to Greece and resulting discrimination.

Outcome of the war

Great Britain, Greece, and Turkey agreed in 1959 to the London Three Power Treaty which, while allowing for Cypriot independence as of August 16, 1960, also constitutionally provided for Turkish-Cypriot participation in politics and administration. The Greek-Cypriot Makarios became President, the Turkish-Cypriot Kücük was Vice President. As guarantors of the treaty Great Britain, Greece, and Turkey maintained a right to intervene and stationed a contingent of troops on the island.

Six hundred to 1,000 people died in the conflict.

Notes

1. The EOKA was founded in 1952 by Makarios, among others, and led since October 1954 by the Greek Colonel Georgios Grivas. The EOKA played an important role in the later 1963/64 war (see War Nr. 73).
2. "Vulcan" was a terrorist group founded in 1955 with money from the Turkish government. It changed its name to TMT in December 1956 and after February 1957 began to fight on the British side. TMT was also involved in the 1963/64 war (see War Nr. 73).

Sources

Dischler 1960; *Der Spiegel* 34/1964 (19.8.1964):55; Kadritzke/Wagner 1976; Meinardus 1984; Tenekides 1966; South 8/1986:44 (AAB)

---

43 Hungarian Revolution
24.10.1956 - 10.11.1956[1]; A-1
War ended by: military victory for side B

Warring parties:

Aggressor (Side A): Heterogeneous opposition forces and factions within the Hungarian Army[2]
Aggressed (Side B): Hungary
Intervention favoring side B: USSR

Object and goals of the conflict

In the aftermath of the enormous industrialization introduced by the Soviet Union following the Second World War, Hungary experienced increased urbanization and impoverishment. Dissatisfaction grew among the population as demands for output were raised while living standards remained low and the security policy (AVO) began to harshly repress people.

A further, and decisive, factor was the attempt by Prime Minister Imre Nagy from 1953 to 1955 to introduce a liberalized version of communism. The Stalinist faction foiled such attempts, and he was neutralized and seemed to have lost political credibility in the eyes of the population, especially after he was made Prime Minister again at the beginning of the disturbances on October 24, 1956.

A mass demonstration with 100,000 participants on October 24, 1956, to show solidarity with Polish workers quickly turned into a demonstration against Stalinism and the presence of Soviet troops in Hungary and raised demands for free elections. Official institutions were occupied or besieged soon thereafter, and gun battles finally erupted.[3] Soviet troops intervened, only to soon withdraw again. Part of the Hungarian Army confronted the rebels, and Budapest was bombed. Negotiations with the Soviet government brought a new cadre of party and government leadership, including many anti-Stalinists, but the revolt continued to expand and soon encompassed battles with AVO units across the land. On Octber 31, 1956, Prime Minister Nagy stunningly announced Hungary's neutrality and consequent withdrawal from the Warsaw Pact while party leader János Kádár negotiated with Moscow. On November 4, Soviet troops returned with reenforcements from the border regions where they had been withdrawn, and descended on Budapest. Heavy fighting between armed resistance groups and parts of the Hungarian army continued until November 10/11, 1956.

Outcome of the war

Nagy was deposed and eventually killed, Hungary remained a member of the Warsaw Pact, and Soviet hegemony was maintained. Despite damning criticism Western powers did not intervene, underscoring the reciprocal recognition of spheres of influence. Participants in the revolt were severely repressed, though after the consolidation of power Kádár began cautiously to implement small steps toward political and economic liberalization. At least 32,000 people died in the war.

Notes

1. It is not possibl to determine exactly when the fighting ended. The Soviet Union claimed that the revolt was over on November 4, 1956, but armed resistance continued in individual regions. By November 10/11 the revolt seems to have been crushed.
2. A heterogeneous opposition movement consisting of anti-Stalinist communists, worker's councils, social democrats, the small landholders party, the nationalist peasant party, and conservative to reactionary groups, as well as parts of the Hungarian army.
3. It is not clear who started shooting first.

Sources

Gosztony 1988; Piribauer 1957 (PT)

---

73 Cyprus II
21.12.1963 - 11.8.1964; B-2
War ended by: Third-party mediation (UN)

Warring parties:

Aggressor (Side A): Greece / Cyprus / Ethniki Organosis Kiprion Agoniston (EOKA; National Union of Cypriot Fighters)[1]
Aggressed (Side B): Türk Mukavement Teskilati (TMT)[2] / Turkey

Object and goals of the conflict

Following Cyprus' independence in 1960 there arose persistent confrontations between the Greek-Cypriot majority and the Turkish-Cypriot minority. The minority made such intensive use of their constitutional veto power that they endangered the ability of the communities to function effectively. In response, President Makarios suggested 13 constitutional changes. These plans met with escalating demonstrations, and the underground movements on both sides (EOKA for the Greeks, TMT for the Turks) began armed attacks with the support of the respective troops stationed by their "motherlands" on the island. EOKA sought annexation to Greece (a policy known as Enosis), while the TMT fought for a policy of Taksim, or division of the island. A front line emerged across the middle of Cyprus.

Outcome of the war

The war led to a de facto division of the island and movement toward the Turkish goal of Taksim (division) by partially separating the two populations. British troops stationed on the island tried to stop the fighting but were unsuccessful. A cease-fire was achieved only by a decision of the U.N. Security Council to station U.N. troops (UNIFICP) along the demarcation line. The opposing parties remained armed, however, and in the following years they found opportunities for limited confrontations. Makarios began to build a Cypriot national guard in order to implement state power, though its officers belonged to the Greek army. A military coup in Greece on April 21, 1967, introduced a period of detente until 1974 (see War Nr. 116). The victim figures range from 60 to 500 dead.

Notes

1. The EOKA had fought previously against the British for the independence of Cyprus (see War Nr. 38).
2. The TMT had fought in the first Cyprus war on the British side (see War Nr. 38).
3. Greece, Turkey and Great Britain all maintained an intervention right in Cyprus based on the London Three Power Treaty of 1959.

Sources

Der Spiegel Nr. 34 (19.8.1964):56; Hahn 1982; Kadritzke/Wagner 1976; Meinardus 1984; Nowacki 1982; Tenekides 1966 (AAB)

---

106 Spain (Basque Country)
7.6.1968[1] - 1979; AB-2 / B[2]-2
War ended by: fighting fell under the threshold of war

Warring parties:

Aggressor (Side A): Euskadi ta Askatasuna (ETA; Basque Country and Freedom)[3]
Aggressed (Side B): Spain

Object and goals of the conflict

ETA's original goal of Basque secession from Spain and the establishment of an independent state is today shared only by a militant portion of the ETA. In the end phase of Franco's dictatorship (1974) the ETA split into military and political-military factions. The latter forswore armed confrontation and turned into the Euskadi ko Ezkerra (EE), while the former continued the armed struggle. The military faction in turn also produced a political party, the Herri Batasuna Party, which acts as the legal political arm of the militant ETA.

Outcome of the war

While the Franco regime responded to the autonomy claims of the ETA with military means, the democratically elected Spanish government of 1977 tried to find a political solution to prevent Basque secession from Spain. Since 1975 Basque language and culture are recognized and since 1979 they have received wide-ranging political autonomy. The granting of political autonomy undermined the support for armed confrontation. The formerly unbridgable gulf between the Basque population, the Church and the ETA no longer exists. Nonetheless there has not been a complete halt to violent acts. The economic decline of the region and the corresponding social consequeneces continues to provide the militant ETA with a basis for recruitment.

War and terrorist actions took approximately 600 lives (*Der Spiegel*, 25.9.89), with 90 percent of those casualties occurring after 1976. ETA's violence has spurred economic decline, for Basque entrepreneurs have often been the target of kidnapping, extortion and attacks.

Notes

1. The first armed confrontation between the ETA and police was on June 7, 1968.
2. The anti-regime component was removed after the establishment of a democratic Spain in June 1977.
3. An organization founded by Basque intellectuals on July 31, 1959. The phase of armed confict began in 1967, with the first fighting in 1968.

Sources

Blaschke 1984; *Der Spiegel* 25.9.89; Gilmour 1985; Herzog 1979; Herzog 1987; Lang 1988; Sullivan 1988; Waldmann 1981; Waldmann 1988; Waldmann 1989b (WR/UB)

105 Northern Ireland
since 14.8.1969[1]; B-2

WARRING PARTIES:

Aggressor (Side A): Great Britain[2] / Ulster Defense Association (UDA)/ Ulster Freedom Fighters (UFF)[3] (since 1971) / Ulster Volunteer Force (UVF)/Protestant Action Force (PAF)[4] (since 1969)
Aggressed (Side B): Irish Republican Army/Provisionals (IRA)[5] (since 27.6.1970) / Irish National Liberation Army (INLA)/Irish Republican Socialist Party (IRSP)[6] (?? - 1986)

OBJECT AND GOALS OF THE CONFLICT

The current war in Northern Ireland is part of the 800-year-old Irish-British conflict over rule of the island. The society of Northern Ireland has been repeatedly segmented, from the bloody repressions of Irish uprisings, colonialism, anglicization attempts, English settlement waves, and last but not least the decline in the textile and harbor industries which sustain the Northern Irish economy. In the 1922 Irish war of independence the northern provinces fell under British control, a semi-autonomous part of the United Kingdom. The drawing of borders resulted in a population that was two-thirds Protestant and one-third Catholic, which resulted in widespread discrimination against Catholics at all levels of social and political life.

Both Protestants and Catholics, however, had substantial underclasses, despite a concentration of wealth among Protestants. Competition between the underclasses developed into confessional mobilization of each side. Historical conflicts between the two religions further strengthened the opposing camps. On August 12, 1969 Protestants in Derry (Londenderry) stormed the Catholic quarter known as the Bogside, and the inhabitants erected barricades. Two days later the Northern Irish Prime Minister called on Britain for help. At this time the IRA had not yet adopted the role it eventually came to play: simultaneously the guerrilla army and defenders of the Catholic population.

OUTCOME OF THE WAR

The war continually reproduces the structures which give it life. Social life has become militarized, even while the people grow tired of war. Neither the British nor the IRA have the ability for complete military victory over the other, the former is too powerful, the latter too popular and fluid.

Many negotiations came to naught, starting with the first on November 11, 1992, between representatives of moderate Catholic nationalists, the Irish Republic, Great Britain, Unionists, and radical Unionists. The paramilitary organizations were excluded from the talks, as was Sinn Fein, the political wing of the IRA. Before and after the 1992 negotiations the IRA continued sporadic terrorist attacks in England, mostly bombs set off in London and other English cities. Only in the late 1990s did negotiations produce largely effective cease-fires, and Sinn Fein gain the legitimacy necessary to participate in talks. On April 10, 1998 British Prime Minister Tony Blair presided over a historic agreement signed by all parties to the conflict. The agreement created new governing structures, ended the Irish Republic's claims to Northern Ireland, and kept the province within the United Kingdom. There is hope that this time the negotiations will be ultimately successful in ending what seemed to be an endless war, but it is too early to tell. Northern Ireland remains among the poorest regions in Europe, and the long war has economically ruined the country, and this presents additional challenges to healing the wounds of the divided society.

Approximately 3,000 persons were killed in the war.

Notes

1. British troops were called in by the Northern Irish Prime Minister on August 14, 1969. As a starting date for the war, however, this is problematic, since the IRA as such had been fighting since 1919. Only after they split into the Provisional IRA (PIRA) and Official IRA did the PIRA enter the new war on June 27, 1970. Previously the Catholic population in the north had organized their self-defense on their own, as the IRA was immobilized by internal disputes.
2. In the first years of the war the British army numbered about 40,000 soldiers. As the war dragged on they were reduced to 10,000, though special units for fighting terrorism, such as the Special Air Service (SAS), were introduced by the British. A Northern Irish unit of the British Army was also created as the "Ulster Defense Regiment (UDR), with 6,000 men. They worked together informally with paramilitary Protestant groups, as did the 12,000-man-strong Royal Ulster Constabulary (RUC), the Northern Ireland police who were also Protestant dominated.
3. The UDA is the ony paramilitary organization in Northern Ireland with legal status. They are an umbrella organization for diverse splinter groups, and boast a membership of approximately 50,000. They also operate informally with British securitiy forces. The UFF operates as hit squads in its ranks. The UDA regards the UVF as competitors.

4. An illegal paramilitary organization which focuses primarily on Catholic civilians, and includes the hit squads PAF in its ranks. The UVF is the oldest paramilitary organization (since 1913), and is a loyalist Protestant force with approximately two to three thousand members.
5. Founded in 1919 as the alliance of various armed Republican groups in the aftermath of the Easter Rebellion of 1916. They fought the British Army in the war of independence, and thereafter split into two groups. One part became the Southern Irish Army, the other continued the civil war against Great Britain in the North as the IRA. Later the IRA split again into the Provisional (PIRA) and Official IRA (OIRA). The Official IRA ended armed struggle, but the PIRA fought on. Their political representation is the Sinn Fein Party.
6. They emerged from the OIRA after it laid down its arms. In 1986 they split into different factions which fought each other. The result was two meaningless organizations, the INLA and the Irish People's Liberation Organization (IPLO).

Sources

Coogan 1987; Schulze-Marmeling/Sotschek 1989; Stadler 1979; Waldmann 1989b (PT)

---

116 Cyprus III
20.7.1974 - 16.8.1974; BC-1
War ended by: third-party mediation (U.N.)

Warring parties:

Aggressor (Side A): Turkey
Aggressed (Side B): Cyprus
Side B: Ethniki Organosis Kiprion Agoniston-B (EOKA-B; National Union of Cypriot Fighters)[1]
Intervention favoring side B: Greece

Object and goals of the conflict

In an apparent attempt to divert attention from domestic and economic crises in Greece the military junta instigated a coup against the president of Cyprus, Makarios, on July 15, 1974. The EOKA-B and Greek officers of the Cypriot National Guard carried out the coup. EOKA-B had organized right-wing extremist groups of Greek Cypriots, who had operated underground for years against Makarios. The EOKA-B and the Greek officers wanted Enosis: the annexation of Cyprus to Greece, an issue which was a tinderbox for the relationship between the Greek and Turkish Cypriots.

Makarios fled, and the EOKA leader Sampson took over as president and announced Cyprus' annexation to Greece. In response Turkish troops invaded the island in two offensives with over 40,000 soldiers, but eventually withdrew to the Turkish side of the so-called Attila demarcation line.

OUTCOME OF THE WAR

North of the "Attila Line" became Turkish-Cypriot North Cyprus, consisting of about one-third of the island, while South Cyprus became wholly Greek-Cypriot. Since the war the border has been under U.N. supervision. About 200,000 Greek-Cypriot inhabitants in the northern part of the island fled or were expelled to the south. On February 13, 1975, the north declared itself an autonomous state, and Turkey and the Turkish Cypriots no longer recognized the Greek-Cypriot-led central government (where, incidentally, Makarios had come back to power). So far only Turkey has recognized the autonomy of Northern Cyprus. Ongoing negotiations mediated by the U.N. between the two halves of the island have so far remained without results.

Up to 5,000 people, including 3,000 civilians, lost their lives in the fighting, and over 200,000 persons were displaced.

Note

1. The EOKA was restructured in 1967, necessitating the addition of "-B" to their name (see Wars Nr. 38 and 73). They fought on the Greek-Cypriot side.

SOURCES

Bouthoul/Carrère 1979; Bunge 1980; Choisie 1991; Hahn 1982; Kadritzke/Wagner 1976a; Meinardus 1984; Nowacki 1982; South 8/86:44 (AAB)

---

169 Romania
17.12.1989 - 28.12.1989; A-2
War ended by: military victory for side A

WARRING PARTIES:

Aggressor (Side A): Opposition factions of the Romanian Army
Aggressed (Side B): Romania[1]

Object and goals of the conflict

Nicolae Ceauçescu came to power in 1965 as a reformer, but soon brought all decisive political and economic functions under his and his family's personal control. He used brutal repression to maintain his family's power, especially through use of the secret police known as the Securitate. He further cemented his power by promoting an extreme cult of personality and fanning the flames of Romanian nationalism against national minorities, especially the Hungarians and the Germans.

The regime dealt with problems of legitimacy by successfully distancing itself from its socialist "brother" states and nurturing close relations with China. The relationship with Hungary deteriorated due to Romanian pressure on ethnic Hungarians to assimilate, and eventually the government erected a fence along the Romanian/Hungarian border to prevent a mass exodus of the minority to Hungary.

The perfect repressive apparatus of the state prevented rebellion until December 1989, despite terrible living conditions, forced resettlement of farmers and peasants, and an export policy which kept the country out of debt but destroyed its ability to feed itself. The catalyst for revolt was an otherwise everyday occurrence: the Securitate arrested an oppositional, ethnic Hungarian priest in Timisoara (Temesvar). The population of Timisoara protested the arrest, and despite brutal attempts to quell the unrest it spread across the land instead. The army refused to heed the command to put down the revolt and crossed sides to support the people. A civil war developed between the regular Romanian army and the Securitate. A "National Salvation Front" emerged at the top of the unorganized revolt, consisting mostly of intellectuals and former party members.

Outcome of the war

Although poorly equipped, the army and popular resistance won the war against the Securitate. Nicolae Ceauçescu and his wife were sentenced to death in a rapid trial and executed on videotape. Yet the old elite remained surprisingly in power after elections in 1990 and 1992. The nationalist policies toward the Hungarian, German and other minorities, especially Sinti and Roma (Gypsies), have not improved much. Many still flee the country. Estimates for war dead vary between 1,000 and 6,000.

Note

1. The Securitate, the secret police loyal to Ceauçescu.

SOURCES

Die Zeit 20.3.90, 18.5.90; Gabanyi 1990a; Gabanyi 1990b; Olschewski 1990; NYT 18.12.89, 21.12.89; FR 23.1.90, 22.12.89, 23.12.89, 28.12.89, 27.1.90, 29.1.90, 26.5.90, 24.7.90, 20.6.90, 27.6.90, 5.4.91; *Romanian News* 28.12.89 (Nr.1 der „Neuen Serie“); Tagesspiegel 19.12.89, 21.12.89; FT 23.12.89, 2.1.90, 21.6.90, 23.10.90, 22/23.12.90; taz 29.9.89, 2.1.90, 8.1.90, 23.1.90, 26.1.90, 26.3.90, 31.3.90, 2.4.90; SZ 8.7.89, 20.12.89, 21.12.89, 22.12.89, 23.12.89, 27.12.89, 28.12.89, 29.12.89; *The Guardian* 6./7.1.90 (JPF/JH)

---

200 Yugoslavia (Slovenia, Croatia, Bosnia)
since 26.6.1991; BC-2

WARRING PARTIES:

Aggressor (Side A): Yugoslavia
Aggressed (Side B): Slovenia (26.6.1991 - 3.7.1991) / Croatia[1] (7/1991 - 1/1992) / Bosnia[2] (since 3/1992)
Side A: Serbian Militias und Chetniks[3] (since 7/1991)
Side B: Croatian Militias in Bosnia-Hercegovina (since 3/1992)

OBJECT AND GOALS OF THE CONFLICT

The essential causes underlying the bloody conflict which tore Yugoslavia apart lie in Josip Tito' s ultimately failed attempt to integrate a region overshadowed by historical Serb-Croat antagonism and social, cultural, religious and economic differences. The Yugoslav state arose as an artifical product of the interwar period in response to a power vacuum left by the end of the First World War. It disintegrated during the Second World War, when its constituent members turned on each other with horrific ferocity, but reemerged phoenix-like under Tito, the former leader of the Partisan movement against the Germans. He tried to forge a single national identity out of the six primary national groups (Slovenia, Croatia, Serbia, Bosnia-Hercegovina, Macedonia and Montenegro, plus the semi-autonomous regions of Kosovo and Voivodina within Serbia). A North-South economic divide and transfer payments provided tinder for bitterness between the export-oriented northwest and the poorer southeast. Economic competition soon acquired nationalist colors.

State institutions were also unsuccessful in forcing a Yugoslavian pan-nationalism. The two truly integrated organizations were the Communist Party and the Yugoslav People's Army (JNA), although after Tito's death the army became dominated by Serb officers, a development which foreshadowed the terrible war to come.

As economic problems increased the individual republics maximized their interests at the expense of the national state. The Communist Party also became divided between the reform-minded Croatian and Slovenian factions, who advocated a transformation to pluralism and a market economy, and the Serbian faction, under Tito's successor Slobodan Milosevic, who saw the crisis in terms of a loss of centralized power and sought to hold on to power by falling back on the model of a Serb-dominated, authoritarian state.

A crisis in the ethnic Albanian majority province of Kosovo pushed Yugoslavia into free fall, as Milosevic revoked the regions autonomy (along with that of Voivodina in the north) and began a campaign of repression. This provoked in turn a deterioration of relations between the Serbs and the other Republics, and resulted eventually in Slovenia and Croatia declaring independence from Yugoslavia.

At first Serbia seemed interested in holding Yugoslavia together by force, but soon it became apparent that Serbia wished to establish itself as a sovereign state encompassing regions in other Republics where Serbs lived (Greater Serbia). The JNA became a de facto Serbian army. After a token resistance against Slovenian independence, the JNA attacked Croatia, where great fighting broke out between extreme nationalist militias. After a U.N.-brokered cease-fire took effect (see below) the fighting shifted to Bosnia-Hercegovina in March 1992, where nationalists Croatians, Serbs, and Bosnians (mostly Muslim) fought a long and brutal war against each other in ever-spiraling surges of violence. The aim of all sides was nominally the establishment of homogeneous ethnic states, and especially the Croatian and Serbian exclaves sought annexation to their "mother" countries. The Bosnian government was alone in wanting to maintain an ethnically heterogeneous Bosnia, though not all Bosnians supported this goal, especially after the genocidal tactics employed against them by their fellow Serbs and, to a lesser extent, Croats.

Outcome of the War

The fighting in Slovenia ended on July 3, 1991 with a cease-fire mediated by the European Community and the Treaty of Brioni (July 8) which allowed Slovenia to successfully secede from Yugoslavia.

Solvenia was internationally recognized along with Croatia as a sovereign state in mid-January 1992.

In Croatia, however, the process did not run so smoothly. Small confrontations between Croatian police/National Guard and the JNA broke into open warfare in mid-July 1991, and the Serb-inhabited regions of Slavonia and the Krajina experienced terrible fighting and massacres. The Serbs occupied both regions, which were later subject to a U.N. cease-fire and put under U.N. control (United Nations Protection Forces; UNPROFOR)[4]. The Croats re-occupied the Krajina in a later offensive, but Slavonia has not yet fully reverted to Croat control.

The fighting in Croatia costs at least 10,000 lives, and at least 700,000 were expelled or fled. Both sides practiced "ethnic cleansing," whereby the respective "enemy" ethnic group was expelled or killed in an attempt to create ethnically "pure" regions.

The war in Bosnia-Hercegovina (B-H) erupted after fighting stopped between Croatia and Serbia. In many senses it was a continuation of the same conflict, but assumed even more intense levels of violence since B-H was the most ethnically diverse Republic of them all: 43 percent Muslim,[5] 31 percent Serb and 17 percent Croat.

The Serbs and Croats, as mentioned, wanted to divide up B-H among themselves. The B-H leadership wanted to maintain the country as a sovereign entity with a federal system. The fighting in Bosnia-Hercegovina prompted the international community to intensify its sanctions against Yugoslavia, including a weapons embargo which only helped freeze in the Serbs dominant military advantage. The U.N. introduced a no-flight zone, but the Serbs were successful in "ethnically cleansing" nearly two-thirds of the territory with tactics including concentration camps, mass executions, torture and genocidal policies reminiscent of the Nazis. The Croats, and to a much lesser extent the Bosnian army, also participated in ethnic cleansing and atrocities. Genocidal rape, a Serb policy, resulted in the organized and systematic rape of more than 20,000 women, mostly muslims. The war in B-H resulted in millions of refugees, including three milllion in Croatia alone.

While the Croatians and Bosnians were nominally allied against the Serbs, they often fought each other as well. An attempt by international mediators known as the Vance-Owen plan to divide Bosnia into ten largely autonomous provinces failed. Particularly vicious Serb mortar attacks on the besieged city of Sarajevo motivated NATO to undertake an air campaign against Bosnian Serb targets, and the war began to shift in the Bosnian governments favor. Recognizing this, all parties agreed to a peace plan signed in Dayton, Ohio, in 1995, which had been hammered out by a "Contact Group" consisting of the United States, Russia, Germany, France and Great Britain. Since then an uneasy peace has settled on the war-ravaged country, but tensions remain high and simmer under the watchful eyes of a U.N.-authorized NATO force (IFOR).

More than 140,000 people died in the fighting in Bosnia-Hercegovina.

Notes

1. Croatian National Guard: formed originally as a police unit in 1974, it expanded to integrate the new Croatian army during the war in Slovenia.
2. Croats and Muslims both participated in the territorial defense of Bosnia-Hercegovina against the Serbs.
3. Chechniks: irregular units which see themselves in the tradition of the Partisans of World War II.
4. Thirty-one states participated in UNPROFOR, which consisted of 13,870 people.
5. Muslims were regarded in Yugoslavia as a separate nationality.

Sources

Felder 1992; Furkes/Schlarp 1991; Libal 1991; Maibach/Karajica 1992; Razumkovski 1991; HSFK-Report 1/1992; Frieden und Abrüstung Nr. 40 (1992) (TS/ZK/TF)

---

205 Moldova (Dniester Republic)
3/1992 - 8/1992; B-1
War ended by: third-party mediation (neighboring state[s])

Warring parties:

Aggressor (Side A): Moldova
Agressed (Side B): Armed forces of the Dneister Republic and Cossacks
changing sides: "Third Power"[1] (1992 - 8/1992)
Intervention to end the war: Russia (7/1992 - 8/1992)

Object and goals of the conflict

Moldova's current nationality problems are closely tied to the tumultous history of the region: in 1812 Moldova shifted from the Ottoman Empire to Russian control, and Russians, Ukrainians and other nationalities settled in the area which Romania views as its historical cradle. After World War I the region fell to Romania with one important exception: the strip of territory east of the Dneist river became part of the Soviet Union (as an autonomous republic), and was further settled by Russians. The autonomous regions formed the core of the later Soviet Republic of Moldova, which came into existence after Romania was forced to give up more territory following their defeat in the Second World War. As the Soviet Union began its path toward collapse in late 1988, Romanian nationalist movements emerged in the Republic, and were supported by Romania. In June 1990, Moldova declared itself an independent republic and then, on August 17, 1991, after the failed coup in Moscow, a sovereign state.

The Russian and Gagaus national minorities reacted to the declaration of independence and efforts to unite with Romania by themselves declaring a separate republic within Moldava. The Gagaus minority was quickly quieted by the Moldavan leadership, but they were less successful with the Dneister region separatists. In the so-called Trans-Dneister region lay over 40 percent of Moldova's industry and a majority of Russians and Ukrainians. Soon after the Dneister region held a referendum on independence and its own presidential elections the tensions spilled over into armed conflict. In March of 1992 skirmishes broke out which quickly developed into open war between Moldavan forces and the armed defenders of the "Dneister Republic."

Outcome of the war

After many attempts to negotiate a cease-fire, Russian mediation finally achieved this goal in July 1992 by united units of the Russian 14th Army, Moldovan units and units of the Dneister Republic together into one organization whose job it became to oversee a "peace corridor" between oppossing factions. By the end of July this new force was concentrating on fighting the so-called "Third Power." Since August 1992 there have been no more large military altercations.

The cease-fire agreement was signed by the Russian and Moldovan presidents in the presence of the president of the Dneister region. The agreement reiterated the territorial integrity of Moldova (with no new claims by Russia) and called for upholding human and minority rights. The inhabitants of the Dneister region were to decide their status by vote if Moldova were ever to unify with Romania.

The leadership of the Dneister Republic (mostly Russian former Communists) was able to secure its position and control the area east of the Dneister river, as well as the city of Bendery. They also continued to build their own state and military structures. In this sense they largely achieved their war goals.

Moldova, for its part, gave up on its goal of rapid unification with Romania, and also lost de facto any control over the Dneister region. The conflict has thus not really been solved, although it currently is not expressing itself militarily. In addition to the existing tensions, the region can also be affected by Romania's standing claim to nearby Ukrainian territory.

Note

1. Armed units from Moldava and the Dneister region who sought to torpedo the cease-fire agreement siged in July 1992.

Sources

FAZ, FR und SZ 1992 (TS/CN)

Near and Middle East

179 Iraq (Barzanis)
8/1945[1] - 10/1945; B-1
War ended by: military victory for side B

Warring parties:

Aggressor (Side A): Barzanis
Aggressed (Side B): Iraq
Intervention favoring side B: Great Britain (9/1945)

Object and goals of the conflict

The League of Nations allocated the former Ottoman province of Mossul, with its majority population of Kurds, to Iraq in 1925. Numerous revolts, including the Barzanis rebellion, demanded more autonomy in the face of Iraq's increasing centralization of power. In 1944 the Iraqi government called a temporary cease-fire in order to consolidate its power, and by the summer of 1945 reopened hostilities and militarily defeated the Barzanis.

Outcome of the war

The Barzanis sought refuge in Iran after British air attacks prevented them from possibly prevailing over Iraqi troops and Kurdish tribal governments opposed to the Barzanis had been mobilized. Up to 1,000 people died in the fighting.

Note

1. Starting in April of 1945 the Barzanis clashed with Iraqi police, but it was not until after August 10, 1945, that the fighting between them became regular. On August 25, the army moved in against the Barzanis.

Sources

Farouk-Sluglett/Sluglett 1987; Haraldsson 1966; O'Ballance 1973a (TS)

---

5 Mahabad (Iran)
14.3.1947[1] - 13.4.1947[2]; B-2
War ended by: military victory for side A

Warring parties:

Aggressor (Side A): Iran
Aggressed (Side B): Barzanis

Object and goals of the conflict

During World War II, Iran was occupied and divided between British and Soviet zones. Between these spheres of influence ran a neutral zone, which in January of 1946 Kurds declared to be the independent "Republic of Mahabad." The military backbone of the fledgling republic consisted of about 3,000 Barzanis, who had fled there after losing their war against Iraq the previous year (see War Nr. 179). When Iranian troops entered Mahabad following negotiations in December 1946, the Barzanis attempted to arrange safe transport back to Iraq in exchange for stopping fighting. While the negotiations continued, however, Iranian troops attacked Barzani positions.

Outcome of the war

The Kurdish republic was destroyed, their leadership slaughtered, and Kurdish culture was repressed. The Barzanis fled back to Iraq, from where they fled again to the USSR.

Notes

1. In April and May of 1946 there had been several battles between Kurds and Iranian troops. This confrontation, however, had been settled by a cease-fire arrangement.
2. About 500-800 Barzanis fled at the end of May, 1947, back into Iran from Iraq, and from there to the USSR.

Sources

Arfa 1966; Eagleton 1963; Ibrahim 1983; Kinnane 1964; O'Ballance 1973a; Roosevelt 1984 (TS)

---

17 North Yemen 1948
2/1948 - 14.3.1948[1]; A-2
War ended by: military victory for side A

Warring parties:

Aggressor (Side A): Yemenite tribes and the army faction under Crown Prince Ahmad
Aggressed (Side B): North Yemen

Object and goals of the conflict

Imam Yahya of the ruling North Yeminte Hamid ad-Din family had isolated his country from the outside world and allowed no economic or political development. The opposition against Yahya was very diverse: it included "reformers" who sought the end of Yahya's unconstrained rule, Sunni Muslim opponents of the Shiite ruling class, and Shiite families who themselves laid claim to replacing Yahya as Imam. This loose opposition was mostly successful in the cities and among exiled Yemenites, but made few inroads among the tribes.

Imam Yahya was assassinated on February 17, 1948, after which his competitor Abdulla al-Wazir became Imam. Many exiled Yemenites returned and many took posts in the government and administration. The government was largely inept, however, due to internal power struggles and grave political and religious differences. In this atmosphere it was fairly easy for one of Yahya's sons, the Crown Prince Abdulla al-Wazir, to mobilize tribes in the north against al-Wazir. Part of the justification for the war was al-Wazir's perceived transgression of not atoning for Yahya's death, and also the failure of the central government to present the tribes with enough "gifts" to win their loyalty.

Outcome of the war

Crown Prince Ahmad succeeded in taking power, whereupon he executed all of the former coup leaders who overthrew his father. Ahmad himself became the new Imam on March 15, 1948. He feared revenge, however, from the inhabitants of the capital, Sana, for his troops had sacked the city during their victorious campaign. Ahmad moved the seat of government to Taiz, which led to Sana's economic decline.

During the plundering of Sana approximately 5,000 people died.

Note

1. Sporadic resistance from government troops persisted unti March 19, 1948.

Sources

O'Ballance 1971; Stookey 1978; Wenner 1967 (TS)

---

18 First Arab-Israeli War
15.5.1948 - 7.1.1949; C-2
War ended by: third-party mediation (U.N.)

Warring parties:

Aggressor (Side A): Egypt/ Syria / Lebanon / Jordan / Iraq / Saudi Arabia
Aggressed (Side B): Israel

Object and goals of the conflict

With a two-thirds majority the United Nations decided on November 29, 1947, to divide the former British mandate of Palestine into a Jewish and an Arab section. On May 14, 1948, Ben Gurion proclaimed the state of Israel. The next day, on May 15, troops from Egypt, Transjordan, Syria, and smaller contingents from Iraq, Saudi Arabia and Lebanon attacked the one-day old Israel. The aggressor's goal, supported by the Arab League, was the elimination of the state of Israel and the U.N.-mandated division, which they had previously rejected.

Outcome of the war

Israeli forces successfully repulsed the attack and extended Israeli territory by occupying Galilee, West Jerusalem, parts of West Jordanland and theMediterranean coast. Only Gaza fell under Egyptian administration. The border with Syria was demilitarized, and approximately 700,000 Palestinian refugees fled the land. The remaining parts of West Jordanland became part of Jordan in 1950.

Sources

Baumgarten 1983; Bunzel 1980; Freimark 1981; Herzog 1984; Mejcher 1981; O'Ballance 1981; Tibi 1989 (DJ)

---

44 Protectorate Aden / North Yemen
1955 - 1958[1]; C[2]-2
War ended by: termination of hostilities

Warring parties:

Aggressor (Side A): Great Britain
Aggressed (Side B): North Yemen

Object and goals of the conflict

Great Britain and (North) Yemen had already squabbled over the northern border of the British protectorates of Western and Eastern Aden in the interwar years, and had signed a treaty in 1934. Despite the treaty, however, in the 1950s the protectorates became an issue again. The catalyst for this was Britain's intensified efforts to establish control over the strategically important surrounding area of Aden, which included bringing the sheiks and sultans of the region together into a southern Arab federation. North Yemen saw such a federation as an obstacle to their goal of unification with South Yemen.[3]

Outcome of the war

Although numerous British-Yeminite negotiations came and went without results, the border incidents lapsed by the end of the 1950s. Great Britain continued its goal of forming a federation, which in 1959 emerged as the Federation of the Amirates of the South. Despite this achievement, the British were unsuccessful in preventing an independence movement from gaining steam (see War Nr. 87). Even in the north one could see signs of the revolution to come (see War Nr. 68).

Successful revolutions in both countries failed to settle the border question or the issue of unification, however, leaving plenty of flammable material for future wars.

Notes

1. Already prior to 1955 (and after 1958) there were sporadic border incidents.
2. Although the border issue and the struggle for spheres of influence were the main foci of this war, the rhetoric from North Yemen and the participation of Southern Yemen tribal fighters lend this war a decolonization component.
3. It is not possible to state with certainty whether North Yemen stirred up unrest among the tribes in the Protectorates, or whether the unrest was the first genuine intimation of the independence movement.

Sources

Lackner 1985a; O'Ballance 1971a; Keesing's 14004A, 15296A, 15502A, 15910B, 16087A, 16139D, 16586A (TS)

---

42 Second Arab-Israeli War (Suez War)
29.10.1956 - 6.11.1956; C-2
War ended by: third-party mediation (U.N.)

Warring parties:

Aggressor (Side A): Great Britain (31.10.1956 - 6.11.1956) / France (31.10.1956 -6.11.1956) / Israel
Aggressed (Side B): Egypt

Object and goals of the conflict

The Suez war occurred against a background of pan-Arabic euphoria the decolonization of the Near East. Violent demonstrations against the British had already occurred in Egypt in 1951, resulting in the ousting of Britain's ally, King Farouk. At the 1955 Bandung Conference Egyptian President Nasser announced a policy of non-alliance. Seven days after the British pulled the last of their troops out of the Suez Canal zone on June 19, 1956, Nasser announced the nationalization of the canal. This action provoked a joint attack on Egypt by Great Britain, France, and Israel on October 29 and 31, 1956. The issues for Israel were the status of the Sinai peninsula and the closing of the Gulf of

Akaba and the Suez Canal to Israeli ships. France and Great Britain were involved in a type of colonial rearguard action, in addition to their economic interests in keeping the canal open.

Outcome of the war

The war ended under pressure from the USA and USSR, and a new chapter in Middle East politics began. The imperial power of Great Britain and France came to an end, and the region entered the superpower spheres of the U.S. and USSR. Nasser became cast as a hero and was a charismatic leader for the Pan-Arab Movement.

The Suez Canal remained nationalized, though five percent of its income was given as damages to the expropriated actors and the use of the canal for all nations was guaranteed. British and French troops left and Israeli troops withdrew to the cease-fire line from 1949. U.N. peacekeeping troops were stationed in the Sinai, Gaza, and Gulf of Akaba.

More than 2,000 people died in the war. The vast majority of casualties were Egyptian (about 2,000 compared to 181 Israelis).

Sources

Bowie 1974; Dooley 1979; Fullick/Powell 1979; Glasneck/Timm 1992; Herzog 1984; O'Ballance 1959; Steinbach 1981; Tibi 1989 (DJ/PT)

---

48 Muskat and Oman (Imam-Sultan Conflict)
7/1957 - 1959[1]; B-1
War ended by: military victory for side B

Warring parties:

Aggressor (Side A): Imam of Oman[2]
Aggressed (Side B): Muskat and Oman
Intervention favoring side B: Great Britain[3]

Object and goals of the conflict

The decades-long state of war between the Imam of Oman and the Sultan of Muskat was ended by the Treaty of Sib in 1920. Oman received de facto independence with the help of the British. After the Second World War the conflict flared up again, however, over the question of transferring oil-exploration rights to British Petroleum (BP). The Imam forbade the Sultan from making concessions without his agreement. Under the pretext of protecting oil-exploration teams the Sultan's troops peacefully occupied Oman, with British support. The Imam's brother, however, formed a guerrilla army in neighboring Saudi Arabia over the next few years, with the goal of achieving independence for Oman and the return of the Imam to his former power. Great Britain's concerns in this conflict were its diminishing influence in Arabia and the securing of access to oil.

Outcome of the war

The Sultan extended his rule over all of Oman, which had previously been largely autonomous. The oil concessions remained primarily in British hands.

Notes

1. According to British sources the 1959 uprising was quickly put down. Koszinowski (1976a) und Kelly (1976a) report that the fighting extended over several years.
2. Involving above all tribal fighters from the Bani Riyami and the Bani Hina.
3. British policy in the region shifted along with their interests from protecting sea passages to protecting oil fields: they became more interested in establishing stronger central governments and firming up internal borders (in order to ascertain the property rights to oil fields). For this reason Great Britain was more willing in 1959 than in 1955 and 1957 to intervene militarily in Oman.

Sources

Bierschenk 1989; Keesing's 15709A; Kelly 1976; Koszinowski 1976 (TS)

---

49 Lebanese Civil War 1958
9.5.1958 - 14.10.1958; AB-2
War ended by: third-party mediation (U.N.)

Warring parties:
Side A: Lebanon[1]
Side B: United National Front (UNF)[2]

Object and goals of the conflict

One cannot speak of a unitary goal in this war. Originally the members of the majority Muslim UNF were primarily interested in a more participation in political institutions and the resignation of President

Chamoun. The regional and the international political situations, however, led to an escalation of the internal Lebanese power-struggles.

The internal sources of conflict can be sketched as follows:

1. Dissatisfaction of Muslims against the overrepresentation of Christians in public office due to the proportional representation system; 2. Personal differences between traditional political leaders; 3. President Chamoun's exclusion of these political leaders by manipulating the elections of 1957; and 4. Rumors about an attempt by Chamoun to change the constitution to allow his re-election.

These internal divisions were subsumed by a larger polarization of Arab politics in the context of an Iraqi-Egyptian dispute as well as the East-West Conflict. When Chamoun accepted the Eisenhower Doctrine in 1957 his opponents saw this as taking sides with the "anti-Arab" West and, consequently, was viewed as a violation of the "National Pact of 1943." In this way the struggle between traditional political leaders became a fight between "Nasserism" and "Western integration."

Outcome of the war

The result of the civil war was a restoration of the status quo. Chamoun denied aspirations for re-election, and Fuad Shihab was elected as an acceptable president for all warring parties. He appointed cabinet members from both sides and restored consensus.

The war caused 2,000 to 3,000 deaths and scores of injured. At the height of the fighting the economic losses were pegged at two million dollars per day.

Notes

1. The government side was supported by the greater Syrian, anti-Nasser "Parti Polulaire Syrie (also known as the Parti Populaire Sociale; PPS), by the Katib (Falange) of the Maronite Gemayel, as well by Majid Arslan, the Druse contingent under Kamal Jumblat, and security and paramilitary forces of the police. The army under Fuad Shihab played a sort of buffer role between the polarized camps.
2. Besides the traditional leaders in the UNF (Saib Salam, Abdallah al-Yafi, Rashid Karami (Sunni), Ahmad al-Asad, Sabri Hamadah (Schiite), Kamal Jumblat (Druse), Hamid Faranjiya, Rene Muawad (Maronite) the following parties were also represented: Najjada (Muslim youth movement), Socialist Progress Party of K. Jumblat (PSP), various pro-Nasser parties and the Baath Party. The Communist Party of Lebanon supported the regime's opponents, but never joined the UNF.

Sources

Agwani 1965; B¸ren 1983; Hottinger 1961; Kerr 1961; Qubain 1961; Salibi 1965 (DJ)

---

61 First and Second Iraqi-Kurdish War
16.9.1961 - 10.2.1964[1]; B-1
War ended by: agreement without mediation

Warring parties:
Aggressor (Side A): Iraq
Aggressed (Side B): Barzanis and allied tribes / Kurdish Democratic Party-Iraq (KDP-Iraq) (3/1962 - 10.2.1964)
Intervention favoring side A: Syria (10.6.1963 - ??)

Object and goals of the conflict

After the monarchy fell in Iraq, the new leader General Kassen could barely hold on to power by exploiting differences among his rivals and gaining concessions from the Kurds. After he successfully eliminated the pro-Nasser elements and the Communists he then turned on the Kurds: Kurdish functionaries were arrested, Kurdish civil servants were fired, Kurdish publications were banned, and so forth. The main goals of the Iraqi central government were the maintenance of a unitary Iraqi state and full control over the oil fields which lay in Kurdistan. The Kurdish Democratic Party-Iraq (KDP) demanded autonomy for Kurdistan, Kurdish as an official language, a completely Kurdish army and police within Kurdistan, appropriate participation in the government, cultural sovereignty, and a fairer share of the budget to develop Kurdistan. In 1961, the central government felt strong enough to attack the Kurds militarily.

Outcome of the war

A provisional constitution granted the Kurds new national rights, and offered them amnesty, a lifting of the economic boycott and the rebuilding of destroyed settlements. The negotiations, however, led to no concrete results (see War Nr. 85). The cease-fire split the KPD, and supporters of the Barzani won an ensuing struggle against the party leadership around Ibrahim and the Talaban.

The highest casualties were borne by Kurds who were the target of Iraqi air attacks. In the first six months of the war 3,000 people were killed, 500 villages destroyed and 180,000 people made homeless.

Note

1. This war is sometimes regarded as two wars, because there was a fairly long cease-fire from February to June, 1963, following the successful coup of Arif against Kassem.

Sources

Khalil 1985; Kinnane 1964; O'Ballance 1973a; Keesing's 18535A, 19323A, 20165A; (TS)

---

68 Yemenite Civil War
10/1962 - 1969; A-1
War ended by: third-party mediation (Arab League)

Warring parties:

Aggressor (Side A): Republican Troops
Aggressed (Side B): North Yemen
Intervention favoring side A: Egypt (10/1962 - 10/1967)

Object and goals of the conflict

This war was a conflict about the nature of the political system pitting the maintenance of the traditional, patrimonial Imamat[1] against the erection of a Yemenite Republic. Egypt wanted to emphasize its role as leader of the Arab world, and thereby supported the Republican Troops while advancing its own pan-Arab, anti-royalist policies. Saudi Arabia supported the Imam's troops, since they were concerned about negative spill-over effects on their own traditional system by the Republican movement in Yemen. One should also not underestimate the power of tribal conflicts within Yemenite politics, which continue to play a central role.

Outcome of the war

The Imam's troops relinquished their fight after a reconciliation conference in 1969, and the republican system remained intact. Sporadic fighting continued until 1970.

Note

1. Every Yemenite government requires Islamic legitimation, but the Imamat is specifically connected to the Zaidit Islam which is dominant in the north of Yemen.

Sources

Koszinowski/Menshaui/Meyer 1980; Kruse 1989; O'Ballance 1971; Triebswetter 1979; Wenner 1965 (DJ)

---

87 Protectorate Aden (Southern Yemen War of Independence)
14.10.1963 - 30.11.1967; D-2
War ended by: agreement without mediation

Warring parties:

Aggressor (Side A): National Liberation Front (NLF) / Front for the Liberation of South Yemen (FLOSY)[1] (1966 - 11/1967)
Aggressed (Side B): Great Britain

Object and goals of the conflict

Arabic nationalism grew in the 1950s within the British colony of Aden and its Eastern and Western protectorates. Furthermore, the economic upturn in Aden's strategically important harbor created a powerful union that moved beyond social issues to make political demands, such as an end to the discrimination of workers from the outlying areas or the north, support of Nasser, and a general strike during the French and British occupation of the Suez Canal Zone. Great Britain reacted to the growing sentiments for independence with a proposal for negotiating autonomy within the British-created "Federation of the Amirates of the South."

While FLOSY emerged from the union movement, the more radical NLF had its roots in the Arabic Nationalist Movement (MAN2). Shortly after its formation in February 1963 the NLF began plans for the armed struggle against the British. The NLF had the advantage over FLOSY of belonging to the few Southern Yemenite organizations which commanded supporters both in Aden and in the surrounding area.

Outcome of the war

Great Britain transferred power to the NLF, who had won an internal power struggle against FLOSY, in Novemer 1967, and formal independence occurred on November 30. The closing of the Suez Canal (see War Nr. 95), the withdrawal of British troops and the one-sided development policies of the former colonialist power led to substantial economic problems. Furthermore, the People's Republic of Yemen remained internationally isolated during its first years. Internal contradiction in the NLF had been papered over during the war for independence, and now broke out causing wide policy swings and leading to war-like confrontations between factions.

Notes

1. FLOSY was originally conceived in January 1966 as an Egyptian initiative for an alliance containing the NLF and the Organization for the Liberation of the Occupied South (OLOS). The NLF left the organization in November 1966, however, and after that FLOSY and the NLF engaged in armed struggle against each other.
2. Regarding MAN see War Nr. 88 (PFLOAG) and Nrs. 15 and 108 (PFLP, DFLP).

Sources

Halliday 1990; Koszinowski/Menshaui/Meyer 1980; Lackner 1985; Keesing's 19993A (TS)

---

85 Third Iraqi-Kurdish War
4.4.1965 - 29.6.1966[1]; B-2
War ended by: agreement without mediation

Warring parties:
Aggressor (Side A): Iraq
Aggressed (Side B): Kurdish Democratic Party - Iraq (KDP - Iraq)

Object and goals of the conflict

The Iraqi government delayed implementation of the 1964 cease-fire agreement (see War Nr. 61) until it had adequately consolidated its power. They then proceeded to increase political and economic pressure on the Kurds and to pursue their "Arabization" process in the Kirkuk oil region. In mid-1964 Barzani, who had succeeded in winning the power struggles within the KPD-Iraq, began to form autonomous administrative units in the regions controlled by Kurds. This was met with distinct displeasure by the Iraqi government, who refused to enter into negotiations with Barzani to find an acceptable solution.

Outcome of the war

The Kurds dealt the Iraqi army severe blows, and in May 1966 the new government once again began cease-fire negotiations with the Kurds. On June 29, 1966, the Iraqi Prime Minister al-Bassas put forward a 12-point plan to decentralize administration and thereby guarantee Kurdish national identity. Furthermore Kudish was to be allowed as an official language of instruction, and Kurds were to be proportionately represented in the government, army, and bureaucracy. These promises, however, were never implemented, and led to further conflict (see War Nr. 102).

Note

1. In April 1966 a one-month cease-fire followed the death of Iraqi President Arif.

Sources

Franz 1977; Keesing's 20953A, 21539A; Khalil 1985; Nebez 1972; O'Ballance 1973a (TS)

---

88 Uprising in Dhofar
9.6.1965[1] - 11/1975[2]; AB-1 / A-1[3]
War ended by: military victory for side B

Warring parties:

Aggressor (Side A): Popular Front for the Liberation of Oman (PFLO)4
Aggressed (Side B): Muskat and Oman (ab 1970: Oman)
Intervention favoring side B: Great Britain (1966 - 11/1975) / Iran (11/1973 - 11/1975) / Jordan (?? - 9/1975)

Object and goals of the conflict

In the mid-1950s the Sultan Said bin Taimur moved his seat of government to Salalah, the capital of the Dhofar province. His extreme conservative approach resulted in strict limits on Dhofari freedom of movement, which had serious economic repercussions and led to economic stagnation. In the early 1960s a sprinkling of illegal opposition groups arose, spurred by contacts with pro-Nasser and socialist-revolutionary ideas encountered abroad. In 1964 these organizations joined together calling themselves the "Dhofari Liberation Front" (DLF). On June 9, 1965, they carried out their first military operation. At this stage in their history the DLF sought first and foremost the overthrow of the Sultan and independence for Dhofar. These goals were changed somewhat by Israel's military victory over the Arab states in 1967 and the success of the NLF in South Yemen: power shifted in the DLF toward the wing closest to the leftist Movement of Arab Nationalists (MAN[5]). In 1968 the DLF renamed itself the PFLOAG and added the goal of a social revolution in the Gulf to its existing demands. In 1974 they shortened their name to the PFLO and their goals shifted toward liberating Oman from occupation by foreign troops and the overthrow of the Sultan of Oman.

Great Britain entered the war to stabilize the government of Oman, largely in order to safeguard British strategic interests, above all the al-Masirah air field. Iran was also interested in stabilizing the Oman government so that it could continue to have unfettered passage through the straights of Hormuz. They were also interested in expanding their sphere of influence into the Arabian peninsula. Jordan entered the war in order to protect their Iranian ally from criticism from other Arab countries.

Outcome of the War

The Sultan Said bin Taimur was overthrown by his own son, Qabus bin Said, with British support on July 23, 1970. Qabus followed a double strategy of a massive military build-up coupled with improvement of the province's standard of living. Both had fairly rapid effects. The Oman army, together with the Iranian army, managed to reconquer all of the villages once controlled by the PFLO, whereupon the Sultan decreed the war to be over. In March 1976 he proclaimed an amnesty for the "rebels" (but only after Oman and the Democratic People's Republic of Yemen signed a cease-fire). Iranian troops withdrew from Oman in 1977. By now the fighting power of the PFLO was severely broken, yet small groups of former guerrillas continued sporadic attacks.

Notes

1. The PFLOAG took control of the majority of Dhofar after 1969. There exists only vague information about the level of violence in the preceding years.
2. The PFLO lost the last villages under their control in November, 1975, after which they were reduced to small and ineffectual attacks.
3. British, Pakistanis, Indians and Jordanians all served as commanders or trainers in the Oman army. The largest contingent of soldiers was originally Baluchans, but they were gradually replaced by Omanis. Abu Dhabi sent soldiers to serve in the Oman garrisons.
4. They were the Dhofari Liberation Front (DLF) until 1968, then the Popular Front for the Liberation of the Occupied Arabian Gulf (PFLOAG), and after December 1971 they became the Popular Front for the Liberation of Oman and the Arabian Gulf (PFLOAG), and finally after August 1974 they settled on the Popular Front for the Liberation of Oman (PFLO).
5. See War Nr. 87 for more on the influence of MAN.

Sources

Halliday 1990; Kelly 1976; Keesing's 25180D, 25654A, 26988A, 27483A, 27716C, 28353A, 29804A; AdG 22281A; FAZ 9.10.71, 10.7.75, 15.7.75; FR 30.1.76; SZ 8.10.76; Koszinowski 1976; Page 1985; Peterson 1978; Price 1975; Times 18.10.75, 2.12.75; Spiegel 17/75 (TS)

---

95 Third Arab-Israeli War ("Six-Day War")
5.6.1967 - 10.6.1967; C-2
War ended by: third-party mediation (U.N.)

Warring parties:
Aggressor (Side A): Israel
Aggressed (Side B): Egypt / Syria / Jordan

Object and goals of the conflict

The war, which Israel began on June 5, 1967, was the culmination of escalating tensions since the mid-1960s. The roots of these tensions lay in the Arab-Israeli enmity which arose around Arab opposition to the formation of the Israeli state. After Pan-Arabism had passed its zenith, Nasser attempted to legitimate his leadership of the Arab world by adopting an increasingly radical line toward Israel. There is much speculation about Nasser's motives, but there is little doubt that his firey rhetoric, closing of the Gulf of Akaba and concentration of troops along the Israeli border contributed considerably to the rising tension. The situation was also provoked by the actions of the new Syrian government, which, after coming to power in a 1966 coup, supported Palestinian commando actions carried out from Jordanian territory. It remains speculation whether Israel's preventive strike really forestalled an Arab attack or whether Israel took the opportunity to secure its borders and extend control around its territory.

Outcome of the war

The war ended victoriously for Israel, which in the process occupied West Jordanland, the Gaza Strip, the Sinai (see War Nr. 103), parts of the Syrian Golan Heights as well as East Jerusalem.

The Israeli army suffered 766 deaths, while Egypt lost 10,000 soldiers through fighting or through thirst in the desert of the Sinai Peninsula. Jordan lost 6,000 soldiers (Arnold 1991a:299). The Israeli occupation of the above territories affected approximately 1.3 million Palestinians, about 300,000 of whom fled to East Jordan, while one million remained under Israeli occupation (Flores 1989a:17).

Sources

Arnold 1991; Barker 1981; Brecher 1980; Bulloch 1974; Flores 1989; Herzog 1984; Laqueur 1968; O'Ballance 1972; Rabinovich/Shaked 1978; Steinbach 1981; Tibi 1989 (DJ)

---

165 Israel, Palestine
since 1968; B^1-2

Warring parties:
Aggressor (Side A): Palestinian Liberation Organization (PLO) with its most important organizations: al-Fatah, People's Front for the Liberation of Palestine (PFLP), Democratic Front for the Liberation of Palestine (DFLP), Palestine Liberation Front (PLF) and PFLP-General Command (PFLP-GC); since the second half of the 1980s also Hamas and Jihad Islami
Aggressed (Side B): Israel

Object and goals of the conflict

After the Arab defeat in the 1967 Six-Day War (see War Nr. 95), the PLO began to be controlled by guerrilla organizations, especially Fatah. Fatah's chairman, Arafat, was elected to chair the PLO in 1968. The PLO took up its armed struggle against Israel by leading commando raids from Egyptian, Jordanian and Lebanese territory and by occasionally hijacking aircraft and taking hostages. Since the PLO was militarily too weak to confront Israel directly they adoped terrotist tactics, which served two main purposes: 1. to publicize the Palestinian cause to the world, and 2. to bring the Arab states back to a state of war with Israel. The latter goal, however, has gradually changed over time—especially within the al-Fatah—to forcing Israel to negotiate with the PLO.

March 1978 and June 1982 saw the largest clashes between the guerrilla groups and the Israeli army, all on Lebanese territory. Since 1987 the confrontations have been dominated by a Palestinian uprising ("Intifada") within the occupied territories. In the forefront of this rebellion have been some mass actions (the "War of Stones"), unorganized attacks by individuals, and smaller armed groups emerging from self-defense units of the Intifada. Israel reacted to these actions with the deployment of undercover special units who have been accused of assassinating key Intifada activists.

Since the PLO made negotiations with Israel their official platform after a national council meeting in 1988, the Islamic organizations Hamas and Islamic Jihad have taken up attacks against Israeli security forces, settlers, and terrorist attacks.

Outcome of the war

Since the mid-1970s the PLO attained certain diplomatic successes: in 1974 they attained observer status in the U.N., in 1976 they became a full member of the Arab League, and in November 1988, the PLO declared (from Algeria) a Palestinian state encompassing the West Bank, Gaza Strip and East Jerusalem. The majority of the Palestinian National Council accepted U.N. resolutions 242 and 338, which de facto meant recognition of the state of Israel according to the borders before the 1967 war. Jordan gave up its claim to the West Bank in favor of the Palestinian claim, although Jordan had controlled the territory until 1967 and laid claim to it in the years following.

Since October 30, 1991, there have been peace negotiations under the auspices of the U.S. and USSR (soon to be the Russian Federation) between Israel on the one side and Syria, Lebanon and Jordan on the other. The West Bank Palestinians are included as part of the Jordanian delegation. Israel offers limited Palestinian autonomy, while the Palestinians demand self-determination. Real progress in peace talks led to a strengthening of Islamic Fundamentalist forces among the Palestinians and also to increasing internal differences among the Israelis and Palestinians.

So far the war has caused over 10,000 deaths, 1,000 during the Intifada.

Note

1. The question of Palestine was certainly one of the main causes for the wars between Israel and its Arab neighbors, but this conflict does not count as an interstate war. since we only focus here on those elements regarding Palestinians within the Israeli-occupied territories.

Sources

Alterman 1988; Baumgarten 1983; B¸ren 1982; Bunzel 1980; Diner

1982; Diner 1988; Eisenstadt 1987; Flores 1981; Flores 1984; Flores 1989; Hottinger 1988; Hottinger 1989; Khalidi 1988; K^hler 1981; Koszinowski 1981; Koszinowski 1989; Mejcher 1981 (DJ/PT)

---

98 People's Republic of Yemen
6/1968[1] - 11/1968[2]; A-2
War ended by: military victory for side B

Warring parties:
Aggressor (Side A): Front for the Liberation of South Yemen (FLOSY) / South Arabian League (SAL) (1968 - 1968)
Aggressed (Side B): DPR Yemen

Object and goals of the conflict

The creation of the Democratic People's Republic of Yemen (DPR Yemen) threatened Saudi Arabia's conservative monarchy, and led them to reorganize the Southern Yemenite opposition against the DPR Yemen, above all the recently defeated FLOSY and the SAL. These attempts hit the DPR Yemen during a period of internal difficulties. The radical wing of the NLF took control after the Fourth Party Congress. This led to an internal struggle in which President ash-Shabi, with the support of the army, forced the left wing from the party leadership and the government and put down their attempts at insurrection.

Outcome of the war

Divisions within the Saudi leadership and the heterogeneity of the Southern Yemen opposition resulted in great discrepancies in the intensity of their attacks across the DVP Yemen border, and they were eventually repulsed. Nonetheless they continued to present a persistent threat to the South Yemen central government and were an important tool for Saudi pressure (see Wars Nr. 101 and 114).

The left and moderate wings of the NLF put aside their differences to fight against the Saudi-supported opposition forces. The left wing returned to government and successfully overthrew President ash-Shaabi on June 22, 1969.

Notes

1. Individual attacks by the opposition began already in February 1968.
2. Sporadic attacks by opposition forces continued across the border for some time afterwards.

Sources

Burrowes 1987; Halliday 1990; Koszinowski/Menshaui/Meyer 1980; Lackner 1985; Page 1985 (TS)

---

102 Fourth Iraqi-Kurdish War
3.1.1969 - 11.3.1970; B-2
War ended by: agreement without mediation

Warring parties:

Aggressor (Side A): Iraq / Talabani-Faction of the Kurdish Democratic Party- Iraq
Aggressed (Side B): Kurdish Democratic Party - Iraq (KDP-Iraq)

Object and goals of the conflict

The 12-Point Plan which ended the last set of hostilities between Iraqi's and Kurds (see War Nr. 85) had been barely implemented, and the situation was made worse for the Kurds when in 1968 the right-wing of the Baath Party came to power in a coup. The new al-Bakr regime attempted to divide the Kurds by making concessions and starting negotiations with the Kurds opposed to the KDP. Once the central government felt strong enough it started, once again, a new offensive against the Kurds.

Outcome of the war

The March 11 Agreement gave the Kurds relative autonomy, cultural sovereignty, amnesty, reconstruction aid and conceded a change in the constitution, and the KDP entered the government. As with previous treaties, however, this agreement was only implemented in fits and starts. The government continued to attack the Kurds and force migration from Kurdish settlements as part of the "Arabization" of the oil fields policy (see War Nr. 117).

Sources

Franz 1977; Keesing's 23827A, 23916A; Nebez 1972; O'Ballance 1973a; Vanly 1984 (TS)

---

103 Egypt / Israel
6.3.1969 - 7.8.1970; C-1
War ended by: third-party mediation (U.N.)

Warring parties:
Side A: Egypt
Side B: Israel
Intervention favoring side A: USSR (3/1970 - 7.8.1970)

Object and goals of the conflict

Egypt sought a revision of the results of the third Israeli-Arab war (see War Nr. 95). This was important for Egypt especially in view of the domestic unrest caused by the loss of the war. They also tried to prevent the results of the June ("Six-Day") War from becoming recognized internationally as the status quo.

The USSR was the main arms supplier and "ally" for Egypt, and they attempted to balance their loss of prestige after the Six Day War. They were concerned that any further defeat of Egypt would endanger Soviet influence in the Near East.

Israel was concerned about securing its territorial acquisitions from the Six-Day War. Israel successfully repulsed the Egyptian attacks and later tried to win the war by flying air attacks within Egyptian territory.

Outcome of the war

The war resulted in no change in the territorial situation. There was a relative shift in the balance of power, however, since Israel was forced to acknowledge Egypt's increased strength.

Sources

Avnery 1972; Bar-Simon-Tov 1980; Bulloch 1974; Herzog 1984; Tibi 1989 (DJ)

---

101 DPR Yemen / Saudi Arabia
26.11.1969 - 12/1969[1]; C-2[2]
War ended by: third-party mediation (Arab League)

Warring parties:
Aggressor (Side A): DPR Yemen
Agressed (Side B): Saudi Arabia

Object and goals of the conflict

Wide swaths of border within the Arabian peninsula are undetermined and indeterminable, and therefore provide grist for border conflicts. The unclear border in the region around Wadiah had already led to confrontations between Saudi Arabia and the former colonial power of Great Britain in 1954/55. After the PDR Yemen achieved independence from Great Britain it, too, came into conflict with Saudi Arabia about claims to the region. Rumors of supplies of oil and water around Wadiah sharpened the crisis. Furthermore, the atmosphere between Saudi Arabia and the PDR Yemen were already strained due to Saudi support for the South Yemen opposition (see War Nr. 98).

Outcome of the war

Both sides increased their military presence in the disputed region. Relations between Saudi Arabia and the DPR Yemen remained tense.

Notes

1. Due to the sparse information available about this war it is not possible to ascertain whether fighting continued after December 7, 1969, or which side could be considered the aggressor.
2. An air attack by one or more planes from the Arab Republic of Yemen against Saudi positions seems to have been the unauthorized action of individual pilots. This attack, however, led to increased political tensions between the two countries.

Sources

Halliday 1990; Orient 1970:24, 60 (TS)

---

108 Jordan (Black September)
16.9.1970 - 19.7.1971; AB-1
War ended by: military victory for side A

Warring parties:

Aggressor (Side A): Jordan
Aggressed (Side B): Palestine Liberation Organization (PLO)
Intervention favoring side B: Syria (20.9.1970 - 23.9.1970)

Object and goals of the conflict

The influx of approximately 200,000 Palestinians as refugees in Jordan after the Six Day War of 1967 (see War Nr. 95) caused domestic political tensions to rise. The People's Front for the Liberation of Palestine (PFLP) and the People's Democratic Front for the Liberation of Palestine (PDFLP) sought to topple the Jordanian Monarchy. In northern Jordan they declared a "liberated" zone. The moderate forces in the PLO under Arafat did not challenge the Monarchy, and wanted to continue to use Jordan as a base for their guerrilla attacks against Israel.

The Jordanian government did not want to let itself be drawn into a war with Israel by the PLO, but was unable to prevent PLO attacks. Nor were they able to effectively oppose the threat to their own position of power from the guerrilla groups. Yet when the Jordanian government felt strong enough it decided to confront its opponents with military force. King Hussein received the support of the conservative Bedouin tribes and significant parts of the army, who regarded the Palestinian refugees as interlopers.

Outcome of the war

The first phase of the war in September 1970 focused on breaking the power of the PLO in and around Amman. The USA concentrated Marine forces in the Mediterranean to deter Syria, Iraq or the Soviet Union from coming to the aid of the Palestinians. Israel, who had no interest in overthrowing King Hussein, threatened to intervene if the Jordanian forces were defeated.

In the months following the battles of September 1970 the Jordanian army was primarily concerned with driving the PLO from its positions in northern Jordan. The PLO was focused on demonstrating and maintaining its fighting strength. The PFLP continued to urge the overthrow of King Hussein and the institution of a nationalist people's regime. In July 1971, the Jordanian troops began their final phase of destroying the PLO's bases in Jordan and interned the guerrillas. The PLO moved its bases to Lebanon.

All together 40,000 Palestinians are said to have been killed, and the fighting in Amman cost 1,500 lives.

Sources

Arnold 1991; Bailey 1984; Frangi 1982; FAZ 25.1.71, 1.2.71, 4.6.71; FR 19.7.71; FR 24.11.1993; Hottinger 1970; Kalb/Kalb 1974; Keesing's 23896A, 24059A, 24148A, 24225A, 24364A, 24769A; NZZ 22.1.71, 3.3.71, 17.3.71, 7.4.71, 16.4.71, 15.7.71, 19.7.71, 29.7.71, 24.9.71; O'Ballance 1973b; Snow 1973 (HSa)

---

114 AR Yemen / DPR Yemen
26.9.1972 - 13.10.1972[1]; C[2]-2
War ended by: third-party mediation (Arab League)

Warring parties:

Aggressor (Side A): AR Yemen

Aggressed (Side B): DPR Yemen

Object and goals of the conflict

The southern Yemenite opposition forces used the Arab Republic of Yemen as a base for their operations against the Democratic People's Republic of Yemen, as did Saudi Arabia (see War Nr. 98). A gradual easing of tensions between the two Yemens in 1970 was undermined by increasing border incidents between the southern Yemen opposition and the DPR's army. Both governments accused the other of plotting coups against them. In August 1972, the situation along the border escalated as the southern Yemenite opposition groups joined forces as the "United National Front for South Yemen." The North Yemenite government began to arm tribes and to mobilize its army. Despite efforts at mediation by the Arab League the fighting on the border intensified, and after September 26, 1972, the armies of both countries confronted each other directly.

Outcome of the war

On October 28, 1972, the Prime Ministers of both Yemens signed a treaty in Cairo to unify both countries, presided over by the Arab League. The agreement planned for the withdrawal of all troops from the contested area, the closing of training camps for exile groups and an end to all acts of sabotage. The agreement, however, had little effect, and both sides continued to support opposition forces against their respective neighboring governments (see War Nr. 135).

Notes

1. Both sides complained about the many interruptions in the cease-fire of October 13, 1972.
2. Since there is no information available about the extent of the southern Yemenite opposition's participation in the war, we classify this conflict as a pure interstate war.

Sources

Braun 1981; Keesing's 26417A; Koszinowski 1973; Koszinowski/Menshaui/Meyer 1980; (TS)

---

115 Fourth Arab-Israeli War (October War)
6.10.1973 - 26.10.1973; C-2

War ended by: third-party mediation (U.N.)

Warring parties:
Aggressor (Side A): Egypt / Syria / Iraq / Palestine Liberation Organization (PLO)
Aggressed (Side B): Israel

Object and goals of the conflict

Egypt and Syria sought to recapture territory lost to Israel in the 1967 war (see Wars Nr. 95 and 103), or at least to change the political situation so that restoration of the territories would be possible through negotiation. Israel sought to defend the occupied territories, and if possible to occupy somewhat more. The PLO fought with the goal of conquering all of former Palestine.

Outcome of the war

Israel ended up being forced to withdraw from the Sinai Pass and to return the territory east of the Golan Heights as well as the Kuneitra and Rafid territories to Syria. Later, during negotiations with the USA and Egypt, Israel withdrew from the Sinai and from the East Bank of the Suez Canal. The October War resulted in at least three diplomatic achievements: the Sinai Agreement, the Geneva talks, and the "separate peace" between Israel and Egypt achieved through the Camp David peace talks.

3,000 Israeli, approximately 7,700 Egyptian and 835 Iraqi soldiers were killed in the fighting.

Sources

Aker 1985; Allen 1982; Arnold 1991; Baram 1987; Brecher 1980; Bulloch 1974; Herzog 1984; Rabinovich/Shaked 1978; Steinbach 1981; Tibi 1989 (DJ)

---

117 Fifth Iraqi-Kurdish War
14.3.1974 - 3/1975; B-2[1]
War ended by: military victory for side A

Warring parties:
Aggressot (Side A): Iraq
Aggressed (Side B): Kurdish Democratic Party - Iraq (KDP-Iraq)

OBJECT AND GOALS OF THE CONFLICT

The Iraqi central government had only partially implemented the treaty of March 11, 1970 (see War Nr. 102). One of the main sticking points was the inclusion of the oil regions around Kirkuk in the Kurdish autonomous zone. Rather than resettling expelled Kurds there, as was called for in the agreement, the Iraqi government continued its policy of "Arabizing" the oil regions.

The leader of the KDP-Iraq, Barzani, was encouraged by Iran and the USA to push through his demands for far-reaching autonomy with military means. At the end of 1971 military confrontations sprang up, and by March 11, 1974 the Iraqi government proclaimed parts of the Kurdish region autonomous. The KDP-Iraq, however, was distrustful and rejected this offer of autonomy, and this became the official cause of the war.

OUTCOME OF THE WAR

Iran withdrew its essential support for the Kurds after Iraq made concessions regarding the border around Schatt al-Arab. This led to the defeat of the Kurdish forces. For the first time since 1961 the Iraqi government successfully took control of the entire Kurdish territory. The autonomous status for part of the Kurdish region remained in force on paper, but in reality the power lay in the hands of the Baath Party. The Iraqi government continued its "Arabization" policies (see War Nr. 126) and most of the Kurdish refugees who returned from Iran settled outside of Kurdistan. The Iraqi government also began to de-populate the areas along the border to Turkey and Iran.

At least 16,000 people died in the war.

Note

1. Iranian artillery units attacked in support of the Kurdish side after December 1974, but since there is no available information about the continuity of this support it is impossible to determine whether it counts as intervention.

SOURCES

Deschner 1983; Franz 1977; Keesing's 27053A; Khalil 1985; Sim 1980; Vanly 1984 (TS)

120 Lebanon War
since 13.4.1975; ABC-2

Warring parties:
Aggressor (Side A): Forces Libanaises (FL)[1] / Southern LebaneseArmy (SLA) (since 1985) / Israel (since 16.3.1978)
Aggressed (Side B): Lebanon / National Movement / Palestine Liberation Organization (PLO)
changing sides: Syria (since 1/1976)

Object and goals of the conflict

The civil war in Lebanon developed out of confessional, ethnic-cultural, regional and economic tensions. Up to the 1970s a constitutional measure to insure confessional proportionality guaranteed reasonable stability, but this order began to fall apart as the demographics changed in favor of the Muslims, and after the left wing gained influence and the PLO moved its bases to Lebanon (see War Nr. 108). At the beginning of the war the National Movement under Kamal Jumblat (PSP), together with the PLO, sought to abolish confessional proportionality in the political system and maintain Lebanon's "Arab identity," largely through the expansion of their own power. The PLO wanted to maintain its operational base on Lebanese territory. The Christian militias, mostly Maronites, came together in the FL to counter the "National Movement." The FL wanted to maintain the current political system and Lebanon's "Western character." Syrian troops intervened in 1976 on the side of the National Movement, but then switched sides to avoid a defeat of the FL. Syrian troops have remained in the country since then.

The war led to Lebanon's division into territories controlled by groups competing for power. Political and religious differences were not generally barriers to coalitions, which sprang up between the most unlikely groups. Yet the fighting and massacres led to religious "separation" whereby territories became the exclusive provenance of one or the other religion where previously heterogeneity had been the norm.

Israel intervened in 1978 and 1982 with the goal of destroying the PLO bases, and in 1982 Israeli and Syrian troops clashed. Israel withdrew partially in 1985, but the SLA, financed in part by Israel, and the

Israeli army still control the border area around Galili. The Israeli air force has also carried out "revenge" attacks on PLO and Hezbollah positions in Lebanese territory. Iran sent Revolutionary Guards to support Hezbollah in the Bekka valley, and contingents of U.S. peacekeeping troops (UNIFIL) and U.S. marines were involved to varying degrees during the war.

OUTCOME OF THE WAR

The war essentially led to the compete disintegration of the Lebanese state. Only in 1989 was it possible for a commission from the Arab League (Algeria, Morocco and Saudi Arabia) to work out an agreement to pacify the conflict. In October of that year Lebanese parliamentarians signed a "Charter for National Reconciliation" in Taif, Saudi Arabia. This agreeement contained no great changes regarding confessional proportionality, but it did diminish Christian control in favor of Sunni and Shiite Muslims, and put forth a plan for the withdrawal of Syrian troops. An environment for the successful implementation of the agreement was not at hand, however, until Syria became part of the anti-Iraq coalition in the (Second) Gulf War (see War Nr. 194). As a sort of recompense from the U.S. and Israel, Syria was left to freely break the military back of the Christian faction under General Aoun. The Lebanese parliament also withdrew support for the PLO and with Syrian support they managed to confiscate almost all of the PLO's heavy weaponry and exert control over the country. After this it was only Hezbollah who continued attacks against the SLA and the Israeli army, and Israel attacked their positions in Lebanon in retaliation for both cross-border attacks and terrorist attacks within the occupied territories.

Elections occurred in Lebanon for the first time in twenty years in August and September of 1992, leading to a vicotry for the pro-Syrian and Islamic fundamentalist forces. Many Christians boycotted the elections, claiming that the agreement signed in Taif was not being fully implemented, especially concerning the withdrawal of 40,000 Syrian troops. On the contrary, Syria managed to bind itself even closer to Lebanon. 1,000 Israeli soldiers were withdrawn from southern Lebanon, but relations remained strained because Lebanon insists on Israel's unconditional withdrawal (based on U.N. resolution 425) and Israel insists on the disarming of all militias, especially Hezbollah, the withdrawal of Syrian troops as called for in the Taif agreement, and guarantees for a secure border. A further sticking point is Israeli interest in southern Lebanon water sources.

Approximately 150,000 people died in the civil war (including 14,000 alone during the Israeli invasion), and several hundred thousand Lebanese and Palestinians were driven from their homes by fighting and massacres. Few have been able to return due to the level of destruction and the new religious demographics. A further problem for the Lebanese government is the 300,000 Palestinians living mostly in refugee camps.

Note

1. Organization of Christian militias.

Sources

Agwani 1965; B¸ren 1983; B¸ren 1989; Dawisha 1984; Deeb 1980; Deeb 1984; Haddad 1985; Hajjaj 1977; Hanf 1983; Hanf 1990; Hourani 1981; Hudson 1985; Kliot 1987; Köhler 1975; Rabinovich 1985; Rotter 1986; Salibi 1965; Salibi 1976; Zamir 1980 (DJ)

---

126 Sixth Iraqi-Kurdish War
since 5/1976; BA-3[1]

Warring parties:

Aggressor (Side A): Kurdish Democratic Party - Iraq (KDP-Iraq) / Patriotic Union of Kurdistan (PUK)[2] / Socialist Party of (Iraqi)-Kurdistan (PSK)[3] (since 1979) / Iraqi Communist Party(since 1979)
Aggressed (Side B): Iraq
Intervention favoring side A: Iran[4] (20.7.1983 - 20.8.1988)
Intervention favoring side B: Turkey[5] (26.5.1983 - 2.6.1983)

Object and goals of the conflict

Iraq pursued its "Arabization" policies with renewed vigor after the end of the fifth Iraqi-Kurdish war (see War Nr. 102). In 1976 they began evacuating all Kurdish villages within ten to twenty kilometers of the borders to Iran and Turkey and resettling their occupants. The farmers rose in spontaneous resistance, joined forces with Kurdish organizations, and continued the fight for Kurdish autonomy within a more democratic Iraq.

The Kurdish organizations were often great rivals, especially the

KDP-Iraq and the PUK, and they formed various alliances with Iraqi opposition parties. But in June 1988 the Iraqi Kurdistan Front was created, including the KDO-Iraq, PUK, Iraqi Communist Party, and the PSK.6 The Kurdish resistance movement had been severely weakened by government offensives after the end of the Iran/Iraq war (the first Gulf War, see War Nr. 194) but Iraq's defeat in the (second) Gulf War lent them a new lease on life, and soon they controlled a wide swath of Iraqi-Kurdistan. Yet after the Iraqi central government successfully put down a Shiite uprising in the south (see War Nr. 196) their troops also reconquered most of the Kurdish-controlled territory in the north.

Outcome of the war

Iraq's reconquering of the north in 1991 produced massive flows of refugees. These prodded the U.S. and other members of the anti-Iraq coalition to declare a no-fly zone in northern Iraq and to use their own soldiers to set up "protective" zones for returning refugees. For the first time the U.N. Security Council interfered in the internal affairs of a state with resolution 688 on April 5, 1991. After the withdrawal of allied troops the Kurdish organizations took control of the established protective zones. Yet all the following attempts at peace negotiations between the Iraqi government and the most important Kurdish representatives—Barzani (KDP-Iraq) and Talabani (PUK)—failed on the issue of defining the borders of the Kurdish autonomous region. Again the problem centered on oil interests.

The Iraqis proceeded to impose an economic embargo on the Kurdish region, and occasionally heavy fighting has been reported. The KDP-Iraq and PUK have worked together to build autonomous political structures in the region, but are still not fully in control, as witnessed by their attack on the PKK in northern Iraq in October 1992 (see War Nr. 159). This could also be seen as an attempt by the Kurdish organizations to improve relations with Turkey, who has grave reservations about the existence of an autonomous Kurdish region.

There are no reliable figures for victims of this war. The Iraqi government began to use chemical weapons in April 1987, and the largest poison gas attack on the Kurdish city of Helabja in March 1988, cost 5,000 to 7,000 lives and at least as many wounded. Chemical weapons were used more widely after the end of the Iran/Iraq War (the first Gulf War), and hundred of thousands of Kurds fled to Turkey. The fighting in 1991 produced another wave of refugees. Between 1976 and 1987 over one million Kurds from approximately 250,000 villages were forcibly relocated, and more than 2,000 people "disappeared" in the process. The Kurds are also affected both by the internal economic embargo imposed by the central government and the general embargo against Iraq imposed by the U.N.

Notes

1. Turkish and Iranian troops attacked sporadically (see notes 4 and 5). The Kurds also suffered under allied bombardments during the (second) Gulf War.
2. The PUK is an umbrella organization for the League of Kurdish Workers, the Kurdish Socialist Movement, and supporters of Jalal Talabani. In 1982 the latter two joined to form the Kurdistan Revolutionary Union.
3. The Kurdish Socialist Party emerged in May 1981 from the United Kurdish Socialist Parties, which itself emerged from a split between the KDP and the Kurdish Socialist Movement (until then part of the PUK) in August 1979.
4. While one cannot separate Iran's involvement in the war from their war with Iraq (see War Nr. 148), common Kurdish-Irani operations between 1983 and 1988 had a significant effect on the outcome.
5. Turkey always claimed the PKK as its official goal during attacks into Iraqi territory (see War Nr. 159), but they also repeatedly hit other Kurdish organizations and Kurdish civilians (e.g., in May/June 1983, August 15, 1986, March 4, 1987, and October/November 1992).
6. The Kurdish Socialist Party (Pasok), founded in 1976, and the Democratic People's Party of Kurdistan (PGDK), founded in 1981, also had guerrilla units, but little is known of their size or degree of participation in the war.

Sources

Farouk-Sluglett/Sluglett 1987; Franz 1986; Franz 1987; Hottinger 1987; Jaber 1988; Khafaji 1988; Keesing's 28116, 29241, 30103A, 32687A, 35135, 35162, 35862, 37424; Vanly 1984; Vanly 1986; Schneider 1987 (TS)

---

135 AR Yemen
1978 - 1982; A^1-1
War ended by: third-party mediation (Arab League)

WARRING PARTIES:

Aggressor (Side A): National Democratic Front (NDF)
Aggressed (Side B): AR Yemen
Intervention favoring side A: DPR Yemen (23.2.1979 - 3/1979)

OBJECT AND GOALS OF THE CONFLICT

The moderate wing of the Republicans supplanted the leftist People's Resistance Forces (PRF) during the revolution in North Yemen (see War Nr. 68), and sought to compromise with the opposing, royalist tribes from the North. The Republicans wanted to invite them into the government, yet did not want to give in to their concessions. This attempt by the central government to compromise elicited dissatisfaction within the Republican-sympathizers in the PFP and the NDF, an umbrella organization founded in 1976 and supported by the DPR Yemen. The NDF became the spokesperson for all who were unhappy with the government's policies, and built its power base in the south. In early summer, 1978, they took the leading role in an armed uprising against the government. This increased tensions between the two Yemenite states, which only worsened after the president of the AR Yemen, Ali Abdallah Salih, was murdered and his death blamed on the DPR Yemen. Both states massed their troops on the border and between February 23 and mid-March 1979, engaged in border skirmishes. The Arab League worked out a cease-fire and once again got both states to agree to unite (see War Nr. 114). The DPR Yemen's support for the NDF varied depending on relations between the two states.

OUTCOME OF THE WAR

Both states found common foreign policy ground when they supported the PLO following Israel's invasion of Lebanon. This brought the two Yemens closer together and this time the cease-fire from summer, 1982 held. An amnesty for the NDF led to negotiations between the leaders of the NDF and the AR Yemen, and although no substantial progress was made, the talks did put an end to the military confrontation.

Note

1. The fightings between the AR Yemen and the DPR Yemen across their borders in February/March 1979 could be classified as a short interstate war.

Sources

Braun 1981; Burrowes 1987; Koszinowski/Menshaui/Meyer 1980; Halliday 1990 (TS)

---

141 Afghanistan
since 10/1978; A-1

Warring parties:

Aggressor (Side A): Conservative-traditional resistance movements[1] / Muslim-fundamentalist resistance groups[2]
Aggressed (Side B): Afghanistan
Intervention favoring side B: USSR (12/1979 - 14.2.1989)

Object and goals of the conflict

In 1973, the Afghan Monarchy was overthrown by a coup and replaced by the cautiously reformist, conservative Daoud government. In April the Communist Party overthrew Daoud with the partial support of the military. Their attempts to drive through rapid socialist reforms failed, however, especially because their agrarian reform aimed at dissolving the traditional patrimonial rural relations of production, with the effect that the small farmers rose up in opposition. These rural peasants, together with large property owners, formed the base for an Islamic-conservative alliance against communist rule. The Communist Party itself was divided about foreign policy and strategies for socio-economic and cultural change, and fell into constant disputes. Their leadership changed often. As their ability to take command of domestic problems dissipated, the militant resistance movements grew stronger. In 1979, the Soviet Union sent troops into Afghanistan to overthrow the ruling national-communist government and replace it with a faction of the Communist Party which had previously lost out in the internal power struggles. The Soviet goal was to overcome the political crisis in what was then a friendly communist country, and further to extend their influence into the important geopolitcal region in between the rival People's Republic of China, U.S.-allied Pakistan, and fundamentalist Iran, whose influence threatened to extend into Soviet Central Asia.

Outcome of the war

Soviet troops withdrew from Afghanistan in 1989 as the USSR underwent a fatal internal transformation, the East-West conflict abated and negotiations with the U.S. under U.N. auspices continued. After the Soviets left, their favored president, Najibulla, managed to hang on to power by delicately balancing competing forces and fighting against the guerrilla (Mujehadin) forces. He was aided by internal power struggles within the Mujehadin, which weakened their military effectiveness. When the USSR withdrew its financial and military support from Najibulla in fall, 1991, however, his government's fate was sealed. Najibulla planned to leave office by the end of April 1992, in the framework of a U.N. peace plan. This was the signal for many government troops to desert and find accommodation with the Mujehadin, including Abdul Rashin Dostam, the leader of the militarily strong Uzbeki militia. Together they overthrew Najibulla in mid-April, and Mujehadin organizations came to power in localities throughout the country.

In the capital city of Kabul the situation was very complex, despite a formal transition government led by the Mujehadin. The contradictions among the Mujehadin came to the fore amidst different nationalities and different Islamic sects. These differences led to war-like confrontations: the Shiites fought for more participation in the government, which lay mostly in Sunni hands, while the Pashtune Kekmatjar fought against Tadziki and Uzbeki forces for their traditional power position in Kabul. The intense and escalating battles provoked a massive flight from the city to the surrounding region or Pakistan, above all in July 1992. Along with these power struggles another powder keg consisted of the relative independence of most of the regions vis à vis the central government, and the supra-regional loyalty of different nationalities (mostly the Uzbeks and Tadziks, but also the Belutsha and Pashtuna) (see War Nr. 208). Finally there existed the influence of the rival states with an interest in Afghanistan, most importantly Saudi Arabia, Iran, and Pakistan.

The Afghan economy has been utterly and completely devastated by the war, and the number of refugees has been estimated at two million within the country, and five million who fled abroad (mostly to Pakistan). Millions of land mines pose a constant threat for the civilian population, and have killed over 4,000 people and wounded tens of thousands more. All in all the war has claimed over one million lives.

Notes

1. These include the majority Pashtun organizations Jabha-i-Nejat-i-Melli (National Liberation Front), led by Sibghatullah Mujaddidi, Mahaz-i-Melli-i-Islami (National Islamic Front) under Pir Sayyed Ahmad Gailani

and Harakat-i-Inqilab-i-Islami (Movement for Islamic Revolution), led by Mohammed Nabi Mohammadi.

2. To these belong the majority Tadzik Jamiat-i-Islami (Islamic Society), led by Burhanuddin Rabbani and Ahmad Schah Massud, and the majority Pashtun organization Hizb-i-Islami (Islamic Party), under Gulbuddin Hekmatjar, an offshoot of the same name, led by Yunis Khalis, and Ittihad-i-Islami (Islamic Alliance) under Abdul Rasoul Sayyaf. In addition to these Sunni organizations exists the noteworthy Shiite alliance Hizb-i-Wahdat.

Sources

Braun 1989; Canfield/Shahrani 1984; Gupta 1986; Halliday 1978; Halliday 1980a; Halliday 1988; Kircheisen 1979; Klass 1988; Lewis, B. 1988; Orient-Institut 1981; Roy 1988; Saikal 1989; Scharpf 1988; Trautner 1991; Wiebe 1984; Ziem 1989 (SA/KJG)

---

144 Iran (Kurdistan)
7/1979 - 1988; B-2
War ended by: military victory for side A

Warring parties:

Aggressor (Side A): Iran
Aggressed (Side B): Kurdish Democratic Party-Iran (KDP-Iran) / Komalah[1] (7/1979 - ??)

Object and goals of the conflict

National minorities make up more than 50 percent of the total population of Iran. They fought against the Shah's regime, and hoped that this fall would translate into more national and cultural freedom. After the victory of the Islamic Revolution the Kurds set about developing autonomous administrative structures. This contradicted the clerics concept of an Islamic central state, but they tolerated the development until their government was firmly in power. Once secure, they then attacked the Kurdish autonomy movement with the full force of the military, and reconquered territory previously under Kurdish control. On the Kurdish side there arose two primary organizations: the KDP-Iran, who fought for the Kurdish autonomy in a loose coalition with the Alliance for a Democratic Iran, and the Komalah, who maintained that Kurdish autonomy was only possible in the context of a Marxist revolution.[2]

Outcome of the war

After Iranian troops took control of Kurdish territory the remaining

Kurdish troops were driven into the high mountains, from where only small units were able to form and carry out limited operations. This minimal resistance was even further hampered by Iran's use of the northern regions for bases in the war with Iraq (see War Nr. 148). Internal struggles between the KPD-Iran and the Komalah weakened the Kurdish case even more. Eventually the military operations dropped to almost nothing, and negotiations between the KPD-Iran and the Iranian government failed. In 1989 the General Secretary of the KPD-Iran, Ghassemlou, was murdered during secret negotiations in Vienna. The continued repression of Kurds, coupled with Iran's economic crisis, lead in June and July of 1992 to uprisings in Kurdish towns. Since then there has been increased guerrilla activity, but it has not yet crossed the threshold into war.

Notes

1. Since 1983: The Communist Party of Iran in Kurdistan.
2. The People's Fedayin and Mujehadin also fought in south-east Kurdistan against the central government, with the different goal, however, of overthrowing the regime and transforming Iran. Their interest in an autonomous Kurdistan was only secondary, and for this reason they are not considered here as warring parties.

Sources

Bruinessen 1981; Entessar 1984; Franz 1981; Franz 1986; Ghassemlou 1984; Hottinger 1987; Keesing's 30629A, 31797A, 32690A; Malek 1989; Zabih 1986 (TS)

---

148 Iraq / Iran (Iran-Iraq War, also known as the "First" Gulf War)
22.9.1980[1] - 20.8.1988; C-2[2]
War ended by: third-party mediation (U.N.)

Warring parties:

Aggressor (Side A): Iraq
Aggressed (Side B): Iran

Object and goals of the conflict

Iraq wanted to decide once and for all its border conflict with Iran with a military attack. The border traditionally ran along the Iranian shore of the river Shatt-el-Arab, but in 1975 the Shah's regime had redefined it in the Treaty of Algiers to run down the middle of the river. Iraq had further war goals: they wanted a return of Iran-occupied territory promised them in the Treaty of Algiers, the annexation of the oil-rich Khusistan (Arabistan) province to Iraq, and the return of three islands in the straights of Hormuz, occupied by Iran in 1971, to other Arab states. Iraqi leader Saddam Hussein thought he could manipulate the confusion following the Iranian Islamic Revolution to insure a quick victory. Hussein was also worried that Iran would export its Shiite revolution to Iraq.[3] A swift victory would also help Saddam Hussein secure his leadership position in the Arab world. During the war Iran sought to topple Saddam Hussein and to incite the Iraqi Shiites to rebellion. They also found the war to be a good means of distracting the population from internal political conflicts, such as the demands of local minorities for autonomy.

Outcome of the war

Iran retook the Khusistan Province in May and June 1982, and a lengthy war of attrition followed. Iraq offered numerous cease-fires, and reduced its war goals to the recognition of the Algiers Treaty. Iran, on the other hand, expanded its demands to include Iraq's unconditional surrender, reparation payments, the end of the Iraqi government, the return of expelled Shiites, and international condemnation of Iraq as the aggressor.

In May of 1984 Iraq and Iran intensified their attacks on commercial ships from different nations, and Kuwait officially requested the formation of an international war fleet under U.S. command to protect international shipping, especially oil transports. This opportunity also served to underscore U.S. interests in the region. In 1988 a cease-fire was signed by both sides, which took effect for the most part only after the end of 1990, when Iraq had occupied Kuwait and perceived a growing threat from the U.S. (see War Nr. 194).

The figures for casualties range between 300,000 and 1,500,000. Both parties bought over 27 million dollars worth of heavy weaponry during the eight-year war.

Notes

1. Small border skirmishes happened regularly after April 1990.
2. Twenty-nine Syrian soldiers, including 17 officers, were killed on the Ira-

nian front in December 1981. It remains unclear whether they were involved in the fighting as military advisors or regular troops. In addition there were the following third-party interventions in this war: in June 1981, Israel bombed an Iraqi nuclear power plant which was suspected of being part of a nuclear weapons program; in November 1981, Iran bombed Kuwait; in May and August 1983, Turkish troops attacked Kurdish guerrillas on Iraqi territory to help reduce pressure on Iraqi forces; in February and March 1984, Turkey bombed Kurdish positions in Iran; in February 1984, a U.S. warship fired at Iranian fighter planes and a warship; in June 1984, Saudi Arabian fighter jets shot down an Iranian fighter plane; and following May 1984, Iraq and Iran attacked commercial ships from different nations, resulting in an international force to protect the shipping lanes.

3. In Iraq live 50 percent Shiites, and 50 percent Sunnis, yet 20 percent of the Sunnis are Kurds, so that Shiites constitute the relative majority. The remaining Sunnis are Arabs.

SOURCES

Friedrich Ebert Stiftung 1982; H‚nseler 1982; Keesing's: 31005A, 31517A, 31848A, 32678A, 33056A; Nima 1983; Rondot 1981; SIPRI 1988; Z‚rrer 1981 (BMu)

---

153 Syria 1982

2.2.1982[1] - 2/1982[2]; A-2

War ended by: military victory for side B

WARRING PARTIES:

Aggressor (Side A): al-Jabha al-Islamiyya (Islamic Front)[3]

Aggressed (Side B): Syria

OBJECT AND GOALS OF THE CONFLICT

In 1970, Assad came to power as the result of a coup d'etat. He proceeded to fill nearly all leading positions with Alavite relatives or trusted friends and sought to secularize the state. Both these actions met with resistance from the orthodox forces within the Sunni majority, for they neither recognized Alavites as Muslims nor could they accept the process of seculariztion. The Muslim Brotherhood became the leading opposition force, who counted among its members the hard-hit lower middle class. In 1976 the Brotherhood began assassination attempts against representatives from the state, the ruling Baath party and the secret service. They also bombed state institutions, and after

Syria and the Soviet Union signed a friendship treaty they included Soviet targets in their bombing campaign. The Syrian central government reacted to these attacks by intensifying their repression of the orthodox Muslims.

The industrialization policies of the government formed the background for the biggest battles in the city of Hama. Hama's economy was based traditionally on trade and small craftwork, and the new policy of industrialization greatly disoriented the city. Besides the economic transformations, the Syrian government stoked ethnic tensions by settling Alavites around the Sunni-controlled city. The last straw for the Sunni residents of Hama was a house-to-house search for weapons and secret Muslim Brotherhood meeting places, carried out by the Syrian army.

Outcome of the war

The Muslim Brotherhood was unable to extend uprising to all of Syria, and the rebellion in Hama failed. After this setback the organization was saddled with internal contradictions, especially after the Brotherhood decided to join with other opposition groups to create the "National Alliance for the Liberation of Syria." The military potential of the organization was considerably weakened when they excluded the "militant vanguard" from taking part.

The number of victims is subject to debate, and the numbers range from several hundred to more than 30,000 (Hama had approximately 200,000 inhabitants). The Hama city center was mostly destroyed, and the cost of reconstruction has been estimated at over $500 million.

Notes

1. Since 1976 the Muslim Brotherhood had carried out a series of assassinations and attacks, resulting occasionally in armed confrontations with the police and the army.
2. The Muslim Brotherhood continued to carry out sporadic attacks, but their number never achieved the intensity of the pre-1982 campaigns.
3. There exists no evidence that members of the army went over to the side of the Muslim Brotherhood.

Sources

Batatu 1982; FAZ 25.2.81, 15.2.82, 18.2.82, 7.4.82; FR 25.3.86;

Keesing's 31623A; SZ 17.2.80; Koszinowski 1985; Lawson 1982; Mayer 1983 (TS)

---

159 Turkey (Kurds)
since 15.8.1984; B-2

Warring parties:
Aggressor (Side A): Partiya Karkeren Kurdistan (PKK; Kurdish Worker's Party)[1]
Aggressed (Side B): Turkey

Object and goals of the conflict

Since its inception, the modern Turkish state has refused to recognize the existence of the Kurdish people. Officially Kurds are referred to as "mountain Turks," and any mention of a Kurdish people or use of the Kurdish language is met with severe punishment. In the crisis-filled atmosphere of the 1970s the Kurdish resistance movement, together with the student movement, became increasingly radicalized. In 1980 the military came to power through a coup, and the PKK remained the only Kurdish organization left to reorganize resistance. Since the summer of 1984 they have led military operations against the Turkish army and NATO installations, originally from bases in Iraq. The PKK regards Kurdistan as being colonized by Turkey, and proclaims as its first goal a national-democratic movement led by the proletariat. After the independence of the Turkish-controlled part of Kurdistan is achieved (the northwest region), then the goal moves to forming a united, socialist Kurdistan out of the remaining Kurdish territories which lie today within Syria, Iraq and Iran.

Outcome of the war

The Turkish government has been unable to break the Kurdish resistance despite actions including intensive repression (mass arrests, torture, arbitrary killings, death penalties), the forced resettlement of approximately 100,000 Kurds with the goal of creating a no-man's-land along the southern border, a high Turkish military presence in Kurdistan, and military operations into Iraqi and Iranian territory. The PKK has been able to extend its influence. To the extent that the Turkish government has been forced to concede that the guerrilla movement is difficult to destroy, and that the Turkish army has lost control over some parts of southeast Anatolia, there has been a slight movement towards liberalization—for example, discussion about the Kurdish people is no longer punishable by law in every instance. After the (Second) Gulf War the Turkish government allowed the private use of the Kurdish language.

Since mid-October 1992 the Turkish army began a large offensive in southeast Anatolia and into the U.N. "protection zone" in Iraq, with the goal of finally annihilating the PKK bases. The Turks received support from the Kurdish organizations in Iraq, who feared the PKK's influence into their region and also sought better relations with Turkey in the hopes of establishing an autonomous Kurdish zone in Iraq (see War Nr. 126). This recent Turkish offensive alone, which lasted until the end of November 1992, cost over 3,000 lives.

Note

1. The guerrilla units operate as Hezen Rizgariya Kurdistan (HRK; Kurdish Liberation Units). In 1985 the PKK founded the Eniya Rizgariya Netewa Kurdistan (ERNK; National Liberation Front of Kurdistan).

Sources

Bruinessen 1984; Bruinessen 1988; Franz 1986; Gunter 1988; Heinrich 1989; Hottinger 1987; Vanly 1986 (TS/MS)

---

162 South Yemen 1986
13.1.1986 - 29.1.1986; A-2[1]
War ended by: military victory for side B

Warring parties:

Aaggressor (Side A): DPR Yemen
Agressed (Side B): Left Wing of the Yemenite Socialist Party (JSP)

Object and goals of the conflict

The National Liberation Front (after 1978 the Yemenite Socialist Party) had been able to contain their internal differences during the war of liberation against Great Briatin, but after the independence of South Yemen in 1967 their conflicts broke out into the open (see Wars Nr. 87 and 98). The radical wing pursued the goal of a socialist transformation, while the moderate wing sought economic liberalization and better relations with their Arab neighbors. Both wings maintained close relations to the USSR. Not to be overlooked within the context of internal power struggles is the role of tribes. A tribe became more powerful when its members occupied powerful positions.

The power struggle exploded into the open in February 1985, after the leader of the radical wing, Abdel Fattah Ismail, returned from Soviet exile. By their third party conference the JSP was deeply divided and incapable of functioning. Fighting broke out during a politburo meeting on January 13, 1986, and it remains unclear who started it. It is likely, however, that President Ali Nasir Muhammad attempted a preventive strike to safeguard his position. He was ousted anyway.

OUTCOME OF THE WAR

About 15,000 supporters of Muhammad were arrested, while he himself successfully fled abroad, from where he vainly tried to organize his return. The Soviet Union had warned other countries from interfering in Yemen's internal affairs, which resulted in no foreign states coming to Muhammad's aid. The new Yemenite government continued Muhammad's political course, however, largely under pressure from the USSR. Former Prime Minister Haider Abu al-Attas became the new head of state, since he did not belong to any of the rival factions and his tribe had remained neutral during the war. One result of the war was the unification of the two Yemenite states.

More than 10,000 persons died, including 4,000 members of the JSP.

Note

1. On January 17, 1986, the Aden airstrip was bombed by an unidentified plane. According to unsubstantiated rumors the bomber was either from Oman or Israel.

SOURCES

Burrowes 1989; Hottinger 1986; Koszinowski 1991; Lawson 1986; Pollock 1986; Werle 1986; Keesing's 34393A; SZ 15.1., 27.1, 28.1., 30.1., 1.2., 18.3.1986; NZZ 26.1., 29.1., 1.2., 5.2., 10.2.1986 (TS)

---

174 USSR (Azerbaijan, Armenia); Nagorno-Karabakh
since 20.1.1990[1]; B[2]-2[3]

WARRING PARTIES:

1. Phase

Aggressor (Side A): USSR
Aggressed (Side B): Armenian National Army (ANA)4 / People's Front of Azerbaijan also National People's Front)
2. Phase:
Aggressor (Side A): Azerbaijan
Aggressed (Side B): Armenian Militias from Nagorno-Karabah

OBJECT AND GOALS OF THE CONFLICT

The core of the conflict lies in two areas: the historic tensions between the Islamic Azerbaijani and the Christian Armenians, and their respective transition to independence after the fall of the Soviet Union at the end of 1991. In 1987, 75,000 Armenians sent an open letter to the Soviet General Secretary Gorbachev, asking for the annexation of the Armenian-populated autonomous territory of Nagorno-Karabah, which lay inconveniently within the territory of Azerbaijan. In 1988 pograms erupted between both ethnic groups. The Azerbaijani People's Front cut Nagorno-Karabah's supply lines from Armenia. Over 20,000 Armenians and Russians fled Baku, the capital of Azerbaijan. Fighting erupted along the border of the two republics and attacks began on villages. On January 3, 1990, Azerbaijani nationalists began to tear down the border to Iran to demonstrate their desire to unify with Iranian southern Azerbaijan. At the same time the autonomous republic of Nakhichevan, populated by a majority of Azerbaijanies declared its secession from the USSR. On January 15, 1990, President Gorbachev declared a state of emergency for Armenia, parts of Azerbaijan, Nagorno-Karabah and the border to Iran. Five days later 20,000 Red Army soldiers and special units from the Soviet Ministry of the Interior occupied important posts in Azerbaijan.

OUTCOME OF THE WAR

The Soviet military quickly broke any military resistance in Azerbaijan, but the state of emergency remained in effect until the end of 1991. Two elections secured the Communist Party in power, although participation was only around 20 percent.The Azerbaijani People's Front complained of massive electoral fraud. On August 30, 1991, Azerbaijan declared itself independent of the Soviet Union.

Armenian independence had been declared one year before, on August 23, 1990, introduced its own currency and began building its own army. Armenian resistance against Soviet control continued to grow.

Ninety-five percent of the population voted in a referendum on September 21 for an independent Armenia.

The relations between Armenia and Azerbaijan remained confrontational, especially regarding Nagorno-Karabah. All attempts at negotiations and cease-fires failed. In November 1991 Azerbaijan cut Armenia off from Russian gas by closing the pipelines, and also cut all railway connections. On January 2, 1992 Nagorno-Karabah was put under direct Azerbaijani administrative control, and in response Nagorno-Karabah declared its independence from Azerbaijan. Since January 1992 the fighting intensified tremendously between Armenian militias in Nagorno-Karabah and Azerbaijani troops. In March 1992 Azerbaijan lost the support of troops originally sent by the CIS (former Soviet Union), and in May of that year Armenian militias captured and held a land corridor to Armenia. At the beginning of July the Armenian parliament decided to no longer regard Nagorno-Karabah as part of Azerbaijan, and this led to intensified border fighting between Armenia and Azerbaijan, as well as violence in Nakhichevan. Azerbaijan also attacked the Armenian enclave of Artsvachen, which lay in Azerbaijani territory. Turkey, Iran, the Russian Federation and the CSCE tried without success to negotiate cease-fires. Both Turkey and Iran have threatened to intervene against Armenia if they do not halt their fighting, and both countries have increased their military along the borders.

All the affected regions suffered terrible economic losses, and 15,000 people have lost their lives in the fighting so far. Over 100,000 Azerbaijanis have fled Armenian territories, depopulating almost all of southeastern Azerbaijan. Armenians have also fled from Azerbaijan, where they have been the target of repression and violent persecution.

Notes

1. Confrontations between Armenian Christians and Azerbaijani Muslims started in February 1988. Soviet troops intervened for the first time on January 20, 1990.
2. The war has to be seen in the context of the disintegration of the Soviet Union. The first phase concerned military confrontations between two republics, and the Soviet military attempted to control the situation. After the end of the USSR the war entered a second phase which centered primarily around the disputed Armenian exclave of Nagorno-Karabah in Azerbaijan.
3. Armenia denies that their troops have taken part in the Nakorno-Karabah fighting, but they admit supporting the Armenians there with important

supplies, and that they allowed "volunteers" to go there. Since May 1992 there have also been skirmishes between Armenia and the Autonomous Republic of Nakhichevan, which belongs to Armenia, and occasional incidents along the Armenian-Azerbaijani border. There is too little information available, however, to categorize these confrontations as war or to see them as continuations of existing conflicts.
4. The ANA is the largest of the fifteen Armenian militias.

Sources

Die Zeit 12.4.91; FAZ 14.7.88, 30.4.90; FR 24.8.90, 29.1.92; FT 24.8.90; Huber 1991; Simon 1988a; Saroyan 1990a; *Newsweek* 5.9.90; Spiegel-Serie: 13-15/92; taz 16.1.90, 18.1.90, 22.1.90, 3.5.90, 29.5.90; *Current Digest of the Soviet Press*, Vol XLII, No. 3ff (DS/KJG/ZK)

---

182 Iraq / Kuwait
2.8.1990 - 4.8.1990; C-2
War ended by: military victory for side A

Warring parties:

Aggressor (Side A): Iraq
Aggressed (Side B): Kuwait

Object and goals of the conflict

Iraq coveted Kuwait, and when Kuwait became independent from Great Britain in 1961 Iraq demanded its annexation. Iraq justified its claim to Kuwait by pointing out that Kuwait was a former part of the Ottoman province of Basra. After Britain intervened, however, Iraq recognized Kuwait in 1963. The relations between both countries worsened, however, during Iraq's war with Iran (see War Nr. 148). Kuwait, nonetheless, had provided financial support to Iraq, and after the end of the Iran-Iraq war Iraq demanded that Kuwait forgive all of Iraq's debt. Kuwait was also charged with usurping oil which belonged to Iraq. A further sticking point was that during the war Kuwait had refused to cede certain strategically important islands to Iraq. Tensions rose after Iraq accused Kuwait and the United Arab Emirates of driving oil prices down by overproduction, thereby denying Iraq much-needed income.

Outcome of the war

Iraq quickly occupied Kuwait, and proceeded to plunder the kingdom and annex the country. The U.N. Security Council condemned the occupation and annexation and imposed sanctions against Iraq. The

U.S., in particular, called for an international military force in the Gulf to drive Iraq out of Kuwait and check Iraqi military potential (see War Nr. 194).

Several hundred people, at least, died during Iraq's occupation and subsequent looting.

Sources

Die Zeit; *Financial Times*; Frankfurter Rundschau; Middle East International; S¸ddeutsche Zeitung (TS)

---

202 Georgia (South Ossetia)
12/1990 - 7/1992; B-2
War ended by: third-party mediation (neighboring state[s])

Warring parties:
Aggressor (Side A): Georgia
Aggressed (Side B): South Ossetian Nationalists

Object and goals of the conflict

The conflict in this autonomous region within Georgia began in 1989. South Ossetia sought to unite with the Autonomous Republic of North Ossetia, which belonged to the Russian Federation. Attempts by the Georgian government to declare Georgian as the official language and to repeal minority rights led to a deteriorating relationship between South Ossetia and Georgia. In September 1990, South Ossetia declared itself independent of Georgia, just as Georgia was declaring its independence from the Soviet Union. In December of 1990, Georgia repealed South Ossetia's autonomy, and the conflict escalated into war.

Outcome of the war

The war ended with the stationing of Russian-Georgian-Ossetian peace troops on July 14, 1992. The conflict, however, was not solved and continues on. Voting stations during Georgian parliamentary elections on October 11, 1992, for example, remained closed in South Ossetia.

More than 150 casualties and more than 20,000 Ossetian and 100,000 Georgian refugees were reported.

Sources

Dobson 1975; Report on the USSR 25.10.89, 1.12.89; Kegel 1991; Lang 1962; Meyer 1990; Simon 1991; St^lting 1990; Suny 1988; Suny 1990 (TD/PT)

---

194 Anti-Iraq Coalition / Iraq (Gulf War, also known as "Second" Gulf War)
17.1.1991 - 27.2.1991; C-2
War ended by: military victory for side A

Warring parties:
Aggressor (Side A): United States of America / Great Britain / France / Italy / Egypt / Syria / Saudi Arabia / Kuwait
Aggressed (Side B): Iraq

Object and goals of the conflict

The goal of the conflict was to liberate Kuwait, which Iraq had invaded, occupied and annexed on August 2, 1990 (see War Nr. 182). Iraq made any peaceful return of Kuwait dependent on a repeal of the U.N. resolution condemning the Iraqi invasion and the convening of a Middle East conference to return Israeli-occupied territories. The U.S., as the strongest military power after the end of the cold war, did not want to give in to these demands. They and their allies also regarded the high level of Iraqi armaments, both conventional and non-conventional, as a threat to regional stability, though Iraq's weapons programs had been previously supported by Western countries. The U.S. was concerned about the supply of oil, Arab countries and Israel were concerned about the threat posed by Iraq's desire for regional hegemony. The U.S. was successful in implementing a military solution with the support of its Western and Arab allies. Many attempts at mediation by Arab states and the Soviet Union foundered either on U.S. or Iraqi objections.

Outcome of the war

The U.S.-led anti-Iraq coalition, made up of more than 28 states[1], was militarily successful but did little to solve the political problems of the region. Iraq's military potential was seriously retarded, but Saddam Hussein's dictatorial powers remained intact, despite international condemnations and embargos. Kurdish and Shiite uprisings in the aftermath of the war were quickly quelled by Iraqi troops (see Wars Nr. 126 and 196).

The cease-fire which ended the war was signed by Iraq on April 3, 1991 (U.N. Resolution 687). Its conditions included the destruction, under international control, of Iraqi weapons of mass destruction. The U.N. resolution did not differentiate between civil and military installations, and there has followed prominent and cantakerous conflicts between U.N. inspection teams and the Iraqi government. These conflicts have threatened to esclatate into new military confrontations, since Iraq views the U.N. actions as interference in their sovereign affairs, while the U.N. sees Iraqi intransigence as violation of the terms of the cease-fire.

Exact numbers of casualties are still not available. Estimates vary tremendously. Official Iraqi figures claim 85,000 to 110,00 soldiers and 40,000 civilians were killed in the war. The Iraqi opposition maintains that 150,000 to 250,000 soldiers and 70,000 to 180,000 civilians died. The U.S. published official numbers for the coalition forces, which sustained 343 casualties. In addition 7,000 Kuwaitis and 13 Israelis were killed (the latter by missile attacks from Iraq on Israeli cities).

The direct and indirect consequences of the war are immense. The environment has suffered from oil spills and burning Kuwaiti oil wells, from which over one million tons of soot and sulfur fell onto Kuwait alone. The combined effect of the war and the economic embargo has all but destroyed the Iraqi economy, cutting industrial production by more than half. Even more serious problems have been caused by the contamination of water and sewage systems, which has led to the outbreak of diseases, made worse by the lack of medical supplies. Official Iraqi figures claim more than 100,000 people, including nearly 30,000 children under five years of age, have died directly or indirectly as a result of the embargo.

Note

1. The forces of the anti-Iraq coalition included: Afghanistan, Egypt, Argentina, Australia, Bahrain, Bangladesh, Belgium, Bulgaria, Czechoslovakia, Denmark, Germany, France, Greece, Great Britain, Italy, Canada, Kuwait, Morocco, New Zealand, Holland, Niger, Norway, Pakistan, Poland, Romania, Saudi Arabia, Senegal, Spain, South Korea, Syria, Turkey, USA and the United Arab Emirate. Not all troops sent to the regions

by various countries took part in the fighting. Israel was not directly involved in the fighting, but was attacked by Iraqi missiles.

Sources

Gresh 1990; Krell/Kubbig 1991; Salinger/Laurent 1990 (UB)

---

196 Iraq (Shiites)
since 2.3.1991; A-2

Warring Parties:
Aggressor (Side A): Iraq Shiites[1]
Aggressed (Side B): Iraq

Object and Goals of the Conflict

The Shiites in Iraq have endured decades of repression under the Sunni-majority government in Bagdad. Saddam Hussein's defeat at the hands of the anti-Iraq coalition encouraged the leaders of the Shiite resistance movement to begin a popular uprising.

Outcome of the War

The Shiite rebels hoped in vain for support from either the U.S. or other opposition groups in Iraq. The U.S. was interested neither in a repetition of the Iranian revolution nor further fragmentation of Iraq, and the Iraqi opposition movements were too fragmented to be of any use. Iran also tried to heavily influence the rebellion, and their interference was not welcomed by either Iraqi Shiites or the U.S. Iraqi troops managed to quickly defeat the popular yet isolated uprising.

Sporadic fighting since then remains concentrated in southeastern Iraq, particularly in the swamp areas which serve as safe cover for deserters and opponents of the regime. In August 1992, the U.S. intervened in the conflict by imposing a no-fly zone for all Iraqi aircraft south of the 32nd parallel. The Iraqi government turned to environmental war tactics, and used existing canals to empty the swamps which had been providing cover for the rebels.

More than 800,000 residents of southern Iraq were left homeless during the first phase of the war, and at least 12,000 persons were killed. Some estimates speak of 50,000 to 60,000 victims.

Note

1. According to Radio Teheran the rebels numbered more than 30,000.

SOURCES

MEI 3.5.91, 14.6.91; Martin 1991 (UB/MS)

---

201 Georgia (Anti-Regime War)
since 9/1991; A-2

WARRING PARTIES:

Aggressor (Side A): Oppositional National Movement[1]
Aggressed (Side B): Georgia

OBJECT AND GOALS OF THE CONFLICT:

Zviad Gamzakurdia's "Round Table" alliance won the first round of free elections in Georgia in 1990, and he himself was directly elected president on May 26, 1991. He was unable, however to make much of a dent in the country's worsening economic and social situation. Furthermore, his dictatorial style and brutal nationalities policies gave birth to sworn enemies amongst his own ranks. In 1991 Gamzakurdia seemed to want to ally himself with the leaders of the Moscow Coup in order to defeat the Georgian opposition. This was the last straw, and he lost the support of most of the Georgians who had originally elected him precisely because of his anti-Soviet stance. In September the opposition joined forces, and clashed with police and military forces during demonstrations outside the government palace in Tblissi (Tiflis). This was followed by a rapid escalation into open warfare between the supporters of Gamsakurdia and his opponents.

OUTCOME OF THE WAR:

After four months of fighting, which witnessed the defection of parts of the National Guard and the former Defense Minister Kitovani to the opposition, the opposition prevailed in a deal. Gamzakurdia fled on January 7, 1992 to Armenia. At least one hundred persons died in the fighting. A military council consisting of leading opposition members charged former Prime Minister Sigua with creating a transition government. Former Soviet Foreign Minister Eduard Shevardnadze became the chair of the state council. Coup attempts by Gamzakurdia supporters in April and June of 1992 were put down, yet sympathy for

Gamzakurdia remains strong in Western Georgia. New fighting later broke out in connection with the war in Abkhazia (see War Nr. 207).

Note

1. The oppositional National Movement was made up of more than thirty different groups, united through their opposition to Gamzakurdia. The most important players were the National Democratic Party under Chanturia and their armed wing the "White Knights."

Sources

Dobson 1975; Iberi 1989; Kartveli 1989; Kegel 1991; Lang 1962; Meyer 1990; Report on the USSR 8.8.89, 25.10.89, 1.12.89; Simon 1991; Stölting 1990; Suny 1988; Suny 1990 (TD/PT/TK)

---

208 Tadjikistan
since 8/1992; A-2

Warring Parties:

Aggressor (Side A): Tadjikistan

Aggressed (Side B): Coalition consisting of the Western-oriented Democratic Party, the nationalist movements Rastokhez and Lali Badakhshan, and the Islamic Renaissance Party

Object and Goals of the Conflict:

Tadjikistan is the poorest member of the Community of Independent States (CIS) which replaced the Soviet Union in 1991. The Communist President Nabiyev, elected in November 1991, was unable to control the deepening economic crisis. The continuing crisis spurred on the opposition, which consisted of a diverse coalition of anti-Communists, nationalists, liberals, and Islamic forces. They laid siege for several days to the Parliament with thousands of supporters, and clashed violently with pro-government demonstrators. In early May, Nabiyev allowed a third of the government to consist of opposition members. Clashes continued, however, especially between the Communist government supporters and the Islamic forces. These confrontations awoke latent regional and ethnic tensions. The Communists were most influential in the industrialized north, populated largely by ethnic Russians and Uzbeks, and the city of Kuljan in the south. The opposition was strongest in the traditionally Islamic Garm region. A march by armed citizens and subsequent siege of the government buildings led to Nabiyev's flight on September 1, 1992, and soon thereafter he resigned. An Islamic/national-liberal coalition under Eskanderos took power. The change of government, however, did not end the fighting, and the old Communist alliance was able to force Eskenderos from power on November 19, 1992. The former Communist Rachmanow was elected President, and days later opened an offensive against the Islamic opposition.

Outcome of the War

Russian troops stationed in Tadjikistan were often targets of militias, who stole their weapons. Fighting also broke out along the Afghan border between the government and the Afghani Mujehadin, who supported the Islamic opposition.

Up to December 1992 the war had cost more than 50,000 lives and caused more than 250,000 persons to flee their homes.

Sources

Brown 1993; taz, FR and SZ August through December 1992 (TS/TD)

---

207 Georgia (Abkhazia)
since 14.8.1992; B-2[1]

Warring Parties:

Aggressor (Side A): Georgian
Aggressed (Side B): Abkhazian Militias with support of the Caucasus mountain people and various Autonomous Republics within the Russian Federation

Object and Goals of the Conflict

Abkhazia sought independence since the fall of the USSR. The population consists of approximately 18 percent Abkhazis and 48 percent Georgians. As tensions worsened with the central Georgian government, the Abkhazi Parliament declared independence by reviving the 1925 Abkhazian constitution on July 23, 1992. In doing so they neglected to get the necessary two-thirds majority prescribed by law for such an action. Georgia saw this as a provocation, which was only heightened by the kidnapping of the Georgian Interior Minister in Abkhazia by Gamzakurdia supporters (see War Nr. 201). On August 14, 1992 the Georgian government sent troops into Abkhazia. After a brief round of fighting the Georgian leader Shevardnadze demanded the resignation of the Abkhazian President Ardsindba and the dissolution of his parliament. After this failed to happen the Georgian National Guard attacked anew.

Outcome of the War

The Russian Federation has tried in vain to broker a lasting cease-fire.

Note

1. Russian troops stationed in Abkhazia have been sporadically involved in skirmishes. The Russian Defense Ministry claims that Russian troops only fought in self-defense against Georgian attacks, while the Georgian Defense Ministry claims that Russian troops were siding with the Abkhazi forces.

Sources

SZ, FR and taz July to December 1992 (TS/TK)

---

209 North Ossetia- (Prigorodny District)
since 30.10.1992; B-2

Warring Parties:

Aggressor (Side A): Armed Inguishian Forces
Aggressed (Side B): Autonomous Republic of North Ossetia
Intervention to End the War: Russian Federation (since 31.10.1992)

Object and Goals of the Conflict

Since October 1992 fighting has escalated in the Autonomous Russian Federal Republic of North Ossetia between armed forces of the Inguishian national minority and North Ossetian troops in the Prigorodny district. The Inguschian goal is the incorporation of the district into their already-existing Autonomous Republic.

The catalyst for the territorial conflict was the Moscow decree which rehabilitated repressed peoples in the Russian Federation. Theoretically, at least, this step implied that territorial borders betweeen Autonomous Republics could be changed. The Inguishian Republic had been part of the Autonomous Republic of Chechno-Inguishia, and was recognized by Moscow as independent in June of 1991. Since then they and their brethren in North Ossetia have demanded the annexation of the historically Inguishian district of Prigorodny into the new republic.

Outcome of the War

On December 30, 1992 the Russian Interior Minister sent 3,000 members of their special unit OMON and a commando group under Chisha to North Ossetia. Since then the Russian government has tried in vain to get the Inguishians to back away from their territorial claims. Meanwhile the Inguishians have begun to create a transition government in North Ossetia in preparation for annexation. In November, 1992, Russian President Yeltsin decreed martial law in the district, which was still in effect at the end of 1992. Although the Russian government began to withdrew the OMON troops as soon as ten days after their deployment, the Russian government remains involved in the conflict.[1]

More than 300 people were killed in the fighting up to the end of 1992. 35,000 Inguishians have fled from North Ossetia, and numerous villages have been destroyed in the Prigorodny district.

Note

1. There are two other conflicts which play an important role here: the conflict between the Inguishians and the Chechniyans, and between the Chechnyians and the Russian Federation. Unlike Inguishia, Chechnya is one of the first Republics to want to secede from the Russian Federation.

Sources

Stölting 1990a; APuZ 20.12.1992; International Alert 1992a (CF)

Africa

North Africa

30 Morocco
7.12.1952 - 2.3.1956; D-2
War ended by: agreement without mediation

Warring parties:

Aggressor (Side A): Istiqlal

Aggressed (Side B): France
first B, then A: Berber tribes

Object and goals of the conflict

The Istiqlal party emerged in the context of a anti-colonial struggle and against the background of a search for Arabic unity. Their demands included autonomy, reform, and democratization, and the Sultan counted among their supporters.

France, however, sought to maintain the status quo in which the French protectorate in Morocco was ruled jointly by Morocco and France. In addition to France there were also two other foreign settlements on Moroccan territory: the international zone of Tangiers and the Spanish protectorate in the north.

The Berber Uprising, which led to Sultan Sidi Mohammed Ben Youssef's abdication, served as a precursor to this war. This uprising, however, had been initiated by France, who commanded the loyalty of the Pasha of Marrakesh and the leaders of the surrounding Berber tribes. The Berber tribes, however, switched sides after the loss of their spiritual leader Ben Youssef and after the nationalist movement began to expand. Within this nationalist movement a National Liberation Army was formed, which fought after 1954 against the French.

Outcome of the war

Istiqlal was unable to attain their political goal of founding a republic. Instead Morocco became a constitutional monarchy. Sultan Sidi Mohammed Ben Youssef, who had been exiled by the French to Corsica in 1953, returned to the throne. Morocco did attain national sovereignty and France recognized its right to conduct autonomous diplomacy and raise an army. One important reason for the French concessions was the simultaneous development of the war in Algeria (see War Nr. 35).

Sources

AdG 3782A, 4157A, 4488A, 4575C, 5210B, 5216B, 5260C, 5316A, 5327B, 5385A, 5399A, 5661D, 5755A, 5855E, 6060C; Carmoy 1967; Grimal 1965; Weisenfeld 1980 (NN)

---

35 Algeria
1.11.1954 - 19.3.1962; D-2
War ended by: agreement without mediation

WARRING PARTIES:

Aggressor (Side A): Front de Libération Nationale (FLN) and their Armée de Libération Nationale (ALN), until 1955 as Comité Revolutionaire pour l'Unité et l'Action (CRVA) / Mouvement National Algérien (MNA) (11/1954 - 5/1956)
Aggressed (Side B): France
Side B: Organisation de l'Armée Secrète (OAS) (dates unknown)

OBJECT AND GOALS OF THE CONFLICT

On November 1, 1954, the FLN began an armed struggle against the French colonists in a war of independence. The independence movement was split internally between the FLN and the MNA, who enjoyed the support of Algerian workers in France. In France itself these two organizations engaged in bloody confrontations.[1] At the same time the war in Algeria began to escalate, and France refused to enter into talks with the nationalists. Instead the French reacted with extensive military interventions and counter insurgency measures. By 1961 over 400,000 French soldiers were stationed in Algeria.

During the war, a faction of French settlers in Algeria formed the OAS, which in cooperation with French military units sought to force Algeria to remain part of France through acts of terror. The OAS even continued their attacks a year after a cease-fire was signed betwen the FLN and France in March, 1962. In 1961 there had been a coup attempt by parts of the French army and settlers, but it was defeated by troops loyal to de Gaulle.

OUTCOME OF THE WAR

The war bore partial responsibility for the end of the Fourth French Republic and de Gaulle's ascension to power. Two years of talks following the war finally resulted in ending France's 130- year colonization of Algeria. Algeria gained independence in 1962, but with the heavy baggage of internal and economic strife.

The numbers of casualties range from 250,000 to two million. Over 20,000 villages were destroyed, and whole forests and large portions of agricultural land were devastated. Ninety percent of the over one million French settlers fled Algeria, and more than 700,000 refugees had to be repatriated. Unemployment rose immediatey after the war and reached 70 percent. France's financial outlay for the war was more than 20 million dollars.

Note

1. Until 1958 there was a bitter mini-war between the MNA and the FLN. The MNA ("Messalists") lost its original basis among workers and the rural population to the FLN, split away from the independence movement and eventually faded into obscurity.

Sources

Andrews 1962; Chikh 1981; Droz/Lever 1984; Elsenhans 1974; Entelis 1986; La Guerre 1990; Mahfoud 1988; Meinicke-Kleint 1965; Ministre de l'Information 1984; Nelson 1986; Pieck 1987; Weisenfeld 1980 (KS)

---

40 Sudan (first Sudanese Civil War)
18.8.1955 - 2/1972; B-2
War ended by: third-party mediation (diverse)

Warring parties:
Aggressor (Side A): Anya-Nya
Aggressed (Side B): Sudan

Object and goals of the conflict

In pre-colonial times black southern Sudan served the Arabic/Muslim north as a reservoir for their slave trade. Sudan became a state under British colonial rule, which also had the effect of intensifying the separate development in the north and the south. During the decolonization process the central administration was occupied almost entirely by Arab/Muslim elites from the North. In 1955, south Sudanese military units rose up in protest against the south's systematic disadvantage in trade, bureaucracy, and the military.

Over seventeen years of conflict the southern Sudanese elite and political groups developed a form of Southern Sudanese national consciousness, which transformed into demands for autonomy and even secession. The north continued to defend its economic and political privileges with the aid of the national armed forces.

Outcome of the war

After numerous negotiations met with failure the government was finally able to end the conflict, with the help of external mediation, through the Addis Ababa Agreement of 1972. Southern Sudan received limited regional autonomy, the right to its own elections and political institutions, as well as assistance for the reintegration of approximately one million refugees. More than 10,000 Anya-Nya fighters and the tribal authorities were integrated into the national armed forces, the police and national administration. Nonetheless, essential aspects of the North-South problem in Sudan remained unsolved by the treaty, and war erupted again in the 1980s (see War Nr. 158).

500,000 to 700,000 people lost their lives in the war. Between 1964 and 1972 approximately 695,000 Sudanese fled to the neighboring states of Ethiopia, Zaire, Central African Republic, and Uganda. Within Sudan about 850,000 people were displaced.

Sources

Badal 1976; Beshir 1975; Beshir 1970; Hermann 1976; Keesing's 5319B, 8790B, 1122A, 14729C, 14854A, 16454A, 16973A, 17015A; Malwal 1981; Peter 1989; Sellin 1980; Sellin/Tetzlaff 1982 (SiS/BM)

---

47 Spanish-Morocco (Ifni)
10/1957 - 4/1958; D-2
War ended by: third-party mediation (former colonial power)

Warring parties:

Aggressor (Side A): Ejército de Liberación (EL)
Aggressed (Side B): Spain

Object and goals of the conflict

After France and Spain had granted Morocco independence in 1956 Spain continued to maintain several protectorates on Moroccan territory: the enclaves of Ceuta and Melilla, which belong still today to Spain, and West Sahara, which Spain transferred in 1974 to Morocco and Mauritania (see War Nr. 123).

Sidi Ifni found itself in the border area of northern West Sahara and independent Morocco. At first the Spanish tolerated the independence movement of the indigenous inhabitants. As the EL, however, began to help the Sidi Ifni develop strategies for armed insurrection, Spain attacked the guerrilla camps in October of 1957 with approximately 8,000 Spanish soldiers.

Outcome of the war

Spain was unable to secure a decisive military victory, especially after an offer of French support never came to pass. In April 1958 Spain transferred the northern zone of Ifni to Morocco, although legally Morocco only gained full control over the territory in 1969. The Spanish side sustained 190 casualties, 500 wounded and 80 missing, and the Moroccan side suffered 1,000 dead.

Sources

Aguirre 1980; Prados/Rivas 1991 (UB)

---

57 Tunisia (Bizerta Crisis)
19.7.1961 - 22.7.1961; D-2
War ended by: third-party mediation (U.N.)

Warring parties:

Aggressor (Side A): Tunisia
Aggressed (Side B): France

Object and goals of the conflict

Since Tunisia achieved independence from France in 1956 it had demanded the French withdrawal from their military post on the peninsula of Bizerta, which had served the French as an important base in the Algerian War (see War Nr. 35). On July 19, 1961, after letting numerous ultimatums to the French pass unchallenged, Tunisian President Bourguiba allowed Tunisian troops to block the garrison and symbolically raise the Tunisian flag at the French border marker 233. France reinforced its troops in the base and threatened military action if the blockade were not lifted within 24 hours. Before the deadline was up Tunisian troops fired at French planes and helicopters, and French troops returned fire. Two days later the French air force bombed the city of Bizerta.

Outcome of the war

France offered to withdraw, and on July 20, 1962, the Tunisian government accepted the offer. Diplomatic relations were resumed between the two states, and on July 22 the U.N. Security Council resolved that both sides should stop all fighting. Both sides formally accepted the resolution, but French troops continued to occupy the city of Bizerta until October 15, 1962.

More than 670 Tunisians were killed in the war, and 30 French troops. In all, 1,500 people were injured.

Sources

AdG 9211B, 9249B, 9267B, 9278A, 9303D, 9317H/3; Brockway 1973 (UB/TF)

---

70 Morocco / Algeria ("Tindouf War")
10/1963 - 2/1964[1]; C-2
War ended by: third-party mediation (regional state[s])

Warring parties:
Aggressor (Side A): Morocco
Aggressed (Side B): Algeria

Object and goals of the conflict

In the western part of the Algerian Sahara there exists no firm borderline dividing the desert between Algeria and its neighbor, Morocco. In hopes of creating a "Greater Morocco," a goal shared equally by the Moroccan government and opposition Istiqlal, King Hassan II sent troops into the Tindouf region. The Tindouf was considered strategically important because it is Algeria's sole land connection to West Sahara. Additional motivation for the conflict lay in the region's valuable deposits of minerals and phosphates.

Outcome of the war

The Moroccan troops were forced back by Algeria. The war ended officially through negotiations mediated by Ethiopia. The status quo ante was restored, but it was not until 1972 that Morocco signed a border treaty with Algeria in which it relinquished its claim to Tindouf. The border treaty was finally ratified years later in May 1989, and even then its publication in the Moroccan "Official Bulletin" occurred first on June 22, 1992.

Approximately 1,000 soldiers died in the fighting.

Note

1. The length of the war is reported differently by differnt sources: ORAE gives it as October 1963 through February 1964, while Dingemann places it in 1962, and Small and Singer present the war as taking place from 1963 to 1964.

Sources

AdG 10889B, 10947B; Dingemann 1983; Hassan II 1993; Maalmi

1993; Spencer 1980; Trout 1969(UB)

---

123 West Sahara
18.11.1975[1] - 6.9.1991; D-2 / B[2]-1
War ended by: third-party mediation (U.N.)

Warring parties:
Aggressor (Side A): Frente Popular de Liberación de Seguia el Hamra y Rio de Oro (POLISARIO)
Aggressed (Side B): Spain (18.11.1975 - 28.2.1976)
Side B: Morocco / Mauritania (10.12.1975 - 12.7.1978)
Intervention favoring side A: Algeria (11/1975 - 1987)

Object and goals of the conflict

King Hassan II of Morocco had designs on the West Sahara even when it was a Spanish colony, hoping to include it in a future "Greater Morocco" and supporting the POLISARIO (1973) in hopes of undermining Spanish rule. In 1974 he threatened violent annexation of the colony, and afterwards appealed to the International Court of Justice which had ruled in October 1975 that neither Morocco nor Mauritania had sovereign rights to West Sahara, and that the people of the region should decide the future of their land by referendum. On November 6, 1975, Hassan II organized a march (Green March) of over 350,000 civilians to West Sahara. He then used the domestic political crisis in Spain following Franco's death to finagle a Spanish-Moroccan treaty against the wishes of Algeria and the U.N. The agreement called for Spain to withdraw all its troops by February 1976 and for West Saharan administration to fall to Morocco and Mauritania, in other words, a de facto split of the territory. Immediately following the "Green March" military confrontations broke out between POLISARIO and Moroccan troops.

On the day after the Spanish withdrawal, February 27, 1976, POLISARIO formed an exile government in Algeria and declared the existence of the "Democratic Arab Republic of Sahara." Meanwhile the south of West Sahara was occupied by Mauritanian troops, and Moroccan troops occupied the northern half. For a brief period Algerian troops also took part in the fighting, partially because of ideological reasons in support of the right to self-determination, but more importantly to prevent the conservative Morrocan government from further increasing its power. POLISARIO also received Libyan support until Libya and Morocco drew closer in 1984.

OUTCOME OF THE WAR

POLISARIO concentrated its attacks at first on the weaker Mauritanian army. Military defeat and economic crisis lead to the overthrow of the Mauritanian president and the new military government ended fighting with POLISARIO. On August 5, 1979 Mauritania signed a peace treaty with POLISARIO. The result, however, was that Morocco then occupied the former Mauritanian part of West Sahara.

At the 1981 OAU summit in Nairobi, Morocco declared themselves willing to accept a referendum about the future of West Sahara. It was not until 1988, however, that the U.N. peace plan was accepted by both sides. Since then there have been only sporadic instances of fighting. A complete cease-fire was signed on September 6, 1991 with U.N. mediation. The intensity of the armed conflict has sunk since then under the threshold of war.

The actual referendum to decide whether West Sahara will become independent or remain an autonomous province within Morocco, however, still has not taken place, and the conflict remains therewith unsolved. The main point of contention lies in determining who has the right to vote. POLISARIO insists on numbers from a Spanish census of 1974, while Morocco wants to include those who have moved to the territory since them.

According to POLISARIO, 24,000 people died in the war, other sources report more than 10,000 dead.

Notes

1. The first armed attack of POLISARIO was on May 20, 1973 against a Spanish police station in El Khanga. The threshold to the level of war, however, was broached first in November of 1975.
2. At first (until 1976) a decolonization war against Spain, but since 1975/76 it became an internal war against Morocco and Mauritania. Since the recognition of the Democratic Arabic Republic of Sahara (DARS) by 72 states and DARS acceptance in the OAU (1982) one can also consider this an interstate war (C), at least since 1970.
3. West Sahara became a Spanish colony in 1884 through a decision of the Berlin Conference.

Sources

Barbier 1982; Clausen 1992; Hodges 1983; Hodges 1985; Marquina 1988; Osterud 1989; Seddon 1988 (UB)

---

158 Sudan (South Sudan Conflict, Second Southern Sudan Civil War)
since 9/1983; BA-2[1]

Warring parties:

Aggressor (Side A): Sudanese People's Liberation Movement / Army (SPLM/SPLA; since August 1991 SPLA-Garang faction)[2]
Aggressed (Side B): Sudan
alternating: Sudanese People's Liberation Army (SPLA-Macar faction)[3] (since 8/1991)

Object and goals of the conflict

Since independence in 1956 until 1972 Sudan was in a state of war between the Arabic- and Muslim-dominated central government and the more heterogeneous ethnic and religious groups of the south (see War Nr. 40). The agreement which ended the war had guaranteed limited autonomy for the south, yet this had never been implemented.

Many factors contributed to ignition of the latent conflict between North and South, including: the illegal dismissal of the regional parliament in 1980, attempts to divide the south into three separate provinces (divide and conquer), a canal project disadvantageous to the south, the stationing of southern soldiers in the north contrary to agreement, and the Islamic renaissance of the Numeiri regime which imposed Islamic law (Sharia) on the non-Muslim populations. The demands by the SPLM/SPLA for a new constitutional convention and end to Islamic law were met by successive central governments with violence.

Outcome of the war

The war lead to the collapse of the Numeiri regime in early 1985 and generated a deep domestic instability which doomed the democratic experiment which followed Numeiri's fall. On June 30, 1989 the military under Islamic leadership (al-Bashir) staged a successful coup, and the SPLA attained significant territorial and political benefits. After many negotiation rounds the views of the various banned political organizations converged (with the exception of the Umma Party), and the conflict gained a strong anti-regime component. General al-Bashir's internationally isolated Junta was not prepared to deal with a united southern Sudan and contemplated forcing them to secede. By May 1991, however, the regime declared a general political amnesty, even for members of the armed opposition.

At the end of August 1991, however, the SPLA split into two factions wracked by internecine conflict, and the Sudanese army overcame the rebels. The end of the wars in Ethiopia (see Wars Nr. 63, 122 and 131) worked to the SPLA's disadvantage. The SPLA has slowly collapsed since then into feuding factions while the Sudanese army has attained significant military successes. With the disintegration of the SPLA the civil structures which once provided basic support for the population have deteriorated, and the U.N. ordered the Operation Lifeline Sudan to aid the region.

So far the war has taken over half a million lives and generated massive refugee movement within Sudan (about 3.5 million people). The economy and infrastructure of southern Sudan has been severely incapacitated. In 1992 a famine set in which threatened the lives of over a million people.

Notes

1. Possible participation in the war by Libian advisors in support of the central government remains unproved.
2. The main faction was under the leadership of Colonel John Garang, who sought a unitary, secular and pluralist state with a Muslim north and heterogeneous south. After the SPLA split in August 1991, mostly along tribal lines, the Garang faction consisted mostly of members of the Dinka tribe.
3. After a failed coup attempt in August 1991 against John Garang, a new faction arose around the coup leader Riak Macar. The ensuing split of the SPLA fell along tribal lines, and the new SPLA faction under Macar was mostly supported by the Nuer tribe, and its influence was limited to their settlement area of the upper Nile. The Macar faction supported the Sudanese army against the Garang faction.

Sources

Aquarone 1987; Baya 1988; Beshir 1975; Bleuchot 1990; Eikenberg 1988b; Garang 1987; Gurdon 1984; Jada 1988; Khalid 1987; Lesch 1987; Mattes 1990; Mawut 1986; Niblock 1987; Peterson 1984; Tetzlaff

1984; Tetzlaff 1988; Wirz 1982; Jane's Defence Weekly 13.2.1993; MEI 5.2.1993 (CL/SiS)

---

206 Algeria
since 1/1992; A-2

Warring parties:
Side A: Algeria
Side B: Mouvement islamique armé (MIA) / Groupe islamique armé (GIA) (since 1993)

Object and goals of the conflict

The first democratic elections in Algeria, on December 26, 1991, were won not by a democratic party but by the Islamic front (Front islamique du salut [FIS]). The Algerian military government under General Khaled Nezzar, which was overseeing the transition to democracy, sought to prevent the fundamentalist FIS from legally attaining power. They forced the head of government, Chadli Bendjedid, from power in January 1992 when he appeared willing to cooperate with the FIS, and put a military council (HCE) at the helm with the seventy-three-year- old Boudiaf as president. Boudiaf soon showed conciliatory gestures with the goal of new elections, and on June 29 1992 he was murdered by a member of his security detail.

Outcome of the war

The FIS was forbidden, and their leaders Anführer Abassi Madani and Ali Benhadj were sentenced to twelve years in jail. Martial law was declared and supporters of the FIS were arrested en masse. In response the MIA, under Abdelhader Chebouti, began a guerrilla war against the Algerian military. While support for the Islamic groups seemed initially strong, a long and brutal campaign of heinous civilian massacres and assassinations against representatives of secular Algeria (journalists, women without head-covering, intellectuals, etc.) has eroded their popularity and the Islamic movement itself has split. There are rumors of military complicity in at least some of the massacres. The violence against civilians continues to escalate, and calls for international mediation have so far been met with stiff rejection from the Algerian government as interference in their internal affairs.

At least 500 members of the MIA have been killed, and there have been thousands killed in the war so far.

Sources

Barrin 1993; Mounir 1993; Ramonet 1993 (UB)

West Africa

77 Guinea-Bissau (War of Independence)
1/1963 - 25.4.1974; D-2
War ended by: termination of hostilities

Warring parties:

Aggressor (Side A): Partido Africano da Independéncia da Guiné e Cabo Verde (PAIGC)[1] / Frente da tutta para la Independencia Nacional da Guiné (FLING) (1963)
Aggressed (Side B): Portugal[2]

Object and goals of the conflict

The war sought to end Portugal's century-old colonial rule over Guinea-Bissau and the Cape Verde Islands[3] as well as to reintroduce a historical identity marginalized under colonial repression.

Outcome of the war

The war ended in the national liberation of Guinea-Bissaus and the Cape Verde Islands. This successful decolonization, however, was less a result of PAIGC's military prowness than a consequence of the end of the fascistic regime in Portugal in April 1974 . The war nonetheless cost about 15,000 lives, including 5,000 civilians.

Notes

1. The FARP (Armed Revolutionary People's Force) was the guerrilla arm of the PAIGC, and was founded in 1964 after the PAIGC's first military successes.
2. The Fula and Mandinka tribes fought on Portugal's side. The Portugese secret police "Pide" had already undertaken actions against the PAIGC before the beginning of the war.
3. Open fighting came to the Cape Verde Islands only after the revolution in Portugal, but their national elite had been represented by the PAIGC. The Cape Verdean Amilcar Cabral had been one of the five founders of the PAIGC.

Sources

Braganca/Wallerstein 1982; Cabral 1982; Davidson 1981; Komitee Southern Africa 1974; Gerwerk 1974; Nohlen/Witzel 1978; Rudebeck 1974; Schiefer 1986; Sigrist 1980; Ziegler 1980 (HJ)

---

184 Nigeria (Tiv Uprising, 1964)
8/1964 - 12/1964; AB-2
War ended by: termination of hostilities

Warring parties:
Aggressor (Side A): armed Tiv Militias
Aggressed (Side B): Nigeria

Object and goals of the conflict

The largest non-Muslim people in northern Nigeria are the Tiv, primarily Christian and sympathetic to the United Middle Belt Congress (UMBC). They rose up against the Muslim Haussa-Fulani hegemony and their dominant organ in north Nigeria, the Northern People's Congress (NPC). The Tiv's traditional aversion to central authority and foreign domination found expression in the uprising too. Additionally they were protesting a rise in taxes. Their goals were self-administration and the assertion of their own identity.

Already in August and September of 1960 there was a rash of attacks against NPC functionaries and sympathizers, however the target was more property than persons and there were hardly any casualties. The catalyst for the unrest of 1964, however, was the murder of a clan chief on February 12, 1964. The aftermath saw confrontations with the police and attacks by armed gangs. The anniversary of the UMBC saw the transformation of the conflict into an open rebellion against members of the NPC.[1]

Outcome of the war

The Nigerian Army brought the conflict under control in December 1964. The troops were sent in November, and remained until June 1965. The Tiv was promised more self-administration. Their rebellion, however, added to the increasingly flammable domestic climate, which soon erupted in two military coups during 1966 and, one year later, into the Biafra war (see War Nr. 97).

The uprising caused up to 4,000 deaths and public order in the region collapsed almost entirely.

Note

1. There was no central organization of the rebellion in a narrow sense, however the traditional authority Tarka Tiv provided some direction.

Sources

Anifowose 1982; Dent 1966; Miners 1971; Wirz 1982 (PK)

---

97 Nigeria (Biafra War)
6.7.1967 - 15.1.1970; B-2
War ended by: military victory for side A

Warring parties:
Aggressor (Side A): Nigeria
Aggressed (Side B): Biafra[1]

Object and goals of the conflict

The Biafra War resulted from an already existing power struggle between the leadership of the Muslim Haussa-Fulani, who dominated the central government since independence (in 1960), and the leadership of the christian Igbo from the southeast region of Biafra, who dominated the military and the state bureaucracy. Biafra's oil supplies strengthened the Igbo claim to national dominance, especially since oil was just becoming the primary pillar of the Nigerian economy. Both the Igbo and Haussa-Fulani leadership experienced bloody coups in 1966 (January and July, respectively), and tensions rose between the Igbo region and the central government. War came when the central government divided Nigeria into twelve federal states at the end of May, 1967, in order to remove the Igbo from the administration of the oil-rich regions. The Igbo sought independence for Biafra in an attempt to compensate for their loss of power in Nigeria and to secure their exclusive access to the oil reserves.

Outcome of the war

The war ended with the military defeat of the secessionist Igbo and the integration of Biafra into the Nigerian federal state. The Igbo leadership lost its leading role in the state, bureaucracy, and military.

Up to two million people were killed in the war and related famine, including many children. Material destruction in Biafra was considerable.

Note

1. Only toward the end of the war did the guerrilla troops take the name "Biafran Organization of Freedom Fighters" (BOFF).

Sources

AdG 40/1970; Brown 1985a; Dingemann 1983a; Wirz 1982a; Nafziger 1983a; Phillips 1984a (PM/PK)

---

185 Nigeria (Maitatsine Uprising in Kano/Northern Nigeria)
18.12.1980 - 29.12.1980[1]; B-2
War ended by: military victory for side B

Warring parties:
Aggressor (Side A): Maitatsine Movement[2]
Aggressed (Side B): Nigeria

Object and goals of the conflict

The Maitatsine Movement was founded by the Cameroonian preacher Muhammadu Marwa and functioned as an umbrella movement for the poor and disposessed in Muslim north Nigeria. The movement maintained an armed guard. The first confrontations with state forces occurred already in 1972 in Kano, but in 1979/80 the rebellious activities intensified. Marwa himself was killed during an uprising known as the Kano Riots in December 1980.

Outcome of the war

The Maitatsine Movement continued to exist after Marwa's death and reorganized itself following military defeats and confronted the Nigerian forces again in Maiduguri and Kaduna in October 1982, Yola in March 1984 and Gombe in April 1985. After the defeat of the revolt in Gombe the movement seemed finished. The Maitatsine Movement contributed to the deteriorating domestic political climate and the escalation of organized violence in Nigeria.

The Kano Riots of December 1980 caused 4,177 deaths, according to official sources. Altogether the confrontations cost over 8,600 deaths

(3,350 alone in 1982). The affected areas also sustained substantial material damages.

Notes

1. The discontinuous nature of the confrontations are due to the Maitatsine Movement's need to reorganize after military defeats. The Kano Riots of December 1980 prompted the Nigerian state to react with massive military deployment, including the Air Force. Only this particular confrontation is treated here as war, although the other events are very close to the threshold of war according to our definition.
2 The Maitatsine Movement also operated under the names Yan Tatsine and Kalo Kato.

Sources

Bienen 1986; Hiskett 1987; Isichei 1987; Lubeck 1988 (PK)

---

180 Mali / Burkina Faso (Border War)
21.12.1985[1] - 31.12.1985; C-2
War ended by: third-party mediation (diverse)

Warring parties:

Aggressor (Side A): Mali
Aggressed (Side B): Burkina Faso

Object and goals of the conflict

The object of the war was the Agacher strip of territory in the northeast of Burkina Faso, which had been disputed by both sides for over twenty years. Both agricultural states hoped to ameliorate their difficult economic situations by exploiting the manganese, natural gas, titanium and uranium deposits which both suspected lay hidden there. Mali, driven by the tense interstate and a difficult domestic political situation, decided to shift the old French colonial border to their advantage by military force. Burkina Faso had little military means by which to respond.

Outcome of the war

A cease-fire was signed in only six days due to immediate political intervention by Ivory Coast, Nigeria and Libya. In December 1986 both sides accepted the arbitration of the International Court of Justice in the

Hague, and the territory was divided between both states. Sixty to 100 persons died in the war, and there were a similar number of wounded.

Note

1. There had already been sporadic gun battles on November 25 and December 14, 16 and 18, 1975, but the conflict did not have the character of a war, although there had been casualties. At that time the conflict was kept under control by the mediation of several African states.

Sources

Brownlie 1979; Church/Synge/Hodgkinson 1988; Englebert 1986; AdG 19216; Financial Times; Fraternité Matin; Le Monde; Le Monde Diplomatique (SiS/BM)

---

168 Liberia, Sierra Leone
since 24.12.1989; A-1

Warring parties:

Aggressor (Side A): National Patriotic Front of Liberia (NPFL)[1] / Independent National Patriotic Front of Liberia (INPFL)[2] (since 1990) / United Liberation Movement for Democracy (ULIMO)[3] (since 9/1991)
Aggressed (Side B): Liberia / Sierra Leone (since 3/1991)
Intervention to end the war: Gambia (since 8/1990) / Mali (since 3/1991) / Senegal (10/1991 - 1/1993) / Guinea (since 8/1990) / Sierra Leone (8/1990 - 3/1991) / Ghana (since 8/1990) / Nigeria (since 8/1990)

Object and goals of the conflict

The background to the conflict consists of a litany of economic and political problems in Liberia: the repression of any opposition and independent press, corruption, international isolation, declining production and falling export prices, declining imports due to balance of payment problems and indebtedness, and finally the suspension of World Bank and IMF support.

The economic decline and state terror under the Doe regime in the 1980s presaged the NPFL's invasion of December 1989. In February 1990 the INPFL split, but internal differences did not prevent the capture and execution of Doe in September 1990. August of 1990 also saw the intervention of the Economic Community of Western African States (ECOWAS), a group consisting mostly of anglophone states, above all Nigeria. The official reason for intervention was the growing burden of war refugees, which had reached over 500,000, and dangers posed to their respective citizens living in Liberia, especially by widespread massacres of civilians. The ECOWAS Monitoring Group (ECOMOG) and the INPFL entered a fragile alliance and the INPFL temporarily participated in the interim government set up by ECOMOG. Yet ECOWAS was unable to achieve a peaceful end to the conflict. In the autumn of 1991 former government troops calling themselves ULIMO attacked from NPFL-controlled territory in neighboring Guinea and Sierra Leone. The NPFL had already carried the war into Sierra Leone in the spring of 1991, partially because Sierra Leone participated in ECOMOG's armed forces. In Sierra Leone the NPFL cooperated with the indigenous rebel group RUF (the Revolutionary United Front). This cooperation earned the attention of the Sierra Leonian Army, which together with troops from Nigeria and Guinea as well as the ULIMO fought the NPFL and RUF starting in April 1991. The spread of hostilities to Sierra Leone created favorable conditions for the successful coup in that country in April 1992.

Outcome of the war

Numerous cease-fire agreements came and went, leaving the land divided: up to mid-1992 the NPFL controlled the largest portion of the country, while the INPFL and the ECOMOC troops controlled the area around the capital, Monrovia. The 1991 Yamoussoukro peace plan, worked out by all West African states, was undermined by the massive offensives of ULIMO in late 1991. NPFL seemed to have the military advantage and lay poised to take the capital, when in October 1992 ECOMOG troops successfully repelled them.

Tens of thousands have died in the war, most of them civilians. Over one million war refugees have been created in Liberia and Sierra Leone.

Notes

1. The NPFL financed their activities by exporting tropical hardwood and other raw materials. The buyers included European companies.
2. The INPFL split in February 1990 from the NPFL. The only known reason for this split is the personal rivalry between Prince Johnson (INPFL) and Charles Taylor (NPFL).

3. The ULIMO consists for the most part of former government troops, Muslim Mandingos, and Christian Krahn.

Sources

Daniels 1992; Kappel/Korte 1990; Kappel/Korte 1991; Ruiz/US Committee for Refugees 1992; Schlichte 1990; West Africa 6.-12.4.1992 (KS/PK/CS)

---

191 Senegal (Basse Casamance Conflict)
since 4/1990; B-2[1]

Warring parties:

Aggressor (Side A): Mouvement des Forces Démocratiques de la Casamance (MFDC)[2]
Aggressed (Side B): Senegal

Object and goals of the conflict

The main sources of the conflict lay in two areas: the exploitation of natural resources (mostly timber and fish) by non-local Senegalese in the Casamance region, and the central Senegalese government's general neglect of the local Diola people. The political demands of the MFDC centered on the region's receiving a greater share of the national budget, support for agriculture, industry and infrastructure, as well as political autonomy. The central government reacted by increasing repression of the movement, which only strengthened separatist sentiments. There had already been bloody protests in the early 1980s, and after negotiations failed in 1989 the MFDC proclaimd the Casamance's independence. Since 1985 the MFDC had been assembling a military wing with modern weaponry, and starting in April 1990, their sporadic attacks developed into a guerrilla war. [3]

Outcome of the war

The war was temporarily ended on May 31, 1991, through a cease-fire agreement mediated by Guinea-Bissau. The cease-fire did achieve the withdrawal of all military units, the reduction of Senegalese military presence in the region, and an amnesty for MFDC supporters. Yet the situation remains tense, and since the end of 1991 new military confrontations have been reported. The area in which fighting occurs has expanded to include the entire Guinea-Bissau border region.

The MFDC seems split internally between a radical basis prepared for violence and a leadership interested in negotiation. So far at least 800 people have lost their lives in the fighting.

Notes

1. In mid-May, 1990, the armies of Senegal and Guinea-Bissau clashed on the border in the context of the conflict in Senegal.
2. The MFDC developed in the 1970s as the representatives of the Diola people's interests. The Diola people live in the Basse Casamance, and constitute 5 percent of Senegal's population.
3. The tensions at this time in Senegal were not limited to the Casamance region. There was almost a war with Mauritania, stemming from border skirmishes and reciprocal cross-border pogroms. The sources of conflict were nomadic border crossings and the toleration of Mauritanian resistance groups in Senegal. Political talks avoided an escalation to war, but the situation remains tense.

Sources

Drame 1992; Wegemund 1991 (PK/ML)

---

197 Tuareg (Mali, Niger)
since 7.5.1990[1]; B-2

Warring parties:

Aggressor (Side A): Mouvement Populaire de l'Azawad (MPA) (since 1990) / Front Islamique Arabe de l'Azawad (FIAA) (since 1990) / Front Populaire pour la Libération de l'Azawad (FPLA) (since 1990) / Alliance Révolutionnaire pour la Libération de l'Azawad (ARLA) (since 1991)
Aggressed (Side B): Mali (since 6/1990) / Niger

Object and goals of the conflict

When the French colonialists created borders between their possessions in West and North Africa the lives of the Taureg nomads were substantially altered: 600,000 to 700,000 live in what became Niger, 300,000 to 400,000 in Mali, and 20,000 to 50,000 each in Algeria, Libya and Burkina Faso (formerly Upper Volta). Almost everywhere the Taureg faced discrimination, even when they mixed with the Arabic/Muslim populations. They were hated by the black Africans, who they had formerly enslaved. State pressure against nomadic existence and droughts accelerated the dissolution of traditional Taureg communities. Political dissatisfaction among the Tuareg grew when demands for autonomy among the Ifora-Taureg in Mali remained unfulfilled.[2] While the effect of growing dissatisfaction was temporarily muted by the oil booms of the 1970s and 1980s, during which many Tuareg found employment in Algeria and Libya, the end of the boom meant a return to an uncertain future. The governments in Mali and Niger withheld promised aid for reintegration of returning workers, and this was a sufficient catalyst to bring the long-standing conflict to a boiling point. On May 7, 1990, the Tuareg began a revolt in Niger with an attack on the town of Tchin-Tabaradène. By the end of June the attacks had spread to Mali and the rebel factions demanded autonomous rights for the northern territory of Mali and Niger.

Outcome of the war

The uprising accelerated the collapse of the Traoré regime in Mali, a moderate socialist government which had come to power in 1968, and whose fate was sealed by a coup on March 26, 1991. Despite cease-fire agreements in September 1990 and January 1991 the military conflict did not end. Finally on December 18, 1991 another cease-fire arranged in Mopti (Mali) led to the eventual signing of a peace treaty on April 11, 1992. Under the treaty refugees were repatriated from neighboring states, local self-administration for the Taureg's was guaranteed, and an exchange of prisoners was arranged. Despite these achievements, however, in May 1992 fighting erupted again.

The Niger government met in May 1992 to begin negotiations with the Taureg, which have yet to be continued. Skirmishes continue between the rebels and the army.

The war cost at least 1,000, and possibly more than 2,000 casualties. Many Taureg civilians were murdered by the Army and some black African civilians, and 50,000 Taureg fled from Mali to Mauritania.

Notes

1. The war began in Niger, briefly encompassed Niger and Mali, then shifted to Mali before escalating again in Niger.
2. There had already been an Ifora-Taureg revolt against the Malian government in 1963/64. The reasons for this revolt lay in the economic and political disadvantaging of the Tuareg by the central government,, especially a 300 percent increase in the cattle tax which the government tried to collect by force. From the second half of 1963 to the middle of July 1964 there were sporadic but bloody confrontations.

SOURCES

Du Bois 1964; Klute 1990; Der Spiegel 45/1990:216-218; Jeune Afrique 13.11.91:19-21, 24.1.92:41-43; NZZ 29.12.91; AdG 11003 (PK/MB)

Central Africa

39 Cameroon (Decolonization)
5/1955 - 1963[1]; DA-2[2]
War ended by: military victory for side B

WARRING PARTIES:

Aggressor (Side A): Union des Populations du Cameroun (UPC)
Aggressed (Side B): France (5/1955 - 1.1.1960) / Cameroon (1.1.1960 - 1963)

OBJECT AND GOALS OF THE CONFLICT

Following the socialist-revolutionary model of Guinea, union leader Ruben Um Nyobé led the liberation front UPC in a struggle for independence. France at first fought to retain their colonial rule, then sought a neo-colonial solution by transferring power to a francophone elite.

OUTCOME OF THE WAR

Although under neo-colonial conditions, the independence movement did achieve their goal on January 1, 1960. The post-colonial government remained closely tied to France. The UPC was, however, militarily defeated.

The war took 10,000 to 20,000 lives.

Notes

1. Isolated confrontations continued until about 1970.
2. After independence the Republic of Cameroon, with French military support, continued fighting against the UPC.

SOURCES

Baumann/Illy 1982; Eyongetah/Brain 1974; Illy 1976; Joseph 1977;Joseph 1986; Ki-Zerbo 1981; Kom 1971; Legum 1969; LeVine 1964; LeVine/Nye 1974; LeVine 1990; Mveng 1963; Rubin 1971; Ziemer 1978 (KS)

55 Zaire (Secession and Anti-Regime War)
11.7.1960 - 14.1.1963[1]; AB[2]-1
War ended by: military victory for side B

Warring parties:
Aggressor (Side A): Conféderation des Associations Tribales de Katanga (CONAKAT) / Mouvement National Congolais/Lumumba (MNC/L) (9/1960 - 9/1961) / Mouvement National Congolais/Kalondji (MNC/K), Kasai-Secessionists (9.8.1960 - 13.1.1963)
Aggressed (Side B): Zaire
Intervention favoring side A: Belgium (11.7.1960 - 4.9.1960)
Intervention to end the war: Ethiopia / Ghana[3] / India / Ireland / Italy / Canada / Morocco / Sweden / Sudan / Tunisia (all: 7/1960 - 13.1.1963)

Object and goals of the conflict

Belgium had pursued extremely repressive policies in its Congo colony, later to become Zaire (and later still the Democratic Republic of Congo). Decolonization was achieved only when Belgium was convinced that it could no longer be prevented. Zaire was poorly prepared for independence in June 1960, and centrifugal forces emerged from the country's economic, sociocultural and political heterogeneity which threatened to pull the country apart. The elite of the resource-rich Katanga (Shaba) and Kasai provinces sought to prevent the central government from getting their hands on their copper, cobalt and diamonds, and proposed secession. The central government under Patrice Lumumba tried in vain to restore unity, and in September 1960 he was overthrown (supposedly for alleged pro-communist sympathies) and replaced by the head of the army, Sésé Mobutu. Lumumba supporters gathered in military resistance against Mobutu and in 1961 formed an exile government in Stanleyville (Kisangani) in northeastern Zaire. For a short period of time there existed four competing governments: the central government in Leopoldville (Kinshasa), then the competition in Elisabethville (Lubumbashi/Katanga (Shaba), Bakwanga (Mbuji-Mayi)/ Kasai and Stanleyville (Kisangani)/northeast Zaire.

Outcome of the war

The introduction of U.N. troops in the beginning of 1963 ended the secession. The exile government could not prevail militarily, and in July 1961 Lumumba-supporter Gizenga became the deputy head of government in a compromise cabinet. Just months later, however, he was removed from office, arrested, and later exiled. Lumumba himself was arrested by Mobutu's troops and murdered in early 1961 under unclear circumstances. He would later be declared a "national hero" by Mobutu's regime, once it was firmly in power.

At least 100,000 people died in the war, the economy was devastated and Zaire's unity was tested to its utmost.

Notes

1. The last U.N. troops left Zaire on June 30, 1964.
2. Although the secessionists in Katanga (Shaba) and Kasai (from 1960-63) and the exile government in northeast Zaire (1960-61) developed separately, we treat them as parts of one war because of their situation and effect.
3 A contingent of Ghanese troops had arrived in the Congo before the U.N. decision to send troops. It could not be determined whether or not they ever took part in the fighting (Lefever 1965a: 26).

Sources

Abi-Saab 1978; Chomé 1966; Gendebien 1967; Gérard-Libois 1963; Hoskyns 1965; Kanza 1972; Levefer 1965; Vanderlinden 1981; Wirz 1982; Young 1965 (KS/PK)

---

72 Rwanda (Tutsi Invasion from Burundi, Uganda and Zaire)
20.12.1963 - 1966; A-2
War ended by: military victory for side B

Warring parties:
Aggressor (Side A): Inyenzi Groups of Rwandan Tutsi[1]
Aggressed (Side B): Rwanda

Object and goals of the conflict

Before achieving sovereignty as separate states in 1962, Rwanda and Burundi had been administered as one unit by Belgium under the name Rwanda-Urundi. Even in the pre-colonial period the minority Tutsis had developed separate monarchies from the mostly agricultural Hutus, who made up 80 to 90 percent of the entire population. The Tutsis demanded tribute from the Hutus, and even into the colonial period the Tutsis retained their privileges in the state, society and economy.[2] In 1959 the Hutus rebelled violently against the Tutsis. Although Belgian troops brought the situation under control, in 1961 a Hutu coup succeeded in overthrowing the Tutsi monarchy. In 1959 and 1961 the confrontations between the Hutu and Tutsi bordered on war, and tens of thousands of Tutsi fled to neighboring lands, from where they began a campaign, mostly after 1963, to restore their dominance in Rwanda (in Burundi the Tutsis remained in power).

Outcome of the war

By the beginning of 1964 the Tutsi invasion suffered severe military defeat. They continued to try militarily to restore their dominance, but failed repeatedly. The situation on the ground calmed down finally after a border treaty was signed between Rwanda and Burundi in 1966. The war had the effect of strengthening the Hutu's position in Rwanda.

Burundi was also affected by the war. In October 1965 the Tutsi monarchy was the target of an unsuccessful coup by the Tutsi military, which led to an equally unsuccessful coup this time by the Hutu military. Unrest and massacres following the coups resulted in the murder of almost the entire Hutu elite in Burundi. Finally, in 1966 the Tutsi military was successful in toppling the monarchy, and for years to come the Hutu played no role at all at the level of political or military leadership.

In Rwanda the tensions between Hutu and Tutsi escalated again in 1973 in connection with the war in Burundi (see War Nr. 113). The destabilized domestic situation led to a coup in July, 1973, and the consequent installment of a military government. The exiled Tutsis who were still living in Uganda attempted yet another invasion in October 1990 (see War Nr. 189).

The hostilities following 1963 and the civil strife in 1959 and 1961 caused around 105,000 deaths, of which only 3,000 casualties were from the military. The country's economic and political development was sorely retarded by the conflict.

Notes

1. The Inyenzi troops were organized from Rwandan Tutsi who fled in 1959 and 1961 from Rwanda to Burundi, Uganda and Zaire.
2. Hutus and Tutsis were divided primarily about access to (and exclusion from) economic opportunities, social privileges and political influence. Linguistically and culturally there are few differences between the two groups, though the Hutus are descendents of the physiologically shorter Bantu peoples, while the Tutsi are part of the noticably taller Nilotic group.

Centuries of living together and intermarriage has created familial networks and mixtures.

SOURCES

Baumhögger 1984; Lemarchand 1970; Weinstein 1977; Weinstein/Schrire 1976; AdG 8072, 9438, 11007, 12666; NZZ 3.12.63, 18.1.64, 25./26.1.64. (UP/PK)

---

78 Zaire (Anti-Regime Revolt)
1/1964 - 24.9.1966; A-1
War ended by: military victory for side B

WARRING PARTIES:

Aggressor (Side A): Mulelists (12/1963 - 1966) / Conseil National de Libération (CNL) and his troops, the Armée Populaire de Libération (APL) (15.4.1964 - 1966) / Katang soldiers and mercenaries under Ferdinand Tshimpala (24.7.1966 - 24.9.1966)
Aggressed (Side B): Zaire
Intervention favoring side B: United States of America (15.4.1964 - 5/1966) / Belgium (dates unknown)

OBJECT AND GOALS OF THE CONFLICT

No legitimate central government had been able to establish itself in the aftermath of Zaire's decolonization process, and a power vacuum arose which precluded any straightforward consolidation of centralized power. This situation gave rise to new hostilities:[1]

a. Pierre Mulele led a Maoist guerrilla group from January 1964 to 1966 in revolt, during which they temporarily brought the majority of Kwilu territory under their control.

b. From April until mid-1966 the CNL (nationalist supporters of Lumumba) conducted a campaign during which they temporarily conquered substantial parts of eastern Zaire, including the regional capital Stanleyville (Kisangani).

c. From July to September 1966 a group of rebellious soldiers frm Katanga (Shaba) fought againts the Mobutu regime (who came to power in a 1965 coup) in an attempt to restore the former prime minister Tschombé, who came from Katanga. They occupied Stanleyville (Kisangani) while other rebels invaded from Angola into southeastern Zaire.

OUTCOME OF THE WAR

The fighting further weakened the central government, and on November 24, 1965 the military staged a coup (with alleged support from the CIA) which brought Mobutu to power. The Mobutu regime was able to stabilize the volatile situation through a mix of foreign intervention and the use of mercenaries.

The Kwilu uprising ended in defeat, and Mulele himself went into exile in 1967. Although assured of free passage back to Zaire, when he did return the next year he was arrested and executed on October 9, 1968. The CNL rebels were defeated with foreign help (USA, Belgium), and the military activities of Tshombé's supporters were halted by mercenaries. Later these same mercenaries became a problem for the Mobutu regime (see War Nr. 96).

Note

1. Although the various conflicts within Zaire developed separately, we treat them as parts of one war because of their possible, though unclear, connections, the overall situation and the common outcome.

SOURCES

Beys/Gendebien/Verhaegen 1964; CRISP 1967; Levefer 1965; Vanderlinden 1981; Verhaegen 1969; Wirz 1982; Young 1965; Young/Turner 1985; AdG 12983A (KS/PK)

---

92 Chad (Civil War)
since 6/1966; ABC[1]-2

WARRING PARTIES:

The Front de Libération Nationale (FROLINAT) fought until their disolution in 1977 against troops of the first nominally independent Chad government. Since 1979 President Malloum's Forces Armées Tchadiennes (FAT) had been recruiting in the south for troops to fight against Prime Minister Habré's Forces Armées du Nord (FAN). The FAN was at this time supported by the later Habré-opponent Oueddi, who together with Habré built the transition government Gouvernement d'Unité National Transitoire (GUNT). Since 1980 Oueddi's FAP, with Libyan support, fought against Habré's FAN (Habré became president in 1983), who was supported by French troops. In addition there were battles between Chad and Libyan troops about the Aouzou strip. Since 1989 the Mouvement pour la Salvation Nationale Tchadienne (MOSANAT), which later became the Mouvement Patriotique du Salut (MPS) under Déby's leadership, fought against Habré's troops. After Habré's fall from power in 1990 fighting erupted between his supporters and the new President Déby's troops.

Object and goals of the conflict

FROLINAT formed in June 1966 as an organization representing different exiled groups (though heavily represented by the ethnic groups of the north) with the common goal of overthrowing President Tombalbaye. This occurred on April 15, 1975. Until then the armed conflict in Chad could be seen as a pure anti-regime war, though afterwards it developed into an internal power struggle with many changing governments, which all attempted to consolidate power with the help of a third party, usually France,[2] Libya, or the USA.

Although initially the conflict ran along ethnic north-south lines, the power vacuum following Tombalbaye's death led to a general splintering of groups and parties. FROLINAT dissolved into different groups, including the FAN and the FAP. The shape of the conflict changed several times in the following years. When Habré became president in 1983 after defeating his former ally Oueddi, the domestic situation stabilized somewhat without the war actually coming to an end. The border conflict between Chad and Libya strengthened the Habré regime. Nonetheless, it proved unwilling to implement reforms, and by the end of 1990 rebellious troops under Idriss Déby overthrew Habré. A brief period of quiet ended in the second half of 1991, when supporters of Habré took up arms against the Déby regime. The new government had lost credibility with the population by persecuting the opposition.

Outcome of the war

The war effectively hollowed out Chad's sovereignty. The social, economic and political improvements sought at the beginning of the war were not achieved. Nigeria declared its support for the Déby regime and began to persecute Chad rebels who used bases in Nigeria, as well as Chad refugees who were living in the Nigerian state of Borno. In July 1992 Gabon mediated a peace treaty for the warring parties in Chad, but it is unclear whether this will have any effect on ending the conflict.

At least 21,000 people have died in this war.

Notes

1. The war began in 1966 as a spontaneous uprising and developed from an anti-regime war to a complex internal war (AB) with frequent intervention by Libya and France. Between 1987 and 1989 it was also an inter-state war (between Chad and Libya).
2. France alternately supported the stronger military group within Chad. France always intervened in support of the respective government that was threatened by Libya, but let Déby's pro-Libyan government stand.
3. Libya had often intervened in Chad's internal conflicts. The border war between the two countries over the Aouzou strip from 1987 to 1989 was officially ended by a common peace memorandum.

Sources

Brill 1987; Brill 1988; Meyer 1988; Reinhard 1981; Russel 1982; Wrobèl-Leipold 1986a (PK/MB)

---

96 Zaire (Mercenary Rebellion)
5.7.1967 - 6.11.1967; A-2
War ended by: military victory for side B

Warring parties:

Aggressor (Side A): Mercenaries and parts of the former "Gendarmerie Katangaise"
Aggressed (Side B): Zaire

Object and goals of the conflict

European mercenaries and parts of the former "Gendarmerie Katangaise," who had been integrated into the Zairean army after the war in Katanga (see Wars Nr. 55 and 78), sought the return of former Prime Minister Moise Tschombé from exile in Algeria. One mercenary group under the Belgian Jean Schramme fought from Kisangani (Stanleyville) through to Bukavu, from where he was encircled and driven by Zairean forces into Rwanda, where he was interned until 1968. A second mercenary group under the Frenchman Robert ("Bob") Denard attacked Zaire from Angola, but were quickly repulsed. The Mobutu regime sought to disarm the mercenaries in order to secure its own power position.

OUTCOME OF THE WAR

The war ended with the *status quo ante bellum*. Mobutu stabilized his grip on power. During the fighting there were massacres of the civil population, and the city of Bukavu was destroyed and depopulated. Reports put the death toll of the fighting at 1,020.

SOURCES

Young/Turner 1985; AdG 13360A, 13452 (KS)

---

113 Burundi
29.4.1972 - 6/1973; A-2[1]
War ended by: military victory for side B

WARRING PARTIES:

Aggressor (Side A): Hutu Guerrillas
Aggressed (Side B): Burundi

OBJECT AND GOALS OF THE CONFLICT

The war evolved from the tensions between Tutsis and Hutus which gave rise to the war in Rwanda (see Wars Nr. 72 and 189). Unlike in Rwanda the Tutsis had maintained their dominance over the Hutus through the monarchy. By the mid-1960s, however, their rule had become unstable, and in October 1965 the Tutsi monarchy was the target of an unsuccessful coup by the Tutsi military, which led to an equally unsuccessful coup this time by the Hutu military. Unrest and massacres following the coups resulted in the murder of almost the entire Hutu elite in Burundi. In 1966 the Tutsi military succeeded in toppling the monarchy and secured further Tutsi dominance in Burundi. The Hutus were excluded from political and administrative life, Hutu farmers were the targets of massive socioeconomic discrimination, and by April 1972 the Hutus rebelled against Tutsi dominance, though the rebels ended up being the victims. Hutu refugees fled by the thousands to neighboring lands, where they recruited guerrilla troops which moved on Burundi in March 1973 from Zaire, Rwanda and Tanzania. This invasion provoked a response from the Burundian army which not only destroyed the Hutu Guerrillas but also exterminated nearly every literate Hutu in Burundi through massacres of civilians.

Outcome of the war

The war ended with the military victory of the Burundian army and the even further consolidation of Tutsi dominance. A new constitution in June 1974 forbid distinguishing between ethnic groups in an attempt to erase the Hutu-Tutsi issue. Rwanda was destabilized by the war in Burundi, and the situation there led to a military coup in July 1973, which brought the Habyarimana regime to power.

In August 1988 units of the Burundian Army massacred up to 50,000 Hutus, allegedly in revenge for undocumented mass murder of Hutus against Tutsis. Between the war and the massacres 100,00 to 250,000 people died, including almost the entire literate Hutu population of Burundi. 80,000 to 200,000 people fled as refugees.

Note

1. Zaire temporarily sent troops to Burundi to help bring the situation under control. Tensions arose between Burundi and Tanzania, from where Hutu fighters entered Burundi.

Sources

Baumhögger 1984; Kay 1987; Seitz 1988; Weinstein 1976; Weinstein/Schrire 1976; AdG 32492 (UP/PK)

---

128 Zaire (Shaba I, "80-Day War")
8.3.1977 - 26.5.1977; A-1
War ended by: termination of hostilities[1]

Warring parties:

Aggressor (Side A): Front de Libération Nationale Congolais (FLNC)[2]
Aggressed (Side B): Zaire
Intervention favoring side B: Morocco (4/1977 - 26.5.1977)

Object and goals of the conflict

The background for the war was formed by the Mobutu regime's ruinous policies which devastated the state and the economy; extensive corruption, unscrupulous exploitation of resources, blatant human rights violations, and international credit problems undermined the regime's political legitimacy. The FLNC sought to overthow the regime by conquering Zaire's most important economic zone, the Shaba. Its fighters entered Zaire from Angola. Yet the FLNC's political orientation was far less developed than their military organization, and they had no clear program for Zaire's future should their military plans be successful.

Portions of the population in the affected areas sympathized with the FLNC, above all the Lunda people, whose opposition in the 1960s was the basis for the Katanga secessionist attempt. There were also contacts between the FLNC and other anti-regime organizations, including the Parti de la Révolution Populaire (see War Nr. 175).

OUTCOME OF THE WAR

The status quo ante bellum was restored, but there was increasing pressure on the Mobutu regime to stabilize his rule and the economy through political liberalization and free-market reforms.

The war cost up to 1,000 lives.

Notes

1. After the termination of hostilities the situation remained tense in Shaba. The FLNC was not finished, rather they withdrew in order to reorganize, and returned for the second Shaba war one year later (see War Nr. 133).
2. The FLNC was formed in 1968 by former army officer Nathaniel Mbumba, with its headquarters in Paris. It recruited mostly from former Katanga secessionists who had fled to Angola and fought there for the Portugese colonialists until Angolan independence. The FLNC was never developed a clear political orientation, and was unsuited for broad political mobilization. In matters military, however, they were better organized than the Zairean army, which only prevailed due to external military help from Morocco and Western industrialized states.

SOURCES

Adelman 1978; Callaghy 1984; Gran 1979; Mangold 1979; Nzongola-Ntalaja 1986; Schümer 1984; Young 1978; Young/Turner 1985 (PK)

---

133 Zaire (Shaba II)

12.5.1978 - 30.5.1978; A-1

War ended by: military victory for side B

WARRING PARTIES:

Aggressor (Side A): Front de Libération Nationale Congolais (FLNC)[1]

Aggressed (Side B): Zaire

Intervention favoring side B: Belgium (20.5.1978 - 30.5.1978) / France (19.5.1978 - 30.5.1978)

Object and goals of the conflict

As with the previous Shaba War (see War Nr. 128), the Mobutu regime's ruinous policy provided the background for the fighting. Once again the FLNC entered from Angola and sought to take control of Shaba. In part because of Zaire's strategic importance in the East-West conflict and because of Mobutu's pro-Western stance, Zaire's otherwise untenable army received massive military support from Western states which enabled them to fend off the better-trained FLNC.

Outcome of the war

The FLNC was militarily defeated, but the Mobutu regime had to make compromises to please the Western states who saved his derrière: the IMF took control of the central bank, the finance ministry, and customs. Mobutu was forced to sign a peace treaty with Angola with the goal of preventing the FLNC from using that country as a base for further attacks. The creditor countries strengthened Mobutu's position with further economic and military aid. From June 1978 to September 1979 an inter-African force, consisting mostly of troops from conservative francophone states, was stationed in Shaba to stabilize the situation.

Up to 1,000 people died in the war.

Note

1. See note 1 for War Nr. 128.

Sources

Adelman 1978; Callaghy 1984; Gran 1979; Mangold 1979; Nzongola-Ntalaja 1986 Schümer 1984; Young 1978; Young/Turner 1985 (PK)

---

189 Rwanda (Rebel Invasion from Uganda)
since 1.10.1990; AB-1[1]

Warring parties:

Aggressor (Side A): Front Patriotique Rwandais (FPR)
Aggressed (Side B): Rwanda
Intervention favoring side B: Zaire (5.10.1990 - 10/1990)

Object and goals of the conflict

Once again the ethnic conflict between the Hutu and Tutsi people formed the background for this conflict, especially the bloody reversal of Tutsi domination by the installation of a Hutu government in the early 1960s (see War Nr. 72). Hundreds of thousands of Tutsis fled persecution at the hands of the Hutu, and their return was blocked by the Habyarimana regime on the grounds that Rwanda was allegedly overpopulated.

The FPR recruited mostly from the approximately 250,000 Rwandan Tutsi, who either themselves or whose parents and grandparents had fled from Rwanda to Uganda. Many of the FPR fighters had experience serving in the Ugandan National Resistance Army, and had helped Museveni come to power in Kampala in 1986 (see War Nr. 152). The FPR leader Fred Rwigyema had been the deputy defense minister for Uganda (he was killed at the beginning of the war in Rwanda). After an alleged falling out between Rwigyema and Museveni, Rwigyema left his Ugandan position and formed the FPR to topple Habyarimana and to enable the return of Tutsi refugees to Rwanda. In October 1991 the FPR attacked Rwandan troops from Uganda.

The traditional Hutu-Tutsi antagonism had been displaced within Rwanda by a north-south division among the Rwandan elite. The privileged north, from where President Habyarimana hailed, had tense relations with the disadvantaged south. As the economic crisis in Rwanda intensified, largely due to a fall in global coffee prices, corruption, and the deteriorating human rights situation, there arose an environment in which an anti-regime coalition could form across ethnic lines. Hutus gained influence in the FPR leadership, and changed the character of the FPR away from an exclusively ethnic-oriented organization.

Outcome of the war

After an initial open offensive the FPR switched to guerrilla tactics after November 1990. The Habyarimana regime was forced to move toward a democratic posture to secure its position, and was unable to defeat the rebels. In July 1992 cease-fire negotiations began, but to no avail, and there is no end in sight to the conflicts in Rwanda.

Up to 1992 over 5,000 people lost their lives in the fighting.

Note

1. The Mobutu regime in Zaire sent soldiers to Rwanda to fight on the side

of the Rwandan Army, but were withdrawn in a few days after their lack of discipline resulted in attacks against the civilian population. Belgium and France also sent troops, officially to protect strategic installations in Kigali and to evacuate European citizens, but the presence of these troops fortified the position of the Rwandan army against the rebels.

SOURCES

Grohs 1988; Helbig 1990; Hildenbrand 1982; Kübler 1988; Philipp 1976; Weiss 1988; Weiss 1989; Weiss 1990 (PK/ST)

East Africa

12 Madagascar (Decolonization War)
29.3.1947 - 1948; D-2
War ended by: military victory for side B

WARRING PARTIES:

Aggressor (Side A): PANAMA[1] / JINA[1] / Mouvement de la Rénovation Malgache (MDRM)[2]
Aggressed (Side B): France[3]

OBJECT AND GOALS OF THE CONFLICT

The catalyst for war was the manipulation of the 1946 elections by the French colonial government to prevent the electoral victory of the radical MDRM. The colonial government favored the Parti des Deshérités de Madagascar (PADESM). The atmosphere had already been charged by social problems of integrating 15,000 Madagascan mercenaries returning from the Second World War, and Vichy France had drawn popular ire by dictating an unreasonably low price for Madagascan rice.

The first demand was to extend the French civil rights to all people, and not only the Madagascar elite who enjoyed them exclusively until then. The French stubbornly refused to accommodate such demands, however, and popular demands shifted toward complete sovereign independence from France. Starting in March 1947 the nationalist movement rebelled at locations across the country.

OUTCOME OF THE WAR

The rebels were under-equipped and poorly organized, and the French troops defeated them soundly. The MDRM leader was arrested and executed. Marshall Law was imposed and remained in effect until 1950.

Political activity was prohibited until 1955, and the right to vote was re-introduced only in 1956. The MDRM was utterly destroyed.

The war cost 60,000 to 100,000 lives.

Notes

1. Secret organization, anti-colonial, probably a participant in the organization of the revolt.
2. As the most popular liberation organization the MDRM was accused by the colonial government of organizing the revolt. In all likelihood, however, the revolt was organized by the secret organizations PANAMA and JINA. The MDRM rejected the revolt.
3. Senegalese mercenaries fought along with French troops.

Sources

Bertaux 1966; Cadoux 1969; Girrbach/Nohlen 1982; Kent 1962; Nelson 1973; Wachendorfer 1978; Werobel-La Rochelle 1978 (HJ)

---

29 Kenya (Mau-Mau Uprising)
10/1952 - 11/1956; D-2
War ended by: military victory for side B

Warring parties:

Aggressor (Side A): Mau-Mau Resistance Fighters
Aggressed (Side B): Great Britain

Object and goals of the conflict

The Mau-Mau Uprising pitted the economically disadvantaged and politically defenseless African population against the British colonial regime. The catalyst for conflict was the question of land distribution. The Mau-Mau Movement, a secret organization, was unwilling to any longer accept the exclusive use of the most fertile and productive land (the "White Highlands") for European settlers—1 percent of the Kenyan population, mostly white settlers, owned 25 percent of the land for cultivation. Furthermore Africans were widely excluded from commerce in the cities, which was controlled by Europeans and Indians. Unemployment, segregation, and political repression were the hallmarks of the underprivileged. Resistance arose first in the Kikuyu Central Organization (Kikuyu was one of the areas most heavily affected by white settlers), and then in the Kenya Africa Union, whose ties to the Mau-Mau Movemenent remain unclear. In the early 1950s resistance intensified, and the British colonial government responded with more repression. Leading Africans were arrested, blacks were deported en masse from the "White Highlands" and the cities. With this move the escalation of the conflict was clearly visible.

Outcome of the war

The resistance movement was held in check militarily, and there was no immediate change in the status quo. The colonial regime began to make political concessions, however, and eventually granted formal independence to Kenya in 1962. The withdrawal of colonial troops, however, had begun already in 1956 after the termination of hostilities.

The war cost up to 13,500 deaths, incuding approximately 2,000 African civilians and 32 European civilians.

Sources

Bujtenhujs 1973; Hofmeier/Schönborn 1984; Leifer 1977; Ojwando Abuor 1971; Rosberg/Nottingham 1966 (SiS/PK)

---

63 Ethiopia (Eritrea Conflict)
1962 - 5/1991; B-2
War ended by: military victory for side A

Warring parties:

Aggressor (Side A): Eritrean Liberation Front (ELF)[1] (1962 - 1981) / Eritrean Liberation Front-Popular Liberation Forces (ELF-PLF)[2] (1969 - 1988) / Eritrean Peoples Liberation Front (EPLF)[3] (1976 - 5/1991)
Aggressed (Side B): Ethiopia

Object and goals of the conflict

At the end of the nineteenth century Eritrea was colonized by Italy against the wishes and resistance of Ethiopia, which also had designs on the same territory. Eritrea, from its position on the Red Sea, served Italy with agricultural supplies and as a destination for Italian immigrants. In 1935/36 Italy most notably used their colony as a staging and recruiting area for the Italian invasion of Ethiopia. By 1941, however, it was Great Britain who occupied Eritrea, and in 1950 the U.N. acknowledged the region's autonomy, but allocated it to the sovereign state of Ethiopia. Ethiopia undermined Eritrea's autonomy and fully incorporated it into the Ethiopian state in 1962. A militant resistance movement rose up to oppose Ethiopia's annexation and the corresponding economic exploitation and sociocultural discrimination.

In 1962 the ELF began the armed struggle for an independent Eritrea. The ELF split internally, and the resulting EPLF eventually overtook the resistance by the beginning of the 1970s.

Outcome of the war

The EPLF coordinated its offensives, especially after 1988, with the Tigray Peoples Liberation Front (TPLF) (see War Nr. 122). Together they were instrumental in the overthrow of the Mengistu regime in Addis Abbaba in 1991 and the dissolution of the Ethiopian army. Eritrea won the right to self-determination by referendum, and after a vote in May 1993 Eritrea became independent.

Hundreds of thousands of people died in the war, and agriculture, trade, transportation and supplies suffered severe set-backs in Eritrea.

Notes

1. The ELF, founded in 1960, lost its dominant position to internal squabbles. Militarily the ELF played no further role after it lost a confrontation with the rival EPLF in 1981, but politically they attempted to form an alliance to take part in possible negotiations with the Ethiopian government in 1988/89.
2. Formed in 1969 out of internal conflict in the ELF, the ELF-PLF worked with two other splinter groups from the ELF (the PLF2 and the Ala Group), who later united and formed the core of the future EPLF.
3. A confrontation within the ELF-PLF led Osman Saleh Sabbe (one of the ELF founders) to withdraw his small fighting group to the Sudan, leaving behind a majority who proceeded to reject the ELF. This group, under Ramadan Mohammed Nur and Isagas Afeworki, became the EPLF in 1976.

Sources

Eikenberg 1988a; Eikenberg 1989; Eikenberg 1990a; Eikenberg 1990b; Kaplan 1988; Markakis 1987;Matthies 1981; Matthies 1989; Schröder 1989 (SiS/PK)

---

71 Ogaden Uprising (Ethiopia 1963/64)
8/1963 - 12/1964; BC-2[1]
War ended by: military victory for side B

Warring parties:

Aggressor (Side A): Ogaden Liberation Front[2]

Aggressed (Side B): Ethiopia
Side A: Somalia (6.2.1964 - 30.3.1964)

Object and goals of the conflict

In 1963 the Somali nomads living in Ethiopia formed the Ogaden Liberation Front as a "liberation army" and as a "provisional revolutionary government." They demanded that the Ethiopian government repeal tax laws and grant them autonomy. The autonomy was to guarantee the nomads unfettered access to Somalia, where they were dependent upon their traditional watering holes and pasture. Despite the emphasis on autonomy and the importance of Somalia, secession or annexation were not explicit goals of the Ogaden uprising. In the context of the conflict Ethiopia and Somalia came to blows between February 6 and March 31, 1964, which was settled through mediation from the OAU.

Outcome of the war

The cease-fire between Ethiopia and Somalia of March 30, 1964, also required Somalia to end its support of the Ogaden Liberation Front. In the second half of 1964 Ethiopian troops won back control of Ogaden territory. The Ethiopian government attempted to defuse the crisis by improving water, education and health care for the nomads. Despite this the following years saw sporadic acts of sabotage against the central government, and in 1975 war broke out again (see War Nr. 127).

At least 2,000 people lost their lives in the war.

Notes

1. There is no proof for the rumors that either Somali troops fought with Ogaden units, or that Israeli secret service specialists served with Ethiopian anti-guerrilla units (see Matthies 1977:205-207).
2. The connections between the Ogaden Liberation Front and the later Western Somali Liberation Front (WSLF) are unclear.

Sources

Matthies 1977 (SiS/PK)

---

75 Kenya (Shifta War)
11/1963 - 10/1967; B-2
War ended by: third-party mediation (OAU)

Warring parties:
Aggressor (Side A): Shifta Guerrillas (Kenya)[1]
Aggressed (Side B): Kenya

Object and goals of the conflict

The Guerrilla's goal was the annexation of the "Northern Front District" (NFD) of Kenya to Somalia. Nomadic Somalis lived in the NFD as a result of colonial British borders, and they sought secession from Kenya both because of their economically disadvantaged position and their sociocultural affinity to Somalia. Since the mid-1900s the Somalis had lived in the NFD area. This Somalian consciousness played an important role alongside economic considerations. The Kenyan government was concerned with maintaining the territorial integrity of their state's colonial borders.

Outcome of the war

The status quo ante bellum remained in place, and the NFD's loyalty to the Kenyan government remained questionable. The Kenyan government did make some economic and social concessions to the region.

The war resulted in 4,000 to 6,000 casualties.

Note

1. The Somali government supported the separatists logistically, financially, with propaganda and weaponry. Whether Somalia sent "volunteers" to fight with them is unclear.

Sources

AdG 10471H, 10434B, 13500A/10; Matthies 1973; Matthies 1977; Michler 1991 (KS/PK)

---

112 Uganda / Tanzania (I)
8/1971 - 7.10.1972; AC-2[1]
War ended by: third-party mediation (OAU)

Warring parties:
Aggressor (Side A): Uganda
Aggressed (Side B): Tanzania
Side B: Uganda People's Army[2] (9/1972 - 7.10.1972)

Object and goals of the conflict

Tensions rose between Tanzania and Uganda after Idi Amin successfully overthrew the centrist, socialist-oriented government of Milton Obote on January 25, 1971[3] and replaced him with a military government. The Tanzanian Nyere government refused to recognize the dictatorial Amin regime and gave Obote shelter and support. Tanzania regarded Amin's push as a British-Israeli attempt to make Uganda dependent on Western interests. Repeated border violations by Ugandan troops served two purposes: to force Tanzania to recognize the Amin regime, and to provide distraction from unrest in Uganda.

Tanzania supplied units when Obote's guerrilla troops invaded Uganda in the hope of overthrowing Amin.

Outcome of the war

The war ended with Amin's troops victorious over the exiled Ugandans. The peace treaty of Mogadishu, mediated by Somalia, ended all hostilities by both states and their support for subversive groups who sought to undermine the respective neighboring state. Neither state achieved their main goals, however. The Amin regime did not succeed in forcing Tanzania to recognize their legitimacy, and Nyere was unable to reinstall Obote as Ugandan president. The conflict itself was not settled, and later flamed up again (see War Nr. 134). Up to 1,500 people died in the war.

Notes

1. In September 1972 Libya sent 400 soldiers to Uganda, but it remains uncertain whether they actively fought on the side of the central government.
2. The Uganda People's Army were the guerrilla troops loyal to deposed President Milton Obote.
3. Obote himself had come to power by overthrowing King König Kabaka Mutesa II. It is doubtful whether reports of up to 10,000 dead in fighting between the army and the King's personal guard are true, more likely they are propaganda. In the fighting between the army and security forces during Amin's coup 250 people were reported killed.

Sources

Apter 1961; Avirgan/Honey 1982; Füsser 1989; Gertzel 1974; Halbach 1973; Karugire 1980; Keesing's: 18393A, 20656A, 21360A,

21429A, 24450A, 24757A, 24849A, 25022A, 25543A, 26062A; Kiwanuka 1979; Low 1962; Mamdami 1976; Mazrui 1975; Mittelmann 1975; Omara-Otunnu 1987; Posner 1978; Sathyamurthy 1986
(AH)

---

122 Ethiopia (Tigray Conflict)
1975 - 5/1991; B-2 / BA-2
War ended by: military victory for side A

Warring parties:
Aggressor (Side A): Tigray Peoples Liberation Front (TPLF)[1] / Tigray Liberation Front (TLF)[2] (1975 - 1976) / Ethiopian Peoples Revolutionary Party (EPRP) / Ethiopian Democratic Union (EDU)[3] (1977 - 5/1991) / Ethiopian Peoples Democratic Movement (EPDM)[4] (11/1980 - 5/1991)
Aggressed (Side B): Ethiopia

Object and goals of the conflict
The resistance groups fought for more just economic and political participation for the northern Tigray region. Their goals were a more balanced distribution of resources and regional autonomy, eventually also seccession. When the TPLF and EPDM merged they acquired a more national, pan-Ethiopian orientation in place of a regional and ethnic disposition.

Outcome of the war
The EPRDF became the dominant power after the central government fell. A national conference agreed on a charter which gave the peoples of Ethiopia the right to self-determination. Ethiopia itself was divided into fourteen regions and ERDF troops assumed responsibility for national defense. This situation, however, gave rise to new conflicts between the ethnic groups in Ethiopia (see War Nr. 131).
Over 20,000 people died in the Tigray fighting.

Notes

1. The TPLF was originally formed at the University of Addis Abbaba in 1974 as the Tigray National Organisation (TNO). They changed their name to the TPLF in 1975.
2. Probably founded in 1974 from student organizations. After fleeing to Eritrea they found support from the Eritrean Liberation Front (ELF), but were defeated in 1976 by the TPLF.

3. Active most likely since 1977. They were a conservative, monarchist group defeated by the central government in 1978 and driven into the Sudan by the TPLF. They are rumored to have cooperated in 1990 with Ethiopian government troops.
4. Originally founded in 1980 as a splinter group from the EPRP, they merged in 1988 with the TPLF, whose pan-Ethiopian ideas appealed to them. The new organization was called the Ethiopian People's Revolutionary Democratic Front (EPRDF).

Sources

Brüne 1990; Clapham 1990; Eikenberg 1989; Eikenberg 1990a; Eikenberg 1990b; Eikenberg 1991; EPRDF 1991; Markakis 1987; Matthies 1987; Matthies 1989; Matthies 1990a (SiS/PK)

---

127 Second Ogaden War (Ethiopia)
1975 - 1984; BC-1[1]
War ended by: third-party mediation (other third-party states)

Warring parties:
Aggressor (Side A): Somalia (7/1977 - 3/1978) / Western Somali Liberation Front (WSLF)[2]
Aggressed (Side B): Ethiopia
Intervention favoring side B: Cuba (22.12.1977 - 1984)

Object and goals of the conflict

The WSLF fought against the Ethiopian central government with the goal of splitting the Somali-populated Ogaden from the Ethiopian state. They received support from Somalia, and during the second phase of the war (July 1977-March 1978) the war escalated to an interstate conflict between Ethiopia and Somalia. Somalia sought the direct annexation of the Ogaden. After Somalia was militarily defeated, the war turned into a lengthy military confrontation between WSLF guerrillas and the Ethiopian state. The more successful the guerrillas were, the more they shied away from the goal of secession and sought instead more autonomy and the right to self-determination.

Outcome of the war

Somalia was defeated and the territorial status quo ante bellum was restored. The WSLF was slowly worn down by their continuous conflict with Ethiopia, and were unable to acheive their minimum goal of autonomy. Approximately 109,000 people died in the war, many in the interstate phase between Somalia and Ethiopia, and around 840,000 refugees fled Ogaden for Somalia.

Notes

1. Up to 1,500 military advisors from the USSR and GDR supported the regime in Addis Abbaba after 1977, and over 400 remained in 1988. They provided advice on technical matters and military leadership, and possibly took part in the direct overseeing of units and as pilots (Matthies 1987a; ACR; Hofmeier 1989a).
2. The WSLF seems to have cooperated at times with the Somali Abo Liberation Front (SALF), but the degree and continuity of such cooperation is uncertain (see ACR, et.al. 1979/80).

Sources

Hofmeier 1989; Matthies 1987; Laitin/Samatar 1987 (SiS)

---

131 Ethiopia (Oromo-Conflict)
since 1976[1]; AB-2

Warring parties:

Aggressor (Side A): Oromo-Liberation-Front (OLF)[2] / Islamic Front for the Liberation of Oromiya (IFLO)[3] (since 4/1978)
Aggressed (Side B): Ethiopia

Object and goals of the conflict

Oromo territory was conquered by Ethiopia at the end of the nineteenth century. The Oromo fell under the cultural, political and economic dominance of the Amharen (who ruled Ethiopia). In part the Oromo assimilated, but Amharic dominance irked portions of the Oromo population who resisted the socioeconomic discrimination and autocratic centralism of the Ethiopian government. These issues led to repeated confrontations (1941-1943 and 1963-1970), none of which, however, ever seriously endangered the power of the state. When the Oromo attempted to form a civil opposition in the 1960s and early 1970s they were met with repressive military measures. These led to the formation of a guerrilla army, which began its armed struggle against the central government in 1976. In 1990 political demands for self-determination came to the fore.

OUTCOME OF THE WAR

The Oromo uprising contributed to the collapse of Mengistu's regime and to the military victory of the Ethiopian People's Revolutionary Democratic Front (EPRDF) and Eritrean Peoples Liberation Front (EPLF) over the Ethiopian army in May 1991 (see Wars Nr. 122 and 63).

After Mengistu fell the OLF participated in the transition government. Soon, however, serious conflicts arose, and the dominant EPRDF oversaw the formation of a sister party, the People's Democratic Organization of the Oromo. The OLF and IFLO boycotted the regional elections, and the OLF withdrew its ministers from the transition government. Fighting erupted between the OLF guerrillas and the national guard (formed by the EPRDF). The latter was regarded as a new occupation army. It seemed that the Ethiopian government had chosen an authoritarian pacification policy to deal with the Oromo problem.

Amnesty International reported torture, extra-legal executions, and disappeared prisoners in the rural areas, and many prominent Oromos were said to be held in secret prisons. By the end of 1992 up to 20,000 OLF members were apparently imprisoned.

Notes

1. The exact date of the start of the war is impossible to determine. The OLF guerrilla actions started in 1976, but there is no available information about the quality (intensity, continuity, etc.) of their actions in the first few months.
2. The OLF was founded in 1974 and reactivated in 1976. In 1978 it split into two organizations due to internal differences, but both retained the name OLF.
3. The IFLO emerged in 1978 from one of the OLF splinters under the first OLF leader Abdul Karim Ibrahim. They retained the name OLF until 1986.

SOURCES

Bitima 1983; Brüne 1990; Clapham 1990; Eikenberg 1988a; Eikenberg 1989; Eikenberg 1990a; Eikenberg 1990b; Eikenberg 1991; Markakis 1987; Matthies 1987; Matthies 1989; Matthies 1990b; Zitelmann 1990 (SiS/PK/CL)

---

134 Uganda/Tanzania (II)
10/1978 - 2.6.1979; AC-1

War ended by: military victory for side B

WARRING PARTIES:
Aggressor (Side A): Uganda
Aggressed (Side B): Tanzania
Side B: Uganda National Liberation Army (UNLA)[1] (1/1979 - 2.6.1979)
Intervention favoring side A: Libya

OBJECT AND GOALS OF THE CONFLICT

As in the previous war between Uganda and Tanzania in 1971/72 (see War Nr. 112) the issues were Tanzania's refusal to recognize Idi Amin as the legitimate ruler of Uganda and the increasing instability of the Amin regime. Ugandan exile troops had ensconced themselves in Tanzania, from where they sought to topple the Amin regime. Uganda provoked Tanzania by occupying the disputed Kagera point, which had been an issue in the first war. After Tanzania came to see the OAU as incapable of effectively settling the conflict, they sent troops not only to counter Uganda's territorial aggression but to help the exiled Ugandan troops in their quest to overthrow Amin.

OUTCOME OF THE WAR

Amin was successfully toppled, and after the war the UNLF formed a provisional government. After two interim presidents (Lule and Binaisi), P. Muwanga came to power through a coup-like manuever by UNLF military council, dominated by the Ugandan People's Congress (UPC). Disputed elections followed in 1980, and Obote returned to power. The anti-Amin alliance persisted less than a year after Amin fell and soon disintegrated due to internal disagreements.

The post-Amin period brought no successful domestic political stabilization, and the country remained in dire need of rebuilding. The war also drew substantial economic and social consequences in Tanzania. A further problem was the continued presence of Tanzanian troops in Uganda, who remained to guarantee internal security. Their brutal behavior, however, led to a deterioration in relations between the two states. The situation in Uganda continued to deteriorate and by 1981 the country convulsed into civil war (see War Nr. 152).

Nearly 10,000 persons died in this round of war, and Tanzania bore direct costs in excess of $250 million. The war had catastrophic economic consequences for Tanzania and Uganda.

Note

1. This was the army of the anti-Amin alliance known as the Uganda National Liberation Front (UNLF). The UNLF was founded by representatives of eighteen exiled groups during the so-called Moshi Conference in Tanzania from March 23-25, 1979. The UNLA consisted primarily of former loyal ex-Obote troops, who had fled to Tanzania after Amin's coup.

Sources

Avirgan/Honey 1982; Halbach 1973; Hansen et al. 1988; Hilgers 1991; Keesing's 28798B, 28904A, 28928A, 29306A, 29669A, 29837A, 30279A, 30589A, 30711A, 31049A; Kiwanuka 1979; Mamdami 1983; Mittelmann 1975; Omara-Otunnu 1987; Sathyamurthy 1986 (AH)

---

152 Uganda
2/1981 - 5/1992; A-2
War ended by: military victory for side B

Warring parties:

1st Phase (until 1986): National Resistance Movement/Army (NRM/A)[1] and Uganda Freedom Movement (UFM)[2] (1982 - 1983) against Uganda

2nd Phase: Uganda Freedom Movement (UFM)[2] (4/1987 - 1988) / Holy Spirit Movement/Batallion[3] (1986 - 10/1987) / Uganda People's Democratic Movement/Army (UPDM/A)[4] (1986 - 6/1988) / Uganda People's Front/Army (UPF/A)[5] (1987 - 5/1988) and numerous smaller groups.

Object and goals of the conflict

In the first phase (until mid-1995) resistance arose in the north against the Obote government. The catalysts were electoral manipulation, human rights violations and the alleged sell-out of the country to Western interests. The rebellion sought political, social and economic reforms. Even after the Obote government was toppled in July 1985, armed resistance continued against the repressive military council led by Okello. By the end of January 1986 NRA troops succeeded in overthrowing the government and Museveni took control.

In the second phase a motley crew of troops, including soldiers from the previous Amin, Obote and Okello regimes, tried to overthrow Museveni and take power for themselves. In the midst of all this the religious "Holy Spirit Movement" rose up, making use of increasing economic, sociocultural and political alienation among the Ugandan population, especially in the north.

Outcome of the war

The Museveni government (1986) improved the human rights situation in Uganda and developed democratic structures at the grass-roots level. The situation remained unstable, however, due not least of all to the myriad militant groups which threatened the government. Uganda's lopsided economic dependence on world coffee prices also increased the country's susceptibility to crisis. Despite IMF programs the economy remains chaotic, undermined by a large black market, smuggling and bartering.

At the end of May 1992, the NRA declared victory over the security problems in the north and east of the country. The potential for conflict remains, however, especially in the debates over a new constitution, the introduction of a multiparty system and the charges by Rwanda that the NRA supported the Rwanda Patriotic Front rebels (see War Nr. 189).

One serious consequence of the war has been an outbreak of AIDs unparalleled in scale. It is estimated that one million people are HIV positive out of a population of 16 million. The war killed up to 520,000 people, and the economy collapsed entirely. Since 1986 it has recovered only gradually.

Notes

1. Previously the People's Revolutionary Army (PRA), before that the Front of National Salvation (FRONASA). They were a leftist splinter group from Obote's Uganda People's Congress (UPC), and cooperated with the Uganda Patriotic Movement (UPM). They were Obote's main opponent between 1981 and 1986, and after conquering Kampala they became the main part of Museveni's army.
2. Formerly the Save Uganda Movement (SUM); they stand politically close to the Democratic Party (DP) under the leadership of former interior minister Kayiira. Originally supporters of Museveni, they shifted in 1987 to the opposition.
3. They probably emerged from parts of the UPDAM under the leadership of the Voodoo priest A. Lakwena. Since 1990 they have called themselves the United Democratic Christian Movement (UDCM).
4. Under John Okello's leadership, who was one of Museveni's prime enemies. It formed originally from members of the Acholi ethnic group.
5. Led by Charles Jesus Ogirot and oriented against Museveni. Primarily consisted of members of the Langi ethnic group.

6. It is possible that during Obote's second round as President more people died than under Idi Amin's dictatorship. There are claims of up to 500,000 civilian casualties during revenge actions by Obote's troops in the area vacated by retreating NRA rebels (Luwero-Dreieck).

Sources

Entrup 1988; Furley 1987; Hansen et al. 1988; Hofmeier 1988; Hofmeier 1989; Keesing's 31049A, 31844A, 32895A, 33508A, 33763, 33890, 34023A, 34535A, 34720, 34729, 35491A, 35621, 36419, 36455, 36968, 37396, 37602, 37765, 37890; Nabudere 1980; Omara-Otunnu 1987; Sathyamurthy 1986; IHT 9.10.92; NZZ 6.8.92; ARB Jg. 1992 (AH/PK/AB)

---

166 Somalia (Civil War)
since 5/1988; A-2 / AB-2

Warring parties:

Aggressor (Side A): Somali National Movement (SNM; Isaaq Clan) / Somali Salvation Democratic Front (SSDF; Darod/Majertain Clan) / Somali Patriotic Movement (SPM; Ogadeni/Darod Clan)[1] (since 8/1989) / United Somali Congress (USC; Hawiye Clan), Fraktion unter Mohammed Farah Hassan "Aidid" (Unterclan Abgal) (since 10.10.1989) / United Somali Congress (USC; Hawiye Clan), Fraktion unter dem Interimspräsidenten Somalias Ali Mahdi Mohammed (Unterclan Habr Gedir) (since 1991) / Somali Democratic Movement (SDM; Rahanweyn Clan) (since ??) / Somali Democratic Alliance (SDA; Gadabursi Clan) (since ??) / Somali National Alliance (SNA)[2] (since 12.8.1992) / Southern Somali National Movement (SSNM; Dir Clans in Southern Somalia) (since ??), Somali National Front (SNF; Darod-Marehan)[3] (since ??) and other organizations.
Aggressed (Side B): Somalia

Object and goals of the conflict

A 1969 military coup under Siyad Barre repealed Somalia's democratic constitution, which had been enacted after independence in 1960. By the mid-1970s terrible drought and global economic crisis undermined attempts by Barre's socialist-oriented military government to institute economic reforms and increase literacy. After the loss of the Ogade war in 1977-78 (see War Nr. 127), armed opposition groups formed with bases in neighboring states, from where they sporadically attacked Somalia. In May 1988 these attacks escalated into war with the goal of unseating Barre and installing a democratic regime with mixed economic reforms.

Barre fled on January 27, 1991, but his absence gave rise to bloody power struggles between rival resistance groups, particularly after November 1991 and primarily in the south and in the capital, Mogadishu. Clans which had not yet been involved in the fighting entered the fray to carve out spheres of influence. On May 18, 1991, the former British-administered northern region (British Somali-land) declared independence, but so far has not received any international recognition.

Outcome of the war

The war resulted in the complete dissolution of any centralized governmental control. Local and regional power struggles reigned across the landscape. Barre fought with the SNF and the rest of his army to restore his regime, but his attempt was foiled in April 1992 by Aidid's faction in the USC. Nigeria granted Barre political asylum on May 17.

In March 1992, the U.N. mediated a cease-fire between Ali Mahdi and Aidid, and four months later the first U.N. observers arrived followed by 500 unarmed Pakistani U.N. soldiers (UNOSOM I). On December 3 the U.N. security council authorized military intervention under chapter VII of the U.N. Charter in order to create a secure environment for the distribution of humanitarian aid. Six days later the U.N. began to deploy the United Nations Unified Task Forces (UNITAF), which consisted of 25,000 U.S. soldiers and 12,000 soldiers from 20 other nations.

The war and the ensuing famine cost approximately 300,000 lives, and over one million Somalis fled to neighboring countries.

Notes

1. One faction, primarily from Ethiopia, belonged to the SNA under Ahmed Omar Jess, the other faction was led by Adan Abdullahi Nur "Gabiyow" and consisted of members mostly from southern Somalia.
2. An alliance of the USC (Aidid faction), SDM, SPM, and SSNM.
3. The SNF was allied with Siyad Barre, and led by his son-in-law Mohamed Siad Hersi "Morgan."
4. U.S. and other troops became invovled after 1993 in the hostilities.

Sources

Bongartz 1991a; Brüne/Matthies 1990a; Eikenberg 1989a; Eikenberg 1990a; Greenfield 1989; Hussein 1992a; Laitin/Samatar 1987a; Lewis 1988; Markakis 1987a; Matthies 1989a; Matthies 1990c; Mavrolean 1992; Nuscheler 1992; Samatar 1988; Watson 1992 (SiS/PK/US)

---

190 Djibuti
since 10/1991; AB-2

Warring parties:

Aggressor (Side A): Front pour la Restauration de l'Unité et de la Démocratie (FRUD)
Aggressed (Side B): Djibuti

Object and goals of the conflict

Since independence in 1977 unrest had plagued Djibuti, mostly because of the socioeconomic and political discrimination against the Afar. The Afar blamed the Issas for securing the most influential posts with the help of the former colonial power, France, while aiming to prevent the Afar from gaining equal rights. In August 1991 several guerrilla organizations formed together into the FRUS, and took up the fight together against the government.

Outcome of the war

Although the government tried to blame the conflict on foreign intervention, President Gouled made political concessions (announced on November 27, 1991) in order to calm the situation. A cease-fire held from the end of February 1992 through early autumn, but the fighting resumed after the FRUD boycotted both a constitutional referendum on September 4, 1992, and subsequent parliamentary elections.

Sources

Monitordienst der Deutschen Welle 19.10., 13.11., 19.11. u. 28.11.1991; FR 12.10.1991; FAZ 14.11.1991; *Le Monde*

Southern Africa

62 Angola
since 4.2.1961; D-1 / ABC-1 / A-2[1]

Warring parties:
1. Phase (until 11.11.1975): Movimento Popular de Libertaçao de Angola (MPLA)[2] / União Nacional pela Independência Total de Angola (UNITA)[3] (since 13.3.1966) / Frente Nacional de Libertação de Angola (FNLA)[4] (since 3/1961) against Portugal
2. Phase: União Nacional pela Independência Total de Angola (UNITA)[3] / Frente Nacional de Libertação de Angola (FNLA)[4] (until 1978) / Frente de Libertação do Enclave de Cabinda (FLEC)[5] (until 1978 und ab 1991) against Angola
Intervention favoring side MPLA/Angola: Cuba[6] (7.11.1975 - 20.6.1991)
Intervention favoring side FNLA/UNITA: Zaire (5/1975 - 2/1976) / South Africa (14.10.1975 - 1/1989)

Object and goals of the conflict

The war began in 1961 as an anti-colonial uprising by rival organizations against Portugal. The collapse of the fascist Salazar regime in Portugal in April, 1974, set off an internal power-struggle for the control of Angola. It was decided in favor of the MPLA, who were helped by intervening Cuban troops. Together they fought during 1975/76 against units of the FNLA and UNITA, as well as against mercenaries and regular troops from Zaire and South Africa.

It was under the MPLA, on November 11, 1975, that Angola became independent. This event was followed by a war to destabilize the MPLA, fought by UNITA and South Africa, and until 1978 by the FNLA. The FNLA wanted to build a pro-Western government dominated by the Bakongo people of northern Angola, while UNITA strove for a pro-Western government under leadership of the Ovimbundu people. In the 1980s UNITA broadened its base and its ethnic character dissipated, as recruits came from dissatisfied Angolans of all kinds and also coerced supporters from within the territories which they controlled. UNITA propounded a free-market economic order. South Africa had two primary reasons for intervening: the first was to fight the socialist government in Luanda, which was opposed to Apartheid, as well as their Cuban allies. The second reason was to eliminate safe areas for the Namibian liberation organization SWAPO, who were using bases in southern Angola.

Outcome of the war

The first phase of the war ended on November 11, 1975 with Angola's independence under the MPLA. UNITA and South Africa's destabilization campaign, however, forced the MPLA to the negotiation table. South Africa suffered serious defeats in battle which strengthened their willingness to negotiate and to withdraw from Angola. Namibia's later independence in 1990 and the eventual collapse of the Apartheid regime are in part connected to South Africa's unsuccessful military engagement in Angola.

In 1988 and 1991 treaties were signed sealing the withdrawal of Cuban and South African troops and guaranteeing democratic elections. On June 20, 1991 a cease-fire took effect, after which the domestic political situation seemed to calm down. After the MPLA won the election in 1992, however, UNITA started fighting again. Pitched battles were concentrated in cities and around economic targets.

In the oil rich enclave of Cabinda the FLEC started military activity after 1991, supported by UNITA. This indicated that an old FLEC goal of an independent Cabinda was not dead. Since summer 1992 the fighting between separatists and government troops has increased in intensity.

Several hundred thousand people have died in the more than thirty years of war. More than two million persons became refugees in their own land, while several hundred thousand fled abroad. The material cost of the war is tremendous: the Angolan government spent between 1980-88 alone between $27 and $30 million. Hunger and illness became widespread as the economy slipped further into ruin.

Notes

1. Until November 11, 1975 the conflict was a decolonization war (D), afterwards a destabilization war (ABC), and since November 1992 it has been an anti-regime war (A).
2. The MPLA was founded in 1956 from different groups with the goal of a diverse, multiethnic state of Angola. They were led by Agostinho Neto from 1962 to 1979, and thereupon by José Eduardo dos Santos. Their

success in gaining power in 1975 was due in large part to Cuban and Soviet assistance, the former with troops and the latter with weapons. After they came to power their politics became increasingly Marxist-Leninist, and the Reagan administration supported UNTIA's and South Africa's attempts to destabilize the MPLA. In the 1980s the MPLA adopted a more pragmatic approach, including negotiations to end the fighting, an opening to the West and a willingness (at the cost of internal dissent) to work with the IMF and World Bank.

3. UNITA rose in 1966 as an anti-colonial organization opposed to Portugese rule. Jonathan Savimbi, who formerly held an important position in the FLNA, became its leader, and directed the war against the MPLA after 1975. Although pro-Western, their political, economic and social programs were never very clear.
4. The FNLA was formed in 1957 under the name of "The Popular Union of North Angola" (UPNA), also briefly known as the UPA. In 1962 they changed their name to FNLA. From the start they were the most militantly anti-Portugese organization. Their support came from the Bakongo people who lived not only in Angola but also in western Zaire and in the Republic of Congo. After 1975 they also fought against the MPLA, and received help not only from the U.S. but also from China and Zaire. Their leader, Holden Roberto, was distantly related to the Zairian dictator Mobutu. After 1978, however, the FLNA was no longer able to function effectively as a military organization, because Mobutu withdrew his support and, most importantly, access to FLNA bases in Zaire, from where they carried out attacks in Angola. This sudden change was the result of a treaty between Zaire and the Angolan government to reciprocally end support for respective rebels in their territory. This came about in the context of the second Shaba War (see war Nr 133).
5. The FLEC was founded in 1963 with the goal of gaining independence for the oil-rich enclave of Cabinda. Although they cooperated with the FLNA, they were militarily weak and were essentially put out of commission by MPLA troops in March 1976. Somewhat surprisingly they began attacks again after June 1991, with the same goal of Cabinda's secession from Angola.
6. Cuban soldiers intervened on the side of the MPLA. Their numbers ranged from 18,000 troops at the beginning of the conflict in 1975 to 50,000 troops at the end of the 1980s. Since 1989 their troops were being withdrawn.

SOURCES

Becker 1988; Davies 1979; Decke 1981; Heimer 1979; Marcum 1969; Marcum 1978; Offerman 1988; Ohlson 1979; Rone 1989; Somerville 1986; Wolfers/Bergerol 1983; Young 1990 (PK/ST)

81 Mozambique (War of Independence)
25.9.1964 - 7.9.1974; D-1
War ended by: agreement without mediation

Warring parties:
Side A: Frente de Libertação de Moçambique (FRELIMO)
Side B: Portugal
Intervention favoring side B: Rhodesia 1 (1967 - 9/1974) / South Africa 2 (1969 - 9/1974)

Object and goals of the conflict

FRELIMO's goal was the full liberation of Mozambique and the installment of a national democratic alliance. After political opposition failed to achieve any results, the FRELIMO turned to military means to oust the Portugese colonialists in 1964.

Outcome of the war

White settler groups attempted to create a "Rhodesian solution" by arranging independence under white rule, but they met with failure. FRELIMO successfully pushed their claim to sole representation and took control of the government. The People's Republic of Mozambique was declared in June 1975. Approximately 30,000 people died in the war.

Notes

1. In 1972 Rhodesians replaced the Portugese in fighting FRELIMO and the armed units of ZANU and ZAPU in the northwest of the country (see War Nr. 91).
2. South African forces participated mostly in the defense of the Cabora-Bassa dam.

Sources

ARB; Braganca/Wallerstein 1982; Henriksen 1983; Isaacman/Isaacman 1983; Meyns 1979; Mondlane 1970; Munslow 1983 (KS)

---

91 Rhodesia ("Chimurenga II")
28.4.1966 - 21.12.1979; A-1
War ended by: third-party mediation (diverse)

Warring parties:

Aggressor (Side A): Zimbabwe African National Union (ZANU)1 / Zimbabwe African People's Union (ZAPU)2 (8/1967 - 21.12.1979) / Patriotic Front (PF) (9.10.1976 - 21.12.1979)
Aggressed (Side B): Rhodesia
Intervention favoring side A: Mozambique (1978 - 21.12.1979)
Intervention favoring side B: South Africa (?? - 21.12.1979)

Object and goals of the conflict

The ZANU and ZAPU sought to end the Smith regimes white minority rule and wanted to institute an African majority government and a just social order. ZAPU pursued a double strategy of military attacks and political negotiation. ZANU, on the other hand, focused exclusively since 1975 on guerrilla warfare, and became the main opponent of the Smith regime.

Outcome of the war

The war led to the end of both the white minority government and the Rhodesian state, which had never gained international recognition. In its place arose Zimbabwe on April 18, 1980, and with it a black majority government. International pressure forced the government to make concessions to the white minority for the first few years, above all in the form of disproportional representation in parliament (see War Nr. 160).

The war cost over 30,000 lives and at least 10,000 wounded. 250,000 to 300,000 people fled to Zambia, Mozambique and Botswana, while 20 percent of the population suffered from malnutrition due to effects of the war.

Notes

1. Its military arm was the Zimbabwe African National Liberation Army (ZANLA).
2. Its military arm was the Zimbabwe Peoples Revolutionary Army (ZPRA).

Sources

Astrow 1983; Baumhögger 1985; Cilliers 1985; Evans 1981; Martin/Johnson 1981; Semin-Panzer 1985; Thompson 1985 (SiS)

93 Namibia (War of Independence)
26.8.1966 - 22.12.1988; B-2
War ended by: third-party mediation (U.N.)

Warring parties:
Aggressor (Side A): South West Africa People's Organization (SWAPO)
Aggressed (Side B): South Africa

Object and goals of the conflict

Since 1966 SWAPO fought for Namibian independence from South Africa, which had adminstered the former German colony (known as Southwest Africa) since a 1920 mandate from the League of Nations. In 1966 the U.N. withdrew the mandate, recognized the territory as Namibia and legitimized SWAPO's right to resist the illegal occupation of their country. South Africa attempted to retain control by military means, and to extend Apartheid into Namibia. In 1975 the war spread to include South African actions against SWAPO bases in Angola.

Outcome of the war

The war ended in 1988 in the framework of a regional peace plan based on the 1978 U.N. resolution number 435. Namibia became fully independent on March 21, 1990, and SWAPO's founder, Sam Nujoma, was elected president. The only deep-water harbor in Namibia, Walvis Bay, was annexed by South Africa in 1977 and remains a sticking point.

At least 20,000 people died in the fighting, and the entire northern part of the country was laid waste during the war. 80,000 people fled the war to Angola and Zambia.

Sources

Ansprenger 1984; Baumhögger 1989; Bley 1968; Bley 1989; Doro/Legum 1989; Drechsler 1966; Ermacora 1981; Grabendorf 1979; Hubrich/Melber 1977; Hutcheson/Moorsom 1989; IDAF 1980; IDAF 1982; Katjavivi 1988; K¸hne 1989; Liberation Support Movement 1978; Manning 1989; Moorsom 1984; Pisani 1985; SWAPO 1976; SWAPO 1981 (PK)

---

119 Mozambique (Destabilization Campaign)
since 1975; AC-2 / A-1

Warring parties:
Aggressor (Side A): Resistência Nacional Moçambiquana (RENAMO)1 (since 1978) / Rhodesien (1975 - 21.12.1979)
Aggressed (Side B): Mozambique
Intervention favoring side A: South Africa (1979 - ??)
Intervention favoring side B: Nigeria (dates unknown) / Tanzania (1985 - 12/1988) / Zimbabwe (since 1982) / Malawi (since 4/1987)

Object and goals of the conflict

The white minority government in Rhodesia and South Africa regarded the success of the FRELIMO in Mozambique as a potential threat to their power (see War Nr. 81). Their fear was based not only on the example set by FRELIMO but by the protection given by FRELIMO to other black liberation movements who sought shelter in Mozambique, above all the ZANU and ZAPU (see War Nr. 91) and ANC (see War Nr. 125). In 1975 and 1979, respectively, Rhodesia and South Africa began military attacks against military and civilian targets in Mozambique to destabilize FRELIMO. As part of this campaign the Rhodesian government along with rebels in Mozambique developed the terrorist organization RENAMO, which continued to exist after the war ended with South African support and took on a life of its own.

Outcome of the war

The war weakened the Mozambique economy and forced FRELIMO to open itself towards the West. In 1984 the IMF and World Bank gained access to the orginally Marxist-Leninist state. RENAMO remained only a military, and not a political, opponent for FRELIMO. In 1984 South Africa and Mozambique signed the Nkomati treaty, in which they promised to stop harboring rebels opposed to the respective governments. Mozambique halted support for the ANC, but South Africa continued to further RENAMO in violation of the agreement. Only after the end of the East-West conflict did South Africa begin to distance itself from RENAMO. Nonetheless Mozambique entered into negotiations with RENAMO and a cease-fire was signed on October 4, 1992. The agreement included the pacification of the country under U.N. observation. Since then, however, the cease-fire has been repeatedly broken.

The war brought hunger and epidemics and cost almost one million lives, including many children. Several million inhabitants became refugees, and a substantial percentage of the population are dependent on food aid for survival.

Note

1. RENAMO also functioned under the name "Angolan National Resistance Movement" (MNR).

Sources

ARB 1975-92; Baumhögger 1985; Diederichsen 1990; Gersony 1988; Grest 1989; Hanlon 1984; Hoik 1989; Marsh 1981; Martin/Johnson 1981; Reichel 1980; Reid Daly/Stiff 1982; Semin-Panzer 1985; Young 1990 (PK/SG)

---

125 South Africa (Anti-Apartheid War)
since 1976; A-2 / AB-2

Warring parties:
Aggressor (Side A): African National Congress (ANC)
Aggressed (Side B): South Africa
alternating: Inkatha (Zulu-Movement) (since 1984)

Object and goals of the conflict

The main ANC goals were to eliminate the Apartheid regime and to build a non-racist society under democratic majority government. In the early 1960s the ANC began its resistance, but was quickly repressed by South African troops. Their sporadic attacks increased between 1961 and 1976, until the massacre of over 200 unarmed black students in Soweto on June 16, 1976, pushed the confrontation over the threshold of war.

With the advent of the de Klerk government in 1989 the possibility for a negotiated settlement arose. The ANC suspended its armed struggle on August 6, 1990. Increased tensions between the white and black population accompanied the gradual loosening in the South African hardline position. Through the 1980s bloody confrontations in the Natal province and in some of the so-called homelands have taken on warlike attributes. The conservative Inkatha under Buthelezi's leadership also added to the violence, along with so-called "vigilantes," supported by the South African state, who sought to weaken and split the ANC by terrorist means.

Outcome of the war

Growing international pressure against the Apartheid system, military defeat in Angola in 1988 (see War Nr. 62) and internal problems led first to a peace treaty with Namibia (see War Nr. 93) and eventually a dismantling of Apartheid under de Klerk. The ANC was not in a position to succeed militarily, and acquisced to negotiations to end Apartheid. Whether the situation will become more peaceful, however, is unclear, since the ANC and Inkatha (and the Pan-Africanist Congress [PAC]) continue to fight each other, and there still exist armed white extremists.

The war of the ANC against the Apartheid regime cost at least 4,000 lives. The clashes between the ANC and Inkatha since the mid-1980s has caused more than 10,000 deaths.

Sources

Berger/Godsell 1988; Brewer 1989; Davenport 1977; Frankel 1988; Hanf/Weiland/Vierdag 1978; Hirson 1979; Lodge 1983; Lucius 1981; Mahncke 1989; Noller 1977; Rotberg 1980; SAIRR 1990; Study Commission 1981; Wilson/Thompson 1971; Woods 1978 (PK/SG)

---

181 Zambia (Mushala Uprising)
14.1.1976 - 20.11.1982; A-2
War ended by: military victory for side B

Warring parties:

Aggressor (Side A): Mushala Army1
Aggressed (Side B): Zambia

Object and goals of the conflict

Adamson Mushala, a politically ambitious leader, used the socioeconomic hardships of the northwest province to recruit supporters in his struggle against the Kaunda regime. His ethnic mobilization strategy limited the war to the northwest, and Mushala's troops concentrated their attacks against government institutions. Their goal was to destabilize the Kaunda regime and its one-party system. Mushala tried unsuccesfully to gain the support of South Africa. His troops eventually succumbed to the better-supplied and -outfitted Zambian army. The Kaunda regime rejected any negotiations and sought to utterly eradicate Mushala's movement.

Outcome of the war

The war had little effect on Zambia's political system. If anything it strengthened the trend toward an autocratic state. From a regional perspective the Mushala uprising negatively impacted the development of the northwest province. Ethnic tensions also increased in the region.

Note

1. The Mushala army consisted of a core of 50-100 people, padded by new recruits. Mushala attempted to create a parallel political organization, but both the military and political wings were so focused on him that after his death the movement collapsed.

Sources

Papstein 1981; Peters/Tetzlaff/Wegemund 1991; Wele 1987 (CP)

---

160 Zimbabwe (Matabele Conflict)
1/1983 - 5/1988; A-2
War ended by: agreement without mediation

Warring parties:

Aggressor (Side A): Dissidents1
Aggressed (Side B): Zimbabwe (1/1983 - 3/1988)

Object and goals of the conflict

This conflict grew out of competition between the Zimbabwe African National Union (ZANU), led by Robert Mugabe, and the Zimbabwe African People's Union (ZAPU) led by Joshua Nkoma. On the surface the conflict was also a confrontation between the two largest ethnic groups, the Shona (77 percent) and the Ndebele (18 percent). Differences between these two organizations had been miminized during their common struggle against the white minority government of Rhodesia (see War Nr. 91). ZAPU felt itself unfairly disadvantaged after it won only twenty seats in the first elections in independent Zimbabwe, while ZANU won fifty-seven. Differences also arose in the context of insufficient agricultural and social reforms, the South African attempts at destabilization, long-standing internal disputes and the orientation of ZANU toward a one-party state. The warring parties had a wide range of goals from destabilization to social reforms to more power.

Outcome of the war

On December 22, 1987, ZANU and ZAPU successfully concluded negotiations. The war ended with the granting of a general amnesty for ZAPU fighters in May 1988. In 1990 ZANU became the de facto unity party by incorporating ZAPU into the new ZANU-PF (Patriotic Front). In the March 1990 elections they achieved 117 out of 120 seats.

Estimates for casualties range from several thousand to tens of thousands.

Note

1. "Dissidents" was the government's designation for anyone during the conflict who caused unrest or pursued resistance. This term was used indiscriminately without regard for political orientation. There was also the so-called Super-ZANU, which consisted of militant groups trained in South Africa by the former Rhodesian secret service for sabotage acts in Zimbabwe.

Sources

Hodder-Williams 1983; Kreile 1990; LCHR 1986; Mandaza 1986; Martin/Johnson 1981; Martin/Johnson 1986; Meyns 1987; Peters 1991; Reichert 1984; Ripken/Prein 1990; Stoneman 1988; Stoneman/Cliffe 1989 (SiS)

Asia

South and Southeast Asia

3 Indonesia
29.9.1945 - 27.7.1949; D-2
War ended by: third-party mediation (U.N.)

Warring parties:

Aggressor (Side A): Indonesian Liberation Movement
Aggressed (Side B): Great Britain[1] (29.9.1945 - 1946) / Netherlands[2] (1946 - 27.7.1949)

Object and goals of the conflict

The Netherlands lost control of their colony when the Japanese Army occupied the Malaysian archipelago in December 1942. The Japanese were initially received by the Indonesian people as liberators, but later the occupation provoked an Indonesian independence movement led by Sukarno and Hatta. Facing military defeat in the war as a whole, the Japanese promised Indonesia independence in 1944, just about the time that the allies were deciding on a status quo ante bellum solution for "Dutch-Indonesia" in anticipation of the end of the war. Aware of this, Sukarno and Hatta accelerated the preparations for Indonesian independence, and on August 17, 1945 they declared the independent Republic of Indonesia.

It was not until September 29, 1945, that British troops landed on Java to disarm the remaining Japanese Army. From the very beginning there were armed clashes between the Indonesian Liberation Movement and the allies, which soon escalated to larger confrontations (including the "slaughter of Surabaya"). Great Britain decided to recognize Indonesian independence de facto, and refused to restore Dutch control. By the end of 1946 Great Britain had removed itself from the conflict.

The Netherlands tried to restore their old power, starting in 1946, by destabilizing the Indonesian Republic. They established new Indonesian states to challenge the Republic's claims to sole representation (e.g., the Sultanat Jojakarkta). Indonesian territory was divided into a smaller Indonesian Republic and a much larger Dutch-Indonesian region, but this did not diminish the Republic's influence. The Dutch did not want to accept this division as the status quo, and in July 1947 sent troops into Republican-controlled areas of Java and Sumatra.

Outcome of the war

International pressure, especially from the U.S., forced the war to draw to a close with a cease-fire on July 27, 1949. Dutch troops withdrew, and on December 27, 1949 an Indonesian state was declared within the borders of the former Dutch colony. A temporary exception was the exclusion of West-Papua (see War Nr. 64).

At least 5,000 people died in the fighting, and recent sources have placed the death toll among the Indonesians at much higher numbers.

Notes

1. British troops left Indonesia at the end of 1946. Indian units made up part of the British Commonwealth troops.
2. Dutch Colonial troops KNIL.

SOURCES

Dahm 1966; Dahm 1978; Hall 1964; Kahin 1959; Kubitschek/Wesser 1981; Schiller 1955 (HW)

---

10 Philippines (Luzon)
1946[1] - 1954[2]; A-2
War ended by: military victory for side B

WARRING PARTIES:

Aggressor (Side A): Hubkong Bayan Laban sa Hapon (also: Hukbalahap; HUK; Anti-Japanese People's Army)[3]
Aggressed (Side B): Philippines

OBJECT AND GOALS OF THE CONFLICT

During the Second World War the HUK had fought the large property owners, who collaborated openly with the Japanese occupation power. The HUK sought to do away with the feudal social relations which were a legacy of Spanish control. After the war ended the promised new order failed to materialize, as there was little official retribution against collaborators, no land reform, and so forth. Furthermore, the 1946 elections gave more power to the large property holders, while the HUK was denied representation in Parliament. Unrest developed as dissatisfaction grew, and eventually they escalated into armed confrontations.

OUTCOME OF THE WAR

The war ended as HUK's popular support evaporated and state military forces defeated them in battle. The revolutionary peasant movement suffered severe setbacks, but was not entirely eliminated. The previous socioeconomic structure of the Philippines as a quasi-U.S. colony was continued after independence in 1946. A resettlement program from the island of Luzon to the southern Mindanaos decreased tensions somewhat, but this action created future conflicts of its own (see War Nr. 199).

At least 9,000 people, including upwards of 5,000 civilians, died in the fighting.

Notes

1. There exists only sparse and contradictory information about confronta-

tions between the HUK and the U.S. Armed Forces in the Far East (USAFFE), which were supposed to have taken place starting in 1945.
2. The rest of the HUK continued ever-diminishing military actions up into the 1960s, which at times devolved into purely criminal activities.
3. The left wing of an anti-Japanese guerrilla movement during the Second World War, which received postwar support from the rural population. Their leadership was communist-socialist oriented.

Sources

Europe-Archiv März 1947; Keesing's 498A, 2590A; NZZ 16.4.50, 25.2.54, 18.5.54; Die Welt 13.7.50; Lachica 1971; Liebermann 1968; Reinknecht 1980a; Seekins 1984; Small/Singer 1982:228; South 8/ 1986:45; Tan 1977; Thürk 1980; Weggel 1973 (UP)

---

9 Indochina (first Indochina War)
9.3.1946 - 1.6.1954; D-2
War ended by: third-party mediation (International Conference)

Warring parties:

Aggressor (Side A): France
Aggressed (Side B): Khmer Vietminh (later: Khmer Rouge)[1] / Vietminh[2]
Side B: Pathet Lao[3]

Object and goals of the conflict

France attempted to restore its colonial control over Indochina after the Japanese were defeated in 1945. They were opposed by the national unity fronts of three Indochinese countries, which were led by the respective communist movements. The main goal of the Vietnamese national liberation movement was national independence, which tended to take priority over the systemic reforms sought by the Communist Party. [4] Most of the fighting took place in the north of Vietnam, where Vietminh forces proclaimed the Democratic Republic of Vietnam on September 2, 1945. They bore the brunt of the fighting and the new responsibility, along with the communist movement, which was well-established in the area. As time went on the war expanded to involve the neighboring countries who were also fighting against the French.

Outcome of the war

France was resoundingly and sorely defeated at the battle of Dien Bien Phu in 1954, and thereafter a solution to the conflict was hammered out during cease-fire talks at the Geneva Indochina Conference.

Vietnam was divided into a communist northern and a Western-oriented southern part. The two parts were supposed to be later unified through elections. In Laos the communists were promised two provinces under the condition that they would form a coalition government with the Royalists and later hold elections. In Cambodia the communist movement was promised no land, but was allowed to take part in the free elections planned for the following year. The social conflicts within the respective states, however, remained unresolved and provided flammable material for future wars (see Wars Nr. 41, 99 and 132).

Notes

1. The main military resistance in Cambodia came from the communist movement. They took the lead in the resistance about 1950, and were pro-Vietnamese until they began to be dominated by the Maoist Pol Pot faction starting in 1963.
2. Short for Vietnam Doc Lap Dong Minh Hoi (League for Vietnamese Independence), which was founded in 1941 from the KPIC as part of the National Unity Front against French colonial rule. They formed the core of troops in Ho Chi Minh's Democratic Republic of Vietnam (DRV) and later the regular troops for North Vietnam (see War Nr. 41). A less-used term for the general communist movement in Vietnam is Vietcong, which is short for Viet Nam Cong Sam (Vietnamese Communists).
3. Similar to Vietnam and, to a lesser degree, Cambodia, the communist movement offered the most militant resistance to French colonial rule in Laos. In 1950 the communists took over the leadership of the anti-colonial movement formed in 1945 between nationalists and communists, which after 1950 became dominated by the communists and best known under the designation Pathet Lao (Land of the Lao).
4. Although all three Indochinese states had declared independence at the end of the Japanese occupation, we can still regard this as one war because of the similarity not only in colonial history but the identical goals of war and participants for all concerned. France was pursuing the restoration of its colony Indochina, which from its view consisted of all of these territories. The Vietnamese-dominated Communist Party of Indochina (KPIC) understood itself as a leading force in the struggle for the independence of all of Indochina.

Sources

Benz/Graml 1981; Dingemann 1983; Draguhn/Schier 1987; Thee 1978 (RtH)

13 Pakistan / India (First Kashmir War)
22.10.1947 - 1.1.1949; A-2 / AC-1
War ended by: third-party mediation (U.N.)

WARRING PARTIES:
Aggressor (Side A): Paschtuns[1]
Aggressed (Side B): Kashmir
Side A: Troops of Azad-Kashmir (Azad-Kashmiri-Forces) (10/1947 - 1.1.1949)
Intervention favoring side A: Pakistan (5/1948 - 1.1.1949)
Intervention favoring side B: India (27.10.1947 - 1.1.1949)

OBJECT AND GOALS OF THE CONFLICT

Under British colonial rule Kashmir had retained its own organizational structure as a fiefdom. After the subcontinent was divided into India and Pakistan in August 1947 Kashmir became a *de jure* independent state with the understanding that it would later join either India or Pakistan, as was the case with all 554 fiefdoms. Despite a population consisting of 85 percent Muslims, the Hindu Maharaja sought to keep both options open, and in the process the Muslim Independence Movement was suppressed. On October 22, 1947, however, Pashtuns infiltrated from Pakistan under orders from Pakistani officers and attempted to force Kashmir to annex itself to Pakistan. However, most of the Kashmir troops, of whom the majority were Muslim, deserted and reorganized in the Punjab in support of the newly declared „Azad (free) Kashmir." Following this the Maharaja called upon India for military help, who in turn said they were only ready to aid the Maharaja if Kashmir would join India. One day later Indian forces intervened in the war.

Under pressure from the Muslim League and their own Muslim citizens, the Pakistani government decided to come down on the side of Azad Kashmir. At first Pakistani intervention was prevented by the British threat to withdraw all of their officers from the Pakistani army. British officers played an important role in the early Pakistani army. Yet all this changed when India opened a spring offensive in 1948 which torpedoed all existing negotiations regarding Kashmir. Pakistan was now facing a new wave of refugees, and India threatened to close the vital flow of water to Pakistan from the Indus river. In May Pakistani troops entered the fighting on the side of the rebels, mostly in northern Kashmir. Brines (1968:78) sees the Pakistani move as an attempt to grab as much territory as possible before the foreseen division of the region between them and India. Frey (1978:37), however, interprets the Pakistani intervention as defensive in nature.

Outcome of the war

Kashmir was divided as a result of the war. Two-thirds went to India, which later became the state Jammu and Kashmir and retains a special status, while the remaining third became the Azad Kashmir, administered de facto by Pakistan. Several smaller pockets of territory were annexed directly to Pakistan. A referendum about the future of Kashmir had been requested by the U.N. and promised by the Indian Prime Minister Nehru but has yet to take place.[2]

At least 3,000 people lost their lives in the fighting.

Notes

1. Also known as "Pathans."
2. See also Wars Nr. 84, 82, 161, as well as Wars Nr. 188, 111 and 15.

Sources

Barnds 1972; Brines 1968; Butterworth/Scranton 1976; Chaudry 1976; Choudhury 1968; Frey 1978; Ganguly 1988; Gupta 1967; Korbel 1976; Lamb 1966b; Lamb 1968; Menon 1956; AdG 1299, 1641, 1761 (JPF)

---

20 Malay Alliance ("Malayan-Emergency")
1948 - 2/1960; DA-2
War ended by: military victory for side B

Warring parties:

Aggressor (Side A): Malayan Communist Party (MCP)
Aggressed (Side B): Great Britain[1] / Malay Alliance

Object and goals of the conflict

The guerrillas sought an independent and communist Malaysia under MCP leadership. They recruited mostly among ethnic Chinese. Great Britain supported an independent Malaysia, but wanted a Western-oriented system.

Outcome of the war

The war ended with the total defeat of the MCP and successful implementation of the British goal of a Western-oriented independent Malaysia.

More than 11,000 persons were killed in the war.

Note

1. The British colonial troops included Gurkha units and recruits from Fiji, Kenya and Rhodesia as well as troops from Australia and New Zealand.

Sources

Pye 1956; Short 1975; Stubbs 1989; Tillman 1967 (HW)

---

21 Myanmar (Burma)[1]

since 1948; AB-2

Warring parties:

Aggressor (Side A): Burmese Communist Party (BCP) (1948 - 4/1989) / Ethnic minority groups under the umbrella of the Democratic Alliance of Burma (DAB)[2]
Aggressed (Side B): Burma
Often changing sides: Warlords[3] (since ??)

Object and goals of the conflict

Since achieving independence on January 4, 1949 no less than twenty different ethnic and political rebel groups fought in ever-changing alliances against the central government in Rangoon. The communist parties and guerrilla organizations fought an anti-regime war for the control of state power, while the ethnic minorities such as the Karen, Kachin, Wa or Shan fought for autonomy or secession. Today most ethnic groups have at least one armed rebel organization, whose strength fluctuates between twenty and several thousand fighters.

The origins of this war reach back into the pre-colonial time of the Burmese kings, whose claim to the throne consistently led to conflicts with the other ethnic groups. Later the British colonialists separated the country into "inner" and "outer" Burma, with the former under direct, the latter under indirect British administration. This separation also served as a determining factor in the development of ethnic diversity and tensions between the Burmese, who retained their dominant position in "outer" Burma, and the other respective ethnic groups.

During the Second World War the British Army recruited guerrilla troops from among the Karen and the Kachin to fight the Japanese, who had occupied the entire country. After the war these groups formed the military core of the ethnic rebel movement. The Burmese independence movement, which had at first taken the Japanese side, managed to become the dominant power in Burma during the war. When the war ended members of the Anti-Fascist People's Freedom League (AFPFL) occupied the key military and administrative positions. Internal differences mounted, however, and the communists were removed from all important posts. The communists—above all the BCP—split irreparably from the AFPFL when they could not agree on the formulation of the independence treaty. The AFPFL became the central government, and set out to repress the communists, who reacted by taking up arms.

The national minorities tried to guarantee their autonomy during the negotiations for independence with Great Britain. The Shan, Kayah, Kachin and other peoples received various forms of promises, but the Karen were unable to secure independent status. Shortly after Burmese independence they began an armed struggle for an independent Karen state. The other ethnic groups were fairly quiet until the early 1960s, when secession movements became popular as the Burmese government broke promises to the Shan and Kayah and declared Buddhism the state religion. As tensions rose the military took power on March 2, 1962. Rigid repression and official emphasis on the unity of the Burmese state set most of the ethnic groups into rebellion against the military government.

Alongside these groups the so-called Warlords also played an important political and military role. They recruited originally among the Kuomintang troops who had fled China after their defeat there in the Chinese Civil War. The Warlords primarily followed economic interests, including control over the profitable opium industry. While they cooperated now and again with the various warring parties, their interests often led them to direct confrontation with those resistance groups who either opposed or, as with the BCP or the Wa, were competitors for the opium business.

Outcome of the war

The heterogeneity of the participants and their varied and sundry goals does not bode well for a foreseeable end to the conflict. Since

1976 the NDF has attempted to coordinate the military and political elements of the ethnic opposition groups. In 1988 the NDF responded to another military coup by forming the DAB, which included exiled Burmese organizations. The BCP, who tried to focus on class conflict rather than ethnic differences, collapsed in 1989 under the weight of internal contradictions and since April 1989 has ceased to be a factor in the war.

The National League for Democracy (NLD), whose General Secretary, Aung San Suu Kyi, received the 1991 Nobel Peace Prize, won parliamentary elections in 1990, but has been prevented from taking power by the ruling military government, known by their acronym, SLORC (State Law and Order Restoration Council). SLORC has allowed work on a new constitution under its direction, but in the meantime has also launched successful military attacks against ethnic groups. In 1992 SLORC eased up on some particularly visible repressive policies (such as martial law) in order to avoid U.N. sanctions, but once the dry season began in fall, 1992, the army continued its military actions. The government's strategy centers on trading cease-fire agreements for limited concessions in bilateral talks between individual ethnic groups. The rebels, however, want negotiations to include representatives from all groups and to take place openly under U.N. auspices.

There exists no reliable data about the costs of the war, though economic activity has diminished considerably, as has foreign investment. Burma belongs to the group of the most undeveloped states. Numbers of war dead vary between 100,000 and 300,000, though exact numbers are hard to come by.

Notes

1. Burma renamed itself Myanmar in 1989.
2. The DAB was founded in 1988 by the National Democratic Front (NDF). While it includes exile groups it does not include all the ethnic minorities. Among the most important guerrilla organizations are the Karen National Union (KNU) / Karen National Liberation Army, the Shan State Army (SSA) and Kachin Independence Army (KIA).
3. The Warlords consist largely of former Chinese nationalists troops (see War Nr. 7). Their guerrilla groups include the Shan United Army (SUA) and the Thai Revolutionary Council (TRC).

Sources

Bray 1992; Burger 1987; Fleischmann 1981; Leach 1972; Lehman

1967; Marks 1978; Pogrom 1981; Silverstein 1980; Silverstein 1981; Steinberg 1984; Taylor 1979; Weidemann 1991; Wiant 1984 (HW/JM)

---

15 India / Hyderabad
13.9.1948 - 17.9.1948; C-2
War ended by: military victory for side A

Warring parties:
Aggressor (Side A): India
Aggressed (Side B): Hyderabad[1]

Object and goals of the conflict

Hyderabad, like Kashmir, faced the choice of joining India or Pakistan after the British relinquished colonial control over the sub-continent in 1947. Hyderabad had a Muslim leader, Nizam, yet the population was 87 percent non-Muslim, including 81 percent Hindus. Nizam decided to reject both annexation options and declared national sovereignty for his province, the largest of the 554 Indian fiefdoms which were ordered to join one or the other of the new states.

India, however, refused to accept Hyderabad's independence. Rather, India demanded that the banned Hindu Congress Party be allowed to operate, that social reforms be instituted, and, most importantly, that Hyderabad be integrated into the Indian Union.

Tensions between Hyderabad and India were made worse by internal unrest within the province, including terrorist acts against Hindus, and communist activity against the local regime. Negotiations about the region's status failed, and in May 1948, India began to station troops around Hyderabad. On September 13, 1948 they invaded in a "police action" to "restore order," as the official Indian explanation puts it.[2]

Outcome of the war

The war ended with Hyderabad's annexation to the Indian Union. Nearly 2,000 people died in the conflict.

Notes

1. Hyderabad's troops were supported by the Muslim paramilitary organization Razakar and, according to some, by the Pashtuns (see Menons 1956:371).
2. See also the First Kashmir War (War Nr. 13).

SOURCES

AdG 1483, 1534, 1597, 1635, 1637, 1638; Brines 1968; Chaudry 1976; Ganguly 1988; Menon 1956 (JPF)

---

23 Indonesia (South Molukk)
7/1950 - 12/1950; B-2
War ended by: military victory for side A

WARRING PARTIES:
Aggressor (Side A): Indonesia
Aggressed (Side B): Republic of South Molukk (RMS)

OBJECT AND GOALS OF THE CONFLICT

After achieving independence Indonesia remained a federal "United States of Indonesia" with sixteen internal states. A broad nationalist movement sought to create a unitary state as quickly as possible. States on the periphery, however, feared being discriminated against by any central government based in Java. The South Molluk territory rejected inclusion in a unitary Indonesian state and declared its own goal of sovereign statehood. On April 25, 1950 the Republic of South Molukk (RMS) was declared on the island of Ambon. Yet there was to be no peaceful solution. The U.N. "Commission for Indonesia" (UNCI) and the Netherlands tried unsuccessfully to mediate the dispute. In July 1950 the Indonesian army launched a massive attack against RMS troops, who consisted mostly of repatriated Molukk soldiers from the former Dutch Colonial Army (KNIL). The Indonesian troops suffered high losses, but were able to prevail by December 1950.

OUTCOME OF THE WAR

The secession movement was eradicated, and the restoration of a unitary state was secured. Approximately 17,500 residents of Ambon sought exile in the Netherlands. The rest of the RMS-sympathizers withdrew to the island of Ceram, from where they carried out limited attacks until 1963, when their leader Soumokil was captured and executed.

SOURCES

Dahm 1978; Decker 1957; Heinzelmeir 1976; Nöbel 1975; Prins 1959 (MV)

24 Nepal 1950
11.11.1950 - 1/1951; A-2
War ended by: military victory for side B

WARRING PARTIES:
Aggressor (Side A): Nepalese Congress Party
Aggressed (Side B): Nepal

OBJECT AND GOALS OF THE CONFLICT

Armed troops from the Nepalese Congress Party sought to topple the Rana regime led by King Tribhuvan. One hundred years of Rana rule had left the country in economic ruin, and all political and military leadership positions were occupied by members of the Rana family. The rebels were influenced by contacts with foreign intellectuals and by support from the Democracy Movement in neighboring India.

Outcome of the war

Nepal instituted a constitutional monarchy. The governmental system followed closely the model set forth by India. There were many attempts to rule the country according to parliamentary and democratic principles, but they all remained without effect. In the first four years alone there were nine different cabinets and coalitions (see War Nr. 65).

SOURCES

AdG 2668, 2679; Blaikie/Cameron/Seddon 1980; Shaha 1975 (BMu)

26 India (Nagas)
1954 - 8/1964; B-2
War ended by: third-party mediation (Church)
Warring parties:
Aggressor (Side A): "Federal Army" of the "Federal Republic of Nagas"[1]
Aggressed (Side B): India

Object and goals of the conflict

In the northern part of the mountains between Indian Assam and Burma live the Nagas people, of whom the majority are Christian. In 1965 there were approximately 600,000 Nagas, organized into about twenty tribes, with nearly every village speaking its own dialect of the Nagas language, which belongs to the Tibetan-Burmese family of languages. Their use of Latin script, however, made it easier for their language to adapt to British colonial rule, and as a result English was far more widespread among the Nagas than elsewhere in India.

After the end of British rule the Nagas did not receive their desired independence. Their settlements on the Indian side (others lay in Burma) were allocated to the Indian state of Assam. After all negotiations for independence were exhausted the Nagas took up a campaign of civil disobedience, which the Indian state sought to destroy by sending in paramilitary units and regular army troops in 1954. It was in this context that an underground army was formed among the Nagas, calling itself the "Federal Army" after the "Federal Republic of Nagas,"which was declared on March 22, 1956. This underground army proceeded to confront Indian troops.

Outcome of the war

One result of the war was the founding of an Indian federal state called Nagaland in December 1962. Although this was a concession to the Nagas, who stopped fighting several months after a cease-fire was signed on May 24, 1964, both sides viewed the political talks as merely tactical maneuvers. The cease-fire proceeded to fail, and at the end of the 1960s fighting broke out again between the Nagas and Indian security forces (see War Nr. 172).

At least 1,560 persons were killed in the fighting.

Note

1. Militant movement for Nagas autonomy.

Sources

AdG 10057/10936; FR 5.2.1988; Indo-Asia 1/1965: 19; Jacobs 1990; Maxwell 1973; Nag 1965; NZZ 13.11.66, 3.5.70; Times 28.10.64, 31.12.64 (JPF)

---

41 Vietnam (Second Indochina War)
1957 - 29.4.1975; A-1/AC-1
War ended by: military victory for side A

Warring parties:

Side A: Front National de Libération (FNL)[1]
Side B: South Vietnam

Intervention favoring side A: North Vietnam (4.8.1964 - 29.4.1975) / People's Republic of China (dates unknown)
Intervention favoring side B: United States of America (4.8.1964 - 28.1.1973) / South Korea (1.5.1965-28.1.1973) / Philippines (1.10.1966 - 28.1.1973) / Australia (7.2.1965-20.12.1972) / New Zealand (7.2.1965 - 1971)

Object and goals of the conflict

The FNL guerrillas, supported by North Vietnam, the USSR and the PRC, fought in South Vietnam against the U.S.-supported South Vietnamese government and intervening U.S. troops. The struggle was for the political control of South Vietnam. The South Vietnamese regime[2] had prevented the planned elections for national unification, as called for by the Geneva Indochina Conference which had ended the first Indochina war in 1954 (see War Nr. 9). Since then the South Vietnamese Vietminh guerrillas and troops from North Vietnam (the "Democratic Republic of Vietnam") fought for a communist transformation of society and for unification with the North.

After the Gulf of Tonkin incident in 1964, when North Vietnamese war ships allegedly fired upon U.S. ships, the war escalated dramatically. The U.S. began to heavily bomb North Vietnam in an attempt to halt Northern support for the Southern Vietnamese guerrillas.

By the mid-1960s the war spread to encompass military conflicts in Laos and, in 1970, Cambodia (see Wars Nr. 76 and 99).

Outcome of the war

The conflict ended in 1975 with the FNL victorious over the Southern Vietnamese government troops. The U.S. had withdrawn its troops under intense domestic pressure following the negotiation of a ceasefire in 1973. Vietnam was then united under communist rule. Communists in Laos and Cambodia rode the coattails of North Vietnam's victory and won the conflicts in their respective lands.

The worst consequences of the war for the entire regions was the enormously high number of civilian and military casualties and wounded, including children. Postwar development was severely crippled in Vietnam due to international isolation via an economic embargo (until 1991), rivers of refugees, and massive destruction and defoliation caused by U.S. bombardments.

56,221 U.S. soldiers died in the war, and 303,605 were wounded. The war cost the U.S. approximately $135 to $140 billion. The losses and eventual retreat from Vietnam had a long-lasting demoralizing effect on the U.S. military. Estimates for the total number of war dead are in excess of two million.

Notes

1. Founded on December 20, 1960 as a united front for more than twenty different opposition groups under communist leadership. Better known under the name Vietcong (short for Vietnamese Communists—Viet Nam Cong San), the FNL became superfluous after unification. North Vietnamese took over most of the administration in the South, in part because many Vietminh leaders had been caught and executed by the South Vietnam military after the Tet Offensive of 1968.
2. The corrupt and repressive regime of the Catholic Diem Clan was overthrown in a bloody coup in 1963 after it increased persecution of Buddhists. It was replaced, however, by an equally repressive regime.

Sources

Benz/Graml 1981; Buro 1988; Chen 1973; Chomsky 1972; Draguhn/Schier 1987; Leitenberg 1973; Thee 1978; Weggel 1987; Will 1989; Wood 1974 (RtH)

---

52 Indonesia (Civil War)
3/1958 - 7/1961; AB-1
War ended by: military victory for side B

Warring parties:

Aggressor (Side A): Various Rebel Armies
Aggressed (Side B): Indonesia
Intervention favoring side A: United States of America (3/1958 - 6/1958)

Object and goals of the conflict

The indisputable dominance of Java in Indonesian political and economic life led in 1957 to the creation of several different rebel movements ("Permesta"[1] in Sulawesi, the former Celebes; "Palembang-Charter"[2] in Sumatra). Despite differing goals these groups shared a common interest in breaking Java's central control over Indonesia. A further development adding to growing tension was the expropriation of Dutch interests by the central government following a defeat in the U.N. referendum regarding the West Irian questions (see War Nr. 130).

In February 1958, the guerrillas declared themselves the "Revolutionary Government of the Republic of Indonesia" (PRRI) and the conflict escalated to civil war. The Revolutionary Government was viewed as secessionist and the military went into action against them almost immediately in March 1958. The military was supported, at first secretly, by the U.S., the Philippines, and Taiwan, since they fought in the name of anti-communism and promised to prevent the further rise of the popular Indonesian Communist Party (PKI). Later the government received direct military support from abroad (following the politically embarrassing shooting down of a U.S. pilot in May 1958).

Outcome of the war

The rebels had no chance of military success, especially after the government received outside help while they did not. Already prior to 1958 the government had significantly weakened the rebels, though guerrilla actions continued until 1961. Negotiations led to the gradual surrender of the rebels, who received amnesty. The rebellion's goal was not achieved, rather the central government became stronger and extreme positions gained currency throughout the country, especially those of radical nationalists, supporters of the President, and the Communist Party.

The number of war dead is estimated between 6,000 to 14,000.

Notes

1. Permesta: They called for a general war after the May 2, 1957 massacre in Sulawesi. They sought reform and not secession. Martial law was declared in Eastern Indonesia.
2. Palembang-Charter: This was made public by officers in Sumatra in September 1957 and demanded military reform, decentralization and autonomy for the islands.

Sources

Harvey 1977; Heinzelmeir 1976; Mahajani 1970; Nöbel 1975 (MV)

---

34 Laos (First Laotian Civil War)

1.7.1959 - 16.6.1961; A-1

War ended by: third-party mediation (International Conference)

Warring parties:

Phase 1. (until 8.8.1960):
Side A: Pathet Lao
Side B: Laos[1]
Intervention favoring side A: North Vietnam
Phase 2:
Side A: Pathet Lao / Neutrals under Kong Le
Side B: Laos / Meo Guerrilla (also known as the Hmong)[2] (1960 - 16.6.1961)
Intervention favoring side A: North Vietnam
Intervention favoring side B: Thailand (1960 - 16.6.1961)

Object and goals of the conflict

At the Geneva Indochina Conference of 1954 (see War Nr. 9) the Pathet Lao were recognized as a political and military power along with the royal Laotian government. The Pathet Lao were allocated two northern provinces for a transition period, after which both provinces were to be reintegrated into the Laotian state through elections.

In 1958 a coalition government with Pathet Lao participation was formed, yet it was overthrown in July and August of that year by an anti-Pathet Lao coup. In 1959 fighting escalated into war between the Pathet Lao, supported by North Vietnamese units, and royal government troops.

The second phase of the war began with a coup by Colonel Kong Le against the government on August 9, 1960 in Vientiane. Kong Le's faction is known as neutral, but they fought until the end on the side of the Pathet Lao and the North Vietnamese. The remaining royal troops were led by the conservative military with U.S. support. In the context of their „containment" policy the U.S. was not prepared to accept a neutral Laos with a coalition government which included communists. Thai units intervened on the side of the conservative military, and the U.S. began training guerrillas in 1960 from among the Meo (Hmong) to fight the Pathet Lao.

Outcome of the war

On June 23, 1962 a coalition of conservatives, neutrals and communists was formed. This cooperation notwithstanding, the communists and conservatives continued to wrestle over state power. The U.S.-trained Meo guerrillas and the war in Vietnam (see War Nr. 41) further poisoned the atmosphere. Peace lasted only a short time, and in early 1963 the second Laotian Civil War broke out (see Wars Nr. 41, 76 and 183).

Notes

1. The royal government troops split after Kong Le's coup on August 9, 1960. The conservative faction continued to be known as "government troops," even though Kong Le controlled the capital and regarded this conservative faction as rebels, rather than the government.
2. Parts of the Meo, with U.S. backing and training, fought against the Pathet Lao, while other Meo factions remained neutral or even supported the Pathet Lao.

Sources

AdG 4678A, 7917B, 7927A, 8587A, 8741B, 8850A, 9092D, 9122B; Butterworth/Scranton 1976; Draguhn/Schier 1987; Weggel 1983; Weggel 1986 (WW)

---

65 Nepal 1962
1/1962 - 2/1962; A-2
War ended by: military victory for side B

Warring parties:
Aggressor (Side A): Nepalese Congress Party
Aggressed (Side B): Nepal

Object and goals of the conflict

The Nepalese Congress Party sought to end the rule of King Mahendra, who had dissolved the first elected government of Nepal after only eighteen months and had banned all political parties. He considered the democratic experiment a failure and declared the restoration of the monarchy with its traditional Panchayati council system (see War Nr. 24).

Outcome of the war

The rebels surrendered in resignation, disappointed by the results of a "new" constitution which was supposed to improve the situation. The King retained almost absolute power in the resulting constitutional monarchy, despite some elections.

100 - 114 persons died in the fighting.

Sources

AdG 9702, 9818; Blaikie/Cameron/Seddon 1980; Keesing's 19024f.; Shaha 1975 (BMu)

---

64 Indonesia (West-Papua)
14.1.1962 - 18.8.1962; C-2
War ended by: third-party mediation (U.N.)

Warring parties:
Aggressor (Side A): Indonesia
Aggressed (Side B): Netherlands

Object and goals of the conflict

Indonesia attempted to annex the territory of West Papua by infiltration. For the Netherlands West Papua became a matter of prestige: they did not want to let the young Republic of Indonesia, which thirteen years earlier had fought for its independence against the Dutch, push them out of West Papua (see War Nr. 3). The Netherlands also wanted to maintain rights of access to West Papua's rich natural resources.

Outcome of the war

Many Indonesian landings on West Papua failed to stir up support among the population. Diplomatic pressure by the U.S. and, later, mediation by the U.N. finally led to an end of the war. As part of the diplomatic compromise the United Nations set up a Temporary Executive Authority (UNTEA) to oversee the administration of West Papua in October 1962. On May 1, 1963 West Papua was formally integrated into Indonesia. Although the population was supposed to vote in a referendum in 1969 on their status, West Papua had already been fully incorporated into Indonesia under the new name „West Irian" (see War Nr. 130).

At least 120 soldiers died in the fighting.

Sources

Dahm 1966; Dahm 1978; Keesing's 9977A; Kubitschek/Wesser 1981; Lagerberg 1979; Polley 1991; Suter 1980 (UP/HW)

---

66 People's Republic of China / India
20.10.1962 - 20.11.1962; C-2
War ended by: military victory for side A

WARRING PARTIES:
Aggressor (Side A): People's Republic of China
Aggressed (Side B): India

OBJECT AND GOALS OF THE CONFLICT

The border between the PRC and India was the cause of this dispute, particularly the so-called McMahon line in the East and the Aksai Chin region in the West. This particular border dates from British colonial times and, for the most part, marked only on the map rather than along the territory itself. During the 1950s China renovated the connecting road between Sinkaing and Tibet and expanded its influence across the entire area. Indian border patrols of the territory during that time were few and far between. During negotiations it became clear that China was interested in exchanging Aksai Chin for recognition of the McMahon line, but India rejected any such deal.

Tensions rose between the two nations as border skirmishes occurred in 1959 and 1962, and as the Tibetan leader, the Dalai Lama, operated in exile from India. On October 20, 1962 Chinese forces crossed the McMahon line and ventured deep into Indian territory. The overwhelmed Indian army concentrated on repulsing the attack from the East, and was taken unawares by a Chinese invasion from the West. The Chinese forces then soon withdrew from the East, but occupied the Western territory.

OUTCOME OF THE WAR

China declared a cease-fire which India accepted de facto. During the second Kashmir war of 1965, however, (see War Nr. 84) Chinese troops provoked confrontations along the border between China and the Indian protectorate Sikkim. India's military defeat spurred great increases in military spending in the following years, and the army swelled from 500,000 soldier in 1962 to 825,000, and was outfitted with modern weaponry. Special units were formed for operations in the high mountains.

The confrontation with China also led to new alliances: in 1971 India signed a friendship and mutual assistance treaty. Meanwhile in 1963 Pakistan had already found a new ally in China, following the motto „the enemy of my enemy is my friend." Pakistan had signed an agreement ceding China 4,500 square kilometers of territory in the north of Pakistani-controlled Kashmir. Since India claims all of Kashmir, this agreement was not acceptable. Here the Sino-Indian war had a direct connection to the Indian-Pakistani conflict at its most flammable point, Kashmir (see the five Indian-Pakistani wars Nr. 13, 82, 84, 111, 161).

Relations between India and China were frozen for years, since Chinese troops still occupied the Aksai Chin in the West. The cease-fire of 1962 had only been possible due to mediation by six non-allied countries (the Colombo Conference). Border negotiations between the two countries began only in 1981.

Almost 2,000 people died in the fighting.

Note

1. Chinese troops did not fully withdraw behind the McMahon line until January 15, 1969.

Sources

AdG 7634, 10212, 10235, 10251, 10296; Choudhury 1978; Der Spiegel 8/1971; FEER 26.11.87: 50/51; Fisher/Rose/Huttenback 1963; Ganguly 1988; Indo-Asia 1/1966: 15/16; Lamb 1964; Lamb 1966a; Lamb 1968; Maxwell 1972; Nieh 1971; Sandhu 1988; Simon 1979; South (May) 1987:80-83 (JPF)

---

67 Brunei
12/1962 - 12/1962; A-1
War ended by: military victory for side B

Warring parties:
Aggressor (Side A): North Kalimantan National Army (NKNA)
Aggressed (Side B): Brunei
Intervention favoring side B: Great Britain

Object and goals of the conflict

The British government had intended for its protectorate, Brunei, to be absorbed into the new Malaysian federation, whose strong anti-communism and ethnic composition met British expectations for the region. The political opposition in Brunei, however, had other ideas, namely the uniting of Brunei with neighboring territory (sold by the Sultan in the nineteenth century) and then the building of a sovereign, independent state. The popularity of this concept was borne out by the electoral success of A.M. Azahari (People's Party), who was also the leader of the NKNA. This electoral victory, however, earned neither him nor his party a real say in decisionmaking, which remained the provenance of the Sultan and the British.

Outcome of the war

The NKNA suffered military defeat, annulling declarations of independence.[1] The People's Party was banned in Brunei and a state of emergency was imposed. Yet despite the now utter lack of internal opposition, Brunei did not end up joining Malaysia, most probably because it would have brought economic disadvantages regarding income from the oil industry.

Note

1. An interstate war was briefly a possibility as tensions escalated along the Malay-Indonesian border, but Great Britain prevented war by increasing its military presence (see War Nr. 74).

Sources

aero 177:4929-4933; Bedington 1978; Keesing's 10323; Reinknecht 1980b (UP)

---

76 Laos (Second Laotian Civil War)
31.3.1963 - 4/1973; A-1
War ended by: agreement without mediation

Warring parties:

Aggressor (Side A): Pathet Lao[1]
Aggressed (Side B): Laos[2] / Neutrals under Kong Le[3] / Meo (Hmong) Guerrillas[4] (1965 - 4/1973)
Intervention favoring side A: North Vietnam
Intervention favoring side B: United States of America (5/1964 - 4/1973) / Thailand (1968 - 4/1973) / South Vietnam (6/1964 - 4/1973)

Object and goals of the conflict

The Second Laotian Civil War was a systemic conflict (see the first Laotian Civil War, Nr. 34). The war escalated and became a second theater for the Vietnam War (see War Nr. 41). U.S. intervention (in the guise of bombing, recruiting of special units, the sending of advisors and secret operations) was directed primarily toward preventing the North Vietnamese from using Laos as a route to supply guerrillas in South Vietnam.

After the first Laotian Civil War a provisional coalition government had been formed in June 1962, but by March 1963 fighting broke out between the Pathet Lao and Neutralist troops (under Colonel Kong Le). Most probably the Pathet Lao wanted to improve its position at the cost of the Neutrals.

The Neutrals received hardly any support from the right-wing military at this point, and had been largely driven from their positions by the Pathet Lao. Many Neutrals switched sides and joined the Pathet Lao. In April 1964 individual officers fought heavily in the capital, Vientiane, and this battle weakened the right. In the course of the conflict the right-wing and the Neutrals merged into one force.

Outcome of the war

On February 22, 1973, a peace treaty was signed between the right/Neutral forces, who formed the government, and the Pathet Lao. In April of the following year another coalition government was formed, and in 1975 the Pathet Lao took power peacefully and declared the People's Republic of Laos.

Notes

1. The Pathet Lao is the Laotian communist movement, and their political organization is the Neo Lao Hak Xat.
2. The right-wing controlled the army, though they were weakened during internal fighting in April 1964.
3. The Neutrals provided the prime minister during the war, but they did not control the army.
4. Individual Meo clans fought against the Pathet Lao. Meo guerrillas were supplied and financed by the U.S. Other Meos remained neutral or supported the Pathet Lao (see Wars Nr. 34 and 183).

Sources

AdG 9733A, 9865A, 9913A, 10548A, 11346B, 11685D/5, 11745E, 13363A/16, 14942A, 15497A, 17092B, 17688C, 18178A, 18188A, 18349A, 18389A, 18458B, 18621A; Butterworth/Scranton 1976; Draguhn/Schier 1987; LPDR 1984; Weggel 1983; Weggel 1986 (WW)

---

74 Indonesia / Malaysia ("Confrontation")
12.9.1963[1] - 11.8.1966; C-1
War ended by: agreement without mediation

Warring parties:

Aggressor (Side A): Indonesia / Diverse Rebel Groups[2] (9/1963 - 8/1966)

Aggressed (Side B): Malaysia

Intervention favoring side B: Great Britain[3] (9/1963 - 11.8.1966) / Australia (9/1963 - 11.8.1966) / New Zealand (9/1963 - 11.8.1966)

Object and goals of the conflict

Indonesia perceived the formation of Malaysia, with its planned British presence, as a long-term threat to its regional dominance. Indonesia sought to close foreign military bases in the region (the NEFO Doctrine[4]) and to secure regional hegemony. In the summer of 1963 talks between Indonesia, Malaysia and the Philippines in Manila seemed to present an acceptable solution. Indonesia, however, was caught by surprise by Malaysia's declaration of independence, and viewed this as a violation of the treaty. Indonesia thenceforth pursued the strategy of „confrontation" with the goal of crushing Malaysia. This also served to distract attention away from internal Indonesian problems. Indonesia imposed an economic blockade on Malaysia, and starting in September 1963, regular Indonesian troops took part in infiltrating and commanding guerrilla units in Sarawak and Sabah, and later in Malaya.

The intensity of the „confrontation" varied considerably until 1966, but it never developed into massive military actions. From the very beginning British support for Malaysian troops prevented the possibility of Indonesian military success.

Outcome of the war

In 1965, Indonesia's domestic situation became extremely fragile, as a coup failed in September and General Suharto succeeded President Sukarno in March 1966. Suharto's rise to power was followed by the total eradication of the Indonesian Communist Party (PKI), largely by mass killings of hundreds of thousands.

With the economic situation in Indonesia bordering on catastrophe, there was no stomach left for a continuation of the „confrontation" policy. The Djakarta Agreement on August 11, 1966 ended the policy and recognized Malaysia in its borders. The agreement also ended Malaysia's international isolation (it had left the U.N. between January 1965 and September 1966).

The numbers of war dead vary. According to the British Defense

Ministry 740 people died, including 150 members of Commonwealth units and 590 Indonesians. According to other reports the number of dead reached 1,000 (South 8/1986:45).

Notes

1. The month in which Indonesian troops clearly entered the conflict is considered the beginning of the war. It is probable that Indonesian troops were involved previously, but there is no proof.
2. Various rebel groups including Chinese guerrillas from Sarawak trained by Indonesia.
3. The main contingent of commonwealth troops consisted of British soldiers.
4. The NEFO Doctrine of the early 1960s pitted the "New Emerging Forces" (a synonym for developing countries) against the established industrialized nations their "mechanisms of repression." Indonesia was to play a major role in this.

Sources

Dahm 1978; Heinzelmeir 1976; Mackie 1974; Mahajani 1970; Nöbel 1975; South 8/1986 (MV)

---

178 Vietnam (FULRO)
20.9.1964 - 10/1992; B-2
War ended by: termination of hostilities

Warring parties:

Side A: Force unifiée pour la libération des races opprimées (FULRO)[1]
Side B: North Vietnam

Object and goals of the conflict

FULRO fought for the autonomy and/or independence of approximately thirty mountain tribes (the "Dega") in the Vietnamese, Cambodian and Laotian highlands. After the French withdrew, the guerrillas focused on Vietnamese settlers and soldiers. During the Vietnam War (see War Nr. 41) FULRO supported the United States' troops, who in turn trained and armed them. They hoped for autonomy after the end of the war. After the U.S. retreated from Indochina, however, FULRO was left to its own fate. Between early 1976 and late 1978 the FULRO leadership carried out a bloody "cleansing" within their own ranks. These internal purges resulted probably from rivalries among the ethnic groups, and led to the substantial weakening of the organization. In 1979 FULRO moved its headquarters to the impenetrable jungle on the other side of the Cambodian border.

Outcome of the war

The only success for the guerrillas lay in the delay of the Vietnamese economic development of the highlands. Since 1991 the Vietnamese central government attempted to improve the quality of life in the highlands and to integrate the tribes into Vietnamese society. When Cambodia was put under control of the United Nations Transitional Authority in Cambodia (UNTAC) in the aftermath of the conflict there (see War Nr. 60), FULRO was regarded as a foreign force and subject to U.N. disarmament policy. In October 1992 the last FULRO fighter was flown to Phnom Pen by UNTAC, and it remains unclear whether or not there remain any operative guerrilla groups in the highlands.

There are no exact figures for casualties of the war. 250,000 of the three-million Montagnards are said to have died up to 1975. From 1975 to 1979, FULRO lost 8,000 of their 10,000 fighters, who were either killed, executed, or imprisoned.

Note

1. FULRO was founded in 1964 as a resistance organization for the approximately thirty mountain tribes in the Indochinese highlands. Their first recorded military action was on September 20, 1964 against South Vietnamese soldiers and settlers.

Sources

Asiaweek 7.12.86; Der Spiegel 27.8.84; FAZ 10.10.79; FEER 8.10.82, 11.7.91, 23.4.92; Südostasien aktuell Jan. 1990:102, Nov. 1991:561 (KJG/SW)

---

130 Indonesia (West Irian—West Papua)
since 1965[1]; B-2

Warring parties:

Aggressor (Side A): Organisasi Papua Merdeka (OPM; Organization for a Free Papua)[2]
Aggressed (Side B): Indonesia

Object and goals of the conflict

Sine 1965 the OPM, a small armed liberation army, fought with the support of the indigenous peoples against the annexation of West Papua by Indonesia. Indonesia had renamed the area West Irian, and in 1973 created the name Irian Jaya. The area was originally a Dutch colony, but ended up being administered by and finally incorporated into Indonesia in 1969 (see War Nr. 64).

Due to the overwhelming force of the Indonesian army, the OPM resorted mostly to small operations such as sabotage or ambushes. At first the OPM was supported by the related indigenous peoples in neighboring Papua-New Guinea, who offered the guerrillas a safe haven. Indonesia pressured Papua-New Guinea, however, to tighten their border and officially reject any support for the rebels. In the early 1990s the Indonesian army succeeded in nearly wiping out the OPM, although they continue to reorganize themselves anew.

Outcome of the war

The OPM was unable to change the nature or extent of Indonesian rule in West Papua. The region did not share in Indonesia's rapid economic growth, and an estimated 700,000 Indonesian settlers have moved there from the Indonesian mainland. The known number of war victims swings widely between 30,000 and 150,000 casualties.

Notes

1. The relatively sparse information available does not allow us to date the beginning of the conflict exactly, especially since Indonesia tries repeatedly to play down or deny the existence or extent of the conflict.
2. The OPM was probably founded in 1963 in Holland by exiles from West Papua, since there was no opportunity for political organization after Indonesia took over the region's administration. The Gerakan Papua Kemerdekaan (GKP) is part of the OPM.

Sources

Bell/Feith/Hatley 1985; Kroef 1985; Premdas 1985; Savage 1978; Suter 1980 (UP/MMT)

---

82 India / Pakistan (Rann-Kutch)
9.4.1965 - 1.7.1965; C-2
War ended by: third-party mediation (former Colonial Power)

Warring parties:
Side A: India
Side B: Pakistan

Object and goals of the conflict

A limited war erupted from skirmishes along the disputed Rann-Kutch border between India and Pakistan. The Rann is a salt desert, which is transformed into a sea during the monsoon season from May until late autumn. Pakistan claims that the border runs through the middle of the sea, as is customary in international law. India, however, claims the border along the west of the Rann, which is the former Kutch fiefdom (currently the Pakistani province of Sindh).

The conflict took shape as oil was rumored to be found in the disputed area. The latent Pakistani-Indian conflict in the surrounding region, mostly in the form of skirmishes, was consciously extended by Pakistan into the Rann in order to test India's defensive forces. By the following August the Second Kashmir War (see War Nr. 84) broke out.

Outcome of the war

The British oversaw border negotiations between Pakistan and India between 1966 and 1968, with the result that 90 percent of the Rann went to India and 10 percent to Pakistan.

At least 450 soldiers were killed in the fighting.

Sources

AdG 11947; Barnds 1972; Brines 1968; Butterworth/Scranton 1976; Choudhury 1968; Indo-Asia 3/1965: 206-209; Indo-Asia 4/1965: 302/303; Lamb 1966b; Lamb 1968 (JPF)

---

84 Pakistan / India (second Kashmir War)
5.8.1965 - 23.9.1965; BC^1-2
War ended by: third-party mediation (U.N.)

Warring parties:
Aggressor (Side A): Mujahids (Muslim Guerrillas)[2]
Aggressed (Side B): India
Side A: Pakistan (16.8.1965 - 23.9.1965)

Object and goals of the conflict

On August 5, 1965, Mujahids infiltrated the Indian part of Kashmir and led guerrilla attacks against the Indian security forces. The Indians responded on August 16 by sending their troops into Pakistani-controlled area beyond the agreed-upon cease-fire line. While at first the official Pakistani reaction was limited, by the beginning of September the fighting extended the whole length of the 2,000 kilometer-long border between (West) Pakistan and India, and involved heavy tank battles.

Pakistan intended to liberate Kashmir from India, whom they perceived to be weak. India had previously refused to hold a referendum in Kashmir as Pakistan, with the support of the U.N., had demanded. India wanted to avoid losing Kashmir, and also sought to restore their self-confidence after losing the Sino-Indian war in 1962.

The Second Kashmir War was the zenith, so far, of the conflict between the two states since their founding in 1947. Other confrontations included Rann-von-Kutch[3] and the violation by Indian troops of the cease-fire line in the Kargil region.

Outcome of the war

The war came to the brink of major escalation when China threatened to intervene in support of Pakistan. The United Nations managed to negotiate a cease-fire before this came to pass. China most likely threatened intervention in order to prevent India from opening a second front in East Pakistan.

The status quo ante bellum (that is, the status quo following the First Kashmir War) was restored during a Soviet-sponsored conference in Tashkent on January 10, 1966. At least 20,000 people died in the war.

Notes

1. The "B" component comes from the Kashmiri's struggle against the Indian occupation.
2 The Mujahids recruited partially from Kashmiri Muslims. Whether they came from the Pakistani or Indian part of Kashmir is unknown.
3. See War Nr. 82, also Wars Nr. 13, 111, 161, 188 and 66.

Sources

AdG 12050, 12070, 12076; Barnds 1972; Brines 1968; Chaudry 1976; Choudhury 1968; Der Spiegel 37/1965, 39/1965; Frey 1978; Ganguly 1988; Gupta 1968; Hunck 1966; Indo-Asia 4/1965: 296-301; Indo-Asia 1/1966:7-12; Lamb 1966b; Lamb 1968 (JPF)

89 Thailand
7.8.1965 - 1980; A-2
War ended by: fighting fell below the threshold of war

WARRING PARTIES:
Side A: Thailand
Side B: Communist Party of Thailand (CPT)

OBJECT AND GOALS OF THE CONFLICT

The forbidden Communist Party of Thailand (CPT) thrived against a background of the wars in Indochina, a massive U.S. military presence, Chinese support and, last but not least, the unpopular repressive Thai military regime. In addition to overthrowing the government, the guerrillas sought the removal of U.S. forces, withdrawal from SEATO, and the liberation of the peasants from their feudal conditions. The guerrilla actions spread across the whole land and also targeted U.S. military bases. In 1973 the CPT set up administration for the first "liberated territory." Non-military opportunities, however, soon presented themselves for the Communists under the Kriangsak government's liberalization policies (1973-75). Discussions and disputes about strategy weakened the CPT, even as their popular basis broadened to include workers and students in addition to the rural population. The CPT was also weakened by an ideological split caused largely by events in Southeast Asia after the end of the Vietnam War: Vietnam's occupation of Cambodia (see War Nr. 132) and the Chinese-Vietnam War (see War Nr. 142) complicated the regional picture, and they also lost operational bases in the Moscow-oriented countries of Laos and Cambodia.

OUTCOME OF THE WAR

After hostilities ceased in 1980 the situation among the Thai population remained unchanged. The CPT has all but disappeared, despite forty years of party history and more than twenty years of armed struggle. After the Thai government declared an amnesty in 1980 most of the fighters left the jungle for the city and gave up the struggle. There was, however, no formal end to the war.

Over 3,000 people lost their lives in the fighting.

SOURCES

Keesing's 12850E, 11744E, 15332A, 16705A, 17092B, 20782E, 20532D, 21353B; Blanchard 1959; Lovelance 1971; Morell/ Samudavanija 1981; Schipper 1987; Wit 1969; Wyatt 1984 (SJ)

90 India (Mizos)
1.3.1966 - 31.7.1980; B-2
War ended by: agreement without mediation

Warring parties:
Aggressor (Side A): Mizo National Front (MNF)[1] / Buddhist Chakmas (21.7.1966 - ??)
Aggressed (Side B): India

Object and goals of the conflict

The MNF came into existence after a 1959 famine brought the political and social situation in the Indian federal state of Assam to a head. The indigenous Christian Mizo[2] people prepared a military uprising to counter their perceived repression by the Indian government. Their ultimate goal was complete independence from India and unification with Mizos regions in neighboring Burma and East Pakistan. On July 21, 1966, the Buddhist Chakmas people of Bangladesh joined the fighting on the side of the MNF. In 1973 the MNF set up their headquarters in the rebel Bangladeshi region of Chittagong Hill Tracts (see War Nr. 177), although they later moved it to the mountains of Assam. A peace treaty signed in 1980 only partially ended the hostilities, which continued below the threshold of war until another peace treaty was negotiated on June 25, 1986.

Outcome of the war

In 1971, the Indian government created the autonomous territory of Mizoram in order to weaken the MNF's demands for independence. Mizoram became an Indian federal state in February 1988. Its first prime minister, Laldenga, was the leader of the MNF.

At least 1,500 people died in the fighting.

Notes

1. The MNF was the Mizo's militant organization in their struggle for autonomy. The characterization Mizo National Army (MNA) has also been used, though rarely, to characterize the armed wing of the MNF. Recent literature mentions the existence of a militant Mizo movement called the Mizo Liberation National Front (MLNF).
2. The Mizo are also known as the Lushai.

Sources

AdG 12815, 27741; Akhtar 1988; Maxwell 1973; FEER (19.Jan) 3/

1988:18; NZZ 9.9.88; FR 5.2.88; Indo-Asia 2/1966:117/118 (JPF)

---

99 Cambodia
17.1.1968 - 17.4.1975; AC-1
War ended by: military victory for side A

Warring parties:
Side A: Khmer Rouge[1] / Front Uni National du Kampuchea (FUNK; Kampuchean National Unity Front)[2] (23.3.1970 - 17.4.1975)
Side B: Cambodia
Intervention favoring side A: North Vietnam (3/1970 - 17.4.1975)
Intervention favoring side B: United States of America (30.4.1970 - 15.8.1973) / South Vietnam (30.4.1970 - 29.6.1970)

Object and goals of the conflict

After achieving independence in 1953 Prince Sihanouk sought to unite all the diverse political currents within his Sangkum Party, founded in 1955. The Sangkum party represented a traditionalist, Buddhist-inspired Khmer socialism. The dire economic situation and widespread corruption led him to bring the communists into his government to balance out the old elite. Their presence only made matters worse, however, as the old elite and the right opposed the communists so vociferously that many of them went underground in the mid-1960s and emerged in 1968 as the Khmer Rouge. The Khmer Rouge launched moderate attacks against the government from small "liberated territories" within Cambodia. In 1970 Sihanouk was ousted by the right-wing General Lon Nol, who actively began to fight the Vietnamese and Cambodian communists. The new regime repudiated Sihanouk's neutral foreign policy and entered the war in Vietnam on the side of the South Vietnamese and the U.S. (see War Nr. 41). The U.S. and South Vietnam carried their military activity into Cambodia to cut support lines from the North Vietnamese over the Ho Chi Minh Trail to the Vietminh in the south.

After his fall Sihanouk allied himself in 1970 with the Khmer Rouge, and together they formed the FUNK. His presence lent the guerrilla movement legitimacy in the eyes of the traditionalist rural population.

Outcome of the war

The end of the Vietnam War also led to the Khmer Rouge's victory in Cambodia. The Cambodian government was overthrown, and Pol Pot's terrorist "revolutionary regime" came to power. The Communists also rose to power in the same year in Vietnam and Laos.

There are no exact figures of casualties, but enormous numbers of civilians and soldiers died in the war and large amounts of land were defoliated by massive U.S. bombardment.

Notes

1. The Khmer Rouge are a Maoist, pro-Chinese, anti-Vietnamese, and nationalist Cambodian communist faction, who recruited their leaders from intellectuals who had studied in France and thereby had avoided working with the Vietnamese-influenced Cambodian Communist Party.
2. FUNK was formed on March 23, 1970, as an alliance between the deposed Prince Norodom Sihanouk and the Khmer Rouge. Together they sought the overthrow of Lon Nol. The "Forces Armées Populaires de Libération Nationale du Kampuchea"(FAPLNK: People's Army for the National Liberation of Kampuchea) was first formed in May of that same year. After the fall of Lon Nol's government the Khmer Rouge became the sole power in Phnom Penh and FUNK vanished as an organization.

Sources

Aschmoneit/Werning 1981; Benz/Graml 1981; Buro 1988; Draguhn/Schier 1987; Kiernan 1980; Kiernan 1986; Shawcross 1980; Vickery 1986 (RtH)

---

172 India (Nagas)
1969 - 6.11.1975; B-2
War ended by: agreement without mediation

Warring parties:

Aggressor (Side A): "Federal Army" of the "Federal Republic of Nagas"[1]
Aggressed (Side B): India

Object and goals of the conflict

This second Naga War is a direct continuation of the first (see War Nr. 26), including the same demands for an independent Naga state. The 1964 cease-fire had enabled political talks between the two sides, but led to no convergence in their positions. The Indian side still demanded that Naga have the status of a protectorate, similar to Sikkim, while the Naga side wanted a sovereign state, similar to Bhutan. Military confrontations started to build again as the talks continued in vain. By the end of the 1960s the fighting had returned to pre-cease-fire levels.

Outcome of the war

Pakistani support for the Naga forces dried up after the end of the Bangladesh War (see War Nr. 111), and this led to India exerting complete control over the region. In 1972 India withdrew its troops and left Naga in the hands of the police and paramilitary units. Naga came to resemble a police state, and the status quo ante remained after the signing of the Shillong cease-fire on November 6, 1975.

Note

1. Movement for Naga autonomy.

Sources

Anand 1980; Economic and Political Weekly 4.10.1969: 1583/84; FEER 5.12.75; Maxwell 1973; Misra 1978; NZZ 3.5.70, 22.9.72 (JPF)

109 Philippines (NPA)
since 1970[1]; A-2

Warring parties:

Aggressor (Side A): New People's Army (NPA)
Aggressed (Side B): Philippines

Object and goals of the conflict

After the Philippines achieved independence in 1946 the communist, anti-Japanese "Hukbalahap People's Army" (HUK) fought with the support of the rural population against the new Philippine government. Their goals were land reform and a transformation of social relations. In 1954 HUK was violently repressed with the help of the U.S. (see War Nr. 10), and thereafter they founded the New People's Army (NPA) together with urban intellectuals. The NPA operated as the armed wing of the new, Maoist-oriented Communist Party of the Philippines (CPP), who sought to overthrow the government and implement a socialist social order.

The NPA rose quickly from obscurity to posing a serious threat to the government. Their cause was helped by mass rural poverty and crassly unequal distribution of resources. By the mid-80s the NPA had approximately 25,000 members who operated across the Philippines. Their organization was decentralized and concentrated on small attacks against military and state institutions. The CPP/NPA controlled some territories, where they sought to improve the quality of life for the peasants and farmers. They never managed to cooperate, however, with the other main militant opposition, the Moro National Liberation Front, a Muslim guerrilla army in the south (see War Nr. 199). Both organizations, however, were tolerant of each other's existence in their respective territories.

Outcome of the war

Marshall law was declared in 1972, and the government under Ferdinand Marcos and his family drove the country deeper into repression, replete with electoral manipulation, political assassination, and the liberal use of the military in support of maintaining the economic relations, which benefited mostly Marcos and foreign investors. When the Marcos regime finally gave way peacefully to a new government under Corazon Aquino, she attempted to engage the NPA with a policy of "national reconciliation." This included amnesty, dialogue, and a cease-fire, which was signed in November 1986. Yet the government was not prepared to accept many of the guerrilla's demands, including recognition of their "liberated" territories, bringing the NPA into a coalition government, land reform, and the breaking of Philippine ties with the U.S. Opposition to dialogue with the rebels grew both among former Marcos factions in the military and moderate left-wing politicians. Soon the cease-fire arrangement expired, and both sides resumed their military operations.

Several thousand people are killed on a yearly basis in the fighting, with a total so far of over 50,000 lives lost.

Note

1. Although reliable information about the NPA's military activity exists only after 1970, it is probable that there had already been armed confrontations between them and Philippine security forces just weeks after their formation in March 1969.

Sources

Aktionsgruppe Philippines 1986; Bello 1984; Cooke 1984; FEER 13.3.71; FAZ 15.9.84; Hansen 1991; Keesing's 24186A, 18503A;

Kessler 1989; Luther 1981; Nemenzo 1984; Reinknecht 1980a; Rühland 1982; Seekins 1984; Siemers 1986 (UP)

199 Philippines (Mindanao)
since 1970[1]; B-2

WARRING PARTIES:
Aggressor (Side A): Moro National Liberation Front (MNLF) and their military organization the Bangsa Moro Army (BMA) / Moro Islamic Liberation Front (MILF)[2] (since 1978)
Aggressed (Side B): Philippines

OBJECT AND GOALS OF THE CONFLICT

The Muslim population of Mindanaos and the Sulu Islands found their way of life threatened by the increasing settlement of Christian farmers in their region. Between 1948 and 1960 over one million, and from 1966 to 1976 about three million new Christian settlers arrived in Mindanao and the Sulu Islands. With them came the "Philippenization" of the entire administrative structure and the acquisition of land by the north Philippine elite. Development projects and modernization generally excluded the Mindanaos, who were pushed to the side.

The MNLF was formed in 1969 with the goal of fighting for the independence of the traditional Muslim territories of Mindanao and the Sulu islands from the Philippines. The historical ties to the Catholic-majority Philippines had been traditionally weak. The common struggle for independence papered over differences within the Muslim camp, especially between traditionalists and democratic-socialist oriented factions. At first surprisingly successful, the BMA suffered a terrible defeat at the hands of Philippine forces in 1974 which led to a deep split within the Muslim organization. In 1978 the MLF split entirely from the MNLF.

OUTCOME OF THE WAR

The MNLF/BMA could rely to a certain extent on assistance from Islamic countries, both diplomatic and military. A conference of Islamic foreign ministers forced Marcos to negotiate in 1975/76 with the Mindanaos, who were promised greater autonomy. This promise, however, was never carried out by the government in Manila.

After Marcos fell, another round of negotiations (mediated by the

Organization of Islamic Conference) produced a cease-fire in September 1986. Yet the new Aquino government would not accept the rebels' demands for territorial autonomy, and the Mindanaos rejected the government's offer of more cultural and religious freedom. A referendum was held in the southern Muslim provinces regarding the government's proposals. Suffering from an MNLF boycott the government was only successful in setting its agenda in three of the thirteen provinces. The MNLF/BMA continues to wage its guerrilla war.

The fighting was said to have claimed between 50,000 and 100,000 lives up through 1980 alone.

Notes

1. It is difficult to say with any certainty when exactly the fighting between Christians and Muslims reached an intensity at the level of war. In any case, however, the designation "war" applies to the confrontations in 1970.
2. They split from the MNLF in 1978.

Sources

Ahmad 1982; Lacar 1988; Luther 1981; Molloy 1988; Noble 1981; Werning 1983 (UP)

111 Bangladesh War
25.3.1971 - 17.12.1971[1]; B-2 / B-1 / BC[2]-2
War ended by: military victory for side B

Warring parties:

Aggressor (Side A): Pakistan
Aggressed (Side B): Diverse Guerrilla Movements, often called "Freedom Fighters" (the largest was the Mukti Fauj [„People's Army"]); also Police forces and Paramilitary units (East Bengal Regiment, East Pakistan Rifles, „Ansar")
Side B: India (3.12.1971 - 17.12.1971)
Intervention favoring side B: India (6/1971 - 2.12.1971)

Object and goals of the conflict

The background for this war was the discrimination on all fronts against the Bangladeshi population of East Pakistan by West Pakistan, as well as general domestic political instability. In addition the situation was framed by the rivalry between India and Pakistan since their formation in 1947, which had already led to three wars (see Wars Nr. 13, 84 and 82).

For over twenty years tensions had grown steadily between the two parts of Pakistan, separated from each other by more than 1,800 kilometers. Between the national elections of December 1970 and March 1971 the situation worsened markedly. The secessionist, leftist Awai League won overwhelmingly in East Pakistan, polarizing the parliament along East-West lines. Despite a brief window of opportunity for negotiations, the point of no return for confrontation was reached when the strongest West Pakistani opposition party of Z.A. Bhutto forced the military to forbid the opening ceremony of parliament in March 1971. The Awami League called for massive civil disobedience, which paralyzed public life in East Pakistan. Y. Kahn, the leader of both the Pakistani state and military, cut off negotiations with the Awami League on March 25, and on March 26 the Awami leader M. Rahman was arrested. West Pakistani military units moved into East Pakistan, and an Awami exile government was formed in neighboring India. On the same day as Rahman's capture the exile government declared the new state of Bangladesh. The decisive force in the eventually successful secession was the guerrilla movement.

The secessionist movement in East Pakistan corresponded with India's claims to regional hegemony, and they used the opportunity to pressure the West Pakistani military in East Pakistan by training rebels and supporting the exile government. After up to ten million refugees had fled into India, however, the willingness arose to get directly involved in the fighting. In June 1971 paramilitary units of the Indian Border Security Force (BSF) pushed deep into East Pakistani territory in support of the guerrillas. This escalation provoked an air attack by the Pakistani air force against Indian targets on December 3, which brought about massive battles on the border to both East and West Pakistan.

Outcome of the war

Former East Pakistan became the state of Bangladesh, with over 75 million inhabitants. The internal consolidation of the new state, however, was slow in coming due to internal power struggles, economic crises and natural disasters. In West Pakistan, now known as Pakistan, the military government was forced to resign, and Z.A. Bhutto took power. The military, however, retained enormous influence on Pakistani politics. The emergence of Bangladesh shifted the South Asian balance of power towards India's favor.

At least 300,000 persons died in the fighting.

Notes

1. The war ended in the East on December 16, 1971, by the capitulation of West Pakistani units. The Pakistani army on the West Pakistani front accepted a cease-fire one day later.
3. The first phase was a war of secession against the West Pakistani military starting in March 1971. The second phase consisted of increasing Indian intervention in support of the rebels from June to December 2, 1971; the third phase was interstate war starting on December 3, 1971, with the continuation of secessionist components in East Pakistan.

Sources

Ahmed 1978; Ahmed 1983; Barnds 1972; Brogan 1989; Brown 1972; Choudhury 1973; De Silva/May 1991; Ganguly 1988; Jackson 1975; Linde 1982; Maaß 1989; Schanberg 1971; Talbot 1972 (JPF)

110 Ceylon (JVP Uprising)
5.4.1971 - 5.6.1971; A-1
War ended by: military victory for side B

Warring parties:

Aggressor (Side A): Janatha Vimukthi Peramuna (JVP; People's Liberation Front)
Aggressed (Side B): Ceylon
Intervention favoring side B: India (1971 - 1971)

Object and goals of the conflict

On April 5, 1971, the Maoist People's Liberation Front JVP clashed with the "leftist" Bandaranaike government, which up to a year before had been their supporters. The uprising drew the support of unemployed Singhalese youth with primary school education who grew up during the failed Ceylonese attempt at economic development. The Ceylon economic model was unable to fulfill the expectations it had generated, especially among the Singhalese who felt they would benefit most from the redistribution efforts.

Outcome of the war

The deployment of the Indian air force was decisive for the successful defeat of the rebellion. The Ceylon government had called upon India for assistance. Estimates run up to 10,000 dead during the fighting, including 5,000 JVP supporters. The JVP was banned, and 25,000 interred supporters received a "Buddhist" re-education and were amnestied by 1977.

Sources

AdG 16243, 16410; Alles 1979; Bonn 1971; Der Spiegel 17/1971; Goonetileke 1975; Meyer 1977; Rösel 1994; Tremayne 1988; Wilson 1979 (JPF)

---

177 Bangladesh (Chittagong Hill Tracts)
since 1973; B-2

Warring parties:

Aggressor (Side A): Shanti Bahini (SB; Friedensarmee)[1]
Aggressed (Side B): Bangladesh

Object and goals of the conflict

The Buddhist mountain peoples[2] fought against the Bangladesh government for two main reasons: first, they sought the restoration of special minority rights which they had enjoyed under the colonial government, and second, they wanted a stop to the settlement of Bengalis in the relatively sparsely settled area known as the Chittagong Hill Tracts (CHT). More than half the army (about 50,000 soldiers) supported the new settlers against the militant factions of the mountain peoples.

The dreadfully poor social and economic situation in Bangladesh led to the new migration by Bengalis to the CHT.[3] The conflict was further stoked by traditional ethnic and religious opposition between the Muslims and the Buddhists.

The CHT also gained in importance for the Bengali economy, since oil lay underground and the region could also produce hydro-electric power for Chittagong's ports. The agrarian existence of the mountain people was also seriously encroached upon by the growing of rubber and teak export crops in the CHT.

Outcome of the war

The minorities have been unable to restore their sovereignty.[4] The change of government from the dictator Ershad to the political neophyte Khaleda Zia did little to change the situation, and military still controls many aspects of administration at the local level. The rebels and the government began negotiations in November 1992, but it remains unclear how realistic the rebel's goals of autonomy are, given the demographic shift in the CHT over the years. The military confrontations are currently limited to sporadic attacks, assassinations, and occasionally still massacres.

There are wildly varying estimates of the numbers of victims, since the region is not accessible for observers. Figures swing between 1,500 and 200,000 war dead. Around 70,000 people have fled the CHT and sought refuge from the army in India.

Notes

1. The SB is the military wing of the Porbottya Chattagram - Jana Sanghati Samiti (PC-JSS, CHT People's Solidarity Association).
2. These mountain peoples are mostly Buddhist, and only rarely Hindus or Christians (Mey 1992:48).
3. Today about 480,000 mountain people and 311,000 Bengalis live in the CHT.
4. Twelve traditional peoples live in this mountain region: the Chakma, Taungchengya, Marma, Tripura, Brong, Mru, Khumi, Bawm, Panghkua, Mizo, Khyang and Sak (Mey 1992:48).

Sources

ai 1986; ai 1989; Bertocci 1984; Böttger 1975; Felber/Weidemann 1987; Heinz 1984; Hess 1972; Kulke/Rothermund 1982; Linde 1982; Ludwig 1985; Mey 1979; Mey 1980; Mey 1986; Mey 1988; Mey 1992; Nebelung 1986; OCCHT 1986; Opitz 1988; Pathak 1988; Pathak 1989; Serajul Islam 1987; Serajul Islam 1988; Uhlig 1977; WWIR 1987; Zingel 1983 (HW/JM)

---

171 Pakistan (Beluchistan)
3/1973 - 7/1977; B-2[1]
War ended by: military victory for side B

Warring parties:
Aggressor (Side A): Baluchistan People's Liberation Front (BPLF)
Aggressed (Side B): Pakistan

Object and goals of the conflict

The source of the conflict was Pakistan's economic, social and political neglect of the Beluchistan province. The BPFL had a traditionalist tribal base, but was politically inspired in part by Marxist intellectuals and students (the Belutschistan Students Organization; BSO). They fought for a multinational, socialist Pakistan in which Beluchistan would exist as an equal and largely independent territory. A minority within Beluchistan sought full independence as a sovereign state comprising all Belutchans, some of whom lived beyond Pakistan's borders, especially in eastern Iran. This secessionist trend convinced the Shah of Iran to support the Pakistani government in its war against the ethnonationalist Beluchistan movement. As the war dragged on secession, rather than autonomy within a multinational Pakistan, became the primary goal of the BPLF.

Outcome of the war

The Beluchistan rebels were unable to achieve their goals. On the contrary, the four-year-long war served to restore confidence and legitimacy to the Pakistani military, who had suffered humiliating defeat at the hands of Indian forces in 1971. The military managed to destabilize the civilian government of Z.A. Bhutto, and the subsequent military regime of Zia pursued a development program for the region which did nothing to change the discrimination against the local population. The war cost nearly 10,000 lives.

Note

1. Iranian attack helicopters and pilots participated sporadically in attacks by the Pakistani air force against the guerrillas. Iran was concerned that the goal of a national Belutchan state would include parts of eastern Iran, where many Belutchans lived.

Sources

Amin 1988; Harrison 1981; Harrison 1987 (PK)

---

183 Laos (First Meo [Hmong] Guerrilla War)
1975 - 1979; B-1
War ended by: military victory for side A

Warring parties:
Side A: Laos
Side B: Meo Guerrillas
Intervention favoring side A: Vietnam

Object and goals of the conflict

In 1975, the Pathet Lao took power in Laos and declared the People's Republic of Laos. Part of the Meo (also known as Hmong) Guerrillas who had opposed the Pathet Lao during the civil war (see War Nr. 76) decided to continue the fight against persecution.

The goal of the Pathet Lao and their Vietnamese allies had been to break the remaining resistance within Laos. This goal became more relevant as rumors arose that China was supplying the Meo guerrillas with weapons. Laos' relations with China had reached a low point following the tensions between Vietnam and China (see War Nr. 142).

Outcome of the war

Meo resistance was broken by the use of massive military force. Sporadic guerrilla actions, however, were still reported until 1979. By 1990 the resistance had reorganized itself such that the intensity of the fighting escalated above the threshold of war (see War Nr. 192).

At least 40,000 people died in the fighting. Some sources estimate that more than 300,000 people fled to Thailand. If this number is correct it would mean that approximately ten percent of the Laotian population fled the country. This flow of refugees can also be attributed to the Pathet Lao's seizure of power.

Sources

AdG 19780A, 21923C; Crossette 1987; Karniol 1990; Rosett 1990; Seagrave 1983; Simonitsch 1988; Urban/Urban 1982; Weggel 1980; Weggel 1981; Weggel 1986 (WW)

---

121 Indonesia (East Timor)
since 11.8.1975; A-2 / B^1-2

Warring parties:
Aggressor (Side A): União Democratica Timorese (UDT) (11.8.1975 - 14.9.1975) / Indonesia (since 7.12.1975)
Aggressed (Side B): Frente Revolucionaria de Timor Leste Independiente (FRETELIN)[2]

OBJECT AND GOALS OF THE CONFLICT

Portugal had withdrawn its colonial troops from East Timor after the revolutionary events of April 1974. This led to a power vacuum soon filled by new political organizations. The socialist-oriented FRETELIN sought rapid decolonization and independence for East Timor, which the Portuguese had ruled for over one hundred years. The conservative UDT cooperated at first with FRETELIN, but then attempted to exclude them from power by staging a violent coup on August 11, 1975. Portuguese troops helped FRETELIN prevail, however, and they took over the government after a short civil war. On November 28, 1975, FRETELIN declared the regions independence as the Democratic People's Republic of East Timor. On December 7, 1975, Indonesian armed forces invaded. FRETELIN fled inland, from where they organized resistance for three years against the Indonesian troops, though they steadily lost control over their own territory.

OUTCOME OF THE WAR

Although the Indonesian Republic incorporated East Timor in July 1976 as their twenty-seventh province, the U.N. General Assembly considered East Timor to be still technically under Portuguese administration. In 1980, FRETELIN reorganized itself as a small guerrilla army which continues to fight against the overwhelming superiority of the Indonesian troops. In November of 1992 the Indonesian military captured FRETELIN's leaders.

At least 20,000 people were killed in this war, and human rights groups charge that more than 200,000 people were murdered by the Indonesian occupation forces in the first years of the war. A massacre by Indonesian soldiers on November 12, 1991 resulted in at least 50 deaths and 90 missing persons, according to official figures.

Notes

1. It is possible to view the time between the declaration of East Timor's independence and the occupation by Indonesian troops as an interstate war.
2. Since 1979 the leadership of the socialist-oriented FRETELIN has been weakened in the East Timor resistance movement. FRETELIN continues to exist, however, as a guerrilla movement, mostly in the eastern part of the country.

Sources

South East Asia Chronicle: Aug. 80, June 84; Jolliffe 1978; Kubitschek/Wesser 1981; Ludwig/Horta 1985; Suter 1980; SOAa: May 1992, July 1992, January 1993; Pazifik-Informationsstelle 1991 (UP/DSf)

---

132 Cambodia (Third Indochina War)
since 12/1975; C-2 / A-1[1]

Warring parties:

Phase 1: (until 16.1.1979):
Aggressor (Side A): Cambodia
Aggressed (Side B): Vietnam
Side B: Front d'Union Nationale du Kampuchea pour le Salut National (FUNKSN)[2] (25.12.1978 - 16.1.1979)
Phase 2:
Side A: Khmer Rouge / Armée Nationaliste Sihanoukienne (ANS)[3] (since 1979) / Cambodian People's National Liberation Front (KPLNF)[4] (since 1979)
Side B: Cambodia
Intervention favoring side B: Vietnam (until 9/1989)

Object and goals of the conflict

The origins of this war span a wide variety of issues. The interstate conflict, which was the catalyst for war, focused on competing territorial claims. A further apple of discord was the rivalry between Vietnam, who sought regional hegemony, and Cambodia, who insisted on its independence. Beneath these power-political issues existed historical enmities between socially and culturally diverse peoples, who were further alienated by ideological differences propounded by their respective nationalist-communist governments. These ideological differences were themselves partially the result of the Sino-Soviet conflict being carried into the Vietnam-Cambodian war.

These interstate conflicts determined the nature of the internal war in Cambodia. The roots of the internal conflict, however, go back to the disfigured development process which began under French colonialism, and resulted in the dependence of the rural population on the urban elite. This urban-rural opposition intensified after independence as the war in Indochina insured continued domestic instability. This opposition gradually transformed into the framework of class conflict, and by 1968 the Khmer Rouge fought a civil war against the Sihanouk government. After a series of dramatic turns of events, including Sihanouk's ouster by the right-wing Lon Nol, followed by the overthrow of Lon Nol by the combined forces of the Khmer Rouge and their former adversary, Sihanouk, the Khmer Rouge finally took complete control of Cambodia in 1975 (see War Nr. 99). To institute their notion of social reform and to secure their position—above all against their former ally, Sihanouk—the Khmer Rouge created a regime of terror nearly unparalleled in modern history. The leadership around Pol Pot sought to eliminate the unequal relationship between urban and rural Cambodia by creating a nationalist, pseudo-communist system in which subsistence agriculture formed the structure of the state. To this end all non-agricultural sectors of the population were to be re-educated and/or exterminated. The irrational and grisly policies of Pol Pot led to genocide against his own people and to the invasion of Vietnam.

Two years of inconsequential border skirmishes and a failed offensive around New Years 1977/78 led the Vietnamese to plan a massive offensive on December 25, 1978. They were supported by units of the pro-Vietnamese FUNKSN. This time they were successful, and Pol Pot was forced to flee from office on January 16, 1979. Vietnam proceeded to install a friendly government, just as they had in Laos in 1975. China was greatly displeased by Vietnam's victory, and reacted by attacking Vietnam in February 1979, (see War Nr. 142) to release pressure on the Khmer Rouge by tying up the Vietnamese military along their northern border.

Since January 1979, the Khmer Rouge has fought a guerrilla war against both the government and the Vietnamese occupation army. They were supported by the royalist resistance groups KPNLF (under the former Prime Minister Son Sann) and the ANS (under Prince Sihanouk). In 1982 the former ideological arch enemies came together in a shaky alliance against the Vietnamese, calling themselves the "Coalition Government for Democratic Kampuchea."

Outcome of the War

When Soviet aid dried up (a consequence of Gorbachev's political reforms) Vietnam was forced to withdraw its troops in September of 1989. The guerrilla factions began to fight among themselves, and the return of civil war seemed predetermined. The warring parties, wary of the devastating consequences of another round of war, agreed to find a political solution and began talks under the auspices of ASEAN (continued later by other great powers). In 1990 the arduous negotiations produced agreement on a U.N. peace plan, which called for a U.N. controlled transition to a democratically legitimated government which would represent all four warring parties.

The fighting continued, however, even as 22,000 "blue helmets" of the United Nations Transitional Authority in Cambodia (UNTAC) were stationed there in 1991 to prepare the ground for elections. The U.N. presence was unable to prevent the assassination of candidates and U.N. personnel (in 1992, 56 U.N. troops were killed) or massacres against the Vietnamese.

Fear of the return of the Khmer Rouge, yet the need to somehow constructively bind them to the political process presented an insoluble problem, even after the elections had taken place. This was one of the reasons for the failure of the U.N. to carry out its plan of disarming the guerrillas by early 1993. The Khmer Rouge forbid the U.N. access to the approximately 20 percent of Cambodia under their control, and were unwilling to part with their weapons. The other groups were unwilling to give up their arms as long as the Khmer Rouge kept theirs. The Khmer Rouge were also able to institute a lucrative trade in tropical hardwoods and precious stones by smuggling them across the Thai border, thereby securing enough money to continue their struggle.

Since 1945 Cambodia has seen almost constant war, including the Pol Pot era and the almost twenty-year-long civil war. The result has been approximately two million dead and the rank of Cambodia among the poorest countries in the world. Current problems facing Cambodia include the repatriation of more than 370,000 refugees from Thailand, the 190,000 homeless within the country itself, and the salvaging of almost total economic, agricultural and infrastructural ruin.

Notes

1. During the dry-season offensive Vietnamese and Cambodian troops clashed with Thai border troops.
2. FUNKSN was a pro-Vietnamese unity front against Pol Pot's dictatorship. It was founded before the second Vietnamese invasion and consisted of former Khmer Rouge and Vietnam-oriented communists. They diminished in importance after Pol Pot fell from power.

3. The ANS is a royalist guerrilla army made up of supporters of the former king and president Prince Sihanouk. They emerged in 1981 from three guerrilla organizations (MOULINAKA, Khleang Moeung and Oddar Teus).
4. The KPNLF can be characterized as an anti-Communist guerrilla movement. They were formed on March 5, 1979, from five different anti-Communist and anti-Vietnamese partisan groups and were led by the former Cambodian Prime Minister Son Sann.

Sources

Aschmoneit/Werning 1981; Barth/Terzani 1980; Buro 1988; Chabraborti 1985; Chanda 1986; Die Zeit 3.4.92; Der Spiegel 1.6.92, 17.5.93; Draguhn/Schier 1987; FAZ 13.4.92; FEER 27.5.93; FR 27.5.93; Heinzig 1987; Kiernan 1980; Kiernan 1986; Kiernan/Boua 1982; Schier 1986a; Schier 1986b; Vickery 1984; Vickery 1986; Weggel 1987; Weggel 1988b; Will 1987; Will 1989 (RtH)

---

142 People's Republic of China / Vietnam ("Education Campaign")
17.2.1979 - 1988; C-2
War ended by: termination of hostilities

Warring parties:

Aggressor (Side A): People's Republic of China
Aggressed (Side B): Vietnam

Object and goals of the conflict

The Chinese attack on Vietnam was a direct reaction to Vietnam's invasion of Cambodia (see War Nr. 132). It was a temporally and territorially limited border conflict that served two purposes: to underscore China's hegemonic interests in the region, and to relieve some pressure on China's Khmer Rouge allies in their war against Vietnam.

Outcome of the war

The Chinese military withdrew on March 19, 1979. They were not able to weaken the Vietnamese position in China, yet the threat of a two-front war had long-term economic consequences for Vietnam. To finance a larger army Vietnam had to milk its economy, already impoverished from many wars (see Wars Nr. 9, 34, 41, 76, 99, 132, 178, and 183). Occasional artillery fire was reported along the border until early 1988. Since 1989 the border has been reopened for trade. Regional competition between Vietnam and China has led, however, to infrequent military confrontations between the two states over the Spratly Atolls in the South China Sea.

Sources

Duncanson 1979; Draguhn/Schier 1987; Will 1989 (RtH)

---

155 India (Punjab)
since 7/1982; B-2

Warring parties:

Aggressor (Side A): Khalistan Liberation Force (KLF)[1] / All India Sikh Student's Federation (AISSF)[2] / Khalistan Commando Force (KCF)
Aggressed (Side B): India

Object and goals of the conflict

The end of British colonial rule in 1947 resulted in a division of India and Pakistan which cut across the Punjab. The region's historical metropolis, Lahore, remained in Pakistan. The flight of millions of Hindus and Sikhs from West Punjab altered the demographic character of the Indian Punjab. In 1966 the Indian Punjab was divided again, on the basis of an apparently manipulated referendum. Today approximately 60 percent of the inhabitants of the Punjab are Sikhs.

In 1980, the Sikhs rebelled against their economic, political and religious discrimination under the Hindus. In 1982 the unrest escalated to civil war, in which the Sikhs sought a sovereign independent state called "Khalistan." The charismatic leader of the Sikh extremists, Bindranwale, made his headquarters in the Golden Temple of Amritsar, which played an essential role in the identity of the militant Sikhs. When the Indian army stormed this religious site in July 1984, killing a large number of rebels and demolishing the temple, widespread outrage brought the militants significant support. This event set the context for the assassination of Indian Prime Minister Indira Gandhi by one of her Sikh bodyguards four months later. Her death incited pogroms across all of northern India, in which more than 3,000 Sikhs were murdered.

The new Prime Minister Rajiv Gandhi desired negotiations with Longowal, the leader of the moderate Sikh party Akali Dal. Their attempts at talks collapsed, however, under the opposition of both extremist Sikhs and opponents of dialogue within Gandhi's own party. In 1987 radical Sikhs assassinated Longowal and forced the Akali-Dal government to resign, thus signaling an end to any possible negotiations. India reacted by increasing its troop strength in the region and by taking direct administrative control of the Punjab. The conflict contained additional explosive potential since the Punjab shared a border not only with Pakistan but also with Kashmir (see Wars Nr. 161 and 188).

Outcome of the war

In 1991, the conflict in the Punjab reached its zenith (so far): more than fifteen people died daily in confrontations. Approximately thirty rebel groups grew in power during this year, controlling the border to Pakistan, collecting taxes and instituting the Punjabi as the sole language. Seventy-five months of massive repression by the central government had ended with elections in February 1992, which calmed the revolt. Many of the hundreds of thousands of refugees began to return, and, unlike in Kashmir, the majority of the Sikhs did not identify with the goals of the uprising or its means. In general, the population supported the view of the Akali Dal party, who condemned the terrorist actions since the end of the 1980s as criminal.

The success of Indian units and the regional Sikh police was due in part to their ruthlessness. Amnesty International and Asia Watch denounced the fact that instead of taking prisoners the troops killed more and more people. Torture was a daily occurrence in the prisons, and since the mid-1980s more than 10,000 persons have been jailed without being formally charged. The Indian army also erected a 500-kilometer-long illuminated fence along the Pakistani border to prevent any infiltration and smuggling of weapons.

Notes

1. The Panthic Committees directed the actions of the two largest groups, the Khalistan Liberation Force (KLF) and the Khalistan Commando Force (KCF). Both committees were the religious-ideological leadership for the rebellion and directed four or five groups from their headquarters, which most probably lay in Pakistan.
2. The AISSF was the first of the extremist Sikh groups which came out for an independent Punjab. They led a dual existence between legal contacts with the Akali Dal party and connections to underground groups. Parts of the AISSF were apparently involved in militant activity.

Sources

Asian Survey 12/87, 11/83, 3/86, 2/84, 10/86, 2/85; Bombwall 1986;

De Silva/May 1991; FR 26.7.85, 3.5.86, 25.6.86, 14.7.86, 14.2.87, 27.3.87, 3.4.87, 13.5.87, 30.12.91; FT 10.5.91; Gill/Singhal 1984; Gupta 1985; Government of India 1986; India Today 31.10.84; Jeffrey 1986; Kapur 1986; Kapur 1987; KAS-Auslandsinformationen 5/88; Maaß 1988; Nayar 1966; Schermerhorn 1978; Tagesspiegel 26.7.90; taz 8.1.91; The Economist 7.5.88, 14.1.89; Thomas 1986 (JPF)

---

157 Sri Lanka (Tamil Conflict)
since 7/1983; B-2 / AB-1 / B-1

Warring parties:

Aggressor (Side A): Liberation Tigers of Tamil Eelam (LTTE; also "Tamil Tigers") / Eelam Revolutionary Organisation of Students (EROS) / other small Tamil guerrilla organizations / Janatha Vimukhti Peramuna (JVP; People's Liberation Front) (29.7.1987 - 11/1989)
Aggressed (Side B): Sri Lanka
Side B: Citizen's Volunteer Force (CVF)[1] (1987 - 1989) / Tamil National Army (TNA) (1989 - 3/1990) / Various death squads (1987 - 1989)
first A, then B: Eelam People's Revolutionary Liberation Front (EPRLF) (7/1983 - 29.7.1987) / People's Liberation Organization of Tamil Eelam (PLOTE) (since ??)
Intervention favoring side B: India[2] (15.9.1987 - 24.3.1990)

Object and goals of the conflict

The Sri Lankan Tamils had been a privileged group under British colonialism (1796-1948), for they had a generally high level of education and mobility. After independence, however, the independent state of Ceylon (which later became Sri Lanka) was led by Singhalese, who resented the Tamil's former status. The new constitution of 1972 cemented the leading role of the Singhalese, who also reformed the education system to counter the traditional advantages of the better-education Tamils. The educational reforms, in particular, led to the mobilization of Tamil students, who became the core of the emerging militant Tamil movement of the 1970s. The two ethnic groups became more and more estranged, and events came to a violent head in July 1983 when an island-wide pogrom against Tamils caused over 2,000 deaths and led to a massive exodus. Since that time forth the intensity of conflict has remained above the threshold of war.

As the war developed so did internal power struggles between the

Tamil guerrilla organizations. The victors were the Tamil Tigers who sought secession for the north and northeast regions of the island, where most of the two million Tamil people live, into a socialist-oriented sovereign state called Tamil Elam. The other organizations were cast aside, and now fight with government units against the Tigers.

The southern Indian Tamil Nadu, barely forty kilometers across the sea from Sri Lanka, became the logistical center for the guerrillas and a site of refuge for hundreds of thousands of refugees. The Indian government made three ill-fated attempts at mediation (in 1984, 1985 and 1986), whereupon they decided it was time to intervene militarily in the conflict. A peace treaty between India and Sri Lanka allowed for the stationing of Indian "peacekeeping forces" in the northern and eastern parts of the island. The peace treaty also envisioned patching together the Tamil settlements into a broadly autonomous administrative unit, but this plan met with opposition.

The moderate EPRLF accepted the plan, but the Tamil Tigers soon came to repudiate it, and by 1987 Indian troops and Tamil Tigers became enveloped in a guerrilla war. India, despite the deployment of over 70,000 soldiers and special anti-guerrilla units, was militarily unsuccessful in its contest with the Tigers, and by March 1990 had withdrawn their troops at the Sri Lankan government's behest.

The division of the island, as envisioned by the failed peace treaty, had also caused much unhappiness among the Singhalese because it resurrected traditional fears against both India and the Tamils. The Singhalese nationalist JVP began terrorist attacks against the government and the ruling United National Party. Under these terror tactics the economy and administration teetered at the brink of ruin, and they attempted to stop the terrorism by responding in kind. The government succeeded in November 1989, but the JVP uprising is estimated to have cost up to 30,000 lives. Three years afterwards more than 10,000 persons were reported to still be held in prison with no formal charges.

Outcome of the war

Negotiations between the Sri Lankan government and the Tamil Tigers had begun prior to the Indian withdrawal, though they never really managed to bring the two sides any closer together. The cease-fire did, however, give the guerrillas the opportunity to take over the newly-formed northeast province from Indian troops. The Muslim minority group in the east was not prepared to submit to the Tigers, and it gave up its neutrality in 1990 and formed self-defense groups in cooperation with the government. The war continues, although the Tamil Tigers have been pushed back to the northern peninsula of Jaffna and the surrounding jungle. The Indian government cracked down as well on their presence in Tamil Nadu.

The military stalemate reflects the economic division of Sri Lanka in the north, controlled by the LTTE and the rest of the island, dominated by the Singhalese. The north has been economically wrecked by the war and by economic embargoes. The region's population has declined from 1.7 million people before 1983 to 800,000 in 1992, and the rice fields have been reduced by 49 percent. There is no electricity, no water or medication, and little food. The rest of the island, however, has experienced an economic upturn since the end of the JVP uprising, due mainly to foreign investment and tourism.

Considerably more than 50,000 people have died in the war; India lost 1,500 soldiers. Within Sri Lanka there are over 600,000 refugees, and 150,000 more have been forcibly returned to Sri Lanka from India.

Notes

1. The former Tamil guerrilla groups were re-outfitted by the Indian Peace-Keeping Forces and re-named the Citizen's Volunteer Force (CVF). This group cooperated with the new EPRLF provincial council. They fought on the side of the Indian units in the post-1987 war against the Tamil Tigers. The CVF was succeeded by the TNA, which was later crushed by the LTTE after the EPRLF administration collapsed in March 1990.
2. Indian fighter planes had already escorted transport planes which parachuted supplies to Jaffna in June 1987.

Sources

DÜI 1987; DÜI 1989; De Silva/May 1991; FEER 9.11.89, 23.11.89, 5.4.90, 12.9.91; FEER Asia Yearbook 1991; Hellmann-Rajanayagam 1986; Hellmann-Rajanayagam 1991; IDR 5/1988, 1/1990, 6/1990; Keller 1986; O'Ballance 1989; Obeyesekere 1984; Ponnambalam 1983; Roberts 1979; Rösel 1987; Silva 1984a; Silva 1984b; Spencer 1990; Tremayne 1987; Tremayne 1988; Tagesspiegel 23.8.90; Wilson 1979; Wilson 1988 (JPF)

---

161 Pakistan / India (Ziach Glacier)
4/1984 - 1989; C-2

War ended by: fighting fell under the threshold of war

Warring parties:
Aggressor (Side A): Pakistan
Aggressed (Side B): India

Object and goals of the conflict

In the Karakorum region of northern Kashmir lies a 4,500-square-kilometer region of snow-covered mountains and towering glaciers. This forbidding territory had been left out of the agreements concerning the end of the two Kashmir wars between Pakistan and India and the ensuing cease-fire line (see Wars Nr. 13 and 82). This war is a successor-war to the earlier Kashmir conflicts, and is a limited territorial war for strategic and—more importantly—political purposes. Both governments used this war to document their respective claims to Kashmir, to secure their own rule domestically, and as a sort of "pressure valve" to keep their interstate rivalry active while avoiding a wider war. It is also true that this region of Kashmir was becoming more important from an environmentally strategic perspective, since it contains tremendous water resources. Since 1989 the military activity has fallen below the threshold of war. Since 1990 this conflict was entirely overshadowed by the civil war within the Indian part of Kashmir (see War Nr. 188).

Outcome of the war

The military confrontations continue, though under the threshold of war, to India's strategic advantage (they control the most important mountain passes). It retains potential for escalation back to war within the tense context of the overall Kashmir conflict.

Altogether the war resulted in over 1,000 deaths, though nearly 80 percent of them are attributable to the inhospitable climate rather than battle.

Sources

Albright/Zamorra 1989; Armed Forces 5/1989; Betz 1986; Betz 1989; Braun/Ziem 1989; Chaudry 1976; Desmond 1989; De Silva/May 1991; Draguhn/Hofmeier/Schönborn 1989; FAZ 1.10.87; FR 9.11.84, 9.1.86; FT 1.10.87; Janes Defence Weekly 14.11.87, 30.4.88, 17.2.90; Khan 1989; Ganguly 1988; Mohapatra 1986; Munro 1989; NZZ 10.10.87; Pasricha 1985; Rothermund 1984; SZ 13.8.85; *The Economist* 12.10.85, 3.10.87; *The Times of India* 24.9.89 (RtH)

187 Pakistan (Sind Province)
since 11/1986; B-2

Warring parties:
Aggressor (Side A): Jiye Sind Mahaz (Jeeya-Sind-Movement)
Aggressed (Side B): Pakistan
Side A: Muhajir Qaumi Movement (MQM) / Paschtun drug and weapons mafia their armed militias (since 1986)

Object and goals of the conflict

The division of British India into India and Pakistan in 1947 produced torrents of refugees in both directions. Approximately eight million Hindus and Sikhs were driven from Pakistan, while a similar number of Muslims fled from India. In the two largest cities of the Sind Province—Karachi and Hyderabad—Urdu-speaking Muhajirs became the majority. They occupied key posts in trade, industry, administration and the military, and achieved national influence. The original Sind population found themselves pushed to the side in their own province, and resented their second-class status. The Jeeya-Sind Movement demanded proportional representation of Sindhis in administration and in the army, a reduction in the percentage of "foreigners" in Sind, and autonomy for the province. Following the example of Bangladesh (see War Nr. 111) radical Sind nationalists went further and sought to create the independent state of Sindhu Desh. In May 1988 the Jeeya-Sind Movement transformed into the Sind National Alliance (SNA). Within the Movement there were ties to criminal gangs as well as dissatisfied police of Sindhi descent. It is often impossible to prove a direct connection, however, between Sindhi attacks against Muhajirs and the Jeeya-Sind Movement.

The Sindhi-Muhajir conflict has been overshadowed by the conflict between Muhajirs and Pashtuns. Hundreds of thousands of Pashtuns have descended on Karachi and other Sind cities since the beginning of the Afghan war. Most merely sought survival in a somewhat more economically and politically secure environment. Some, however, formed the hard core of a militant criminal organization which took control of the growing trade in drugs and weapons. While they supported the Afghan rebels against the government in Kabul, they were also having a devastating effect in Sind, where their activities resulted in rising drug use, proliferation of guns, crime and violence. The Mujahirs began to organize against Pashtun influence, and formed the MQM in 1984. This organization advocated a stronger Mujahir identity, recognition of Mujahirs as an ethnic "sub-nationality" and a reduction in Pashtun, Punjabi and Sindi influence in Muhajir cities.

Karachi and Hyderabad have borne the brunt of increasing violence between Sindhi, Mujahirs and Pashtuns, which has escalated since 1986.

Outcome of the war

The constant violent confrontations in Sind contributed significantly to both the Zia and Bhutto regimes' loss of power. The Sharif government came to power in 1990, and in May 1992 sent government troops to the province to impose order. "Operation Blue Fox," as the intervention was known, succeeded in calming the situation, but no political solution has been found. In January 1993 the violence began to rise again.

Many thousands of people have lost their lives in the war.

Sources

Ali 1990a; Ali 1990b; Amin 1988; Harrison 1987; Kennedy 1991; Lamb 1992; Nahost Jahrbuch 1987; Syed 1988; Wirsing 1988; Zirnig 1988 (PK)

---

164 Thailand / Laos
3.11.1987 - 19.2.1988; C-2
War ended by: agreement without mediation

Warring parties:

Aggressor (Side A): Thailand
Aggressed (Side B): Laos

Object and goals of the conflict

The Thai-Laotian border has long lacked a clear demarcation, a situation often resulting in border skirmishes. In November 1987, the Thai military attacked Laotian troops in the disputed area. Both sides reacted with stubbornness and intransigence, yet carefully sought to prevent the conflict from escalating to involve, for example, the approximately 50,000 Vietnamese troops stationed in Laos.

Outcome of the war

Cease-fire negotiations successfully ended the fighting on February

17, 1989. Follow-up negotiations quickly improved relations between the two states.

500 to 700 soldiers died in the fighting.

Sources

FEER 7.1.88, 28.1.88, 3.3.88, 24.3.88; Financial Times 18.2.88; Pretzell 1988; Sarasin 1985; Weggel 1988a; Weggel 1988b (WW)

---

167 Papua New Guinea (Bougainville)
since 2/1989; B-2

Warring parties:
Aggressor (Side A): Bougainville Revolutionary Army (BRA)[1]
Aggressed (Side B): Papua New Guinea

Object and goals of the conflict

The 8,800-square-kilometer island of Bougainville was the site of a huge copper and gold mine owned by an Australian subsidiary of the British firm "Riotinto Zinc" (TRZ). They extracted ore worth U.S.$5 billion, and in the process decimated the environment of the entire island. The founder and leader of the BRA, Francis Ona, demanded that the mine be shut, a reforestation plan, U.S.$11.5 million for environmental damage and compensation for the displaced islanders, and a referendum about the future status of the island. Ethnically and geographically the island belongs to the Solomon Islands, and the Papua-New Guinea government feared the specter of secessionism and its spread to the main island, where mining concerns accounted for a substantial part of the economy (half of the nation's exports consist of copper).

Outcome of the war

The government soldiers were outfitted with modern weapons and trained in brutal tactics. The rebels were spared no mercy, and 5,500 villagers and suspected sympathizers were interned in camps. Captured rebels were tortured and arbitrarily executed. The mine closed down during the course of the fighting (in May 1989), and 2,000 workers and technicians were laid off. The closing had grave effects on the country's economy, since the mine accounted for 40 percent of Papua-New Guinea's total exports and 17 percent of its income.

Despite their military advantages, the government troops were eventually driven off the island and the BRA took power. On May 17, 1990, the BRA declared the independent Republic of Bougainville, whereupon the central government imposed an economic embargo on the island, which effectively shut it off from the outside world. In May 1992, government troops landed again on the island. Negotiations for a peaceful solution to the conflict continue.

The fighting and embargo have resulted in approximately 6,000 deaths.

Note

1. The BRA initially began fighting with bows and arrows, homemade shotguns, and mines made from bombs left from the Second World War. Later they also acquired modern weapons.

Sources

Der Spiegel Nr.11/1990; FR 19.1.90, 25.1.91; SZ 31.1.90 (KJG)

---

203 India (Assam)
1990 - 1992; B-2
War ended by: fighting fell under the threshold of war

Warring parties:

Aggressor (Side A): United Liberation Front of Assam (ULFA)[1]
Aggressed (Side B): India

Object and goals of the conflict

This war's origins lay in the economic disparity between the indigenous peoples and the Bengali settlers in the underdeveloped state of Assam, located in India's turbulent northeast.[2] The profits from Assam's lucrative oil and tea sectors were siphoned by the central government, leaving little for local development. Tensions between the Assam people and the newly arrived Bengalis began at the end of the 1970s, but escalated by 1983 into massacres of Bengalis. The Indian government barely managed to bring the situation under control in 1985 with promises of resettling the Bengalis, but by 1988 violence once again reared its head. The underground organization ULFA (which emerged out of the All Assam Students Union, or AASU) took up the fight against the Indian central government and for an independent Assam. In 1990 the confrontations reached a level above the threshold of war.

OUTCOME OF THE WAR

The ULFA devised a parallel government in most of Assam which financed itself by "taxing" the owners of the tea plantations. Tea production sank, however, as growers refused to pay protection money to the ULFA, who in turn attacked the tea operations. By November 1990 the Indian government put Assam under direct Indian administration, and simultaneously began a five-month military campaign against the rebels. In 1991 the Indian army staged another offensive, which was so successful that the ULFA declared a unilateral cease-fire. The UFLA broke apart into a moderate and a militant wing. The moderates began negotiations with the Indian government under Rao, and the activity of the militant wing receded below the threshold of war by 1992.

At least 300 people died in the fighting.

Notes

1. Founded in April 1979 as a Maoist organization. Occasionally referred to as the military arm of the Asom Gana Parishad (AGP) Party.
2. See also Wars Nr. 26, 90, 172 und 173.

SOURCES

Ahmed 1991; Asian Survey (Berkeley) 10/1984: 1056-1068, 11/1986: 1184-1206; De Silva/May 1991; FAZ 29.11.91; FEER 8.11.90:44/45; FT 10./11.11.90, 19.2.91; Handke 1982; Hindustan Times 17.9.91; India Today 15.12.91, Indo-Asia 1/1969, 1/1988; 15.1.92, 31.3.92; Neues Deutschland 30.12.91; NZZ 7.12.90, 18.9.91, 12.1.92; SZ 29.11.90; The Telegraph 23.2.91 (JPF/SA)

---

188 India (Kashmir)
since 1/1990; B-2

WARRING PARTIES:

Aggressor (Side A): Jammu and Kashmir Liberation Front (JKLF) / Hizbul-Mujaheddin / Jammu and Kashmir People's League / many other organizations
Aggressed (Side B): India

Object and goals of the conflict

India's central government launched a Kashmir policy which hollowed out the special status assigned to the region in the Indian constitution. The lack of political representation, economic discrimination and religious differences had already prompted the Kashmiri Muslims, who formed the majority, to seek secession from India. In 1987 unrest began to grow in the Indian part of Kashmir, and by 1990 it had escalated to civil war. The fighting was concentrated in the Kashmir Valley, where most of the seven million Kashmiris lived.

Over 100 militant Kashmiri movements sprung up, but they essentially represented three strands. The JKLF, who dominated at the beginning of the war, sought an independent, secular Kashmir including the territory under Pakistani rule. This view was shared by the majority of the Kashmiris. As the JKLF withered under Indian attacks, however, pro-Pakistani Muslim fundamentalist groups such as the Hizbul-Mujaheddin gained influence. A third group of non-fundamentalist pro-Pakistani organizations, such as the Jammu and Kashmir People's League, represent a third development, but have had little effect.

This conflict was one of India's most dangerous internal rebellions because of the threat of escalation to interstate war with Pakistan (see previous Indian-Pakistani Wars Nr. 13, 84, 161, 82 and 111). India accused Pakistan of aiding the rebels in Kashmir. The chronic tensions between the two rivals rose again in the early 1990s with ominous implications—only pressure by the United States seems to have averted a possible nuclear confrontation between the two states.

Outcome of the war

India's reaction to the rebels has been dominated by repression. The state of Jammu and Kashmir has been under direct Indian administration since the unrest began in 1990. The nature of the anti-terror laws in the state amount de facto to martial law. Nearly 500,000 of India's 1.2 million soldiers are stationed in Kashmir. Paramilitary groups such as the Border Security Force are notorious for their brutality. The uprising is estimated to have caused up to 20,000 deaths, with 30,000 missing, and 15,000 interned in Indian jails.

Sources

Albright/Zamorra 1989; Braun/Ziem 1989; Chaudry 1976; De Silva/May 1991; FAZ 13.2.90; FEER Nr.20, 18.5.89; FR 6.3.90, 14.4.90, 18.5.90, 7.7.90, 23.8.90; FT 26.1.90, 26.5.90, 6.1.92; Ganguly 1988;

Ganguly 1990; Lamb 1991; Newsweek 30.4.90, 28.8.90, 10.5.93; SZ 12.4.90, 29.8.90; Taylor 1991; IDR 8/1990, 4/1991; NZZ 19.8.89, 14.9.89, 11.1.90, 1.2.90, 13.4.90; taz 13.6.91 (JPF)

---

192 Laos (Second Meo (Hmong) Guerrilla War)
1/1990 - 7/1992; B-1
War ended by: termination of hostilities

WARRING PARTIES:
Side A: Laos
Side B: Lao United Front (LUF) / Ethnic Liberation Organization of Laos (ELOL)
Intervention favoring side A: Vietnam

OBJECT AND GOALS OF THE CONFLICT

Ever since their resistance movement was crushed in 1979 by the Laotian government and Vietnamese troops (see War Nr. 183), the mountain peoples of Laos continued to carry out isolated attacks and guerrilla actions. Since 1981 there have been attempts to unite the resistance movement, which consists of more than two different groups, under the United Lao National Liberation Front (ULNLF). Their military backbone is the ethnic Hmong group, who live in the Laotian highlands. In mid-1985, however, the ULNLF split into the ELOL and the LUF. The otherwise latent guerrilla conflict exploded into war when Laotian and Vietnamese forces launched a massive offensive in January 1990. Portions of the guerrillas were forced into Thailand, while smaller groups continued to operate in Laos, although individually and with no organization.

The ELOL is a Hmong nationalist organization which fights for the "self-determination of the ethnic minorities and recognition of their culture." The LUF, on the other hand, is more interested in establishing a broad anti-Vietnamese coalition. Both factions contest the other's claim to be the sole representative of Hmong interests.

OUTCOME OF THE WAR

The fighting had to stop after the Hmong guerrillas lost their logistical and financial support from Thailand and exiled Laotians. Original support from Thailand changed into hostility, as the Thai security forces began to work with the Laotians against those Hmong operating out of Thailand.

Note

1. "Hmong" is the official name of the highland Laotians, who are known in China as the Miao or the Meo. The Hmong are one of the approximately 102 ethnic groups living in Laos. Like the Kurds, they are distributed across several countries. About two million Miao live in China and 500,000 in the mountains of Laos, Vietnam and Thailand.

Sources

Castella 1991; Country Report 1992; Crossette 1987; Karniol 1990; Rosett 1990; Simonitsch 1988; Weggel 1988a; Weggel 1990a; Weggel 1990b; Weggel 1992a; Weggel 1992b (RtH/WW)

---

198 Indonesia (Aceh)
since 5/1990; B-2

Warring parties:

Aggressor (Side A): Movement to Disrupt the Public Order (GPK)
Aggressed (Side B): Indonesia

Object and goals of the conflict

In the western Indonesian province of Aceh on the island of Sumatra live approximately 3.2 million people who belong to a fundamentalist strain of Islam. This is in marked contrast to the majority of Muslims in Indonesia (87 percent of the total population), who belong to moderate denominations. Conflicts in Aceh were reported during the colonial period, and starting in the 1950s the population opposed Indonesian attempts at assimilation. Despite concessions which gave the region a degree of cultural and religious autonomy, dissatisfaction with rule from Java reached a boiling point in the middle of the 1980s. A major factor in the unrest is probably the relative wealth of natural resources in Aceh—despite having only 2 percent of the population, Aceh produces 30 percent of the national oil and gas exports. It also contains mineral and forest resources. The conflict is fueled by resentment against the central government and the crass economic contradictions which leave the region relatively poor. The citizen's movement for a free Aceh (Aceh Merdeka) has formulated political demands, but the civil war is carried out by the GPK, who themselves formulated a straightforward goal: to kill settlers and soldiers from Java.

Outcome of the war

Many of the Aceh fled to Malaysia in 1990/91. Currently the Indonesian government continues its operations under the official guise of combating drugs. They destroy poppy plantations and use violence against the village populations. The real purpose of the Indonesian military actions, however, lie in tracking down the guerrillas.

At least 2,000 people have died so far in this war.

Sources

FEER 29.8.91; FEER Asia-Yearbook 1991; FT 10.2.89, 2.8., 24./25.11.90, 19.3., 1.5.91; SOAa 1/1991; taz 24.7.90, 3.7.91, 6.7.91 (JPF/UP/MMT)

East Asia

7 China (Chinese Civil War)
15.3.1946 - 21.4.1950; A-2
War ended by: military victory for side B

Warring parties:

Side A: China
Side B: Guerrillas of the Chinese Communist Party[1]

Object and goals of the conflict

The end of the Second World War also saw the end of the tenuous anti-Japanese alliance, formed in 1937 between the Chinese Communist Party and the Kuomintang government (the Nationalist People's Party) under Chiang Kai-shek. The nationalist government had tried to destroy the communist movement even before the Japanese invasion, yet they narrowly escaped annihilation and fled to the rural areas of the north and northeast. After the end of the war the Communist Party, with the support of the rural population, began to fight the government. Attempts at mediation by the U.S. failed, due in part to the massive level of distrust between the warring parties, and eventually the country became divided into zones of respective influence. The Kuomintang's attack on Communist-occupied cities in 1945 demolished any prospects of reconciliation.

OUTCOME OF THE WAR

The Communist Party succeeded in defeating the Kuomintang's troops, who were forced to retreat from the mainland to the island of Taiwan (Formosa). The result was the formation of two Chinese states: the communist People's Republic of China (PRC) on October 1, 1949, and in Taiwan the Republic of China (ROC), under U.S. military protection, on March 1, 1950.[2] The ROC's ruling Kuomintang regards itself as the legal successor to the Chinese Republic of 1911. Both states refuse reciprocal diplomatic recognition and have talked in terms of militarily conquering the other. The resulting tensions led to three military conflicts between 1949 and 1958, all concerning small islands in the Formosa Straits which belong to Taiwan, but are claimed by China (see Wars Nr. 32, 36 and 50).

The deaths during the civil war numbered far above one million, and were accompanied by tremendous destruction, famine, epidemics and floods of refugees.

Notes

1. Some troops deserted the Kuomintang and fought with the guerrillas. After the Communist Party came to power they turned the guerrilla army into the national armed forces (the People's Liberation Army of the People's Republic of China).
2. Earlier literature refers to the Chinese Republic of 1911 as National China. Taiwan refers to itself as the "Republic of China."

SOURCES

Chaffee 1967; Chassin 1965; Kaminski 1971; Keesings Zusammenfassungen der Kriegsergebnisse in den Jahren 1946-50; Kindermann 1980; Meyer 1982; Thorton 1973 (RtH)

---

25 Korean War
25.6.1950 - 27.7.1953; C-1
War ended by: agreement without mediation

WARRING PARTIES:

Side A: North Korea
Side B: South Korea / United States of America (5.7.1950 - 27.7.1953) / Canada / Colombia / Great Britain / Netherlands / Belgium / Luxembourg / France / Greece / Ethiopia / South Africa / Turkey / Thailand / Philippines / Australia / New Zealand (all: 950 - 27.7.1953)

Intervention favoring side A: People's Republic of China (26.11.1950 - 27.7.1953)

Object and goals of the conflict

Both communist North Korea and capitalist South Korea sought to expand their political system across the whole Korean peninsula. The North Koreans allegedly attacked South Korea, after which the U.S. and fifteen other states under the auspices of the U.N. came to South Korea's defense. An instrumental consideration in the U.N. action was to prevent the expansion of communism in East Asia. U.N. troops, that consisted mostly of the U.S. contingent, occupied most of the country, when the People's Republic of China suddenly intervened on the side of North Korea to repulse what they saw as the encroachment of capitalist countries.[1]

Outcome of the war

The status quo ante bellum was restored, which amounted to a seemingly permanent division of the country along the 38th parallel. The adversarial regimes were each strengthened in their uncompromising positions, and the Korean War heightened the global East-West conflict.

850,000 to 2,000,000 people died in the fighting, which destroyed 43 percent of Korea's industrial infrastructure and 33 percent of its housing. All sides violated international law, including the execution of prisoners of war and civilians. There were also unsubstantiated rumors about experiments on prisoners of war and the use of bacteriological weapons.

Note

1. The Korean War first experienced a phase in which the front shifted from the middle of the country to the southern tip, then up to the northern border, and then back to the middle. The second phase began in May 1951 and was characterized by bloody trench warfare.

Sources

Berger 1957; Brecher/Moser/Wilkenfeld 1988; Dingemann 1983; Encyclopaedia Britannica 1959, Bd.13:487; Encyclopaedia Britannica 1976, Bd.10:513; Fehrenbach 1963; Gödeke /Stuckmann/Vogt 1983; Jones 1950; Leckie 1962; Pfeffer 1953; Ridgway 1967 (BHu)

22 People's Republic of China / Tibet
7.10.1950 - 22.10.1950; C-2
War ended by: military victory for side A

WARRING PARTIES:
Aggressor (Side A): People's Republic of China
Aggressed (Side B): Tibet
Side B: Khambas[1]

OBJECT AND GOALS OF THE CONFLICT

The People's Republic of China attempted to militarily decide the disputed status of Tibet to its own advantage. Tibet's international legal status was disputed and unclear, and China asserted its rights to the territory based on claims from the era of the Chinese emperors.

Outcome of the war

The Chinese army quickly crushed the small Tibetan armed forces and their support from the Khambas tribe. Internationally isolated, militarily inferior and with large parts of the country under occupation, a Tibetan governmental delegation signed a treaty in 1951 which integrated Tibet as an "autonomous region" into the People's Republic of China. This treaty was signed without the participation of the Tibetan head of state, the Dalai Lama.

The deeply traditional and religious Tibetan people were prodded into resistance by the Chinese military occupation, economic exploitation, establishment of a socialist order and increased settlement of Chinese. Their dissatisfaction has insured future conflict (see War Nr. 33).

Note

1. Tribal fighters from the northeast of Tibet.

SOURCES

Keesing's 7634, 7636; AdG 2645B, 2649B; Ludwig 1989; NZZ 31.10.50, 29.5.51; Peissel 1973 (RtH)

---

33 Tibet / People's Republic of China
8/1954 - 22.3.1959; AB-2
War ended by: military victory for side B

Warring parties:
Side A: Tibet
Side B: People's Republic of China

Object and goals of the conflict

The Chinese military attempted to enforce a socioeconomic reform program which met with the resistance of the traditional rural population of Tibet.[1] The expropriation of land for Chinese settlers, attempts at collectivization and even the collection of systematic information (such as the counting of cattle) led to considerable tensions. After a few tentative battles the Chinese increased their repression and the situation became politically polarized. Many sectors of the population who until then had remained relatively passive, such as monks and civil servants who had been left alone by the Chinese, came to support complete independence from China.[2]

Outcome of the war

The war ended with military victory for the Chinese troops. The Dalai Lama fled to India, whereupon the Pantchen Lama was instituted by the Chinese government as the head of Tibet. The Chinese government's attempts at reform and integration, however, failed. The brutal Chinese repression forced thousands of Tibetans to flee over the border to Nepal, Bhutan and India. Despite the stabilization of the Chinese military position in Tibet the mountain tribes remain unpacified, and they have undertaken sporadic attacks against Chinese troops in recent years, especially in western Tibet.

Notes

1. China occupied Tibet in November 1950 (see War Nr. 22). The "Peking Treaty" was signed six months later, in May 1951. Tibet retained formal autonomy over internal affairs, and the Dalai Lama became the formal leader again in May of 1952, but he repudiated the treaty and never returned.
2. In March 1959 the Tibetan government renounced the Peking Treaty.

Sources

Ginsburgs/Mathos 1964; Keesing's 2654, 26496, 7636; Ludwig 1989; Peissel 1973 (RtH)

32 People's Republic of China / Taiwan (First Quemoy Crisis)
3.9.1954 - 25.9.1954; C-1
War ended by: Termination of hostilities

Warring parties:
Side A: People's Republic of China
Side B: Taiwan
Intervention favoring side B: United States of America (9/1954 - 25.9.1954)

Object and goals of the conflict

The nationalist Kuomintang Chinese troops had foiled a communist Chinese takeover of these coastal islands in 1949, even though as the Kuomintang government was being driven from the mainland (see War Nr. 7). The ground for future conflicts was lain by competing claims to legitimate Chinese rule in general and over the islands in specific. The nationalist Chinese, now ensconced on Taiwan, occupied the small coastal islands, and stationed a large number of their troops on them. The People's Republic of China (PRC) perceived the strong Taiwanese military presence as a threat to the mainland harbor cities of Amoy and Fuzhou, and the PRC also interpreted extensive weapons exports from the U.S. to Taiwan as evidence of an impending attack. In a form of preemptive strike, the Chinese People's Liberation Army (PLA) set out to conquer the islands. The PRC was also motivated to attack because nearly two-thirds of the 21,000 PLA Chinese soldiers taken prisoner in the Korean war had defected to Taiwan, and the PRC also wanted to test the resolve of the United State's commitment to Taiwan's defense. The attack was also an opportunity to demonstrate the new PRC's regional power.

The Taiwanese Chinese government considered the stationing of large numbers of troops on the islands an appropriate reaction to the threat they perceived from the PRC. Taiwan was understandably concerned by China's efforts to restore the territorial status of the Chinese empire, which was one of the primary claims used to legitimize the invasion of Tibet in 1950 (see War Nr. 22). If these claims were accepted internationally, as they were to a significant degree regarding Tibet, then China could claim that any conflict in the territory of the former empire, including Taiwan, was an internal affair and should not involve outside powers.

Outcome of the war

The war did not change the status quo. Taiwan carried out massive bombardments of the Chinese mainland along the coast, and the PRC troops suddenly held their artillery fire with no explanation.

On December 2, 1954 the U.S. and Taiwan signed a mutual defense treaty, which included Taiwan and the southwestern Pescadore Islands, but did not directly include the coastal islands. Yet the U.S. did not exclude the possibility of coming to Taiwan's defense if a PRC attack on the islands of Quemoy and Matsu signified a Chinese invasion of Taiwan.

Sources

China Handbook 1955; Keesing's 4726A; Kindermann 1977; Rudolph 1986 (RtH)

---

36 People's Republic of China / Taiwan (Da Chen and Nan Ji island groups)
18.1.1955 - 22.1.1955; C-2
War ended by: termination of hostilities

Warring parties:
Side A: People's Republic of China
Side B: Taiwan

Object and goals of the conflict

The conflict was a continuation of the competing claims by Taiwan and the PRC to control the coastal islands (see Wars Nr. 7 and 32). The PRC attempted anew to secure their claim by military means.

Outcome of the war

The PRC was partially successful in conquering some of the territory: the People's Liberation Army (PLA) took control of the Yijiangshan island, part of the Da Chen island group, after air and sea battles with national Chinese (Taiwanese) forces. Taiwan retreated from the rest of the Da Cen and Nan Ji islands in favor of tightening the front elsewhere, since the relatively low strategic importance of the islands made a pitched battle in their defense nonsensical. The U.S. helped organize the withdrawal of Taiwanese troops. The PLA suffered high losses during the fighting.

Sources

Chaffee 1967; Keesing's 4984, 5013D; Kindermann 1977; Kindermann 1980; Rudolph 1986 (RtH)

---

50 People's Republic of China / Taiwan (Second Quemoy Crisis)
23.8.1958 - 25.10.1958; C-2
War ended by: termination of hostilities

Warring parties:

Side A: People's Republic of China
Side B: Taiwan

Object and goals of the conflict

This was the third and largest attempt of the People's Republic of China to take the coastal islands of Quemoy and Matsu by force (see Wars Nr. 7, 32 and 36). Air and sea battles around the northern island of Matsu in early August turned out to be diversionary maneuvers, and on August 23, 1958 the PLA began a massive offensive to capture Quemoy and its surrounding smaller islands. Their tactics changed, however, when the conquest of the first island, Dong Ding, failed in the face of Taiwanese military resistance. The PLA then laid siege to the island, hoping to force the Taiwanese troops to surrender. This attempt also faltered. If Taiwan had lost Quemoy and possibly Matsu, then they would have lost almost half of their ground troops, in addition to an almost unbearable loss of prestige. There was also fear that such a loss would have consequences leading to possible Chinese invasion of Taiwan.

The timing of the PRC's offensive and its continuation after initial failures gave rise to rumors that Beijing was not really interested in pursuing its stated military goals. Speculations centered on the PRC's desire to test both the loyalty of the USSR at a time of increasing Sino-Soviet tension, and the commitment of the U.S. to Taiwan. It seemed sensible to test the Superpower's level of interest in the "Chinese Civil War" at a time when "peaceful coexistence" had become a diplomatic watchword, and when the U.S. had recently become mired in the Lebanon Crisis (see War Nr. 49). Beijing would not have been opposed to a worsening of relations between Washington and Moscow.

Outcome of the war

The military confrontations again led to no change in the status quo.

After the termination of hostilities the military activity was limited to artillery salvos from Mainland China against Quemoy, which were fired on odd days only. In Beijing the view prevailed that there was no military solution and that Taiwan had to be tolerated as a "second China."

The PLA again suffered substantial casualties.

Sources

Gödeke/Stuckmann/Vogt 1983; Keesing's 7272, 7286; Kindermann 1977; Kindermann 1980; Rudolph 1986 (RtH)

Appendices

Chronological List of Wars, 1945-1992

1944

Greek Civil War (Phase 1)
War Nr. 1
War Type: A
3.12.1944 - 12.2.1945

1945

Spain 1945-1950
War Nr. 4
War Type: A
1945 - 1950

Iraq (Barzanis)
War Nr. 179
War Type: B
8/1945 - 10/1945

Indonesia
War Nr. 3
War Type: D
29.9.1945 - 27.7.1949

1947

Paraguay
War Nr. 11
War Type: A
3/1947 - 12.8.1947

Mahabad (Iran)
War Nr. 5
War Type: B
14.3.1947 - 13.4.1947

Madagascar (Decolonization War)
War Nr. 12
War Type: D
29.3.1947 - 1948

Pakistan / India (First Kashmir War)
War Nr. 13
War Type: A/AC
22.10.1947 - 1.1.1949

1946

Philippines (Luzon)
War Nr. 10
War Type: A
1946 - 1954

Bolivian Revolution
War Nr. 8
War Type: A
1946 - 4/1952

Indochina (First Indochina War)
War Nr. 9
War Type: D
9.3.1946 - 1.6.1954

China (Chinese Civil War)
War Nr. 7
War Type: A
15.3.1946 - 21.4.1950

Greek Civil War (Phase 2)
War Nr. 6
War Type: A
6/1946 - 9.10.1949

India / Hyderabad
War Nr. 15
War Type: C
13.9.1948 - 17.9.1948

Costa Rica (Coup Attempt)
War Nr. 16
War Type: A
10.12.1948 - 24.12.1948

1948

Myanmar (Burma)
War Nr. 21
War Type: AB
Since 1948

Malaysian Federation ("Malayan-Emergency")
War Nr. 20
War Type: DA
1948 - 2/1960

North Yemen 1948
War Nr. 17
War Type: A
2/1948 - 14.3.1948

Costa Rica
War Nr. 14
War Type: A
12.3.1948 - 19.4.1948

First Arab-Israeli War
War Nr. 18
War Type: C
15.5.1948 - 7.1.1949

1949

Colombia (Violencia)
War Nr. 19
War Type: A
21.5.1949 - 10.5.1957

1950

Korean War
War Nr. 25
War Type: C
25.6.1950 - 27.7.1953

Indonesia (South Molukks)
War Nr. 23
War Type: B
7/1950 - 12/1950

People's Republic of China / Tibet
War Nr. 22
War Type: C
7.10.1950 - 22.10.1950

Nepal 1950
War Nr. 24
War Type: A
11.11.1950 - 1/1951

1952

Kenya
(Mau-Mau Uprising)
War Nr. 29
War Type: D
10/1952 - 11/1956

Morocco
War Nr. 30
War Type: D
7.12.1952 - 2.3.1956

1954

India (Nagas)
War Nr. 26
War Type: B
1954 - 8/1964

Guatemala (Operation Success)
War Nr. 31
War Type: A
18.6.1954 - 1.7.1954

Tibet / People's Republic of China
War Nr. 33
War Type: AB
8/1954 - 22.3.1959

People's Republic of China / Taiwan (First Quemoy Crisis)
War Nr. 32
War Type: C
3.9.1954 - 25.9.1954

Algeria
War Nr. 35
War Type: D
1.11.1954 - 19.3.1962

1955

Aden Protectorate / North Yemen
War Nr. 44
War Type: C
1955 - 1958

Costa Rica (Second Coup Attempt)
War Nr. 37
War Type: A
11.1.1955 - 21.1.1955

People's Republic of China / Taiwan (Da Chen and Nan Ji Island Groups)
War Nr. 36
War Type: C
18.1.1955 - 22.1.1955

Cyprus I
War Nr. 38
War Type: DB
1.4.1955 - 24.2.1959

Cameroon (Decolonization)
War Nr. 39
War Type: DA
5/1955 - 1963

Sudan (First Sudanese Civil War)
War Nr. 40
War Type: B
18.8.1955 - 2/1972

1956

Hungarian Revolution
War Nr. 43
War Type: A
24.10.1956 - 10.11.1956

Second Arab-Israeli War (Suez War)
War Nr. 42
War Type: C
29.10.1956 - 6.11.1956

Cuban Revolution
War Nr. 45
War Type: A
12/1956 - 1/1959

1957

Vietnam (Second Indochina War)
War Nr. 41
War Type: A-1/AC-1
1957 - 29.4.1975

Muskat and Oman (Imam-Sultan Conflict)
War Nr. 48
War Type: B
7/1957 - 1959

Spanish Morocco
War Nr. 47
War Type: D
10/1957 - 4/1958

1958

Indonesia (Civil War)
War Nr. 52
War Type: AB
3/1958 - 7/1961

Lebanese Civil War 1958
War Nr. 49
War Type: AB
9.5.1958 - 14.10.1958

People's Republic of China / Taiwan (Second Quemoy Crisis)
War Nr. 50
War Type: C
23.8.1958 - 25.10.1958

1959

Laos (First Laotian Civil War)
War Nr. 34
War Type: A
1.7.1959 - 16.6.1961

1960

Zaïre (Secession and Anti-Regime War)
War Nr. 55
War Type: AB
11.7.1960 - 14.1.1963

1961

Angola
War Nr. 62
War Type: D/ABC/A
Since 4.2.1961

Cuba (Bay of Pigs Invasion)
War Nr. 58
War Type: A
17.4.1961 - 20.4.1961

Tunisia (Bizerta Crisis)
War Nr. 57
War Type: D
19.7.1961 - 22.7.1961

First and Second Iraqi-Kurdish War
War Nr. 61
War Type: B
16.9.1961 - 10.2.1964

1962

Guatemala
War Nr. 56
War Type: A
1962 - 1968

Ethiopia (Eritrea Conflict)
War Nr. 63
War Type: B
1962 - 5/1991

Nepal 1962
War Nr. 65
War Type: A
1/1962 - 2/1962

Indonesia (West Papua)
War Nr. 64
War Type: C
14.1.1962 - 18.8.1962

Yemenite Civil War
War Nr. 68
War Type: A
10/1962 - 1969

People's Republic of China / India
War Nr. 66
War Type: C
20.10.1962 - 20.11.1962

Brunei
War Nr. 67
War Type: A
12/1962 - 12/1962

1963

Guinea-Bissau (War of Independence)
War Nr. 77
War Type: D
1/1963 - 25.4.1974

Venezuela
War Nr. 53
War Type: A
2/1963 - 1967

Laos (Second Laotian Civil War)
War Nr. 76
War Type: A
31.3.1963 - 4/1973

Ogaden Uprising (Ethiopia 1963/64)
War Nr. 71
War Type: BC
8/1963 - 12/1964

Indonesia / Malaysia ("Confrontation")
War Nr. 74
War Type: C
12.9.1963 - 11.8.1966

Morocco / Algeria ("Tindouf War")
War Nr. 70
War Type: C
10/1963 - 2/1964

Aden Protectorate (Southern Yemenite
War of Independence)
War Nr. 87
War Type: D
14.10.1963 - 30.11.1967

Kenya (Shifta War)
War Nr. 75
War Type: B
11/1963 - 10/1967

Rwanda (Tutsi Invasion from Burundi, Uganda and Zaïre)
War Nr. 72
War Type: A
20.12.1963 - 1966

Cyprus II
War Nr. 73
War Type: B
21.12.1963 - 11.8.1964

1964

Zaïre (Anti-Regime Uprisings)
War Nr. 78
War Type: A
1/1964 - 24.9.1966

Colombia (FARC)
War Nr. 60
War Type: A
Since 27.5.1964

Nigeria (Tiv Uprising 1964)
War Nr. 184
War Type: AB
8/1964 - 12/1964
Vietnam (FULRO)
War Nr. 178
War Type: B
20.9.1964 - 10/1992

Mozambique (War of Independence)
War Nr. 81
War Type: D
25.9.1964 - 7.9.1974

1965

Indonesia (West Irian / West Papua)
War Nr. 130
War Type: B
Since 1965

Colombia (ELN)
War Nr. 79
War Type: A
Since 7.1.1965

Third Iraqi-Kurdish War
War Nr. 85
War Type: B
4.4.1965 - 29.6.1966

India / Pakistan (Rann-von-Kutch)
War Nr. 82
War Type: C
9.4.1965 - 1.7.1965

Dominican Republic
War Nr. 83
War Type: A
25.4.1965 - 3.9.1965
Peru (Guerrilla Uprising of the MIR
and the ELN 1965)
War Nr. 86
War Type: A
6/1965 - 1/1966

Uprising in Dhofar
War Nr. 88
War Type: AB/A
9.6.1965 - 11/1975

Pakistan / India (Second Kashmir War)
War Nr. 84
War Type: BC
5.8.1965 - 23.9.1965

Thailand
War Nr. 89
War Type: A
7.8.1965 - 1980

1966

India (Mizos)
War Nr. 90
War Type: B
1.3.1966 - 31.7.1980

Rhodesia ("Chimurenga II")
War Nr. 91
War Type: A
28.4.1966 - 21.12.1979

Chad (Civil War)
War Nr. 92
War Type: ABC
Since 6/1966

Namibia (Liberation and War of Independence)
War Nr. 93
War Type: B
26.8.1966 - 22.12.1988

1967

Third Arab-Israeli War ("Six Day War")
War Nr. 95
War Type: C
5.6.1967 - 10.6.1967

Zaïre (Mercenary Rebellion)
War Nr. 96
War Type: A
5.7.1967 - 6.11.1967

Nigeria (Biafra War)
War Nr. 97
War Type: B
6.7.1967 - 15.1.1970

1968

Argentina
War Nr. 104
War Type: A
1968 - 1977

Israel, Palestine
War Nr. 165
War Type: B
Since 1968

Cambodia
War Nr. 99
War Type: AC
17.1.1968 - 17.4.1975

People's Republic of Yemen
War Nr. 98
War Type: A
6/1968 - 11/1968

Spain (Basque Country)
War Nr. 106
War Type: AB/B
7.6.1968 - 1979

1969

India (Nagas)
War Nr. 172
War Type: B
1969 - 6.11.1975

Fourth Iraqi-Kurdish War
War Nr. 102
War Type: B
3.1.1969 - 11.3.1970

Egypt / Israel (War of Attrition)
War Nr. 103
War Type: C
6.3.1969 - 7.8.1970

El Salvador / Honduras ("Soccer War")
War Nr. 100
War Type: C
14.7.1969 - 18.7.1969

Northern Ireland
War Nr. 105
War Type: B
Since 14.8.1969

DPR Yemen / Saudi Arabia
War Nr. 101
War Type: C
26.11.1969 - 12/1969

1970

Philippines (NPA)
War Nr. 109
War Type: A
Since 1970

Philippines (Mindanao)
War Nr. 199
War Type: B
Since 1970

Jordan (Black September)
War Nr. 108
War Type: AB
16.9.1970 - 19.7.1971

1971

Bangladesh War
War Nr. 111
War Type: B/B/BC
25.3.1971 - 17.12.1971

Ceylon (JVP Uprising)
War Nr. 110
War Type: A
5.4.1971 - 5.6.1971

Uganda / Tanzania (I)
War Nr. 112
War Type: AC
8/1971 - 7.10.1972

1972

Burundi
War Nr. 113
War Type: A
29.4.1972 - 6/1973

AR Yemen / DPR Yemen
War Nr. 114
War Type: C
26.9.1972 - 13.10.1972

1973

Bangladesh (Chittagong Hill Tracts)
War Nr. 177
War Type: B
Since 1973

Pakistan (Belutchistan)
War Nr. 171
War Type: B
3/1973 - 7/1977

Fourth Arab-Israeli War (October War)
War Nr. 115
War Type: C
6.10.1973 - 26.10.1973

1974

Colombia (M-19)
War Nr. 137
War Type: A
1/1974 - 8.3.1990

Fifth Iraqi-Kurdish War
War Nr. 117
War Type: B
14.3.1974 - 3/1975

Cyprus III
War Nr. 116
War Type: BC
20.7.1974 - 16.8.1974

1975

Ethiopia (Tigray Conflict)
War Nr. 122
War Type: B/BA
1975 - 5/1991

Second Ogaden War (Ethiopia)
War Nr. 127
War Type: BC
1975 - 1984

Mozambique (Destabilization Campaign)
War Nr. 119
War Type: AC/A
Since 1975

Laos (First Meo Guerilla War (Hmong))
War Nr. 183
War Type: B
1975 - 1979

Lebanon War
War Nr. 120
War Type: ABC
Since 13.4.1975

Indonesia (East Timor)
War Nr. 121
War Type: A/B
Since 11.8.1975

West Sahara
War Nr. 123
War Type: D/B
18.11.1975 - 6.9.1991

Cambodia (Third Indochina War)
War Nr. 132
War Type: C/A
Since 12/1975

South Africa (Anti-Apartheid War)
War Nr. 125
War Type: A/AB
Since 1976

1976

Ethiopia (Oromo Conflict)
War Nr. 131
War Type: AB
Since 1976

Zambia (Mushala Uprising)
War Nr. 181
War Type: A
14.1.1976 - 20.11.1982

Sixth Iraqi-Kurdish War
War Nr. 126
War Type: BA
Since 5/1976

1977

Zaïre (Shaba I, "80 Day War")
War Nr. 128
War Type: A
8.3.1977 - 26.5.1977

Nicaraguan Revolution
War Nr. 129
War Type: A
10/1977 - 19.7.1979

AR Yemen
War Nr. 135
War Type: A
1978 - 1982

1978

Zaïre (Shaba II)
War Nr. 133
War Type: A
12.5.1978 - 30.5.1978

Afghanistan
War Nr. 141
War Type: A
Since 10/1978

Uganda / Tanzania (II)
War Nr. 134
War Type: AC
10/1978 - 2.6.1979

1979

People's Republic of China / Vietnam ("Education Campaign")
War Nr. 142
War Type: C
17.2.1979 - 1988

Iran (Kurdistan)
War Nr. 144
War Type: B
7/1979 - 1988

1980

Guatemala
War Nr. 146
War Type: A
Since 1980

Peru (Sendero Luminoso)
War Nr. 147
War Type: A
Since 17.5.1980

Iraq / Iran ("First" Gulf War)
War Nr. 148
War Type: C
22.9.1980 - 20.8.1988

Nigeria (Maitatsine Uprising in Kano/Northern Nigeria)
War Nr. 185
War Type: B
18.12.1980 - 29.12.1980

1981

El Salvador
War Nr. 140
War Type: A
10.1.1981 - 1.2.1992

Peru / Ecuador
War Nr. 149
War Type: C
28.1.1981 - 2.2.1981

Uganda
War Nr. 152
War Type: A
2/1981 - 5/1992

Nicaragua (Contra War)
War Nr. 151
War Type: A
4/1981 - 1990

1982

Syria 1982
War Nr. 153
War Type: A
2.2.1982 - 2/1982

Falkland/Malvinas War (Argentina / Great Britain)
War Nr. 154
War Type: C
25.4.1982 - 15.6.1982

India (Punjab)
War Nr. 155
War Type: B
Since 7/1982

1983

Zimbabwe (Matabeleland Conflict)
War Nr. 160
War Type: A
1/1983 - 5/1988

Sri Lanka ("Tamil Conflict")
War Nr. 157
War Type: B/AB/B
Since 7/1983

Sudan (South Sudan Conflict, Second Sudanese Civil War)
War Nr. 158
War Type: BA
Since 9/1983

Grenada Invasion (Operation Urgent
Fury)
War Nr. 156
War Type: C
24.10.1983 - 1.11.1983

1984

Pakistan / India (Ziak Glacier)
War Nr. 161
War Type: C
4/1984 - 1989

Turkey (Kurds)
War Nr. 159
War Type: B
Since 15.8.1984

1985

Mali / Burkina Faso (Border War)
War Nr. 180
War Type: C
21.12.1985 - 31.12.1985

1986

South Yemen 1986
War Nr. 162
War Type: A
13.1.1986 - 29.1.1986

Suriname (Jungle War)
War Nr. 163
War Type: B
21.6.1986 - 7.6.1989

Pakistan (Sind Province)
War Nr. 187
War Type: B
Since 11/1986

1987

Peru (MRTA)
War Nr. 204
War Type: A
Since 11/1987

Thailand / Laos
War Nr. 164
War Type: C
3.11.1987 - 19.2.1988

1988

Somalia (Civil War)
War Nr. 166
War Type: A/AB
Since 5/1988

1989

Papua New Guinea (Bougainville)
War Nr. 167
War Type: B
Since 2/1989

Romania
War Nr. 169
War Type: A
17.12.1989 - 28.12.1989

USA / Panama
War Nr. 170
War Type: C
20.12.1989 - 24.12.1989

Liberia, Sierra Leone
War Nr. 168
War Type: A
Since 24.12.1989

1990

India (Assam)
War Nr. 203
War Type: B
1990 - 1992

India (Kashmir)
War Nr. 188
War Type: B
Since 1/1990

Laos (Second Meo Guerrilla War (Hmong))
War Nr. 192
War Type: B
1/1990 - 7/1992

USSR (Azerbaijan, Armenia); Nagorno-Karabakh
War Nr. 174
War Type: B
Since 20.1.1990

Senegal (Basse-Casamance Conflict)
War Nr. 191
War Type: B
Since 4/1990

Indonesia (Aceh)
War Nr. 198
War Type: B
Since 5/1990

Tuareg (Mali, Niger)
War Nr. 197
War Type: B
Since 7.5.1990

Iraq / Kuwait
War Nr. 182
War Type: C
2.8.1990 - 4.8.1990

Rwanda
(Rebel Invasion from Uganda)
War Nr. 189
War Type: AB
Since 1.10.1990

Georgia (South Ossetia)
War Nr. 202
War Type: B
12/1990 - 7/1992

1991

Anti-Iraq Coalition / Iraq
("Second" Gulf War)
War Nr. 194
War Type: C
17.1.1991 - 27.2.1991

Iraq (Shi'ites)
War Nr. 196
War Type: A
Since 2.3.1991

Yugoslavia
(Slovenia, Croatia, Bosnia)
War Nr. 200
War Type: BC
Since 26.6.1991

Georgia
(Anti-Regime War)
War Nr. 201
War Type: A
Since 9/1991

Djibuti
War Nr. 190
War Type: AB
Since 10/1991

1992

Algeria
War Nr. 206
War Type: A
Since 1/1992

Moldavia (Dneister Republic)
War Nr. 205
War Type: B
3/1992 - 8/1992

Tadjikistan
War Nr. 208
War Type: A
Since 8/1992

Georgia (Abkhazia)
War Nr. 207
War Type: B
Since 14.8.1992

North Ossetia (Prigorodny District)
War Nr. 209
War Type: B
Since 30.10.1992

Wars, 1945-1992, Ordered by War Type

War Type: A

Greek Civil War (Phase 1)
War Nr. 1
3.12.1944 - 12.2.1945

Spain 1945-1950
War Nr. 4
1945 - 1950

Philippines (Island Luzon)
War Nr. 10
1946 - 1954

Bolivian Revolution
War Nr. 8
1946 - 4/1952

China (Chinese Civil War)
War Nr. 7
15.3.1946 - 21.4.1950

Greek Civil War (Phase 2)
War Nr. 6
6/1946 - 9.10.1949

Paraguay
War Nr. 11
3/1947 - 12.8.1947

North Yemen 1948
War Nr. 17
2/1948 - 14.3.1948

Costa Rica
War Nr. 14
12.3.1948 - 19.4.1948

Costa Rica (Coup Attempt)
War Nr. 16
10.12.1948 - 24.12.1948

Colombia (Violencia)
War Nr. 19
21.5.1949 - 10.5.1957

Nepal 1950
War Nr. 24
11.11.1950 - 1/1951

Guatemala (Operation Success)
War Nr. 31
18.6.1954 - 1.7.1954

Costa Rica (Second Coup Attempt)
War Nr. 37
11.1.1955 - 21.1.1955

Hungarian Revolution
War Nr. 43
24.10.1956 - 10.11.1956

Cuban Revolution
War Nr. 45
12/1956 - 1/1959

Laos (First Laotian Civil War)
War Nr. 34
1.7.1959 - 16.6.1961

Cuba (Bay of Pigs Invasion)
War Nr. 58
17.4.1961 - 20.4.1961

Guatemala
War Nr. 56
1962 - 1968

Nepal 1962
War Nr. 65
1/1962 - 2/1962

Yemenite Civil War
War Nr. 68
10/1962 - 1969

Brunei
War Nr. 67
12/1962 - 12/1962

Venezuela
War Nr. 53
2/1963 - 1967

Laos (Second Laotian Civil War)
War Nr. 76
31.3.1963 - 4/1973

Rwanda (Tutsi Invasion from Burundi, Uganda and Zaïre)
War Nr. 72
20.12.1963 - 1966

Zaïre (Anti-Regime Uprisings)
War Nr. 78
1/1964 - 24.9.1966

Colombia (FARC)
War Nr. 60
Since 27.5.1964

Colombia (ELN)
War Nr. 79
Since 7.1.1965

Dominican Republic
War Nr. 83
25.4.1965 - 3.9.1965

Peru (Guerrilla Uprising of the MIR
and the ELN 1965)
War Nr. 86
6/1965 - 1/1966

Thailand
War Nr. 89
7.8.1965 - 1980

Rhodesia ("Chimurenga II")
War Nr. 91
28.4.1966 - 21.12.1979

Zaïre (Mercenary Rebellion)
War Nr. 96
5.7.1967 - 6.11.1967

Argentina
War Nr. 104
1968 - 1977

People's Republic of Yemen
War Nr. 98
6/1968 - 11/1968

Philippines (NPA)
War Nr. 109
Since 1970

Ceylon (JVP Uprising)
War Nr. 110
5.4.1971 - 5.6.1971

Burundi
War Nr. 113
29.4.1972 - 6/1973

Colombia (M-19)
War Nr. 137
1/1974 - 8.3.1990

Zambia (Mushala Uprising)
War Nr. 181
14.1.1976 - 20.11.1982

Zaïre (Shaba I, "80 Day War")
War Nr. 128
8.3.1977 - 26.5.1977

Nicaraguan Revolution
War Nr. 129
10/1977 - 19.7.1979

AR Yemen
War Nr. 135
1978 - 1982

Zaïre (Shaba II)
War Nr. 133
12.5.1978 - 30.5.1978

Afghanistan
War Nr. 141
Since 10/1978

Guatemala
War Nr. 146
Since 1980

Peru (Sendero Luminoso)
War Nr. 147
Since 17.5.1980

El Salvador
War Nr. 140
10.1.1981 - 1.2.1992

Uganda
War Nr. 152
2/1981 - 5/1992

Nicaragua (Contra War)
War Nr. 151
4/1981 - 1990

Syria 1982
War Nr. 153
2.2.1982 - 2/1982

Zimbabwe (Matabeleland Conflict)
War Nr. 160
1/1983 - 5/1988

South Yemen 1986
War Nr. 162
13.1.1986 - 29.1.1986

Peru (MRTA)
War Nr. 204
Since 11/1987

Romania
War Nr. 169
17.12.1989 - 28.12.1989

Liberia, Sierra Leone
War Nr. 168
Since 24.12.1989

Iraq (Shi'ites)
War Nr. 196
Since 2.3.1991

Georgia (Anti-Regime War)
War Nr. 201
Since 9/1991

Algeria
War Nr. 206
Since 1/1992

Tadjikistan
War Nr. 208
Since 8/1992

War Type: AB

Myanmar (Burma)
War Nr. 21
Since 1948

Tibet / People's Republic of China
War Nr. 33
8/1954 - 22.3.1959

Indonesia (Civil War)
War Nr. 52
3/1958 - 7/1961

Lebanese Civil War 1958
War Nr. 49
9.5.1958 - 14.10.1958

Zaïre (Secession and Anti-Regime War)
War Nr. 55
11.7.1960 - 14.1.1963

Nigeria (Tiv Uprising 1964)
War Nr. 184
8/1964 - 12/1964

Uprising in Dhofar
War Nr. 88
9.6.1965 - 11/1975

Spain (Basque Country)
War Nr. 106
7.6.1968 - 1979

Jordan (Black September)
War Nr. 108
16.9.1970 - 19.7.1971

Äthiopen (Tigray Conflict)
War Nr. 122
1975 - 5/1991

Indonesia (East Timor)
War Nr. 121
Since 11.8.1975

South Africa (Anti-Apartheid War)
War Nr. 125
Since 1976

Ethiopia (Oromo Conflict)
War Nr. 131
Since 1976

Sixth Iraqi-Kurdish War
War Nr. 126
Since 5/1976

Sri Lanka ("Tamil Conflict")
War Nr. 157
Since 7/1983

Sudan (South Sudan Conflict, Second Sudanese Civil War)
War Nr. 158
Since 9/1983

Somalia (Civil War)
War Nr. 166
Since 5/1988

Rwanda (Rebel Invasion from Uganda)
War Nr. 189
Since 1.10.1990

Djibuti
War Nr. 190
Since 10/1991

War Type: ABC

Chad (Civil War)
War Nr. 92
Since 6/1966

Lebanon War
War Nr. 120
Since 13.4.1975

War Type: ABCD

Angola
War Nr. 62
Since 4.2.1961

War Type: AC

Pakistan / India (First Kashmir War)
War Nr. 13
22.10.1947 - 1.1.1949

Vietnam (Second Indochina War)
War Nr. 41
1957 - 29.4.1975

Cambodia
War Nr. 99
17.1.1968 - 17.4.1975

Uganda / Tanzania (I)
War Nr. 112
8/1971 - 7.10.1972

Mozambique (Destabilization Campaign)
War Nr. 119
Since 1975

Cambodia (Third Indochina War)
War Nr. 132
Since 12/1975

Uganda / Tanzania (II)
War Nr. 134
10/1978 - 2.6.1979

War Type: AD

Malaysian Federation ("Malayan-Emergency")
War Nr. 20
1948 - 2/1960

Cameroon (Decolonization)
War Nr. 39
5/1955 - 1963

War Type: B

Iraq (Barzanis)
War Nr. 179
8/1945 - 10/1945

Mahabad (Iran)
War Nr. 5
14.3.1947 - 13.4.1947

Indonesia (South Molukks)
War Nr. 23
7/1950 - 12/1950

India (Nagas)
War Nr. 26
1954 - 8/1964

Sudan (First Sudanese Civil War)
War Nr. 40
18.8.1955 - 2/1972

Muskat and Oman (Imam-Sultan Conflict)
War Nr. 48
7/1957 - 1959

First and Second Iraqi-Kurdish War
War Nr. 61
16.9.1961 - 10.2.1964

Ethiopia (Eritrea Conflict)
War Nr. 63
1962 - 5/1991

Kenya (Shifta War)
War Nr. 75
11/1963 - 10/1967

Cyprus II
War Nr. 73
21.12.1963 - 11.8.1964

Vietnam (FULRO)
War Nr. 178
20.9.1964 - 10/1992

Indonesia (West Iran/West Papua)
War Nr. 130
Since 1965

Third Iraqi-Kurdish War
War Nr. 85
4.4.1965 - 29.6.1966

India (Mizos)
War Nr. 90
1.3.1966 - 31.7.1980

Namibia (Liberation and War of Independence)
War Nr. 93
26.8.1966 - 22.12.1988

Nigeria (Biafra War)
War Nr. 97
6.7.1967 - 15.1.1970

Israel, Palestine
War Nr. 165
Since 1968

India (Nagas)
War Nr. 172
1969 - 6.11.1975

Fourth Iraqi-Kurdish War
War Nr. 102
3.1.1969 - 11.3.1970

Northern Ireland
War Nr. 105
Since 14.8.1969

Philippines (Mindanao)
War Nr. 199
Since 1970

Bangladesh (Chittagong Hill Tracts)
War Nr. 177
Since 1973

Pakistan (Belutschistan)
War Nr. 171
3/1973 - 7/1977

Fifth Iraqi-Kurdish War
War Nr. 117
14.3.1974 - 3/1975

Laos (First Meo Guerrilla War (Hmong))
War Nr. 183
1975 - 1979

Iran (Kurdistan)
War Nr. 144
7/1979 - 1988

Nigeria (Maitatsine Uprising in Kano/Northern Nigeria)
War Nr. 185
18.12.1980 - 29.12.1980

India (Punjab)
War Nr. 155
Since 7/1982

Turkey (Kurds)
War Nr. 159
Since 15.8.1984

Suriname (Jungle War)
War Nr. 163
21.6.1986 - 7.6.1989

Pakistan (Sind Province)
War Nr. 187
Since 11/1986

Papua New Guinea (Bougainville)
War Nr. 167
Since 2/1989

India (Assam)
War Nr. 203
1990 - 1992

India (Kashmir)
War Nr. 188
Since 1/1990

Laos (Second Meo Guerrilla War (Hmong))
War Nr. 192
1/1990 - 7/1992

USSR (Azerbaijan, Armenia); Nagorno-Karabakh
War Nr. 174
Since 20.1.1990

Senegal (Basse-Casamance Conflict)
War Nr. 191
Since 4/1990

Indonesia (Aceh)
War Nr. 198
Since 5/1990

Tuareg (Mali, Niger)
War Nr. 197
Since 7.5.1990

Georgia (South Ossetia)
War Nr. 202
12/1990 - 7/1992

Moldavia (Dneister Republic)
War Nr. 205
3/1992 - 8/1992

Georgia (Abkhazia)
War Nr. 207
Since 14.8.1992

North Ossetia (Prigorodny District)
War Nr. 209
Since 30.10.1992

War Type: BC

Ogaden Uprising (Ethiopia 1963/64)
War Nr. 71
8/1963 - 12/1964

Pakistan / India (Second Kashmir War)
War Nr. 84
5.8.1965 - 23.9.1965

Bangladesh War
War Nr. 111
25.3.1971 - 17.12.1971

Cyprus III
War Nr. 116
20.7.1974 - 16.8.1974

Second Ogaden War (Ethiopia)
War Nr. 127
1975 - 1984

Yugoslavia (Slovenia, Croatia, Bosnia)
War Nr. 200
Since 26.6.1991

War Type: BD

Cyprus I
War Nr. 38
1.4.1955 - 24.2.1959

West Sahara
War Nr. 123
18.11.1975 - 6.9.1991

War Type: C

First Arab-Israeli War
War Nr. 18
15.5.1948 - 7.1.1949
India / Hyderabad
War Nr. 15
13.9.1948 - 17.9.1948

Korean War
War Nr. 25
25.6.1950 - 27.7.1953

People's Republic of China / Tibet
War Nr. 22
7.10.1950 - 22.10.1950

People's Republic of China / Taiwan (First Quemoy Crisis)
War Nr. 32
3.9.1954 - 25.9.1954

Aden Protectorate / North Yemen
War Nr. 44
1955 - 1958

People's Republic of China / Taiwan (Da Chen and Nan Ji Island Groups)
War Nr. 36
18.1.1955 - 22.1.1955

Second Arab-Israeli War (Suez War)
War Nr. 42
29.10.1956 - 6.11.1956

People's Republic of China / Taiwan (Second Quemoy Crisis)
War Nr. 50
23.8.1958 - 25.10.1958

Indonesia (West Papua)
War Nr. 64
14.1.1962 - 18.8.1962

People's Republic of China / India
War Nr. 66
20.10.1962 - 20.11.1962

Indonesia / Malaysia ("Confrontation")
War Nr. 74
12.9.1963 - 11.8.1966

Morocco / Algeria ("Tindouf War")
War Nr. 70
10/1963 - 2/1964

India / Pakistan (Rann-von-Kutch)
War Nr. 82
9.4.1965 - 1.7.1965

Third Arab-Israeli War ("Six Day War")
War Nr. 95
5.6.1967 - 10.6.1967

Egypt / Israel (War of Attrition)
War Nr. 103
6.3.1969 - 7.8.1970

El Salvador / Honduras ("Soccer War")
War Nr. 100
14.7.1969 - 18.7.1969

DPR Yemen / Saudi Arabia
War Nr. 101
26.11.1969 - 12/1969

AR Yemen / DPR Yemen
War Nr. 114
26.9.1972 - 13.10.1972

Fourth Arab-Israeli War (October War)
War Nr. 115
6.10.1973 - 26.10.1973

People's Republic of China / Vietnam ("Education Campaign")
War Nr. 142
17.2.1979 - 1988

Iraq / Iran ("First" Gulf War)
War Nr. 148
22.9.1980 - 20.8.1988

Peru / Ecuador
War Nr. 149
28.1.1981 - 2.2.1981

Falkland/Malvinas War (Argentina / Great Britain)
War Nr. 154
25.4.1982 - 15.6.1982

Grenada Invasion (Operation Urgent Fury)
War Nr. 156
24.10.1983 - 1.11.1983

Pakistan / India (Ziak Glacier)
War Nr. 161
4/1984 - 1989

Mali / Burkina Faso (Border War)
War Nr. 180
21.12.1985 - 31.12.1985

Thailand / Laos
War Nr. 164
3.11.1987 - 19.2.1988

USA / Panama
War Nr. 170
20.12.1989 - 24.12.1989

Iraq / Kuwait
War Nr. 182
2.8.1990 - 4.8.1990

Anti-Iraq Coalition / Iraq ("Second" Gulf War)
War Nr. 194
17.1.1991 - 27.2.1991

War Type: D

Indonesia
War Nr. 3
29.9.1945 - 27.7.1949

Indochina (First Indochina War)
War Nr. 9
9.3.1946 - 1.6.1954

Madagascar (Decolonization War)
War Nr. 12
29.3.1947 - 1948

Kenya (Mau-Mau Uprising)
War Nr. 29
10/1952 - 11/1956

Morocco
War Nr. 30
7.12.1952 - 2.3.1956

Algeria
War Nr. 35
1.11.1954 - 19.3.1962

Spanish Morocco
War Nr. 47
10/1957 - 4/1958

Tunisia (Bizerta Crisis)
War Nr. 57
19.7.1961 - 22.7.1961

Guinea-Bissau (War of Independence)
War Nr. 77
1/1963 - 25.4.1974

Aden Protectorate (Southern Yemenite
War of Independence)
War Nr. 87
14.10.1963 - 30.11.1967

Mozambique (War of Independence)
War Nr. 81
25.9.1964 - 7.9.1974

Wars, 1945-1992, Ordered by War Number

War Nr. 1
Greek Civil War (Phase 1)
War Type: A
3.12.1944 - 12.2.1945

War Nr. 3
Indonesia
War Type: D
29.9.1945 - 27.7.1949

War Nr. 4
Spain 1945-1950
War Type: A
1945 - 1950

War Nr. 5
Mahabad (Iran)
War Type: B
14.3.1947 - 13.4.1947

War Nr. 6
Greek Civil War (Phase 2)
War Type: A
6/1946 - 9.10.1949

War Nr. 7
China (Chinese Civil War)
War Type: A
15.3.1946 - 21.4.1950

War Nr. 8
Bolivian Revolution
War Type: A
1946 - 4/1952

War Nr. 9
Indochina (First Indochina War)
War Type: D
9.3.1946 - 1.6.1954

War Nr. 10
Philippines (Island Luzon)
War Type: A
1946 - 1954

War Nr. 11
Paraguay
War Type: A
3/1947 - 12.8.1947

War Nr. 12
Madagascar (Decolonization War)
War Type: D
29.3.1947 - 1948

War Nr. 13
Pakistan / India (First Kashmir War)
War Type: A/AC
22.10.1947 - 1.1.1949

War Nr. 14
Costa Rica
War Type: A
12.3.1948 - 19.4.1948

War Nr. 15
India / Hyderabad
War Type: C
13.9.1948 - 17.9.1948

War Nr. 16
Costa Rica (Coup Attempt)
War Type: A
10.12.1948 - 24.12.1948

War Nr. 17
North Yemen 1948
War Type: A
2/1948 - 14.3.1948

War Nr. 18
First Arab-Israeli War
War Type: C
15.5.1948 - 7.1.1949

War Nr. 19
Colombia (Violencia)
War Type: A
21.5.1949 - 10.5.1957

War Nr. 20
Malaysian Federation ("Malayan-Emergency")
War Type: DA
1948 - 2/1960

War Nr. 21
Myanmar (Burma)
War Type: AB
Since 1948

War Nr. 22
People's Republic of China / Tibet
War Type: C
7.10.1950 - 22.10.1950

War Nr. 23
Indonesia (South Molukks)
War Type: B
7/1950 - 12/1950

War Nr. 24
Nepal 1950
War Type: A
11.11.1950 - 1/1951

War Nr. 25
Korean War
War Type: C
25.6.1950 - 27.7.1953

War Nr. 26
India (Nagas)
War Type: B
1954 - 8/1964

War Nr. 29
Kenya (Mau-Mau Uprising)
War Type: D
10/1952 - 11/1956

War Nr. 30
Morocco
War Type: D
7.12.1952 - 2.3.1956

War Nr. 31
Guatemala (Operation Success)
War Type: A
18.6.1954 - 1.7.1954

War Nr. 32
People's Republic of China / Taiwan (First Quemoy Crisis)
War Type: C
3.9.1954 - 25.9.1954

War Nr. 33
Tibet / People's Republic of China
War Type: AB
8/1954 - 22.3.1959

War Nr. 34
Laos (First Laotian Civil War)
War Type: A
1.7.1959 - 16.6.1961

War Nr. 35
Algeria
War Type: D
1.11.1954 - 19.3.1962

War Nr. 36
People's Republic of China / Taiwan (Da Chen and Nan Ji Island Groups)
War Type: C
18.1.1955 - 22.1.1955

War Nr. 37
Costa Rica (Second Coup Attempt)
War Type: A 11.1.1955 - 21.1.1955

War Nr. 38
Cyprus I
War Type: DB
1.4.1955 - 24.2.1959

War Nr. 39
Cameroon (Decolonization)
War Type: DA
5/1955 - 1963

War Nr. 40
Sudan (First Sudanese Civil War)
War Type: B
18.8.1955 - 2/1972

War Nr. 41
Vietnam (Second Indochina War)
War Type: A-1/AC-1
1957 - 29.4.1975

War Nr. 42
Second Arab-Israeli War (Suez War)
War Type: C
29.10.1956 - 6.11.1956

War Nr. 43
Hungarian Revolution
War Type: A
24.10.1956 - 10.11.1956

War Nr. 44
Aden Protectorate / North Yemen
War Type: C
1955 - 1958

War Nr. 45
Cuban Revolution
War Type: A
12/1956 - 1/1959

War Nr. 47
Spanish Morocco
War Type: D
10/1957 - 4/1958

War Nr. 48
Muskat and Oman (Imam-Sultan Conflict)
War Type: B
7/1957 - 1959

War Nr. 49
Lebanese Civil War 1958
War Type: AB
9.5.1958 - 14.10.1958

War Nr. 50
People's Republic of China / Taiwan (Second Quemoy Crisis)
War Type: C
23.8.1958 - 25.10.1958

War Nr. 52
Indonesia (Civil War)
War Type: AB
3/1958 - 7/1961

War Nr. 53
Venezuela
War Type: A
2/1963 - 1967

War Nr. 55
Zaïre (Secession and Anti-Regime War)
War Type: AB
11.7.1960 - 14.1.1963

War Nr. 56
Guatemala
War Type: A
1962 - 1968

War Nr. 57
Tunisia (Bizerta Crisis)
War Type: D
19.7.1961 - 22.7.1961

War Nr. 58
Cuba (Bay of Pigs Invasion)
War Type: A
17.4.1961 - 20.4.1961

War Nr. 60
Colombia (FARC)
War Type: A
Since 27.5.1964

War Nr. 61
First and Second Iraqi-Kurdish War
War Type: B
16.9.1961 - 10.2.1964

War Nr. 62
Angola
War Type: D/ABC/A
Since 4.2.1961

War Nr. 63
Ethiopia (Eritrea Conflict)
War Type: B
1962 - 5/1991

War Nr. 64
Indonesia (West Papua)
War Type: C
14.1.1962 - 18.8.1962

War Nr. 65
Nepal 1962
War Type: A
1/1962 - 2/1962

War Nr. 66
People's Republic of China / India
War Type: C
20.10.1962 - 20.11.1962

War Nr. 67
Brunei
War Type: A
12/1962 - 12/1962

War Nr. 68
Yemenite Civil War
War Type: A
10/1962 - 1969

War Nr. 70
Morocco / Algeria ("Tindouf War")
War Type: C
10/1963 - 2/1964

War Nr. 71
Ogaden Uprising (Ethiopia 1963/64)
War Type: BC
8/1963 - 12/1964

War Nr. 72
Rwanda (Tutsi Invasion from Burundi, Uganda and Zaïre)
War Type: A
20.12.1963 - 1966

War Nr. 73
Cyprus II
War Type: B
21.12.1963 - 11.8.1964

War Nr. 74
Indonesia / Malaysia ("Confrontation")
War Type: C
12.9.1963 - 11.8.1966

War Nr. 75
Kenya (Shifta War)
War Type: B
11/1963 - 10/1967

War Nr. 76
Laos (Second Laotian Civil War)
War Type: A
31.3.1963 - 4/1973

War Nr. 77
Guinea-Bissau (War of Independence)
War Type: D
1/1963 - 25.4.1974

War Nr. 78
Zaïre (Anti-Regime Uprisings)
War Type: A
1/1964 - 24.9.1966

War Nr. 79
Colombia (ELN)
War Type: A
Since 7.1.1965

War Nr. 81
Mozambique (War of Independence)
War Type: D
25.9.1964 - 7.9.1974

War Nr. 82
India / Pakistan (Rann-von-Kutch)
War Type: C
9.4.1965 - 1.7.1965

War Nr. 83
Dominican Republic
War Type: A
25.4.1965 - 3.9.1965

War Nr. 84
Pakistan / India (Second Kashmir War)
War Type: BC
5.8.1965 - 23.9.1965

War Nr. 85
Third Iraqi-Kurdish War
War Type: B
4.4.1965 - 29.6.1966

War Nr. 86
Peru (Guerrilla Uprising of the MIR and the ELN 1965)
War Type: A
6/1965 - 1/1966

War Nr. 87
Aden Protectorate (Southern Yemenite War of Independence)
War Type: D
14.10.1963 - 30.11.1967

War Nr. 88
Uprising in Dhofar
War Type: AB/A
9.6.1965 - 11/1975

War Nr. 89
Thailand
War Type: A
7.8.1965 - 1980

War Nr. 90
India (Mizos)
War Type: B
1.3.1966 - 31.7.1980

War Nr. 91
Rhodesia ("Chimurenga II")
War Type: A
28.4.1966 - 21.12.1979

War Nr. 92
Chad (Civil War)
War Type: ABC
Since 6/1966

War Nr. 93
Namibia (Liberation and War of Independence)
War Type: B
26.8.1966 - 22.12.1988

War Nr. 95
Third Arab-Israeli War ("Six Day War")
War Type: C
5.6.1967 - 10.6.1967

War Nr. 96
Zaïre (Mercenary Rebellion)
War Type: A
5.7.1967 - 6.11.1967

War Nr. 97
Nigeria (Biafra War)
War Type: B
6.7.1967 - 15.1.1970

War Nr. 98
People's Republic of Yemen
War Type: A
6/1968 - 11/1968

War Nr. 99
Cambodia
War Type: AC
17.1.1968 - 17.4.1975

War Nr. 100
El Salvador / Honduras ("Soccer War")
War Type: C
14.7.1969 - 18.7.1969

War Nr. 101
DPR Yemen / Saudi Arabia
War Type: C
26.11.1969 - 12/1969

War Nr. 102
Fourth Iraqi-Kurdish War
War Type: B
3.1.1969 - 11.3.1970

War Nr. 103
Egypt / Israel (War of Attrition)
War Type: C
6.3.1969 - 7.8.1970

War Nr. 104
Argentina
War Type: A
1968 - 1977

War Nr. 105
Northern Ireland
War Type: B
Since 14.8.1969

War Nr. 106
Spain (Basque Country)
War Type: AB/B
7.6.1968 - 1979

War Nr. 108
Jordan (Black September)
War Type: AB
16.9.1970 - 19.7.1971

War Nr. 109
Philippines (NPA)
War Type: A
Since 1970

War Nr. 110
Ceylon (JVP Uprising)
War Type: A
5.4.1971 - 5.6.1971

War Nr. 111
Bangladesh War
War Type: B/B/BC
25.3.1971 - 17.12.1971

War Nr. 112
Uganda / Tanzania (I)
War Type: AC
8/1971 - 7.10.1972

War Nr. 113
Burundi
War Type: A
29.4.1972 - 6/1973

War Nr. 114
AR Yemen / DPR Yemen
War Type: C
26.9.1972 - 13.10.1972

War Nr. 115
Fourth Arab-Israeli War (October War)
War Type: C
6.10.1973 - 26.10.1973

War Nr. 116
Cyprus III
War Type: BC
20.7.1974 - 16.8.1974

War Nr. 117
Fifth Iraqi-Kurdish War
War Type: B
14.3.1974 - 3/1975

War Nr. 119
Mozambique (Destabilization Campaign)
War Type: AC/A
Since 1975

War Nr. 120
Lebanon War
War Type: ABC
Since 13.4.1975

War Nr. 121
Indonesia (East Timor)
War Type: A/B
Since 11.8.1975

War Nr. 122
Ethiopia (Tigray Conflict)
War Type: B/BA
1975 - 5/1991

War Nr. 123
West Sahara
War Type: D/B
18.11.1975 - 6.9.1991

War Nr. 125
South Africa (Anti-Apartheid War)
War Type: A/AB
Since 1976

War Nr. 126
Sixth Iraqi-Kurdish War
War Type: BA
Since 5/1976

War Nr. 127
Second Ogaden War (Ethiopia)
War Type: BC
1975 - 1984

War Nr. 128
Zaïre (Shaba I, "80 Day War")
War Type: A
8.3.1977 - 26.5.1977

War Nr. 129
Nicaraguan Revolution
War Type: A
10/1977 - 19.7.1979

War Nr. 130
Indonesia (West Irian / West Papua)
War Type: B
Since 1965

War Nr. 131
Ethiopia (Oromo Conflict)
War Type: AB
Since 1976

War Nr. 132
Cambodia (Third Indochina War)
War Type: C/A
Since 12/1975

War Nr. 133
Zaïre (Shaba II)
War Type: A
12.5.1978 - 30.5.1978

War Nr. 134
Uganda / Tanzania (II)
War Type: AC
10/1978 - 2.6.1979

War Nr. 135
AR Yemen
War Type: A
1978 - 1982

War Nr. 137
Colombia (M-19)
War Type: A
1/1974 - 8.3.1990

War Nr. 140
El Salvador
War Type: A
10.1.1981 - 1.2.1992

War Nr. 141
Afghanistan
War Type: A
Since 10/1978

War Nr. 142
People's Republic of China / Vietnam ("Education Campaign")
War Type: C
17.2.1979 - 1988

War Nr. 144
Iran (Kurdistan)
War Type: B
7/1979 - 1988

War Nr. 146
Guatemala
War Type: A
Since 1980

War Nr. 147
Peru (Sendero Luminoso)
War Type: A
Since 17.5.1980

War Nr. 148
Iraq / Iran ("First" Gulf War)
War Type: C
22.9.1980 - 20.8.1988

War Nr. 149
Peru / Ecuador
War Type: C
28.1.1981 - 2.2.1981

War Nr. 151
Nicaragua (Contra War)
War Type: A
4/1981 - 1990

War Nr. 152
Uganda
War Type: A
2/1981 - 5/1992

War Nr. 153
Syria 1982
War Type: A
2.2.1982 - 2/1982

War Nr. 154
Falkland/Malvinas War (Argentina / Great Britain)
War Type: C
25.4.1982 - 15.6.1982

War Nr. 155
India (Punjab)
War Type: B
Since 7/1982

War Nr. 156
Grenada Invasion (Operation Urgent Fury)
War Type: C
24.10.1983 - 1.11.1983

War Nr. 157
Sri Lanka ("Tamil Conflict")
War Type: B/AB/B
Since 7/1983

War Nr. 158
Sudan (South Sudan Conflict, Second Sudanese Civil War)
War Type: BA
Since 9/1983

War Nr. 159
Turkey (Kurds)
War Type: B
Since 15.8.1984

War Nr. 160
Zimbabwe (Matabeleland Conflict)
War Type: A
1/1983 - 5/1988

War Nr. 161
Pakistan / India (Ziak Glacier)
War Type: C
4/1984 - 1989

War Nr. 162
South Yemen 1986
War Type: A
13.1.1986 - 29.1.1986

War Nr. 163
Suriname (Jungle War)
War Type: B
21.6.1986 - 7.6.1989

War Nr. 164
Thailand / Laos
War Type: C
3.11.1987 - 19.2.1988

War Nr. 165
Israel, Palestine
War Type: B
Since 1968

War Nr. 166
Somalia (Civil War)
War Type: A/AB
Since 5/1988

War Nr. 167
Papua New Guinea (Bougainville)
War Type: B
Since 2/1989

War Nr. 168
Liberia, Sierra Leone
War Type: A
Since 24.12.1989

War Nr. 169
Romania
War Type: A
17.12.1989 - 28.12.1989

War Nr. 170
USA / Panama
War Type: C
20.12.1989 - 24.12.1989

War Nr. 171
Pakistan (Belutchistan)
War Type: B
3/1973 - 7/1977

War Nr. 172
India (Nagas)
War Type: B
1969 - 6.11.1975

War Nr. 174
USSR (Azerbaijan, Armenia); Nagorno-Karabakh
War Type: B
Since 20.1.1990

War Nr. 177
Bangladesh (Chittagong Hill Tracts)
War Type: B
Since 1973

War Nr. 178
Vietnam (FULRO)
War Type: B
20.9.1964 - 10/1992

War Nr. 179
Iraq (Barzanis)
War Type: B
8/1945 - 10/1945

War Nr. 180
Mali / Burkina Faso (Border War)
War Type: C
21.12.1985 - 31.12.1985

War Nr. 181
Zambia (Mushala Uprising)
War Type: A
14.1.1976 - 20.11.1982

War Nr. 182
Iraq / Kuwait
War Type: C
2.8.1990 - 4.8.1990

War Nr. 183
Laos (First Meo Guerrilla War (Hmong))
War Type: B
1975 - 1979

War Nr. 184
Nigeria (Tiv Uprising 1964)
War Type: AB
8/1964 - 12/1964

War Nr. 185
Nigeria (Maitatsine Uprising in Kano/Northern Nigeria)
War Type: B
18.12.1980 - 29.12.1980

War Nr. 187
Pakistan (Sind Province)
War Type: B
Since 11/1986

War Nr. 188
India (Kashmir)
War Type: B
Since 1/1990

War Nr. 189
Rwanda (Rebel Invasion from Uganda)
War Type: AB
Since 1.10.1990

War Nr. 190
Djibuti
War Type: AB
Since 10/1991

War Nr. 191
Senegal (Basse-Casamance Conflict)
War Type: B
Since 4/1990

War Nr. 192
Laos (Second Meo Guerrilla War (Hmong))
War Type: B
1/1990 - 7/1992

War Nr. 194
Anti-Iraq Coalition / Iraq ("Second" Gulf War)
War Type: C
17.1.1991 - 27.2.1991

War Nr. 196
Iraq (Shi'ites)
War Type: A
Since 2.3.1991

War Nr. 197
Tuareg (Mali, Niger)
War Type: B
Since 7.5.1990

War Nr. 198
Indonesia (Aceh)
War Type: B
Since 5/1990

War Nr. 199
Philippines (Mindanao)
War Type: B
Since 1970

War Nr. 200
Yugoslavia (Slovenia, Croatia, Bosnia)
War Type: BC
Since 26.6.1991

War Nr. 201
Georgia (Anti-Regime War)
War Type: A
Since 9/1991

War Nr. 202
Georgia (South Ossetia)
War Type: B
12/1990 - 7/1992

War Nr. 203
India (Assam)
War Type: B
1990 - 1992

War Nr. 204
Peru (MRTA)
War Type: A
Since 11/1987

War Nr. 205
Moldavia (Dneister Republic)
War Type: B
3/1992 - 8/1992

War Nr. 206
Algeria
War Type: A
Since 1/1992

War Nr. 207
Georgia (Abkhazia)
War Type: B
Since 14.8.1992

War Nr. 208
Tadjikistan
War Type: A
Since 8/1992

War Nr. 209
North Ossetia (Prigorodny District)
War Type: B
Since 30.10.1992

Directory of Wars in which the USA, the USSR/Russian Federation, or the Former Colonial Powers of Great Britain, Netherlands, Belgium, France, Portugal, and Spain Participated

USA

War Nr. 25
Korean War
War Type: C
25.6.1950 - 27.7.1953

War Nr. 31
Guatemala (Operation Success)
War Type: A
18.6.1954 - 1.7.1954

War Nr. 32
People's Republic of China / Taiwan (First Quemoy Crisis)
War Type: C
3.9.1954 - 25.9.1954

War Nr. 41
Vietnam (Second Indochina War)
War Type: A-1/AC-1
1957 - 29.4.1975

War Nr. 52
Indonesia (Civil War)
War Type: AB
3/1958 - 7/1961

War Nr. 76
Laos (Second Laotian Civil War)
War Type: A
31.3.1963 - 4/1973

War Nr. 78
Zaïre (Anti-Regime Uprisings)
War Type: A
1/1964 - 24.9.1966

War Nr. 83
Dominican Republic
War Type: A
25.4.1965 - 3.9.1965

War Nr. 99
Cambodia
War Type: AC
17.1.1968 - 17.4.1975

War Nr. 156
Grenada Invasion (Operation Urgent Fury)
War Type: C
24.10.1983 - 1.11.1983

War Nr. 170
USA / Panama
War Type: C
20.12.1989 - 24.12.1989

War Nr. 194
Anti-Iraq Coalition / Iraq ("Second" Gulf War)
War Type: C
17.1.1991 - 27.2.1991

Great Britain

War Nr. 1
Greek Civil War (Phase 1)
War Type: A
3.12.1944 - 12.2.1945

War Nr. 179
Iraq (Barzanis)
War Type: B
8/1945 - 10/1945

War Nr. 3
Indonesia
War Type: D
29.9.1945 - 27.7.1949

War Nr. 6
Greek Civil War (Phase 2)
War Type: A
6/1946 - 9.10.1949

War Nr. 20
Malaysian Federation ("Malayan-Emergency")
War Type: DA
1948 - 2/1960

War Nr. 25
Korean War
War Type: C
25.6.1950 - 27.7.1953

War Nr. 29
Kenya (Mau-Mau Uprising)
War Type: D
10/1952 - 11/1956

War Nr. 44
Aden Protectorate / North Yemen
War Type: C
1955 - 1958

War Nr. 38
Cyprus I
War Type: DB
1.4.1955 - 24.2.1959

War Nr. 42
Second Arab-Israeli War (Suez War)
War Type: C
29.10.1956 - 6.11.1956

War Nr. 48
Muskat and Oman (Imam-Sultan Conflict)
War Type: B
7/1957 - 1959

War Nr. 67
Brunei
War Type: A
12/1962 - 12/1962

War Nr. 74
Indonesia / Malaysia ("Confrontation")
War Type: C
12.9.1963 - 11.8.1966

War Nr. 87
Aden Protectorate (Southern Yemenite War of Independence)
War Type: D
14.10.1963 - 30.11.1967

War Nr. 88
Uprising in Dhofar
War Type: AB/A
9.6.1965 - 11/1975

War Nr. 105
Northern Ireland
War Type: B
Since 14.8.1969

War Nr. 154
Falkland/Malvinas War (Argentina / Great Britain)
War Type: C
25.4.1982 - 15.6.1982

War Nr. 194
Anti-Iraq Coalition / Iraq ("Second" Gulf War)
War Type: C
17.1.1991 - 27.2.1991

The Netherlands

War Nr. 3
Indonesia
War Type: D
29.9.1945 - 27.7.1949

War Nr. 25
Korean War
War Type: C
25.6.1950 - 27.7.1953

War Nr. 64
Indonesia (West Papua)
War Type: C
14.1.1962 - 18.8.1962

Belgium

War Nr. 25
Korean War
War Type: C
25.6.1950 - 27.7.1953

War Nr. 55
Zaïre (Secession and Anti-Regime War)
War Type: AB
11.7.1960 - 14.1.1963

War Nr. 78
Zaïre (Anti-Regime Uprisings)
War Type: A
1/1964 - 24.9.1966

War Nr. 133
Zaïre (Shaba II)
War Type: A
12.5.1978 - 30.5.1978

France

War Nr. 9
Indochina (First Indochina War)
War Type: D
9.3.1946 - 1.6.1954

War Nr. 12
Madagascar (Decolonization War)
War Type: D
29.3.1947 - 1948

War Nr. 25
Korean War
War Type: C
25.6.1950 - 27.7.1953

War Nr. 30
Morocco
War Type: D
7.12.1952 - 2.3.1956

War Nr. 35
Algeria
War Type: D
1.11.1954 - 19.3.1962

War Nr. 39
Cameroon (Decolonization)
War Type: DA
5/1955 - 1963

War Nr. 42
Second Arab-Israeli War (Suez War)
War Type: C
29.10.1956 - 6.11.1956

War Nr. 57
Tunisia (Bizerta Crisis)
War Type: D
19.7.1961 - 22.7.1961

War Nr. 92
Chad (Civil War)
War Type: ABC
Since 6/1966

War Nr. 133
Zaïre (Shaba II)
War Type: A
12.5.1978 - 30.5.1978

War Nr. 194
Anti-Iraq Coalition / Iraq ("Second" Gulf War)
War Type: C
17.1.1991 - 27.2.1991

Spain

War Nr. 4
Spain 1945-1950
War Type: A
1945 - 1950

War Nr. 47
Spanish Morocco
War Type: D
10/1957 - 4/1958

War Nr. 106
Spain (Basque Country)
War Type: AB/B
7.6.1968 - 1979

War Nr. 123
West Sahara
War Type: D/B
18.11.1975 - 6.9.1991

Portugal

War Nr. 62
Angola
War Type: D/ABC/A
Since 4.2.1961

War Nr. 77
Guinea-Bissau (War of Independence)
War Type: D
1/1963 - 25.4.1974

War Nr. 81
Mozambique (War of Independence)
War Type: D
25.9.1964 - 7.9.1974

USSR/Russian Federation:

War Nr. 43
Hungarian Revolution
War Type: A
24.10.1956 - 10.11.1956

War Nr. 103
Egypt / Israel (War of Attrition)
War Type: C
6.3.1969 - 7.8.1970

War Nr. 141
Afghanistan
War Type: A
Since 10/1978

War Nr. 174
USSR (Azerbaijan, Armenia); Nagorno-Karabakh
War Type: B
Since 20.1.1990

War Nr. 205
Moldavia (Dneister Republic)
War Type: B
3/1992 - 8/1992

Directory of Countries at War, Ordered by Region

Country	Nr.	Intervent.	U.N.	War Number
CENTRAL AMERICA /CARIBBEAN	25	9	0	
Costa Rica	4	1	0	14, 16, 37, 83*
Dominican Republic	1	0	0	83
El Salvador	3	1	0	83*, 100, 140
Grenada	1	0	0	156
Guatemala	3	0	0	31, 56, 146
Honduras	4	3	0	83*, 100, 129*, 151*
Cuba	4	2	0	45, 58, 62*, 127*
Nicaragua	4	2	0	37*, 83*, 129, 151
Panama	1	0	0	170
SOUTH AMERICA	18	2	1	
Argentina	2	0	0	104, 154
Bolivia	1	0	0	8
Brazil	1	1	0	83*
Ecuador	1	0	0	149
Colombia	5	0	1	19, 25**, 60, 79, 137
Paraguay	2	1	0	11, 83*
Peru	4	0	0	86, 147, 149, 204
Suriname	1	0	0	163
Venezuela	1	0	0	53
EUROPE	67	17	12	
Belgium	4	3	1	25**, 55*, 78*, 133*
Bosnia	1	0	0	200
France	11	2	2	9, 12, 25**, 30, 35, 39, 42, 57, 92*, 133*, 194**
Greece	5	1	1	1, 6, 25**, 73, 116*
Great Britain	18	7	2	1*, 3, 6*, 20, 25**, 29, 38, 42, 44, 48*, 67*, 74*, 87, 88*, 105, 154, 179*, 194**
Ireland	1	0	1	55**
Italy	2	0	2	55**, 194**
Yugoslavia	1	0	0	200
Croatia	1	0	0	200
Luxembourg	1	0	1	25**
Moldavia	1	0	0	205

Netherlands	3	0	1	3, 25**, 64
Portugal	3	0	0	62, 77, 81
Romania	1	0	0	169
Sweden	1	0	1	55**
Slovenia	1	0	0	200
Spain	4	0	0	4, 47, 106, 123
USSR/Russian Federation	5	4	0	43*, 103*, 141*, 174, 205*
Hungary	1	0	0	43
Cyprus	2	0	0	73, 116
NORTH AFRICA	22	4	4	
Egypt	7	1	1	18, 42, 68*, 95, 103, 115, 194**
Algeria	3	1	0	70, 123*, 206
Libya	2	1	0	92, 134*
Morocco	4	1	1	55**, 70, 123, 128*
Mauritania	1	0	0	123
Sudan	3	0	1	40, 55**, 158
Tunisia	2	0	1	55**, 57
WEST AFRICA	18	8	1	
Burkina Faso	1	0	0	180
Gambia	1	1	0	168*
Ghana	2	1	1	55**, 168*
Guinea	1	1	0	168*
Liberia	1	0	0	168
Mali	3	1	0	168*, 180, 197
Niger	1	0	0	197
Nigeria	5	2	0	97, 119*, 168*, 184, 185
Senegal	2	1	0	168*, 191
Sierra Leone	1	1	0	168*
CENTRAL AFRICA	9	2	0	
Zaïre	7	2	0	55, 62*, 78, 96, 128, 133, 189*
Cameroon	1	0	0	39
Chad	1	0	0	92
EAST AFRICA	21	1	2	
Ethiopia	7	0	2	25**, 55**, 63, 71, 122, 127, 131
Burundi	1	0	0	113
Djibuti	1	0	0	190
Kenya	1	0	0	75
Rwanda	2	0	0	72, 189
Somalia	3	0	0	71, 127, 166

Country	Nr.	Intervent.	U.N.	War Number
Tanzania	3	1	0	112, 119*, 134
Uganda	3	0	0	112, 134, 152
SOUTHERN AFRICA	17	8	1	
Angola	1	0	0	62
Malawi	1	1	0	119*
Mozambique	2	1	0	91*, 119
Rhodesia	3	1	0	81*, 91, 119
Zambia	1	0	0	181
South Africa	7	4	1	25**, 62*, 81*, 91*, 93, 119*, 125
Zimbabwe	2	1	0	119*, 160
NEAR AND MIDDLE EAST	68	7	4	
AR Yemen	5	0	0	17, 44, 68, 114, 135
Afghanistan	1	0	0	141
Azerbaijan	1	0	0	174
DPR Yemen	5	1	0	98, 101, 114, 135*, 162
Georgia	3	0	0	201, 202, 207
Iraq	12	0	0	18, 61, 85, 102, 115, 117, 126, 148, 179, 182, 194, 196
Iran	5	2	0	5, 88*, 126*, 144, 148
Israel	7	0	0	18, 42, 95, 103, 115, 120, 165
Jordan	4	1	0	18, 88*, 95, 108
Kuwait	2	0	1	182, 194**
Lebanon	3	0	0	18, 49, 120
North Ossetia	1	0	0	209
(Muskat and) Oman	2	0	0	48, 88
Saudi Arabia	3	0	1	18, 101, 194**
Syria	8	2	1	18, 61*, 95, 108*, 115, 120, 153, 194**
Tadjikistan	1	0	0	208
Turkey	5	1	1	25**, 73, 116, 126*, 159
SOUTH AND SOUTHEAST ASIA	77	20	5	
Australia	3	2	1	25**, 41*, 74*
Bangladesh	1	0	0	177
Burma	1	0	0	21
Brunei	1	0	0	67
Hyderabad	1	0	0	15
India	16	4	1	13*, 15, 26, 55**, 66, 82, 84, 90, 110*, 111*, 155, 157*, 161, 172, 188, 203

Indonesia	7	0	0	23, 52, 64, 74, 121, 130, 198
Cambodia	2	0	0	99, 132
Kashmir	1	0	0	13
Laos	5	0	0	34, 76, 164, 183, 192
Malaysian Federation	1	0	0	20
Malaysia	1	0	0	74
Nepal	2	0	0	24, 65
New Zealand	3	2	1	25**, 41*, 74*
North Vietnam	5	4	0	34*, 41*, 76*, 99*, 178
Pakistan	7	1	0	13*, 82, 84, 111, 161, 171, 187
Papua New Guinea	1	0	0	167
Philippines	5	1	1	10, 25**, 41*, 109, 199
Sri Lanka/Ceylon	2	0	0	110, 157
South Vietnam	3	2	0	41, 76*, 99*
Thailand	5	2	1	25**, 34*, 76*, 89, 164
Vietnam	4	2	0	132, 142, 183*, 192*
EAST ASIA	18	3	0	
North Korea	1	0	0	25
South Korea	2	1	0	25, 41*
Taiwan	3	0	0	32, 36, 50
Tibet	2	0	0	22, 33
People's Republic of China/China	10	2	0	7, 22, 25*, 32, 33, 36, 41*, 50, 66, 142
NORTH AMERICA	14	8	4	
Canada	2	0	2	25**, 55**
United States of America	12	8	2	25**, 31*, 32*, 41*, 52*, 76*, 78*, 83*, 99*, 156, 170, 194**
Total	**374**	**89**	**34**	
Number of Countries:	**119**			

Legend: * Participation by Intervention; ** Participation in U.N. Mission

Directory of Casualties and the Costs of War, 1945-1992

War Nr. 1
Greek Civil War (Phase 1)
3.12.1944 - 12.2.1945 (Typ: A-1)
Approximately 15,000 civilians were taken hostage by the EAM/ELAS.

War Nr. 3
Indonesia
29.9.1945 - 27.7.1949 (Typ: D-2)
Total number killed: at least 5,000

War Nr. 6
Greek Civil War (Phase 2)
6/1946 - 9.10.1949 (Typ: A-1)
28,000 children were taken hostage, figures for the dead and wounded range from 44,000 to 158,000. Numbers for refugees fluctuate between 80,000 and 703,000.

War Nr. 7
China (Chinese Civil War)
15.3.1946 - 21.4.1950 (Typ: A-2)
Over one million killed.

War Nr. 8
Bolivian Revolution
1946 - 4/1952 (Typ: A-2)
Total number killed: at least 2,000

War Nr. 10
Philippines (Island Luzon)
1946 - 1954 (Typ: A-2)
Combatants killed: at least 4,000
Civilians killed: at least 5,000
Total number killed: at least 9,000

War Nr. 11
Paraguay
3/1947 - 12.8.1947 (Typ: A-2)
Approximately 28,000 killed and 300,000 (one fourth of the population) became refugees.

War Nr. 12
Madagascar (Decolonization War)
29.3.1947 - 1948 (Typ: D-2)
The war caused 60,000-100,000 casualties.

War Nr. 13
Pakistan / India (First Kashmir War)
22.10.1947 - 1.1.1949 (Typ: A-2/AC-1)
Combatants killed: at least 2,000
Civilians killed: at least 1,000
Total number killed: at least 3,000

War Nr. 15
India / Hyderabad
13.9.1948 - 17.9.1948 (Typ: C-2)
Combatants killed: up to 1,000
Civilians killed: up to 1,000
Total number killed: up to 2,000

War Nr. 17
Nord-Yemen 1948
2/1948 - 14.3.1948 (Typ: A-2)
5,000 persons were killed during the plundering of the capital, Sana.

War Nr. 19
Colombia (Violencia)
21.5.1949 - 10.5.1957 (Typ: A-2)
Total number killed: 126,000 to 200,000
The war cost approximately 200,000 lives. Approximately two million persons became refugees.

War Nr. 20
Malaysian Federation ("Malayan-Emergency")
1948 - 2/1960 (Typ: DA-2)
The war caused more than 11,000 deaths.

War Nr. 21
Myanmar (Burma)
Since 1948 (Typ: AB-2)
Between 100,000 and 300,000 persons were killed.

War Nr. 25
Korean War
25.6.1950 - 27.7.1953 (Typ: C-1)
Total number killed: 854,197 to 2,000,000

War Nr. 26
India (Nagas)
1954 - 8/1964 (Typ: B-2)
Total number killed: at least 1,560

War Nr. 29
Kenya (Mau-Mau Uprising)
10/1952 - 11/1956 (Typ: D-2)
Total number killed: 8.349 to 13.500
The war caused up to 13,500 casualties, including ca. 2,000 African civilians and 32 European civilians.

War Nr. 35
Algeria
1.11.1954 - 19.3.1962 (Typ: D-2)
Combatants killed: at least 250,000
Civilians killed: at least 200,000
Total number killed: 450,000 to 2,000,000
Over 20,000 villages were destroyed. The war cost France approximately $20 billion.

War Nr. 36
People's Republic of China / Taiwan (Da Chen and Nan Ji Island Groups)
18.1.1955 - 22.1.1955 (Typ: C-2)
The People's Liberation Army suffered heavy casualties.

War Nr. 38
Cyprus I
1.4.1955 - 24.2.1959 (Typ: DB-2)
Total number killed: 600 to 1,000

War Nr. 39
Cameroon (Decolonization)
5/1955 - 1963 (Typ: DA-2)
The war caused 10,000-20,000 casualties.

War Nr. 40
Sudan (First Sudanese Civil War)
18.8.1955 - 2/1972 (Typ: B-2)
The war cost 500,000 to 700,000 lives. From 1964 to 1972 approximately 695,000 Sudanese fled to neighboring states and 850,000 became internal refugees.

War Nr. 41
Vietnam (Second Indochina War)
1957 - 29.4.1975 (Typ: A-1/AC-1)
The United States sustained 56,221 deaths, 303,605 wounded and spent approximately $135-140 billion. Estimates for the total number of war dead reach two million.

War Nr. 42
Second Arab-Israeli War (Suez War)
29.10.1956 - 6.11.1956 (Typ: C-2)
The war cost over 2,000 lives (Egypt lost approximately 2,000; Israel 181).

War Nr. 43
Hungarian Revolution
24.10.1956 - 10.11.1956 (Typ: A-1)
Total number killed: at least 32,000
111,000 refugees fled to Austria.

War Nr. 47
Spanish Morocco
10/1957 - 4/1958 (Typ: D-2)
Total number killed: 1,000 to 1,200
On the Spanish side there were 190 dead, 500 wounded, and 89 missing. On the Moroccan side there were approximately 1,000 dead.

War Nr. 49
Lebanese Civil War 1958
9.5.1958 - 14.10.1958
(Typ: AB-2)
The war caused 2,000-3,000 casualties.

War Nr. 50
People's Republic of China / Taiwan (Second Quemoy Crisis)
23.8.1958 - 25.10.1958 (Typ: C-2)
The People's Liberation Army suffered heavy casualties.

War Nr. 52
Indonesia (Civil War)
3/1958 - 7/1961 (Typ: AB-1)
There were between 6,000 and 14,000 deaths.

War Nr. 55
Zaïre (Secession and Anti-Regime War)
11.7.1960 - 14.1.1963 (Typ: AB-1)
There were at least 100,000 casualties, the economy was ruined and the state neared collapse.

War Nr. 57
Tunisia (Bizerta Crisis)
19.7.1961 - 22.7.1961 (Typ: D-2)
670 persons lost their lives on the Tunisian side, and 30 French died. 1,500 people were wounded.

War Nr. 60
Colombia (FARC)
Since 27.5.1964 (Typ: A-2)
Total number killed: 15,000

War Nr. 61
First and Second Iraqi-Kurdish War
16.9.1961 - 10.2.1964 (Typ: B-1)
In the first six months of the war

approximately 3,000 people were killed, 500 villages destroyed and 180,000 persons made homeless by Iraqi air attacks.

War Nr. 62
Angola
Since 4.2.1961 (Typ: D-1/ABC-1/A-2)
Total number killed: at least 600,000
More than two million people became internal refugees, and hundreds of thousands fled abroad. The economy was totally destroyed.

War Nr. 63
Ethiopia (Eritrea Conflict)
1962 - 5/1991 (Typ: B-2)
Total number killed: 250,000 to 2,000,000

War Nr. 64
Indonesia (West Papua)
14.1.1962 - 18.8.1962 (Typ: C-2)
Total number killed: at least 120

War Nr. 65
Nepal 1962
1/1962 - 2/1962 (Typ: A-2)
Combatants killed: at least 86
Civilians killed: at least 14
Total number killed: 100 to 114

War Nr. 66
People's Republic of China / India
20.10.1962 - 20.11.1962 (Typ: C-2)
Total number killed: up to 2,000

War Nr. 70
Morocco / Algeria ("Tindouf War")
10/1963 - 2/1964 (Typ: C-2)
Total number killed: 1,000

War Nr. 71
Ogaden Uprising (Ethiopia 1963/64)
8/1963 - 12/1964 (Typ: BC-2)
The war caused at least 2,000 casualties.

War Nr. 72
Rwanda (Tutsi Invasion from Burundi, Uganda and Zaïre)
20.12.1963 - 1966 (Typ: A-2)
Together with unrest in 1959 and 1961, the war cost approximately 105,000 lives, including 3,000 soldiers.

War Nr. 73
Cyprus II
21.12.1963 - 11.8.1964 (Typ: B-2)
Total number killed: 60 to 500

War Nr. 74
Indonesia / Malaysia ("Confrontation")
12.9.1963 - 11.8.1966 (Typ: C-1)
Total number killed: 625 to 1,000
The British Defence Ministry claims that 740 persons died in all: 150 Commonwealth troops and 590 Indonesians. The magazine

South (8/1986:45) reports over 1,000 deaths.

War Nr. 75
Kenya (Shifta War)
11/1963 - 10/1967 (Typ: B-2)
The war caused 4,000 to 6,000 casualties.

War Nr. 77
Guinea-Bissau (War of Independence)
1/1963 - 25.4.1974 (Typ: D-2)
The war cost approximately 15,000 lives, including 5,000 civilians.

War Nr. 81
Mozambique (War of Independence)
25.9.1964 - 7.9.1974 (Typ: D-1)
The war caused approximately 30,000 casualties.

War Nr. 82
India / Pakistan (Rann-von-Kutch)
9.4.1965 - 1.7.1965 (Typ: C-2)
Combatants killed: at least 450

War Nr. 83
Dominican Republic
25.4.1965 - 3.9.1965 (Typ: A-1)
Total number killed: 2,000 to 3,000

War Nr. 84
Pakistan / India (Second Kashmir War)
5.8.1965 - 23.9.1965 (Typ: BC-2)
Total number killed: at least 20,000

War Nr. 85
Third Iraqi-Kurdish War
4.4.1965 - 29.6.1966 (Typ: B-2)
In 1966 the war cost Iraq approximately $1.8 million per day.

War Nr. 89
Thailand
7.8.1965 - 1980 (Typ: A-2)
Total number killed: at least 3,000

War Nr. 90
India (Mizos)
1.3.1966 - 31.7.1980 (Typ: B-2)
Total number killed: at least 1,500

War Nr. 91
Rhodesia ("Chimurenga II")
28.4.1966 - 21.12.1979 (Typ: A-1)
The war caused ca. 30,000 casualties, and at least 10.000 wounded. 250,000-300,000 people fled to Zambia, Mozambique or Botswana. 20 percent of the population suffered from malnutrition as a result of the war.

War Nr. 92
Chad (Civil War)
Since 6/1966 (Typ: ABC-2)
The war has caused at least 21,000 casualties so far.

War Nr. 93
Namibia (Liberation and War of Independence)
26.8.1966 - 22.12.1988 (Typ: B-2)

At least 20,000 people died in the war. 80,000 persons fled to Angola and Zambia.

War Nr. 95
Third Arab-Israeli War ("Six Day War")
5.6.1967 - 10.6.1967 (Typ: C-2)
Total number killed: at least 19.500
The Israeli's lost 766 soldiers, Egypt lost 10,000 either in battle or to the harsh desert conditions on the Sinai Peninsula. Jordan lost 6,000 soldiers. 300,000 Palestinians fled to East Jordan.

War Nr. 96
Zaïre (Mercenary Rebellion)
5.7.1967 - 6.11.1967 (Typ: A-2)
1,020 persons are estimated to have died.

War Nr. 97
Nigeria (Biafra War)
6.7.1967 - 15.1.1970 (Typ: B-2)
Total number killed: 600,000 to 2,000,000

War Nr. 99
Cambodia
17.1.1968 - 17.4.1975 (Typ: AC-1)
Very high casualty rates among both civilians and military.

War Nr. 100
El Salvador / Honduras ("Soccer War")
14.7.1969 - 18.7.1969 (Typ: C-2)
Total number killed: 3,000

War Nr. 102
Fourth Iraqi-Kurdish War
3.1.1969 - 11.3.1970 (Typ: B-2)
Exact figures are not available, but at least 200,000 Kurds are said to have fled Iraqi air attacks. In 1969 the war cost the Iraqi state 30% of its income.

War Nr. 104
Argentina
1968 - 1977 (Typ: A-2)
10,000 persons were kidnapped and murdered, 20,000-30,000 were imprisoned for political reasons.

War Nr. 105
Northern Ireland
Since 14.8.1969 (Typ: B-2)
The war cost over 3,000 lives to date.

War Nr. 106
Spain (Basque Country)
7.6.1968 - 1979 (Typ: AB-2/B-2)
There were approximately 600 casualties, of which 90% occurred after 1976.

War Nr. 108
Jordan (Black September)
16.9.1970 - 19.7.1971 (Typ: AB-1)
Total number killed: at least 40,000

War Nr. 109
Philippines (NPA)
Since 1970 (Typ: A-2)
Over 50,000 people have died in the war so far, and thousands are killed each year.

War Nr. 110
Ceylon (JVP Uprising)
5.4.1971 - 5.6.1971 (Typ: A-1)
Total number killed: at least 10,000, including 5,000 JVP supporters.

War Nr. 111
Bangladesh War
25.3.1971 - 17.12.1971 (Typ: B-2/B-1/BC-2)
Total number killed: at least 300,000

War Nr. 112
Uganda / Tanzania (I)
8/1971 - 7.10.1972 (Typ: AC-2)
Total number killed: 360 to 1,500

War Nr. 113
Burundi
29.4.1972 - 6/1973 (Typ: A-2)
There were 100,000 to 250,000 casualties, including almost all literate Hutus. 80,000 to 200,000 persons fled to neighboring countries.

War Nr. 115
Fourth Arab-Israeli War (October War)
6.10.1973 - 26.10.1973 (Typ: C-2)
Total number killed: at least 16,000, including at least 3,000 Israeli, 7,700 Egyptian and 835 Iraqi soldiers.

War Nr. 116
Cyprus III
20.7.1974 - 16.8.1974 (Typ: BC-1)
Total number killed: up to 5,000
Over 200,000 persons were forcibly expelled.

War Nr. 117
Fifth Iraqi-Kurdish War
14.3.1974 - 3/1975 (Typ: B-2)
Total number killed: at least 16,000

War Nr. 119
Mozambique (Destabilization Campaign)
Since 1975 (Typ: AC-2/A-1)
Total number killed: up to 950,000
Several million persons became refugees, both internally and abroad.

War Nr. 120
Lebanon War
Since 13.4.1975 (Typ: ABC-2)
Total number killed: at least 150,000
14,000 died during the 1982 Israeli invasion.

War Nr. 121
Indonesia (East Timor)

Since 11.8.1975 (Typ: A-2/B-2)
Total number killed: 20,000 to 320,000

War Nr. 122
Ethiopia (Tigray Conflict)
1975 - 5/1991 (Typ: B-2/BA-2)
Over 20,000 persons were killed.

War Nr. 123
West Sahara
18.11.1975 - 6.9.1991 (Typ: D-2/B-1)
Total number killed: 10,000 to 24,000

War Nr. 125
South Africa (Anti-Apartheid War)
Since 1976 (Typ: A-2/AB-2)
Total number killed: at least 14,000
Since the mid-1980s more than 10,000 persons died in conflicts between the ANC and Inkatha.

War Nr. 126
Sixth Iraqi-Kurdish War
Since 5/1976 (Typ: BA-3)
There are no reliable data for this conflict. At least 5,000-7,000 Kurds were reported killed by an Iraqi poison gas attack in April 1987, with a similar number of wounded. More than 2,000 Kurds „disappeared" during the forced resettlement of more than a million Kurds from approximately 2,500 villages.

War Nr. 127
Second Ogaden War (Ethiopia)
1975 - 1984 (Typ: BC-1)
Approximately 109,000 persons died, most during the interstate conflict between Ethiopia and Somalia in 1977/78. Approximately 840,000 persons fled from Ogaden to Somalia.

War Nr. 128
Zaïre (Shaba I, "80 Day War")
8.3.1977 - 26.5.1977 (Typ: A-1)
Up to 1,000 persons died in the war.

War Nr. 130
Indonesia (West Irian / West Papua)
Since 1965 (Typ: B-2)
Estimates for casualties vary between 30,000 and 150,000.

War Nr. 132
Cambodia (Third Indochina War)
Since 12/1975 (Typ: C-2/A-1)
Approximately two million people died under the Pol Pot regime.

War Nr. 133
Zaïre (Shaba II)
12.5.1978 - 30.5.1978 (Typ: A-1)
The war caused up to 1,000 casualties.

War Nr. 134
Uganda / Tanzania (II)
10/1978 - 2.6.1979 (Typ: AC-1)

Total number killed: 1,830 to 9,320
Tanzania had direct war costs in excess of $250 million.

War Nr. 140
El Salvador
10.1.1981 - 1.2.1992 (Typ: A-2)
At least 75,000 persons were killed.

War Nr. 141
Afghanistan
Since 10/1978 (Typ: A-1)
Total casualties approach one million people. Over two million persons became internal refugees, and more than five million fled abroad. At least 4,000 civilians have been killed by landmines.

War Nr. 146
Guatemala
Since 1980 (Typ: A-2)
Total number killed: at least 45,500
Approximately 100,000 persons have lost their lives due to the repression over the last 30 years. 40,000 persons have „disappeared." 250,000 have fled to neighboring countries. Over one million people were forcibly resettled.

War Nr. 147
Peru (Sendero Luminoso)
Since 17.5.1980 (Typ: A-2)
More than 26,000 persons have died so far in the struggle between the government and the guerrillas.

War Nr. 148
Iraq / Iran ("First" Gulf War)
22.9.1980 - 20.8.1988 (Typ: C-2)
Casualty estimates vary between 300,000 and 1,500,000. Both warring parties purchased heavy weaponry for over $27 million during the war.

War Nr. 149
Peru / Ecuador
28.1.1981 - 2.2.1981 (Typ: C-2)
Total number killed: 9

War Nr. 151
Nicaragua (Contra War)
4/1981 - 1990 (Typ: A-1)
The war cost approximately 60,000 lives.

War Nr. 152
Uganda
2/1981 - 5/1992 (Typ: A-2)
Total number killed: at least 500,000
One consequence of the war has been a massive outbreak of AIDS to over one million people.

War Nr. 153
Syria 1982
2.2.1982 - 2/1982 (Typ: A-2)
Total number killed: 8,000 to 30,000

War Nr. 154
Falkland/Malvinas War (Argentina / Great Britain)
25.4.1982 - 15.6.1982 (Typ: C-2)
Total number killed: 900 to 1,000
The material costs of the war were approximately $2.14 billion for Great Britain and $1.24 billion for Argentina.

War Nr. 155
India (Punjab)
Since 7/1982 (Typ: B-2)
At least 20,000 people were killed.

War Nr. 156
Grenada Invasion (Operation Urgent Fury)
24.10.1983 - 1.11.1983 (Typ: C-2)
Total number killed: up to 100

War Nr. 157
Sri Lanka ("Tamil Conflict")
Since 7/1983 (Typ: B-2/AB-1/B-1)
Casualty estimates exceed 50,000
India lost 1,500 troops, and more than 750,000 persons became refugees.

War Nr. 158
Sudan (South Sudan Conflict, 2. Sudanesischer Civil War)
Since 9/1983 (Typ: BA-2)
The war cost up to half a million casualties, and created approximately 3.5 million refugees.
War Nr. 159
Turkey (Kurds)
Since 15.8.1984 (Typ: B-2)
Total number killed: at least 6,000

War Nr. 160
Zimbabwe (Matabeleland Conflict)
1/1983 - 5/1988 (Typ: A-2)
Total number killed: 1,500 to 25,000

War Nr. 161
Pakistan / India (Ziak Glacier)
4/1984 - 1989 (Typ: C-2)
There were over 1,000 casualties, of which at least 80 percent were due to the harsh environmental conditions and not to fighting.

War Nr. 162
South Yemen 1986
13.1.1986 - 29.1.1986 (Typ: A-2)
Total number killed: 10,000 to 15,000
This includes 4,000 members of the JSP.

War Nr. 163
Suriname (Jungle War)
21.6.1986 - 7.6.1989 (Typ: B-2)
500 casualties were reported, and over 10,000 Maroons fled to French-Guayana. The economy nearly collapsed.

War Nr. 164
Thailand / Laos
3.11.1987 - 19.2.1988 (Typ: C-2)

500 to 700 soldiers died in the fighting.

War Nr. 165
Israel, Palestine
Since 1968 (Typ: B-2)
Over 10,000 persons have died so far, 1,000 during the Intifada.

War Nr. 166
Somalia (Civil War)
Since 5/1988 (Typ: A-2/AB-2)
300,000 persons have died from the war and the resulting famine. Over one million Somalis fled to neighboring countries.

War Nr. 167
Papua New Guinea (Bougainville)
Since 2/1989 (Typ: B-2)
Approximately 6,000 persons have died.

War Nr. 168
Liberia, Sierra Leone
Since 24.12.1989 (Typ: A-1)
Total number killed: at least 20,000
Most of the dead are civilians. More than one million people have become refugees.

War Nr. 169
Romania
17.12.1989 - 28.12.1989 (Typ: A-2)
Casualty estimates range between 1,000 and 6,000.

War Nr. 170
USA / Panama
20.12.1989 - 24.12.1989 (Typ: C-2)
Between 600 and 7,000 persons died in the five-day long war. 15,000 were left homeless, and damages were estimated at $2.2 billion (including U.S. sanctions).

War Nr. 171
Pakistan (Belutschistan)
3/1973 - 7/1977 (Typ: B-2)
Total number killed: 9,000 to 10,000

War Nr. 174
USSR (Azerbaijan, Armenia); Nagorno-Karabakh
Since 20.1.1990 (Typ: B-2)
Approximately 3,000 persons have died in the fighting.

War Nr. 177
Bangladesh (Chittagong Hill Tracts)
Since 1973 (Typ: B-2)
Casualty estimates range widely from 1,500 to 200,000. Approximately 70,000 persons fled to India.

War Nr. 178
Vietnam (FULRO)
20.9.1964 - 10/1992 (Typ: B-2)
There are no reliable data available, but up to 1975 an estimated

250,000 Montagnards have been killed. Until 1979 FULRO lost 8,000 of their 10,000 fighters.

War Nr. 179
Iraq (Barzanis)
8/1945 - 10/1945 (Typ: B-1)
Civilians killed: up to 1,000

War Nr. 180
Mali / Burkina Faso (Border War)
21.12.1985 - 31.12.1985 (Typ: C-2)
Casualties range from 60 to 100 persons, with an equal number of wounded.

War Nr. 182
Iraq / Kuwait
2.8.1990 - 4.8.1990 (Typ: C-2)
Total number killed: 200 to 800

War Nr. 183
Laos (First Meo Guerrilla War (Hmong))
1975 - 1979 (Typ: B-1)
Total number killed: at least 40,000
Up to 300,000 people are said to have fled to Thailand, though these numbers have not been confirmed.

War Nr. 184
Nigeria (Tiv Uprising 1964)
8/1964 - 12/1964 (Typ: AB-2)
Up to 4,000 persons died in the uprising.

War Nr. 185
Nigeria (Maitatsine Uprising in Kano/Northern Nigeria)
18.12.1980 - 29.12.1980 (Typ: B-2)
Total number killed: at least 4,200
The total casualties for all of the Maitatsine confrontations with the government between 1980-1985 are estimated at 8,600.

War Nr. 187
Pakistan (Sind Province)
Since 11/1986 (Typ: B-2)
Total number killed: at least 2,000

War Nr. 188
India (Kashmir)
Since 1/1990 (Typ: B-2)
Total number killed: 6,000 to 20,000

War Nr. 189
Rwanda (Rebel Invasion from Uganda)
Since 1.10.1990 (Typ: AB-1)
At least 5,000 persons died up to 1992.

War Nr. 190
Djibuti
Since 10/1991 (Typ: AB-2)
Total number killed: up to 1,000

War Nr. 191
Senegal (Basse-Casamance Conflict)
Since 4/1990 (Typ: B-2)

Total number killed: at least 800

War Nr. 192
Laos (Second Meo Guerrilla War (Hmong))
1/1990 - 7/1992 (Typ: B-1)
Total number killed: at least 500

War Nr. 194
Anti-Iraq Coalition / Iraq ("Second" Gulf War)
17.1.1991 - 27.2.1991 (Typ: C-2)
Total number killed: 230,000 to 260,000
Exact figures are unavailable. The Iraqi government claims 85,000 to 110,000 soldiers and 40,000 civilians died, though the Iraqi opposition claims that 150,000 to 250,000 soldiers and 70,000 to 180,000 civilians died. The U.S. reports 343 allied casualties. 4,000 to 7,000 Kuwaitis died, and 13 Israelis. There was also tremendous environmental damage due to burning oil fields.

War Nr. 196
Iraq (Shi'ites)
Since 2.3.1991 (Typ: A-2)
More than 12,000 persons died, though some estimates reach as high as 50,000 to 60,000 casualties. More han 800,000 persons were left homeless.

War Nr. 197
Tuareg (Mali, Niger)
Since 7.5.1990 (Typ: B-2)
The war cost 1,000 to 2,000 lives. 50,000 Tuareg fled from Mali to Mauritania.

War Nr. 198
Indonesia (Aceh)
Since 5/1990 (Typ: B-2)
At least 2,000 people were killed.

War Nr. 199
Philippines (Mindanao)
Since 1970 (Typ: B-2)
From 1970 to 1980 alone the war is said to have cost between 50,000 and 100,000 lives.

War Nr. 200
Yugoslavia (Slovenia, Croatia, Bosnia)
Since 26.6.1991 (Typ: BC-2)
In Croatia at least 10,000 persons died, and in Bosnia at least 140,000.

War Nr. 201
Georgia (Anti-Regime War)
Since 9/1991 (Typ: A-2)
At least 100 persons were killed.

War Nr. 202
Georgia (South Ossetia)
12/1990 - 7/1992 (Typ: B-2)
At least 150 people were killed and approximately 20,000 Ossetians and 100,000 Georgians became refugees.

War Nr. 203
India (Assam)
1990 - 1992 (Typ: B-2)
Total number killed: at least 300

War Nr. 205
Moldavia (Dneister Republic)
3/1992 - 8/1992 (Typ: B-1)
Total number killed: at least 1,000

War Nr. 206
Algeria
Since 1/1992 (Typ: A-2)
Total number killed: 400 to 600

War Nr. 207
Georgia (Abkhazia)
Since 14.8.1992 (Typ: B-2)
Total number killed: at least 2,000

War Nr. 208
Tadjikistan
Since 8/1992 (Typ: A-2)
By the end of December, 1992, the war had cost more than 50,000 lives and created 250,000 refugees.

War Nr. 209
North Ossetia (Prigorodny District)
Since 30.10.1992 (Typ: B-2)
Total number killed: at least 300
90 percent of the Ingusitian population (about 35,000 persons) fled from North Ossetia.

Additional Lists

Directory of Wars Ongoing as of December 31, 1992

War	Started	War Type
Central America/Caribbean		
Nr. 146: Guatemala	1980	A-2
South America		
Nr. 060: Colombia (FARC)	27.5.1964	A-2
Nr. 079: Colombia (ELN)	7.1.1965	A-2
Nr. 147: Peru (Sendero Luminoso)	17.5.1980	A-2
Nr. 204: Peru (MRTA)	11/1987	A-2
Europe		
Nr. 105: Northern Ireland	14.8.1969	B-2
Nr. 200: Yugoslavia (Slovenia, Croatia, Bosnia)	26.6.1991	BC-2
North Africa		
Nr. 158: Sudan (South Sudan Conflict, Second Sudanese Civil War)	9/1983	BA-2
Nr. 206: Algeria	1/1992	A-2
West Africa		
Nr. 168: Liberia, Sierra Leone	24.12.1989	A-1
Nr. 191: Senegal (Basse-Casamance Conflict)	4/1990	B-2
Nr. 197: Tuareg (Mali, Niger)	7.5.1990	B-2
Central Africa		
Nr. 092: Chad (Civil War)	6/1966	ABC-2
Nr. 189: Rwanda (Rebel Invasion from Uganda)	1.10.1990	AB-1
East Africa		
Nr. 131: Ethiopia (Oromo Conflict)	1976	AB-2
Nr. 166: Somalia (Civil War)	5/1988	A-2/AB-2
Nr. 190: Djibuti	10/1991	AB-2

Southern Africa		
Nr. 062: Angola	4.2.1961	D-1/ABC-1/
A-2		
Nr. 119: Mozambique (Destabilization Campaign)	1975	AC-2/A-1
Nr. 125: South Africa (Anti-Apartheid War)	1976	A-2/AB-2
Near and Middle East		
Nr. 165: Israel, Palestine	1968	B-2
Nr. 120: Lebanon War	13.4.1975	ABC-2
Nr. 126: 6. Iraqi-Kurdish War	5/1976	BA-3
Nr. 141: Afghanistan	10/1978	A-1
Nr. 159: Turkey (Kurds)	15.8.1984	B-2
Nr. 174: USSR (Azerbaijan, Armenia);		
Nagorno-Karabakh	20.1.1990	B-2
Nr. 196: Iraq (Shi'ites)	2.3.1991	A-2
Nr. 201: Georgia (Anti-Regime War)	9/1991	A-2
Nr. 208: Tadjikistan	8/1992	A-2
Nr. 207: Georgia (Abkhasia)	14.8.1992	B-2
Nr. 209: North Ossetia (Prigorodny District)	30.10.1992	B-2
South and Southeast Asia		
Nr. 021: Myanmar (Burma)	1948	AB-2
Nr. 130: Indonesia (West Irian / West Papua)	1965	B-2
Nr. 199: Philippines (Mindanao)	1970	B-2
Nr. 109: Philippines (NPA)	1970	A-2
Nr. 177: Bangladesh (Chittagong Hill Tracts)	1973	B-2
Nr. 121: Indonesia (East Timor)	11.8.1975	A-2/B-2
Nr. 132: Cambodia (Third Indochina War)	12/1975	C-2/A-1
Nr. 155: India (Punjab)	7/1982	B-2
Nr. 157: Sri Lanka ("Tamil Conflict")	7/1983	B-2/AB-1/B-1
Nr. 187: Pakistan (Sind Province)	11/1986	B-2
Nr. 167: Papua New Guinea (Bougainville)	2/1989	B-2
Nr. 188: India (Kashmir)	1/1990	B-2
Nr. 198: Indonesia (Aceh)	5/1990	B-2

Directory of Wars Waged in 1993 (as of December 31, 1993)

Wars

War	Started	War Type
Latin America		
Guatemala	1980	A-2

Colombia (FARC)	1964	A-2
Colombia (ELN)	1965	A-2
Peru (Sendero Luminoso)	1980	A-2
Peru (MRTA)	1987	A-2
Europe		
Northern Ireland	1969	B-2
Croatia/Serbien	1991	C-2/B-2
Bosnia-Herzegovina	1992	B-1
Africa		
Algeria	1992	A-2
Angola (Unita)	1961	ABC-1/A-2
Ethiopia (Oromo)	1976	AB-2
Djibouti	1991	AB-2
Liberia and Sierra Leone	1989	A-1
Mali and Niger (Tuareg)	1990	B-2
Mozambique	1975	AC-2/A-1
Rwanda	1990	A-1
Senegal (Casamance)	1990	B-2
Somalia	1988	AB-1
South Africa	1976	AB-2
Sudan	1983	BA-2
Chad	1966	ABC-1
Near and Middle East		
Afghanistan	1978	A-1/A-1
Azerbaijan (Nag.-Karabach)	1990	B-2
Georgia (Abkhasia)	1992	B-2
Georgia (Gamsachurdia)	1991	A-2
Iraq (Kurdistan)	1976	BA-1
Iraq (Shi'ites)	1991	A-2
Israel (Palestine)	1968	B-2
Lebanon	1975	ABC-1
Russian Federation (Ingusetia)	1992	B-2
Tadjikistan	1992	A-1
Turkey (Kurdistan)	1984	B-2
Asia		
Bangladesh (Chittagong Hills)	1973	B-2
Burma (Myanmar)	1948	AB-2
India (Kashmir)	1990	B-2
India (Punjab)	1982	B-2
Indonesia (East Timor)	1975	B-2
Cambodia	1975	C-2/A-1

Papua New Guinea (Bougainville)	1989	B-2
Philippines (Mindanao)	1970	B-2
Philippines (NPA)	1970	A-2
Sri Lanka (Tamil)	1983	B-2/
AB-1/B-1		

Armed Conflicts

Latin America	Haiti; Nicaragua
Near and Middle East	Yemen; Russian Federation (Chechenya)
Africa	Egypt; Angola (Cabinda); Burundi; Eritrea; Kenya; Congo; Togo; West Sahara (DARS); Zaïre (Anti-Mobutu); Zaïre (Shaba and Kivu)
Asia	India (Assam); Indonesia (Aceh); Indonesia (West
Irian);	
Laos; Pakistan (Sind)	

Directory of Wars Waged in 1994 (as of December 31, 1994)

Wars

War	Started	War Type
Latin America		
Guatemala	1980 -	A-2
Colombia (FARC)	1964 -	A-2
Colombia (ELN)	1965 -	A-2
Peru (Sendero Luminoso)	1980 -	A-2
Peru (MRTA)	1987 -	A-2
Mexico	1994 -	B-2
Europe		
Northern Ireland	1969 -	B-2
Croatia/Serbia	1991 -	C-2/B-2
Bosnia-Herzegovina	1992 -	B-1
Africa		
Algeria	1992 -	A-2
Angola (Unita)	1961 -	ABC-1/A-2
Angola (Cabinda)	1992 -	B-2
Djibouti	1991 -	AB-2
Liberia and Sierra Leone	1989 -	A-1

Mali	1990 -	B-2
Niger	1990 -	B-2
Rwanda	1990 -	A-1
Somalia	1988 -	AB-1
Sudan	1983 -	BA-2
Republic of South Africa	1976 -	AB-2
Chad	1966 -	ABC-1
Near and Middle East		
Afghanistan	1978 -	A-1/A-1
Azerbaijan (Nag.-Karabach)	1990 -	B-2
Georgia (Abkhazia)	1992 -	B-2
Iraq (Kurdistan)	1976 -	BA-1
Iraq (Shi'ites)	1991 -	A-2
Israel (Palestine)	1968 -	B-2
Yemen	1994 -	B-2
Lebanon	1975 -	ABC-1
Russian Federation (Ingusetia)	1992 -	B-2
Russian Federation (Chechenya)	1994 -	B-2
Tadjikistan	1992 -	A-1
Turkey (Kurdistan)	1984 -	B-2
Asia		
Burma (Myanmar)	1948 -	AB-2
India (Kashmir)	1990 -	B-2
India (Punjab)	1982 -	B-2
Indonesia (East Timor)	1975 -	B-2
Cambodia	1975 -	C-2/A-1
Papua New Guinea (Bougainville)	1989 -	B-2
Philippines (Mindanao)	1970 -	B-2
Philippines (NPA)	1970 -	A-2
Sri Lanka (Tamil)	1983 -	B-2/AB-1/B-1

Armed Conflicts

Latin America	Haiti; Nicaragua
Near and Middle East	Iran (Kurds); Russian Federation (Chechnaya)
Africa	Egypt; Burundi; Kenya; Kongo; Mozambique; West Sahara (DARS); Zaïre (Anti-Mobutu); Zaïre (Shaba and Kivu)
Asia	India (Assam); India (Nagaland); India(Naxalites); Indonesia (West-Irian); Pakistan (Sind)

Directory of Wars Waged in 1995 (as of December 31, 1995)

Wars

War	Started	War Type
Latin America		
Ecuador/Peru	1995 -	C-1
Guatemala	1980 -	A-2
Colombia (FARC)	1964 -	A-2
Colombia (ELN)	1965 -	A-2
Peru (Sendero Luminoso)	1980 -	A-2
Peru (MRTA)	1987 -	A-2
Mexico	1994 -	B-2
Europe		
Croatia/Serbia	1991 -	BC-2
Bosnia-Herzegovina	1992 -	BC-1
Africa		
Algeria	1992 -	A-2
Angola (Unita)	1961 -	A-2
Angola (Cabinda)	1994 -	B-2
Burundi	1994 -	A-1
Chad	1966 -	AB-2
Liberia	1989 -	A-1
Mali	1990 -	B-2
Niger	1990 -	B-2
Rwanda	1990 -	A-1
Senegal (Casamance)	1990 -	B-2
Sierra Leone	1993 -	A-1
Somalia	1988 -	A-2/AB-2
Sudan	1983 -	BA-2
Uganda	1995 -	A-2
Near and Middle East		
Afghanistan	1978 -	A-1
Iraq (Kurdistan)	1976 -	BA-1
Iraq (Shi'ites)	1991 -	A-2
Lebanon	1975 -	ABC-2
Russian Federation (Chechenya)	1994 -	B-2
Tadjikistan	1992 -	A-2
Turkey (Kurdistan)	1984 -	B-2
Asia		
Burma (Myanmar)	1948 -	AB-2

India (Kashmir)	1990 -	B-2
Cambodia	1975 -	A-1
Pakistan (Sind)	1986	B-2
Papua New Guinea (Bougainville)	1989 -	B-2
Philippines (Mindanao)	1970 -	B-2
Philippines (NPA)	1970 -	A-2
Sri Lanka (Tamil)	1983 -	B-2/AB-1/B-2

Armed Conflicts

Latin America	Haiti; Peru (MRTA)
Europe	Northern Ireland
Near and Middle East	Egypt; Iran (Kurds); Israel (Hamas)
Africa	Kenya; Niger (Taureg); Niger (FDR)
Asia	Bangladesh; India (Assam); India (Nagas); Indonesia (West-Papua)

Directory of Colonial Independence

Colony	Year	Current Name
North America	1776	USA
Hispaniola	1804	Haiti
New Granada	1810	Colombia
Paraguay	1811	
General Captaniate Caracas	1811	Venezuela
Rio de la Plata	1816	Argentina
Chile	1818	
El Salvador	1821	
Nicaragua	1821	
General Captinate Guatemala	1821	Divided into Guatemala and Costa Rica
Peru	1821	
Brazil	1822	
New Spain	1823/24	Mexico
Bolivia	1825	
Hispaniola (Eastern Part)	1865	Dominican Republic
Australia	1901*	Australian Federation
Cuba	1902	
Panama	1903	

Colony	Year	Current Name
New Zealand	1907	
South Africa	1910	Republic of South Africa
Iceland	1918	
Bismarck Archipelago	1920	to Australian Federation
German South-West Africa	1920	to South Africa
Kiautschou	1920	to China
Egypt	1922	
Ireland	1922	
Canada	1931*	
Iraq	1932	
New Foundland	1934	1948 to Canada
Indonesia	1945	
Molukken	1950	to Indonesia
South Molukks	1950/51	today to Indonesia

Colony	Year	Current Name
Malakka	1946	Malaysia
Lebanon	1944/46	
Transjordan	1946	Jordan
Syria	1946	
Philippines	1946	
British India	1947	Divided into: India and Pakistan, and since 1971 also into Bangladesh
Burma	1948	
Ceylon	1948	
Korea	1948	Divided into: Republic of Korea (South) and Democratic People's Republic of Korea (North)
Palestine	1948	Israel, East Palestine to Jordan, Gaza to Egypt
Libya	1951	
Indochina	1954	Divided into: Cambodia, Laos, North and South Vietnam (today the Republic of Vietnam)
Tunisia	1956	
Morocco	1956	
Sudan	1956	
Gold Coast	1957	Ghana
West Togo		
Guinea	1958	

Togo (East Part)	1960	
Chad	1960	
Dahomey	1960	Benin
Ivory Coast	1960	
Gabon	1960	
Oubangui-Chari	1960	Central African Republic
French Congo	1960	People's Republic of Congo
Belgian Congo	1960	Zaïre
Cameroon	1960	
Madagascar	1960	
Mauritania	1960	
Niger	1960	
Nigeria	1960	
Upper Volta	1960	Burkino Fasso
Senegal	1960	
Italy and British Somaliland	1960	Somalia
Soudan	1960	Mali
Daman, Diu, Goa	1961	To India
Tanganika	1961	Tanzania
Sierre Leone	1961	
Cyprus	1961	
West Samoa	1962	Samoa-West
Uganda	1962	
Rwanda-Burundi	1962	Divided into: Burundi and Rwanda
Trinidad and Tobago	1962	
Algeria	1962	
Jamaica	1962	
Kenya	1963	
Sabah and Sarawak	1963	To Malaysia
Zanzibar	1963	to Zanzibar
Malta	1964	
Niassa Land	1964	Malawi
North Rhodesia	1964	Zambia
Aden	1965	People's Republic of Yemen
Gambia	1965	
Maldives	1965	
Singapore	1965	
Betchuana Land	1966	Botswana
Guyana	1966	
Basuto Land	1966	Lesotho
Fernando Póo Rio Mundi	1968	Equatorial Guinea
Mauritius	1968	

Colony	Year	Current Name
Nauru	1968	
Swaziland	1968	
(West New Guinea [West Irian])	1969	To Indonesia
Tonga	1970	
Bahrain	1970	
Fidji Islands	1970	Republic of Fidji
Oman	1970	
Liberated Oman	1971	United Arab Emirate
Katar	1971	
Guinea-Bissau	1974	
Grenada	1974	
Mozambique	1975	
Cape Verdean Islands	1975	Cape Verde
Comoros	1975	
São Tomé and Príncipe	1975	
Northeast Guinea	1975	Papua New Guinea
Angola	1975	
Suriname	1975	
(Spanish Sahara	1976	today to Morocco)
(Portugese Timor	1976	to Indonesia)
Seychelle Islands	1976	
French Afar and Issa Coast	1977	
Dominica	1978	
Ellice Islands	1978	Tuvalu
Saint Lucia	1979	
Gilbert Islands	1979	Kiribati
Saint Vincent	1979	
Rhodesia	1980	Zimbabwe
New Hebrides	1980	Vanuatu
British Honduras	1981	Belize
Antigua	1981	Antigua and Barbados
Brunei	1984	
Marshall Islands	1990	

Sources: Meyers Großes Taschenlexikon, Mannheim, Wien, Zürich 1983, Bd 6, S.166f; Harenberg, Schlüsseldaten des 20. Jahrhundert, Dormtund 1993, S. 163, 491, 686; Der Fischer-Weltalmanach '95, S. 427; Der Fischer-Weltalmanach '95, Frankfurt a.M. 1994, Spalte 428.

Directory of States

A. States recognized by the United Nations[1]

	STATE	DATE		STATE	DATE
1.	Egypt	24.10.1945	32.	Peru	31.10.1945
2.	Argentina	24.10.1945	33.	Australia	01.11.1945
3.	Byelorussia (Belarus)	24.10.1945	34.	Costa Rica	02.11.1945
4.	Brazil	24.10.1945	35.	Liberia	02.11.1945
5.	Chile	24.10.1945	36.	Colombia	05.11.1945
6.	China	24.10.1945	37.	Mexico	07.11.1945
7.	CSSR (Czechoslovakia)[2]	24.10.1945	38.	South Africa	07.11.1945
8.	Denmark	24.10.1945	39.	Canada	09.11.1945
9.	Dominican Republic	24.10.1945	40.	Ethiopia	13.11.1945
10.	El Salvador	24.10.1945	41.	Panama	13.11.1945
11.	France	24.10.1945	42.	Bolivia	14.11.1945
12.	Haiti	24.10.1945	43.	Venezuela	15.11.1945
13.	Iran	24.10.1945	44.	Guatemala	21.11.1945
14.	Yugoslavia[3]	24.10.1945	45.	Norway	27.11.1945
15.	Cuba	24.10.1945	46.	Netherlands	10.12.1945
16.	Lebanon	24.10.1945	47.	Honduras	17.12.1945
17.	Luxembourg##	24.10.1945	48.	Uruguay	18.12.1945
18.	New Zealand	24.10.1945	49.	Ecuador	21.12.1945
19.	Nicaragua	24.10.1945	50.	Iraq	21.12.1945
0.	Paraguay	24.10.1945	51.	Belgium	27.12.1945
21.	Philippines	24.10.1945	52.	Afghanistan	19.11.1946
22.	Poland	24.10.1945	53.	Iceland##	19.11.1946
23.	Saudi Arabia	24.10.1945	54.	Sweden	19.11.1946
24.	Syria	24.10.1945	55.	Thailand	16.12.1946
25.	Turkey	24.10.1945	256.	Yemen	30.09.1947
26.	Ukraine	24.10.1945	57.	Pakistan	30.09.1947
27.	USSR[4]	24.10.1945	58.	Burma (Myanmar)	19.04.1948
28.	United States of America	24.10.1945	59.	Israel	11.05.1949
29.	United Kingdom	24.10.1945	60.	Indonesia	28.09.1950
30.	Greece	25.10.1945	61.	Albania	14.12.1955
31.	India	30.10.1945	62.	Bulgaria	14.12.1955
			63.	Finland	14.12.1955

	STATE	DATE
64.	Ireland	14.12.1955
65.	Italy	14.12.1955
66.	Jordan	14.12.1955
67.	Cambodia	14.12.1955
68.	Laos	14.12.1955
69.	Libya	14.12.1955
70.	Nepal	14.12.1955
71.	Austria	14.12.1955
72.	Portugal	14.12.1955
73.	Romania	14.12.1955
74.	Spain	14.12.1955
75.	Sri Lanka	14.12.1955
76.	Hungary	14.12.1955
77.	Morocco	12.11.1956
78.	Sudan	12.11.1956
79.	Tunisia	12.11.1956
80.	Japan	18.12.1956
81.	Ghana	08.03.1957
82.	Malaysian Federation (Malaysia)	17.09.1957
83.	Guinea	12.12.1958
84.	Benin	20.09.1960
85.	Côte d'Ivoire	20.09.1960
86.	Gabun	20.09.1960
87.	Cameroon	20.09.1960
88.	Congo	20.09.1960
89.	Madagascar	20.09.1960
90.	Niger	20.09.1960
91.	Upper Volta (Burkina Faso)	20.09.1960
92.	Somalia	20.09.1960
93.	Togo	20.09.1960
94.	Chad	20.09.1960
95.	Zaïre	20.09.1960
96.	Central African Republic	20.09.1960
97.	Cyprus#	20.09.1960
98.	Mali	28.09.1960
99.	Senegal	28.09.1960
100.	Nigeria	07.10.1960
101.	Sierra Leone	27.09.1961
102.	Mauritania	27.10.1961
103.	Mongolia	27.10.1961
104.	Tanganika (Tanzania)	14.12.1961
105.	Burundi	18.09.1962
106.	Jamaica	18.09.1962
107.	Rwanda	18.09.1962
108.	Trinidad and Tobago	18.09.1962
109.	Algeria	08.10.1962
110.	Uganda	25.10.1962
111.	Kuwait	14.05.1963
112.	Kenya	16.12.1963
113.	Zanzibar[5]	16.12.1963
114.	Malawi	01.12.1964
115.	Malta##	01.12.1964
116.	Zambia	01.12.1964
117.	Gambia#	21.09.1965
118.	Maledives##	21.09.1965
119.	Singapore	21.09.1965
120.	Guyana#	20.09.1966
121.	Botswana	17.10.1966
122.	Lesotho	17.10.1966
123.	Barbados##	09.12.1966
124.	Democratic People's Republic of Yemen[6]	14.12.1967
125.	Mauritius	24.04.1968
126.	Swaziland#	24.09.1968
127.	Equatorial Guinea##	12.11.1968
128.	Fidji#	13.10.1970
129.	Bahrain#	21.09.1971
130.	Bhutan	21.09.1971
131.	Katar#	21.09.1971
132.	Oman	07.10.1971
133.	United Arab Emirate	09.12.1971
134.	Bahamas##	18.09.1973
135.	Federal Republic of Germany	18.09.1973
136.	German Democratic Republic[7]	18.09.1973

	STATE	DATE
137.	Bangladesh	17.09.1974
138.	Grenada##	17.09.1974
139.	Guinea-Bissau#	17.09.1974
140.	Cape Verde##	16.09.1975
141.	Mozambique	16.09.1975
142.	Sao Tome and Principe##	16.09.1975
143.	Papua New Guinea	10.10.1975
144.	Suriname##	04.12.1975
145.	Comoros##	12.12.1975
146.	Seychelles##	21.09.1976
147.	Angola	01.12.1976
148.	Samoa##	15.12.1976
149.	Djibuti##	20.09.1977
150.	Vietnam	20.09.1977
151.	Solomon Islands##	19.09.1978
152.	Dominica##	18.12.1978
153.	St. Lucia##	18.09.1979
154.	Zimbabwe	25.08.1980
155.	St. Vincent and the Grenadines##	16.09.1980
156.	Vanuatu##	15.09.1981
157.	Belize##	25.09.1981
158.	Antigua and Barbuda##	01.11.1981
159.	St. Kitts and Nevis##	23.09.1983
160.	Brunei Darussalam##	21.09.1984
161.	Namibia	23.04.1990
162.	Liechtenstein##	18.09.1990
163.	Democratic People's Republic of Korea	17.09.1991
164.	Estonia	17.09.1991
165.	Latvia	17.09.1991
166.	Lithuania	17.09.1991
167.	Marshall Islands##	17.09.1991
168.	Republic of Korea	17.09.1991
169.	United States of Micronesia##	17.09.1991
170.	Armenia	02.03.1992
171.	Azerbaijan	02.03.1992
172.	Kazakhstan	02.03.1992
173.	Kirgisistan	02.03.1992
174.	Moldavia	02.03.1992
175.	San Marino##	02.03.1992
176.	Tadjikistan	02.03.1992
177.	Turkmenistan	02.03.1992
178.	Uzbekistan	02.03.1992
179.	Bosnia-Herzegovina	22.05.1992
180.	Croatia	22.05.1992
181.	Slovenia	22.05.1992
182.	Georgia	31.07.1992
183.	Slovak Republic	19.01.1993
184.	Czech Republic	19.01.1993
185.	Former Yugoslav Republic of Macedonia	08.04.1993
186.	Eritrea	28.05.1993
187.	Monaco##	28.05.1993
188.	Andorra##	28.07.1993

B. Other States[8]

	STATE	DATE
1.	Democratic Arab Republic of Sahara##, [9]	28.02.1976
2.	Kiribati##	12.07.1979
3.	Nauru##	31.01.1968
4.	Palau##, [10]	
5.	Switzerland	24.10.1648
6.	Taiwan[11]	
7.	Tonga##	05.06.1970
8.	Tuvalu##	01.10.1978
9.	Vatican City##	07.06.1929

Notes:

- \# States with a population of between 500,000 to one million (as of the beginning of the 1990s).
- \#\# States with a population under 500,000 (as of the beginning of the 1990s).

1. Recognition by the United Nations means that the respective state was admitted by the General Assembly upon recommendation by the Security Council.
2. The CSSR ceased to exist as of December 31, 1992. Its successor states were the Slovak Republic and the Czech Republic, both of which joined the U.N. on January 19, 1993.
3. The legal successor to Yugoslavia, the „Federal Republic of Yugoslavia“ consisting of Serbia and Montenegro, is not recognized by the U.N.
4. After the USSR ceased to exist on December 31, 1991, its role in the U.N. was taken over by the Russian Federation.
5. Zanzibar ceased to be a member of the U.N. after it united with Tanganika on April 26, 1964. Today it is part of the United Republic of Tanzania.
6. Member of the U.N. until its unification with North Yemen on May 22, 1990.
7. Member of the U.N. until its unification with the Federal Republic on October 3, 1990.
8. States which are not members of the United Nations.
9. Recognized by 70 states (including 30 OAU states), but annexed by Morocco since 1979.
10. U.S. territory.
11. Not recognized by the U.N. as a sovereign state.

Sources: Fischer Weltalmanach '94; U.N. Information Service Note Nr. 27/Rev. 2 vom 29. Juli 1993; U.N. Handbook 1993
Compiled by: Boris Wilke

Bibliography

Key to News media and journal abbreviations

ACR Africa Contemporary Record, New York, London
AdG Archiv der Gegenwart
AdJ Asian Defence Journal, Kuala Lumpur, Malaysia
APuZ Aus Politik und Zeitgeschichte, Beilage zu Das Parlament, Bonn
ARB Africa Research Bulletin, London
FAZ Frankfurter Allgemeine Zeitung
FEER Far Eastern Economic Review, Hongkong
FR Frankfurter Rundschau
FT Financial Times, London
IDR International Defence Review, London
IHT International Herald Tribune, Frankfurt M.
JDW Jane's Defence Weekly, London
Keesing's: Keesing's Contemporary Archives, London
KAS Konrad Adenauer Stiftung, Auslandsinformationen, Sankt Augustin
MEI Middle East International, London
NZZ Neue Zürcher Zeitung, Zürich
ÖMZ Österreichische Militärische Zeitschrift, Wien
SOAa Südostasien Aktuell, Hamburg
SOAi Südostasien Informationen, Bochum
SZ Süddeutsche Zeitung, München

Works Consulted

Abi-Saab, Georges: International Crisis and the Role of Law. The U.N.-Operation in the Congo 1960-1964, 1978

Adelman, K. L.: Zaire's Year of Crisis, in: African Affairs 77, 306/1978: 36-44

Agee, Philip / Fuchs, Regina / Ernst, Manfred: Generalprobe Grenada, Hamburg 1984

Aguado Sanchéz, Francisco: El maquis en España, Madrid 1975

Aguado Sanchéz, Francisco: El maquis en España: sus documentos, Madrid 1976

Aguilera, Gabriel: Centroamérica: La crisis sin fin, in: Ciencias Sociales 36/1987: 19-27

Aguirre, José Ramón Diego: La guerra del Ifni, Madrid 1980

Agwani, M.S.: The Lebanese Crisis 1958 - A Documentary Study, London/New Delhi 1965

Ahmad, A.: 400 Years War - Moro Struggle in the Philippines, in: Southeast Asia Chronicle (Feb.) 82/1982: 2-22

Ahmed, Moudud: Constitutional Quest for Autonomy: 1950-1971, Wiesbaden 1978

Ahmed, Moudud: Bangladesh: Era of Sheikh Mujibur Rahman, Dacca 1983

Ahmed, Samina: The Politics of Ethnicity in India, in: Regional Studies (Islamabad) 4/91: 22-50

Aker, F.: October 1973. The Arab-Israeli War, Hamden 1985

Akhtar, Shaheen: Mizo Accord: Implications for North-East India, in: Regional Studies (Islamabad) (Vol. 7, Winter) 1/1988-89: 124-170

Aktionsgruppe Philippinen (Hg.): Philippinen - Wenn der Bambus bricht, Hamburg 1986

Alape, Arturo: La Paz, la Violencia: Testigos de Excepción, Bogotá 1985

Albright, David / Zamorra, Tom: India, Pakistan's Nuclear Weapons - All the Pieces in Place, in: Bulletin of the Atomic Scientists, Vol.45, 5/1989: 20-26.

Ali, S.: Sindh Erupts in Wave of Ethnic Killings, in: Far Eastern Economic Review 7.6.1990 (1990a)

Ali, S.: Sindh Violence Eclipsis Kashmir Crisis, in: Far Eastern Economic Review 14.6.1990 (1990b)

Allemann, Fritz René: Macht und Ohnmacht der Guerrilla, München 1974

Allen, P.: The Yom Kippur War, New York 1982

Alles, A.C.: Insurgency - 1971, Colombo 1979

Alterman, E.: Report from the Occupied Territories: Palestinians Speak (Interviews), in: World Policy Journal Vol.5, 3/1988: 519- 542

Amin, Tahir: Ethno-national Movements of Pakistan: Domestic and International Factors, Islamabad 1988

amnesty international (Hg.): Paraguay in den 70er Jahren, (Oktober) 1971
amnesty international (Hg.): "Bangladesh". Widerrechtliche Tötungen und Folter im Chittagong-Gebirge, London 1986
amnesty international (Hg.): Bangladesh Info, Hamburg 1,2,4,5/1989
Anand, Vijay Kumar: Conflict in Nagaland. A Story of Insurgency and Counter-Insurgency, Delhi 1980
Andrews, William G.: French Politics and Algeria: The Process of Policy Formation 1954-1962, New York 1962
Anifowose, R.: Violence and Politics in Nigeria: The Tiv and Yoruba Experience, London 1982
Ansprenger, Franz: Die SWAPO, Mainz 1984
Anthony, Ian: The "Third Tier" Countries: Production of Major Weapons. Arms Industry Limited, Herbert Wulf (Hg.), London: Oxford University Press, 1993: 362-385.
Apter, David E.: The Political Kingdom in Uganda, Princeton 1961
Aquarone, M. Ch.: Les Frontières du Refus - Six Séparatismes Africains, Paris 1987
Arenas, Jacobo: Cese el Fuego. Una Historia Politica de las FARC, Bogotá 1985
Arfa, Hassan: The Kurds. A Historical and Political Study, London 1966
Argueta, Manlio: Un dia en la vida, San José 1982
Arias, Arturo: El movimiento Indígena en Guatemala 1970-1983, in: Camacho, Daniel / Manjívar, Rafael: Movimientos populares en Centroamérica, San José 1985: 62-119
Arnold, Guy: Wars in the Third World since 1945, London/New York 1991
Aschmoneit, Walter / Werning, Rainer (Hg.): Kampuchea, Münster 1981
Astrow, André: A Revolution that has Lost its Way, London 1983
Atkins, G. Pope: Arms and Politics in the Dominican Republic, Boulder (Col.) 1981
Avirgan, Tony / Honey, Martha: War in Uganda. The Legacy of Idi Amin, London 1982
Avnery, A.: The War of Attrition, Tel Aviv 1981
Bacevich, A.J. / Hallums, J.D. / White, R.H.: El Salvador: Una evaluación militar estadounidense, in: CENTROAMERICA / USA, San José 4/1988
Badal, R.K.: The Rise and Fall of Separatism in Southern Sudan, in: African Affairs 75/301, London 1976
Bächler, Günther.: Friedensfähigkeit von Demokratien, Demokratisierung der Sicherheitspolitik und Strukturelle Angriffsunfähigkeit. in: Institut für Friedensforschung und Sicherheitspolitik an der Universität Hamburg (Hg.): Hamburger Beiträge zur Friedensforschung und Sicherheitspolitik, Heft 29, Hamburg 1988.

Bailey, Clinton: Jordans Palestinian Challenge 1948-1983. A Political History, Boulder (Col.) 1984
Bar-Simon-Tov, Y.: The Israeli-Egyptian war of attrition 1969-1970. A case study of limited local war, New York 1980
Baram, Amazia: Ideology and Power Politics in Syria, 1968 - 1984, in: Maoz, Moshe/Yaniv, Auner (Ed.): Syria under Assad, London/Sydney 1987: 125-139
Barbier, M.: Le Conflit du Sahara Occidental, Paris 1982
Barker, A.J.: Der Sechs-Tage-Krieg, Rastatt 1981
Barnds, William J.: India, Pakistan, and the Great Powers, New York/ Washington/ London 1972
Barnet, Richard J.: Intervention and Revolution. The United States in the Third World, New York/Cleveland (Ohio) 1968: 153-180
Barolo, Juan: Der erleuchtete Pfad in die Finsternis, in: Lateinamerika. Analysen und Berichte 9/1985: 111-136
Barrin, Jacques du: L'Algérie démobilisée, in: Le Monde, 9.1.1993
Barry, Tom / Preusch, Deb: The Central American Fact Book, New York 1986
Barth, Ariane / Terzani, Tiziano: Holocaust in Kambodscha, Hamburg 1980
Batatu, Hanna: Syria's Muslim Brethren, in: Merip Reports, Nr. 110, 1982: 12-20
Baumann, Evelyne / Illy, Hans: Kamerun, in: Nohlen, Dieter / Nuscheler, Franz (Hg.): Handbuch der Dritten Welt, Bd.4, Hamburg 1982: 410-428
Baumgarten, H.: Die palästinensische Nationalbewegung vom Anfang des Jahrhunderts bis Mitte der 60er Jahre, in: Bernstein/Hamdan/Schneider (Hg.): Der Palästinakonflikt, Bad Wörishofen 1983: 61-80
Baumgärtner, Ulf / Gabriel, Leo / Gottwald, Gaby / u.a.: D'Aubuisson's neue ARENA, Hamburg 1989
Baumhögger, Goswin: Burundi/Ruanda, in: Hofmeier/Schönborn (Hg.): München 1984
Baumhögger, Goswin: The Struggle for Independence: Documents on the Recent Development of Zimbabwe, Hamburg 1985
Baumhögger, Goswin: Namibia, in: Hofmeier, Rolf (Hg.): Afrika-Jahrbuch 1988, Opladen 1989
Baya, B. K.: Aid at a Stillstand, in: Panos-Institute: War Wounds, Development Costs of Conflict in Southern Sudan, London/Paris/Washington 1988: 17-28
Becker, J.: Angola, Mozambik und Zimbabwe: Im Visier Südafrikas, Köln 1988.
Bedington, Stanley: Malaysia and Singapore, London/New York 1978
Behar, Olga: Las Guerras de la Paz, Bogotá 1985
Béjar Rivera, Héctor: Peru 1965. Aufzeichnung eines Guerrilla-Aufstands, Frankfurt a.M. 1970

Bell, Ian / Feith, Herb / Hatley, Ron: The West Papuan Challenge to Indonesian Authority in Irian Jaya, in: Asian Survey, University of California Press, Vol. 25, 11/1985: 539-556

Bello, W.: Beningo Aquino. Between Dictatorship and Revolution in the Philippines, in: Third World Quarterly 2/1984: 283-310

Benz, Wolfgang / Graml, Hermann: Weltprobleme zwischen den Machtblöcken - Das Zwanzigste Jahrhundert III. (Fischer Weltgeschichte Bd. 23) Frankfurt a.M. 1981

Berger, Carl: The Korean Knot. A Military-Political History, London 1957

Berger, Peter L. / Godsell, Bobby: A Future South Africa, o.O. 1988

Berquist, Charles W.: Labour in Latin America. Comparative Essays on Chile, Argentina, Venezuela and Colombia, Stanford (Cal.) 1986

Berry, Albert R. / u.a. (Hg.): Politics of Compromise: Coalition Government in Colombia. New Brunswick (N.Y.) 1980

Bertaux, Pierre: Afrika. Von der Vorgeschichte bis zu den Staaten der Gegenwart, Frankfurt a.M. 1966

Bertocci, P.: Bangla Desh: Resource Development and Ethnic Conflict - The Case of Chittagong Hill Tracts, in: Asian Thought and Society, Vol.9, 26-27/1984, New York

Beshir, Mohamed Omer: The Southern Sudan, Background to Conflict, Khartoum 1970

Beshir, Mohamed Omer: The Southern Sudan from Conflict to Peace, Khartoum 1975

Betz, Joachim: Krisenherd Südasien. Der indisch-pakistanische Konflikt, in: Hamann, Rudolf (Hg.): Die "Süddimension" des Ost-West-Konflikts. Das Engagement der Supermächte in Krisen und Kriegen der Dritten Welt, Baden-Baden 1986: 173-193

Betz, Joachim: Außen- und innenpolitische Rahmenbedingungen der wirtschaftspolitischen Liberalisierung in Indien, in: Aus Politik und Zeitgeschichte (24.Feb.) 9/1989: 25-38

Beys, J. (Hg.) / Gendebien, P.-H. / Verhaegen, B. (Hg.): Congo 1963, Brüssel 1964

bfai (Bundesstelle für Außenhandelsinformation): Nicaragua: Wirtschaftsdaten, Bonn 1991

Bienen, H.: Religion, Legitamacy, and Conflict in Nigeria, in: The Annuals of the American Academy of Political and Social Science, 483/1986: 50-60

Bierschenk, Thomas: Oil Interest and the Formation of Centralized Government in Oman 1920-1970, in: Orient 1989: 205-219

Biescas, José Antonio / Tuñon de Lara, Manuel: España bajo la dictadura franquista, Barvelona 1980

Bitima, Tamene: Die Oromo - Ein Jahrhundert nationaler Unterdrückung, in:

Steuber, Jürgen / Bitima, Tamene: Die ungelöste nationale Frage in Äthiopien, Frankfurt a.M./Bern 1983
Black, George: Triumph of the People. The Sandinista Revolution in Nicaragua, London 1981
Black, George / u.a.: Garrison Guatemala, New York 1984
Blaikie, P. / Cameron, J. / Seddon, D.: Nepal in Crisis. Growth and Stagnation at the Periphery, Oxford/Delhi/Bombay/Calcutta 1980
Blanchard, Wendell: Thailand, its People, its Society, its Culture, New Haven 1959
Blaschke, Jochen,: Volk, Nation, interner Kolonialismus, Ethnizität. Konzepte zur politischen Soziologie sozialer Bewegungen in Westeuropa, Berlin 1984
Bleuchot, H.: Le Soudan au XIXe Siècle - Le Soudan anglo-égyptien, in: Lavergne, M. (Hg.): Le Soudan Contemporain. De l'invasion turco-égyptienne à la rébellion africaine (1821 - 1889), Paris/Amman 1990
Bley, Helmut: Kolonialherrschaft und Sozialstruktur in Deutsch-Südwestafrika 1894-1914, Hamburg 1968
Bley, Helmut: Namibia nach zwei Jahrzehnten Krieg, in: Vereinte Nationen 2/1989
Bombwall, K.R.: Sikh Identity, Akali Dal and Federal Policy, in: Economic and Political Weekly, Vol.11 (May 17) 20/1986: 888-890
Bongartz, Maria: Somalia im Bürgerkrieg, Ursachen und Perspektiven des innenpolitischen Konflikts, Hamburg 1991
Bonn, Gisela: Revolte auf Ceylon, in: Indo-Asia 2/1971: 160-166
Boris, Dieter / Hiedl, Peter: Argentinien. Geschichte und politische Gegenwart, Köln 1978
Boris, Dieter / Hiedl, Peter / Sieglin, Veronika: Guatemala, in: Boris, Dieter / Rausch, Renate (Hg.): Zentralamerika, Köln 1984: 70-127
Bosch, Juan: Der Pentagonismus oder die Ablösung des Imperialismus, Reinbek 1969
Böttger, B.: 700 Millionen ohne Zukunft? Faschismus oder Revolution in Indien und Bangladesh, Reinbek 1975
Bouthoul, Gaston / Carrère, René: Major Armed Conflicts 1965 - 1. July 1978, in: Peace Research 4/1979
Bowie, R.: International crisis and the role of law, Suez 1956, Oxford 1974
Braganca, Aquino / Wallerstein, Immanuel (ed.): The African Liberation Reader, (Vol.I-III), London 1982
Braun, Dieter: Die Sowjetunion und der "afghanische Knoten" - Verwicklung, Intervention, Herauslösung, in: Senghaas, Dieter (Hg.): Regionalkonflikte in der Dritten Welt - Autonomie und Fremdbestimmung, Baden-Baden 1989: 57-81
Braun, Dieter / Ziem, Karlernst: Die nukleare Komponente eines latenten wie akuten Regionalkonflikts: Indien/Pakistan, in: Senghaas, Dieter (Hg.):

Regionalkonflikte in der Dritten Welt - Autonomie und Fremdbestimmung, Baden-Baden 1989: 273-294
Braun, Ursula: Nord- und Südjemen im Spannungsfeld interner, regionaler und globaler Gegensätze, Bonn 1981
Bray, John: Ethnic Minorities and the Future of Burma, in: The World Today 8/9 1992: 144-147
Brecher, M.: Decisions in Crisis. Israel 1967 and 1973, Berkeley 1980
Brecher, Michael / Moser, Sheila / Wilkenfeld, Jonathan: Crisis in the Twentieth Century. Handbook of International Crises, Oxford 1988
Brewer, John D.: Can South Africa Survive?, London 1989
Brill, Heinz: Libyens Expansionsbestrebungen im Tschad, in: Beiträge zur Konfliktforschung 17/1987
Brill, Heinz: Libyens Außen- und Sicherheitspolitik. Moamar al Gaddafis Motive und Visionen, Baden-Baden 1988
Brines, Russell: The Indo-Pakistani Conflict, London 1968
Brintall, Douglas E.: Revolt Against the Dead, New York 1979
Brockway, Fenner: The Colonial Revolution, London 1973
Brogan, Patrick: World Conflicts. Why and where they are happening, London 1989
Brown, Ashley (Hg.): War in Peace, the Marshall Cavendish Illustrated Encyclopaedia of Post War Conflicts (Bd.6), New York 1985
Brown, Bess: Tajik Opposition to be banned, in: Radio Free Europe / Radio Liberty 14/1993: 9-12
Brown, Norman W.: The United States and India, Pakistan and Bangladesh, Cambridge (Mass.) 1972
Brownlie, J.: African Boundaries, London 1979
Bruinessen, Martin van: Nationalismus und religiöser Konflikt: Der kurdische Widerstand im Iran, in: Berliner Institut für Vergleichende Sozialforschung (Hg.): Religion und Politik im Iran. mardom nameh - Jahrbuch zur Geschichte und Gesellschaft des Mittleren Orients, Frankfurt a.M. 1981: 372-409
Bruinessen, Martin van: The Kurds in Turkey, in: MERIP Reports 121/1984: 6-12
Bruinessen, Martin van: Between Guerrilla War and Political Murder: The Workers' Party of Kurdistan, in: Middle East Report No. 153, 1988: 40-46
Brüne, Stefan: Perestroika in Addis Abeba? Ideologie, Entwicklungskrise und Staat in Äthiopien, in: Brüne, Stefan / Matthies, Volker: Krisenregion Horn von Afrika, Hamburg 1990: 69-89
Brüne, Stefan / Matthies, Volker: Krisenregion Horn von Afrika. Hamburg 1990
Brzoska, Michael.: Militarisierungsdynamik, militante Konflikte und

Entwicklunsdynamik. Arbeitspapiere der Forschungsstelle Kriege, Rüstung und Entwicklung, Nr.40. Universität Hamburg, Hamburg 1990:.

Brzoska, Michael.: Spread of Conventional Weapons Production Technology. in: Arbeitspapiere der Forschungsstelle Kriege, Rüstung und Entwicklung, Nr.76. Hamburg: Universität Hamburg, Hamburg 1994.

Bujtenhujs, Robert: Mau Mau. Twenty Years After, Moutou 1973

Bulloch, J.: The making of war. The Middle East from 1967 to 1973, London 1974

Bunge, Frederica M. (ed.): Cyprus. A Country Study, Washington D.C. 1980

Bunzel, J.: Israel/Palästina: Klasse, Nation und Befreiung im Nahostkonflikt, Hamburg 1980

Büren, Rainer: Ein palästinensischer Teilstaat?, Baden-Baden 1982

Büren, Rainer: Westliche Orient-Politik und das Krisen-Dreieck Libanon- Israel-Palästina, Ebenhausen 1983

Büren, Rainer: Befriedung im Vorderen Orient: Eine "Verlorene Kunst"?, in: Senghaas, Dieter (Hg.): Regionalkonflikte in der Dritten Welt - Autonomie und Fremdbestimmung, Baden-Baden 1989: 167- 180

Burger, Julian: Report from the Frontier: The State of the world's indiginous peoples, London 1987

Burgos, Elisabeth: Rigoberta Menchú. Leben in Guatemala, Bornheim-Merten 1984

Buro, Andreas: Der dreißigjährige Krieg in Vietnam. Ursachen und Perspektiven, in: Gantzel, Klaus Jürgen (Hg.): Krieg in der Dritten Welt, Baden-Baden 1988: 459-489

Burrowes, Robert D.: The Yemen Arab Republic. The Politics of Development, 1962 - 1986, Boulder (Col.) 1987

Burrowes, Robert D.: Oil Strike and Leadership. Struggle in South Yemen: 1986 and beyond, in: Middle East International, 1989: 437-454

Butterworth, Robert Lyle / Scranton, Margaret E.: Managing Interstate Conflict, 1945-1974: Data with Synopsis, Pittsburgh 1976

Cabarrús, P. Rafael Carlos: Génesis de una revolución. Análisis del surgimiento y desarrollo de la organización campesina en El Salvador, México D.F. 1983

Cable, Vincent: "The Football War" and the Central American Common Market, in: Gray, Richard B. (ed.): Latin America and the United States in the 1970`s, Itaca (Ill.) 1971: 194-208

Cabral, Amilcar: Die Theorie als Waffe, Wuppertal 1982

Cadoux, Charles: La république malgache, Paris 1969

Callaghy, T.M.: The State-Society Struggle: Zaire in Comperative Perspective, New York 1984

Camacho, Daniel / u.a.: El fracaso social de la integración centro-americana - capital, tecnología y empleo, EDUCA, Centroamérica 1979

Canfield, R.L. / Shahrani, M.N. (Hg.): Revolutions and Rebellions in Afghanistan. Anthropological Perspectives, Berkley 1984
Carey, John (ed.): The Dominican Republic Crisis 1965. Background Paper and Proceedings of the Ninth Hammarskjöld Forum, New York 1967
Carey, John (ed.) / Carey, (ed.): The Web of Modern Greek Politics, New York/ London 1968
Carías, Marco Virgilio / Slutzky, Daniel: La guerra inútil. Análisis socioeconónomico del conflicto entre Honduras y El Salvador, San José 1971
Carmoy, Guy de: Politiques Etrangères de la France 1944-1966, Paris 1967
Castella, Paul: Bamboo versus Guns. Laos's Hmong resistance movement, in: International Defense Review 5/1991: 397-398
CDHG (Commissión de derechos humanos en Guatemala): Guatemala, in: Terre des hommes (Hg.): Kinder und Krieg in Lateinamerika, Göttingen 1989: 94-159
CENIDH (Centro Nicaraguense de Derechos Humanos): Derechos Humanos en Nicaragua. Informe 1990, Managua 1990
CENIDH (Centro Nicaraguense de Derechos Humanos): Derechos Humanos en Nicaragua. Diciembre 1990 a Febrero 1991, Managua 1991
CEPAL (Comisión Económica para Américan Latina) / (CEPAL/FAO/OIT/ IICA/SIEGA/ OCT/OEA): Tenencia de la tierra y desarrollo rural en Centoamérica, EDUCA, Centroamérica, 1980
Chabraborti, Tridib: Vietnam-Kampuchea Confrontation. A Background Study, in: China Report, New Delhi 21/1985: 143-164
Chaffee, F. u. A.: Area-Handbook for Communist China, Washington 1967
Chamorro, Edgar: Packing the Contras: A Case of CIA Disinformation, New York 1987
Chanda, Nayan: Brother Enemy. The War After the War, San Diego/New York 1986
Chang-Rodriguez, Eugenio (ed.): The Lingering Crisis. A Case Study of the Dominican Republic, New York 1969
Chassin, L.M.: The Communist Conquest of China. A History of the Civil War 1945-49, Cambridge 1965
Chaudry, Mohammed Saeed: Der Kaschmirkonflikt. Seine Ursachen, sein Wesen sowie Rolle und Bemühen der Vereinten Nationen, (3 Bde.), München 1976
Chen, John Hsueeh-ming: Vietnam. A Comprehensive Bibliography, New York 1973
Chikh, Slimane: L'Algerie en armes ou les temps de certitudes, Paris 1981
Chin, Henk E. / Buddingh', Hans: Surinam. Politics, Economics and Society (Marxist Regimes Series), London/New York 1987.

Choisie, Jeanette: Wurzeln und Strukturen des Zypernkonfliktes 1878 - 1990, Diss., Hamburg 1991
Chomé, Jules: Moise Tschombé et l'escroquerie katangaise, Brüssel 1966
Chomsky, Noam: Im Krieg mit Asien, 2 Bde., Frankfurt a.M. 1972
Choudhury, Golam Wahed: Pakistan's relations with India, London 1968
Choudhury, Golam Wahed: The Last Days of United Pakistan, in: International Affairs, 49/1973: 229-239
Choudhury, D.P.: The North-East Frontier of India 1865 - 1914, Calcutta 1978
Church, R.J.H. / Synge, R. / Hodgkinson, E.: Burkina Faso (Länderbeitrag): 288-304, und Mali (Länderbeitrag): 673-686, in: Africa South of the Sahara, London 1988
CIDAMO (Centro de Información, Documentación y Análisis / sobre el Movimiento Obrero Latinoamericano): Carta informativa, México D.F. Mai/ 1980
Cilliers, J.K.: Counter-Insurgency in Rhodesia, London u.a. 1985
CLADDE-RIAL (Centro Latinoamericano de Defensa y Desarme): Programa de Estudios Conjuntos de Relaciones Internacionales de America Latina: La Situación Estratégica de América Latina. Estados unidos y la Seguridad Regional, Santiago, CESOC 1987
Clapham, Christopher: Transformation and Continuity in Revolutionary Ethiopia, Cambridge 1990
Clausen, Ursula: Westsahara: Mißlungene Dekolonisierung und der Freiheitskampf der Sahranis gegen Marokko, in: Hofmeier, Rolf/Matthies, Volker (Hg.): Vergessene Kriege in Afrika; Göttingen 1992: 347-365
Clausewitz, Carl von: Vom Kriege. Hinterlassenes Werk (1832/34), (Hg.) Marie von Clausewitz. Hier zitiert nach dem ungekürzten Text der Erstauflage von 1832-34, Frankfurt am Main / Berlin / Wien: Ullstein, 1980.
Coogan, Tim Pat: The IRA, Glasgow 1987
Cooke, M.: National Security, in: Bunge, Frederica M.: Philippines, A Country Study, Washington 1984: 229-289
Cossías, Tomás: La lucha contra el maquis en España, Madrid 1956
Country Report: Vietnam, Laos, Cambodia. London 3/1992
CRISP (Centre de Recherche et d'Information Socio-Politique): Congo 1966, Les Dossiers du C.R.I.S.P, Brüssel 1967
Crossette, Barbara: East Bloc Events Pierce Temeless World of Laos, in: International Herald Tribune, 25.11.1987
Dahm, Bernhard: Sukarnos Kampf um Indonesiens Unabhängigkeit, Hamburg 1966
Dahm, Bernhard: Indonesien - Geschichte eines Entwicklungslandes (1945-1971). Handbuch der Orientalistik, Bd. 1, Leiden 1978
Daniels, Anthony: Monrovia mon Amour, a Visit to Liberia, London 1992
Davenport, T.R.H.: South Africa, London 1977

Dawydow, Jurij P./ Trenin, Dmitrij W.: Ethnische Konflikte auf dem Gebiet der ehemalihen Sowjetunion, in: Europa-Archiv 7/1993: 179-192
Davidson, Basil: No Frist is Big Enough to Hide the Sky. The Liberation of Guinea-Bissau and Cape Verde, London 1981
Davies, Robert: South African Regional Policy Before and After Cuito Cuanavale, in: Moss, Glenn / Obery, Ingrid (Hg.): South African Review (Johannesburg) 5/1979
Davis, Shelton H.: State Violence and Agrarian Crisis in Guatemala. The Roots of the Indian Peasants Rebellion, in: Diskin, Martin (ed.): Trouble in our Backyard. Central America and the United States in the Eighties, New York 1983: 155-171
Dawisha, Adeed: The Motives of Syria's Involvement in Lebanon, in: Middle East International, Vol. 38, (Spring) 2/1984: 228-236
De Silva, K.M. / May, R.J. (ed.): Internationalization of Ethnic Conflict, London 1991
Decke, B.: A terra é nossa. Koloniale Gesellschaft und Befreiungsbewegung in Angola, Bonn 1981
Decker, Günter: Republik Maluku Selatan - Die Republik der Süd-Molukken, Forschungszentrum für Selbstbestimmungsrecht und Nationalitätenpolitik (Hg.), Göttingen 1957
Deeb, Marius K.: The Lebanese Civil War, New York 1980
Deeb, Marius K.: Lebanon: Prospects for National Reconciliation in the mid-1980s, in: Middle East International, Vol.38, (Spring) 2/1984: 267-283
Dent, M.J.: A Minority Party - The U.M.B.C., in: Mackintosh, J. (ed.): Nigerian Government and Politics, Evanston 1966: 461-507
Department of Peace and Conflict Research, Major Armed Conflicts, (Hg.:) Uppsala University, Uppsala 1988 ff.
Deschner, Günther: Saladins Söhne. Die Kurden - das betrogene Volk, München 1983
Desmond, Edward W.: War on High Ground, in: Time International, (July 17) 29/1989: 18-25
Deutsch, Karl W.: Politische Kybernetik. Modelle und Perspektiven. Freiburg i.Br. (1973), Rombach. (Übers. nach der erw. Neuausgabe von "The Nerves of Government," New York 1966.)
Deutsches Orientinstitut Hamburg (Hg.): Pakistan, in: Nahost Jahrbuch seit 1987
Diederichsen, Telse: Mocambique, in: Hofmeier, Rolf (Hg.): Afrika Jahrbuch 1989, Opladen 1990: 318-325
Dietrich, Wolfgang: Nicaragua. Heidelberger Dritte Welt Studien, Heidelberg 1988
Diner, D.: Keine Zukunft auf den Gräbern der Palästinenser, Hamburg 1982

Diner, Dan: Imperialismus, Universalismus, Hegemonie. Zum Verhältnis von Politik und Ökonomie in der Weltgesellschaft. in: Fetscher / Münkler 1985: 326-360.

Diner, D.: Strukturelemente des Palästinakonflikts, in: Gantzel, K.J. (Hg.): Krieg in der Dritten Welt, Baden-Baden 1988: 351-374

Dingemann, Rüdiger: Bewaffnete Konflikte seit 1945. Zwischenstaatliche Auseinandersetzungen, Befreiungskriege in der Dritten Welt, Bürgerkriege, Düsseldorf 1983

Dischler, Ludwig: Die Zypernfrage, Berlin/Frankfurt a.M. 1960

Dix, Robert: Political Opposition under the National Front, in: Berry / u.a. (Hg.): Politics of Compromise: Coalition Government in Colombia. New Brunswick (N.Y.) 1980: 131-179

Dixon, William J.: Democracy and the Peaceful Settlement of International Conflict. in: Wissenschaftszentrum Berlin für Sozialforschung (Hg.:) FIB Papers, Nr.P 1993/301, 1993

Dobson, Richard B.: Georgia and the Georgians, in: Katz, Zev (ed.): Handbook of Major Soviet Nationalities, New York/London 1975

Dooley, H.J.: The Suez Crisis 1956. A Case Study in Contemporary History, (2 Bde.), London 1979

Dorenwendt, Thomas: Mit Marx und Mohammed. Realsozialistische und traditionale Vergesellschaftung in Usbekistan. Universität Hamburg, Institut für Politische Wissenschaft, Diplomarbeit, Hamburg 1993

Doro, Marion E. / Legum, Colin: Namibia, in: Africa Contemporary Record 1987-1988, New York 1989: B649-B666

Draguhn, Werner: Indien, in: Draguhn, Werner / Hofmeier, Rolf / Schönborn, Matthias (Hg.): Politisches Lexikon Asien, Australien, Pazifik, München 1989: 93-112

Draguhn, Werner / Schier, Peter (Hg.): Indochina - Der permanente Konflikt? Hamburg 1987

Drame, Hassan: Interprétation d'une Crise Centre-Périphérie: Le Cas du Conflit Casamancais, Centre d'Étude d'Afrique Noire (Hg.), Bordeaux 1992

Draper, Theodore: The Dominican Revolt. A Case Study in American Policy, New York 1968

Draper, Theodore: The Dominican Intervention Reconsidered, in: Political Science Quarterly, New York 1971: 1-36

Drechsler, Horst: Südwestafrika unter deutscher Kolonialherrschaft, Berlin 1966

Droz, Bernard / Lever, Eric: Histoire de la guerre d'Algérie 1954-1962, Paris 1984

Du Bois, Victor D.: The struggle for stability in the Republic of Mali, in: West Africa Series, Vol. VII, No. 1: 1-10

DÜI (Deutsches Übersee-Institut, Übersee-Dokumentation) (Hg.):

Kurzbibliographie Sri Lanka: Der Konflikt zwischen Singhalesen und Tamilen, Hamburg 1987
DÜI (Deutsches Übersee-Institut, Übersee-Dokumentation) (Hg.): Kurzbibliographie Sri Lanka: Der Konflikt zwischen Singhalesen und Tamilen. Publikationen 1986-1989, Hamburg 1989
Duncanson, Dennis: Die strategischen Hintergründe des chinesischen Vietnam-Kriegs, in: Europa-Archiv 34/11/1979: 336-344
Dunkerley, James: Der lange Krieg. Diktatur & Revolution in El Salvador, Frankfurt 1986
Durán, Khalid.: Zur inneren Lage des Sudan, in: Vierteljahresberichte der Friedrich-Ebert-Stiftung 108/1987: 151-160
Durán, Khalid: Tragödie Tajikistan, in: Blätter für deutsche und internationale Politik 1/1994: 84-93
Eagleton, William: The Kurdish Republic of 1946, London 1963
Eckardtstein, Maria L. Silva von: Peru-Ecuador-Konflikt, in: Nohlen, Dieter / Boekh, Andreas (Hg.): Pipers Wörterbuch zur Politik, Bd.5 - Internationale Beziehungen, München 1984: 392-393
Ehrke, Michael: Die US-Politik in Zentralamerika: Schwierigkeiten einer Konfliktregelung, in: Hamann, Rudolf (Hg.): Die "Süddimension" des Ost-West-Konfliktes. Das Engagement der Supermächte in Krisen und Kriegen der Dritten Welt, Baden-Baden 1986
Eibl-Eibesfeldt, Irenäus: Krieg und Frieden aus der Sicht der Verhaltensforschung. München / Zürich 1975:.
Eikenberg, Kathrin: Ethnische Konflikte in der Dritten Welt. in Deutsches Übersee-Institut Hamburg (Hg.): Jahrbuch Dritte Welt 1987 - Daten, Übersichten, Analysen, München 1987: 69-83.
Eikenberg, Kathrin: Somalia (Länderartikel); Äthiopien (Länderartikel), in: Institut für Afrikakunde/Hofmeier, Rolf Afrika (Hg.): Afrika-Jahrbuch 1987, Hamburg 1988 (1988a)
Eikenberg, Kathrin: Sudan (Länderartikel), in: Institut für Afrika-Kunde (Hg.): Afrika-Jahrbuch 1987, Hamburg 1988 (1988b): 254-260
Eikenberg, Kathrin: Somalia (Länderartikel); Äthiopien (Länderartikel), in: Institut für Afrikakunde/Hofmeier, Rolf (Hg.): Afrika. Jahrbuch 1988, Hamburg 1989
Eikenberg, Kathrin: Somalia (Länderartikel); Äthiopien (Länderartikel), in: Afrika. Jahrbuch 1989, hrsg. vom Institut für Afrikakunde/Hofmeier, Rolf, Hamburg 1990 (1990a)
Eikenberg, Kathrin: Der Eritrea-Konflikt, in: Brüne, Stefan / Matthies, Volker (Hg.): Krisenregion Horn von Afrika, Hamburg 1990 (1990b)
Eikenberg, Kathrin: Innerstaatliche Kriege in Äthiopien, in: Siegelberg, Jens (Red.): Die Kriege 1985 bis 1990. Analysen ihrer Ursachen, Hamburg 1991

Eisenstadt, S.N.: Die Transformation der israelischen Gesellschaft, Frankfurt a.M. 1987

Elias, Norbert: Über den Prozeß der Zivilisation. Soziogenetische und psychogenetische Untersuchungen. Bd.1: Wandlungen des Verhaltens in den weltlichen Oberschichten des Abendlandes. Bd.2.: Wandlungen der Gesellschaft. Entwurf zu einer Theorie der Zivilisation. Frankfurt am Main: Suhrkamp (1., limitierte Aufl. Basel 1939; 2.Aufl. Bern 1969), 1980

Ellner, Steve: Venezuela's Movimiento al Socialismo, Durham/London 1988

Elsenhans, Hartmut: Frankreichs Algerienkrieg 1954-1962. Entkolonisierungsversuche einer kapitalistischen Metropole: Zum Zusammenbruch der Kolonialreiche, München 1974

Engels, Friedrich (1895). Einleitung zu Karl Marx: Die Klassenk„mpfe in Frankreich 1848 bis 1850, Ausgabe von 1895. Zitiert nach dem vollständigen Abdruck in Marx-Engels-Werke (MEW), Bd.7, Berlin/DDR: Dietz Verlag, 1960, 511-527.

Englebert, P.: La Révolution Burkinabé, Paris 1986

Entelis, John P.: Algeria: The Revolution Institutionalized. Profiles, Nations of the Contemporary Middle East, Boulder (Col.) 1986

Entessar, Nader: The Kurds in Post-Revolutionary Iran and Iraq, in: Third World Quaterly, 1984: 911-933

Entrup, Josef Lütke: Erfahrungen mit Mehrparteiensystemen. Uganda. KAS-Auslandsinformationen (Sep.) 1988

EP (Europa Publications) (ed.): South America, Central America and the Caribbean 1991, London 1991

EPRDF (Ethiopian Peoples Revolutionary Democratic Front): Documents from the First National Congress of the Ethiopian Peoples Revolutionary Democratic Front, o. O. 1991

Ermacora, Felix: Namibia, in: Bayerische Landeszentrale für politische Bildungsarbeit (Hg.), München 1981: 136-142

ERP (Ejército Revolucionario del Pueblo): Revolutionäre Volkskräfte in Argentinien, Dokumente der ERP, Hamburg 1977

Esche, Matthias: Die KP Griechenlands von 1941 - 1949, Hamburg 1982

Evans, Michael: Fighting against Chimurenga. An Analysis of Counter-Insurgency in Rhodesia, 1972-1979, Harare 1981

Ewell, Judith: Venezuela. A Century of Change, London 1984

Eyongetah, T. / Brain, R.: A History of the Cameroon, London 1974

Ezell, Edward C.: Small Arms Today. Harrisburg 1988

Farouk-Sluglett, Marion / Sluglett, Peter: Iraq Since 1958. From Revolution to Dictatorship, London/New York 1987

Favre, Henri: Pérou: Sentier Lumineux et Horizons Obscurs, in: Problèmes d'Amérique Latine 72/1984: 3-27

FEER: Asia Yearbook, Hongkong 1991

Fehrenbach, T.R.: This Kind of War. A Study in Unpreparedness, New York 1963
Feinberg, Richard E. (ed.): Central America: International Dimensions of the Crisis, New York 1982
Felber, R. / Weidemann D. (Hg.): Asien - Kleines Nachschlagewerk, Berlin (Ost) 1987
Felder, Tim: Jugoslawien - Das gescheiterte Experiment - Skizze einer Krise, Kommentierte, aktualisierte Fassung, in: Forschungsstelle Kriege, Rüstung und Entwicklung an der Universität Hamburg (Hg.): Arbeitspapiere, Nr. 53, Hamburg 1992
Ferdowsi, Mir A.: Zum Zusammenhang zwischen Aufrüstung und Krieg in der Dritten Welt - Versuch der Zerstörung einer Legende. in: Österreichische Zeitschrift für Politikwissenschaft, Heft 1/1990, 77-97.
Ferguson, Yale H.: The Dominican Intervention of 1965. Recent Interpretations, in: International Organization (Vol. XXVII), Wisconsin 1973: 517-548
Fetscher, Iring / Münkler, Herfried, (Hg.): Politikwissenschaft. Begriffe, Analysen, Theorien. Ein Grundkurs. Reinbek 1985
Fisher, Margaret W. / Rose, Leo E. / Huttenback, Robert A.: Himalayan Battleground - Sino-Indian Rivalry in Ladakh, London 1963
Fleischmann, Klaus: Arakan: Konfliktregion zwischen Birma und Bangladesh, Hamburg 1981
Flores, Alexander: Die Entwicklung der palästinensischen Nationalbewegung bis 1939, in: Mejcher, Helmut / Schölch, Alexander: Die Palästinafrage 1917-1948, Paderborn 1981: 89-118
Flores, Alexander: Palästinakonflikt, in: Pipers Wörterbuch zur Politik, Bd.5 - Internationale Beziehungen, München 1984: 381-387
Flores, Alexander: Intifada: Aufstand der Palästinenser, Berlin 1989
Frangi, Abdallah: PLO und Palästina. Vergangenheit und Gegenwart, Frankfurt a.M. 1982
Frankel, Philip: State, Resistance and Change in South Africa, Croom Helm 1988
Franz, Erhard: Material zum Kurdenproblem, Hamburg 1977
Franz, Erhard: Minderheiten in Iran. Dokumentation zur Ethnographie und Politik, Hamburg 1981
Franz, Erhard: Kurden und Kurdentum. Zeitgeschichte eines Volkes und seiner Nationalbewegungen, Hamburg 1986
Franz, Erhard: Jelal Talebani, in: Orient, 28. Jg., 1/1987: 5-8
Franz, H.W.: Klassenkämpfe in Spanien heute, Frankfurt a.M. 1975
Freimark, P.: Zum Selbstverständnis jüdischer Nationalität und Staatlichkeit in Palästina, in: Mejcher, Helmut / Schölch, Alexander: Die Palästinafrage 1917-1948, Paderborn 1981: 47-72

Frenz, Helmut / Greinacher, Norbert / Junk, Ulla / Päschke, Bernd: El Salvador. Massaker im Namen der Freiheit, Hamburg 1982
Frey, Hans: Der indisch-pakistanische Konflikt und seine wirtschaftlichen und sozialen Kosten für Pakistan in den Jahren 1958-68, Wiesbaden 1978
Friedrich Ebert Stiftung (Hg.): Horn of Africa. From "Scramble for Africa" to East-West Conflict (Analysen aus der Abteilung Entwicklungsländerforschung 106/107/1983), Bonn 1982
Fronius, Sigrid: Nicht besiegt und noch nicht Sieger. Argentinien und die Entwicklung der peronistischen Arbeiterbewegung, Berlin 1977
Fulbright, J. William: Die Arroganz der Macht, Reinbek 1982
Fuller, Elizabeth: Konflikte im Transkaukasus: Wer könnte vermitteln? in: Europa-Archiv 7/1993: 193-201
Fullick, R. / Powell, G.: Suez: The Double War, London 1979
Furkes, Josip/Schlarp, Karl-Heinz: Jugoslawien. Ein Staat zerfällt, Reinbek 1991
Furley, Oliver: Uganda's Retreat from Turmoil?, in: Conflict Studies 196/1987
Füsser, Wilhelm-Karl: Rebellion in Buganda. Eine Staatskrise im vorkolonialen Afrika, Hamburg 1989
Gabanyi, Anneli Ute: Rumäniens unvollendete Revolution, in: Europa-Archiv 12/1990: 371-380 (1990a)
Gabanyi, Anneli Ute: Die unvollendete Revolution. Rumänien zwischen Diktatur und Demokratie, München 1990 (1990b)
Gabriel, Leo: Aufstand der Kulturen. Konflikt-Region Zentralamerika. Guatemala, El Salvador, Nicaragua, Hamburg 1987
Gallo, M.: Histoire le l'Espagne franquiste, Paris 1969
Gallón Giraldo, Gustavo: Quince Años de Estado de Sitio en Colombia 1958-78, Bogotá 1979
Gallón Giraldo, Gustavo (Hg.): Entre Movimientos y Caudillos, Bogotá 1989
Gambia, V.: The Falklands / Malvinas War. A Model for North-South Crisis Prevention, Boston 1987
Ganguly, Sumit: The Origins of War in South East Asia. Indo-Pakistani Conflicts since 1947, Lahore 1988
Ganguly, Sumit: Avoiding War in Kashmir, in: Foreign Affairs, Winter 1990/91: 57-73
Gantzel, Klaus Jürgen: System und Akteur. Beiträge zur vergleichenden Kriegsursachenforschung. Düsseldorf 1972
Gantzel, Klaus Jürgen: "Gleichgewicht" und "Rüstungswettlauf" - zwei fragwürdige Begriffe in der sicherheitspolitischen Diskussion. in: Peter Lock (Hg.): Frieden als Gegenstand von Wissenschaft - Hamburger Vorträge zur Friedensforschung, Frankfurt a.M 1982: 113-123.
Gantzel, Klaus Jürgen:. Krieg. Pipers Wörterbuch zur Politik, Bd.1, hrsg.v.

Dieter Nohlen und Rainer-Olaf Schultze. München / Zürich 1985: 479-482.

Gantzel, Klaus Jürgen: Tolstoi statt Clausewitz!? Überlegungen zum Verhältnis von Staat und Krieg seit 1816 mittels statistischer Beobachtungen. Kriegsursachen, in: Reiner Steinweg (Red.): Friedensanalysen Bd.21, Frankfurt am Main 1987, 25-97.

Gantzel, Klaus Jürgen, (Hg.): Krieg in der Dritten Welt. Theoretische und methodische Probleme der Kriegsursachenforschung - Fallstudien. in: Schriftenreihe der Arbeitsgemeinschaft für Friedens- und Konfliktforschung (AFK), Bd.XII. Baden-Baden 1988

Gantzel, Klaus Jürgen: Krieg in der Dritten Welt als Forschungsgegenstand. Eine Einführung zum Thema. In ders. 1988: 25-82. (1988a)

Gantzel, Klaus Jürgen: Krieg. in: Dieter Nohlen (Hg.): Wörterbuch Staat und Politik, München / Zürich 1991: 327-330.

Gantzel, Klaus Jürgen: Kriegsursachen: Theoretische Konzeption und Forschungsfragen. in: Krell / Müller 1994:133-156.

Gantzel, Klaus Jürgen / Meyer-Stamer, Jörg (Hg.): Die Kriege nach dem Zweiten Weltkrieg bis 1984 - Daten und erste Analysen. Weltwirtschaft und internationale Beziehungen in: Deutsches Übersee-Institut - Diskussionsbeiträge, Bd.34, München / Köln / London 1986

Gantzel, Klaus Jürgen / Siegelberg, Jens: Krieg und Entwicklung. Überlegungen zur Theoretisierung von Kriegsursachen unter besonderer Berücksichtigung der Zeit seit 1945. Theorien der Internationalen Beziehungen - Bestandsaufnahme und Forschungsperspektiven, in: Volker Rittberger (Hg.): Sonderheft 21 der Politischen Vierteljahresschrift. Opladen 1990: 219-239, (1990a).

Gantzel, Klaus Jürgen / Siegelberg, Jens: Kriege der Welt. Ein systematisches Register der kriegerischen Konflikte 1985 bis 1990. in: Stiftung Entwicklung und Frieden (SEF) (Hg.): Interdependenz - Materialien der Stiftung Entwicklung und Frieden, Nr.1/1990. Bonn 1990, (1990b).

Gantzel, Klaus Jürgen / Schwinghammer, Torsten / Siegelberg, Jens: Kriege der Welt. Ein systematisches Register der kriegerischen Konflikte 1985 bis 1992. in: Stiftung Entwicklung und Frieden (Hg.): Interdependenz - Materialien und Studien der Stiftung Entwicklung und Frieden und des Instituts für Entwicklung und Frieden, Nr.13. Bonn 1992

Gantzel, Klaus Jürgen / Schlichte, Klaus: Das Kriegsgeschehen 1993. Daten und Tendenzen der Kriege und bewaffneten Konflikte im Jahr 1993. in: Stiftung Entwicklung und Frieden (Hg.): Interdependenz - Materialien und Studien der Stiftung Entwicklung und Frieden und des Instituts für Entwicklung und Frieden, Bonn Nr.16/1994

Garang, C.: Südafrikanische Verhältnisse breiten sich aus, in: Blätter des iz3w 146/1987: 8-12

Gelaschwili, Naira: Georgien. Ein Paradies in Trümmern, Berlin 1993
Gendebien, Paul Henri: L'intervention des Nations Unies au Congo 1960 - 64, Paris 1967
Georgel, J.: Il franchismo. Storia e Bilancio, Torino 1965
Gérard-Libois, J.: Secession au Katanga, Brüssel 1963
Gersony, Robert: Summery of Mozambican Refugee Account of Principally Conflict Related Experience in Mozambique, Washington 1988
Gertzel, Cherry: Party and Locality in Northern Uganda: 1945 - 1962, London 1974
Gerwerk, Uta: Guinea-Bissau. Nationaler Befreiungskampf und kollektiver Fortschritt, Stein b. Nürnberg 1974
Ghassemlou, Abdel Rahman: Bericht über Iranisch-Kurdistan, in: Chaliand, Gérard (Hg.): Kurdistan und die Kurden, Göttingen 1984: 183-228
Gilhodes, Pierre: El Ejército Colombiano analiza la Violencia, in: Sánchez, Gonzalo / Peñaranda, Ricardo (Hg.): Pasado y Presente de la Violencia en Colombia, Bogotá 1986: 305-332
Gill, Sucha Singh / Singhal, K.C.: The Punjab Problem: Its Historical Roots, in: Economic and Political Weekly, Vol.19, (April 7) 14/1984: 603-608
Gilmour, David: The Transformation of Spain. From Franco to the Constitutional Monarchy, London 1985
Ginsburgs, G. / Mathos, M.: Communist China and Tibet, The Hague 1964.
Girrbach, Bernd / Nohlen, Dieter: Madagaskar, in: Handbuch der Dritten Welt (Bd.5), Hamburg 1982
Giussani, Pablo: Montoneros. La soberbia armada, Buenos Aires 1984
Glasneck, Johannes / Timm, Angelika: Israel. Die Geschichte des Staates seit seiner Gründung, Bonn/Berlin 1992
Globale Trends 1991, Daten zur Weltentwicklung. (Hg.): Stiftung Entwicklung und Frieden. Düsseldorf 1991
Globale Trends 93/94, Daten zur Weltentwicklung. (Hg.): Ingomar Hauchler / Stiftung Entwicklung und Frieden, Frankfurt am Main 1993.
Goff, Fred / Locker, Michael: The Violence of Domination. US Power and the Dominican Republic, in: Horowitz, Irving Louis / Castro, José de / Gerassi, John (ed.): Latin American Radicalism. A Documentary Report on Left and Nationalist Movements, New York 1969: 249-291
Goldblat, J. / Millán, V.: The Falklands / Malvinas Conflict - A Spur to Arms, Stockholm 1983
Goonetileke, H.A.J.: The April 1971 Insurrection in Ceylon. A Bibliographical Commentary, Leuven 1975
Gosztony, Peter: Der Volksaufstand in Ungarn 1956, in: Militärgeschichtliches Beiheft zur Europäischen Wehrkunde (MGB), hg. vom Militärgeschichtlichen Forschungsamt, Boppard am Rhein 2/1988
Gott, Richard: Guerrilla Movements in Latin America, London 1970
Gott, Richard: Rural Guerrilla in Latin America, Harmondsworth 1973

Government of India (Ministry of External Affairs): India 1986: A Reference Annual, New Delhi (August) 1986

Gödeke, Peter / Stuckmann, Elmar / Vogt, Martin: Kriege im Frieden, Braunschweig 1983

Grabendorf, Wolf: Kubas Politik in Afrika, in: Europa-Archiv 13/1979: 417 ff.

Gran, G. (ed.): Zaire: The Political Economy of Underdevelopment, New York 1979

Greenfield, Richard: Somalia: Barres unholy alliances, in: Africa Report, März-April/1989: 65-68

Gresh, Alain: Ambitions irakiennes, in: Le Monde Diplomatique, Mai 1990

Grest, Jeremy: Mozambique after Machel, in: Moss, Glenn / Obery, Ingrid: South African Review (Johannesburg) 5/1989

Grimal, Henri: La Décolonisation 1919-1963, Paris 1965

Grohs, G.: Rwanda und Burundi, in: Opitz, P.J. (Hg.): Das Weltflüchtlingsproblem, München 1988: 80-83.

Gros, José: Relatos de un guerrillero communista, Paris 1970

Gunson, Phil / Thompson / Chamberlain: The Dictionary of Contemporary Politics of South America, London 1989

Gunter, Michael M.: The Kurdish Problem in Turkey, in: Middle East International, 1988: 389-406

Gupta, B.S.: Afghanistan. Economics and Society. Revolution, Resistance, Intervention, London 1986

Gupta, Dipankar: The Communalising of Punjab, 1980-1985, in: Economic and Political Weekly, Vol.20, (July 13) 28/1985: 1185-1190

Gupta, H.R.: India-Pakistan War 1965, Vol. II, Delhi 1968

Gupta, Sisir: Kashmir. A Study in Indo-Pakistan Relations, Bombay 1967

Gurdon, C.: Sudan at the Crossroads, London 1984

Guzman, German / Fals Borda, Orlando / Umaña Luna, Eduardo: La Violencia en Colombia, Bogotá 1962

Haddad, W.: Lebanon. The Politics of Revolving Doors, Washington D.C. 1985

Haffa, A.I.: Beagle-Konflikt und Falkland-(Malwinen-) Krieg. Zur Außenpolitik der argentinischen Militärregierung 1976 bis 1983, München 1987

Hahn, Barbara: Die Insel Zypern: Die wirtschafts- und politisch-geographischen Entwicklungsprozesse eines geteilten Kleinstaates, Hannover 1982

Hajjaj, Aref: Der Libanesische Bürgerkrieg, seine palästinensischen und innerarabischen Faktoren sowie seine Auswirkungen auf den arabisch-israelischen Konflikt, in: Orient 1/1977

Halbach, A.J.: Die Ausweisung der Asiaten aus Uganda. 7 Monate Aminscher Politik in Uganda, München 1973

Halbach, Uwe: Ethno-territoriale Konflikte in der GUS, in: Bundesinstitut für

Ostwissenschaftliche und Internationale Studien, Köln (Hg.): Bericht 31/1992
Hall, D.G.E.: A History of South-East Asia, London 1964: 804-810
Halliday, Fred: Revolution in Afghanistan, in: New Left Review 112/1980: 3-44 (1980a)
Halliday, Fred: War and Revolution in Afghanistan, in: New Left Review 119/1980: 20-41 (1980b)
Halliday, Fred: Die Kriege Afghanistans, in: Gantzel (Hg.) 1988: 433-457
Halliday, Fred: Revolution and Foreign Policy. The Case of South Yemen 1967 - 1987, Cambridge 1990
Hamilton, Nora / Frieden, Jeffrey A. / Fuller, Linda / Pastor, Manuel Jr. (ed.): Crisis in Central America. Regional Dynamics and U.S. Policy in the 1980's, Boulder (Col.) 1988
Handke, Werner: Regionalismus und Zentralismus in Indien. Wirtschaft und Politik in der indischen Ostregion, Hamburg 1982
Handy, Jim: Gift of the Devil. A History of Guatemala, Toronto 1984
Handy, Jim: Resurgent Democracy and the Guatemalan Military, in: Journal of Latin American Studies 18/1986: 383-408
Hanf, Theodor: Der Krieg in Libanon und die Entstehung der Zweiten Libanesischen Republik, in: Betz, Joachim / Matthies, Volker: Jahrbuch Dritte Welt, Bd.1, München 1983: 101-118
Hanf, Theodor: Koexistenz im Krieg: Staatszerfall und Entstehen einer Nation im Libanon, Baden-Baden 1990
Hanf, Theodor / Weiland, Heribert / Vierdag, Gerda: Südafrika: Friedlicher Wandel?, München 1978
Hanlon, Joseph: Mozambique. The Revolution under Fire, London 1984
Hansen, Holger Bernt / et. al. (Hg.): Uganda now. Between Decay and Development, London 1988
Hansen, Sven: Philippinen. Guerrilla und Revolution. Ursprünge, Entwicklungen und Krise der NDF, Münster 1991
Haraldsson, Erlendur: Land im Aufstand ... Kurdistan, Hamburg 1966
Harding, Colin: The Rise of Sendero Luminoso, in: Miller, Rory (Hg.): Region and Class in Modern Peruvian History, Liverpool 1987: 179-207
Harris, Richard / Vilas, Carlos M. (Hg.): La revolución en Nicaragua. Liberación nacional, democracia popular y transformación económica, Mexico D.F. 1985
Harrison, Selig S.: In Afghanistan's Shadow: Baluch Nationalism and Soviet Temptations, New York 1981
Harrison, Selig S.: Ethnicity and the Political Stalemate in Pakistan, in: Banuazizi, Al / Weimer, Myron (ed.): The State, Religion and Ethnic Politics: Pakistan, Iran and Afghanistan, Lahore 1987: 267-298

Harvey, Barbara S.: Permesta: Half a Rebellion. Monograph Series No. 57, Cornell Modern Indonesia Project, Ithaca (N.Y.) 1977
Hassan II: La Mémoire d'un Roi. Entretiens avec Laurent E., Plon 1993
Heckhorn, Manfred: Die Enkel des Jaguar, Berlin 1983
Heimer, F.W.: Der Entkolonialisierungskonflikt in Angola, München/London 1979
Heinrich, Lothar A.: Die kurdische Nationalbewegung in der Türkei, Hamburg 1989
Heinz, W.S.: Menschenrechte und Minderheiten in Bangladesh, in: Internationales Asienforum, Köln 1984
Heinzelmeir, Helmut: Indonesiens Außenpolitik nach Sukarno 1965 - 1970. Mitteilungen des Instituts für Asienkunde Hamburg 78/1976
Heinzig, Dieter: Sowjetische Asien- und Pazifikpolitik unter Gorbatschow, in: Berichte des Bundesinstituts für Ostwissenschaftliche und Internationale Studien, Köln 1987
Helbig, D.: De Rwanda entre guerre civile et réformes politiques, in: Le Monde Diplomatique, Paris 11/1990: 20
Hellmann-Rajanayagam, Dagmar: The Tamil Tigers in Northern Sri Lanka, Origins, Factions, Programs, in: Internationales Asienforum, 1-2/1986: 63-66
Hellmann-Rajanayagam, Dagmar: Sri Lanka - A Current Perspective, in: Asian Affairs (London) (Vol. 22, Part 3), Oct. 1991: 314-329
Henriksen, Thomas H.: Revolution and Counterrevolution in Mozambique, London 1983
Hermann, Günther: Der Südsudankonflikt. Eine historische Rückschau, in: Pogrom 6/37 (Report Nr.21) Hamburg 1976
Hetmanek, Allen: Islamic Revolution and Jihad come to the former Soviet Central Asia: the case of Tajikistan, in: Central Asian Survey, vol. 12, 3/ 1993: 365-378
Hertoghe, Alain / Labrousse, Alain: Die Koksguerrilla. Der Leuchtende Pfad in Peru, Berlin 1990
Herzog, Werner (Hg.): Terror im Baskenland. Gefahr für Spaniens Demokratie?, Reinbek 1979
Herzog, C.: Kriege um Israel 1948-1984, Frankfurt a.M. 1984
Herzog, Werner (Hg.): Spanien, München 1987
Hess, Peter: Bangla Desh - Tragödie einer Staatsgründung, Frauenfeld/Stuttgart 1972
Hildenbrand, A.: Rwanda. In: Nohlen, Dieter / Nuscheler, Franz (Hg.): Handbuch der Dritten Welt, Bd.5: Ostafrika und Südafrika. Hamburg 1982: 153-161
Hilgers, Andrea: Der Krieg in Uganda, in: Siegelberg, Jens (Red.): Die Kriege 1985 bis 1990 - Analyse ihrer Ursachen, Münster 1991: 156-165

Hippler, Jochen: Menschenrechte und "Politik der Stärke". USA und Lateinamerika, Duisburg 1984
Hirson, Baruch: Year of Fire, Year of Ash, London 1979
Hiskett, M.: The Maitabine Riots in Kano 1980: Assessment, in: Journal of Religion in Africa, 3/1987: 209-223
Hodder-Williams, Richard: Conflict in Zimbabwe: The Matabeleland-Problem, in: Conflict Studies, London 151/1983
Hodges, Tony: Western Sahara. Roots of a Desert War, Westport (Conn.) 1983
Hodges, Tony: At Odds with Self-Determinism. The United States and Western Sahara, in: Bender, Gerald et al. (ed.): African Crisis Areas and U.S. Foreign Policy, Berkley u.a. 1985
Hofmeier, Rolf (Hg.) / Institut für Afrika-Kunde: Afrika-Jahrbuch, Opladen 1988ff
Hofmeier, Rolf / Schönborn, Matthias (Hg.): Politisches Lexikon Afrika, München 1978
Hofmeier, Rolf / Schönborn, Matthias (Hg.): Politisches Lexikon Afrika, 2. Aufl., München 1984
Hoik, David: Mozambique: A Nation in Crisis, London 1989
Hoivik, Tord / Aas, Solveig: Demilitarization in Costa Rica: A Farewell to Arms? in: Journal of Peace Research, Vol.17, 4/1981: 333-351
Holsti, Kalevi J.: Peace and war: armed conflicts and international order 1648-1989. Cambridge / New York u.a. 1991
Hoskyns, Catherine: The Congo since Independence. January 1960 - December 1961, London 1965
Hottinger, Arnold: Zu'ama and Parties in the Lebanese Crisis of 1958, in: Middle East International, Vol.15, (Spring) 2/1961: 127-140
Hottinger, Arnold: Der Machtzuwachs der arabischen Freischärler und ihr Zusammenstoß mit der jordanischen Regierung, in: Europa-Archiv 25, 20/1970: 737-746
Hottinger, Arnold: Der Bürgerkrieg in Südjemen. Machtkämpfe in einer Stammesgesellschaft, in: Europa-Archiv 1986: 311-318
Hottinger, Arnold: Die Kurden im inneren und äußeren Kampf. Revolutionäre und Nationalisten als Minderheiten in vier Staaten, in: Europa-Archiv 6/1987: 161-170
Hottinger, Arnold: Der "Aufstand" der Araber in den besetzten Gebieten, in: Europa-Archiv 11/1988: 299-310
Hottinger, Arnold: Arafats Friedensangebot an Israel, in: Europa-Archiv 2/1989: 39-46
Hottinger, Arnold: Der Sieg der Apparatschiks, in: Europa-Archiv 11/1993: 317-326
Hourani, Albert: The Emergence of the Modern Middle East, Oxford 1981
Huber, Maria: Kein Friede in der Stadt der Winde, in: Die Zeit, Nr.16/1991

Hubrich, H.G. / Melber, Henning: Namibia - Geschichte und Gegenwart, Bonn 1977
Hudson, Michael C.: The Precarious Republic. Political Modernization in Lebanon, Boulder/London 1985
Hunck, J.M.: Der Krieg und die indische Wirtschaft, in: Indo-Asia 1/1966: 43-50
Hussein, Adam M.: Somalia: Militarism, Warlordism or Democracy?, in: Review of African Political Economy, No. 54, 1992: 11-26
Hutcheson, Mac Gregor / Moorsom, Richard: Namibia, in: Strategy Network Int. Ltd., Africa South of the Sahara 1990, London 1989: 729-746
Hünseler, Peter: Der Irak und sein Konflikt mit Iran (Arbeitspapiere zur Internationalen Politik), Bonn 22/1982
Iberi, P.: Moscow, Tiblisi and the Abkhazian Card, in: European Journal of International Affairs 4/1989: 134-143
Ibrahim, Ferhard: Die kurdische Nationalbewegung im Irak. Eine Fallstudie zur Problematik ethnischer Konflikte in der Dritten Welt, Berlin 1983
IDAF (International Defence and Aid Fund) (ed.): Namibia, London 1980
IDAF (International Defence and Aid Fund) (ed.): Apartheid's Army in Namibia. Fact Paper on Southern Africa, London 10/1982
IES (Instituto de Estudio del Sandinismo): ¡ Y se armó la runga! - Testimonios de la Insurrección Popular an Masaya, Managua 1982
Illy, Hans F.: Politik und Wirtschaft in Kamerun. Bedingungen, Ziele und Strategien der staatlichen Entwicklungspolitik, München 1976
Informationsstelle Guatemala (Hg.): Guatemala. Der lange Weg zur Freiheit, Wuppertal 1982
Instituto de Defensa Legal: Peru 1989. En la espiral de Violencia. o.O.
International Alert: Report on Checenia, London 1992
Invernizzi, Gabriele / Pisani, Francis / Ceberio, Jesús: Sandinistas. Entrevistas a Humberto Ortega Saavedra, Jaime Wheelock Róman, Bayardo Arce Castano, Managua 1986
Isaacman, A. / Isaacman, B.: Mozambique. From Colonialism to Revolution, Boulder 1983
Isichei, E.: The Maitabine Risings in Nigeria 1980 - 85: A Revolt of the Disinherited, in: Journal of Religion in Africa 3/1987: 194-208
Jaber, Nadim: Kurdistan: Saddam's final solution?, in: Middle East International, (9.9.88) 333/1988: 3f.
Jackson, Robert: South Asian Crisis. India - Pakistan - Bangla Desh, London 1975
Jacobs, Julian (et al.): The Nagas. Society, Culture and the Colonial Encounter, London 1990
Jada, M. L.: The Four Enemies, in: PANOS, War Wounds, 1988: 125-138
Jeffrey, Robin: What's happening to India?, Basingstoke 1986

Jolliffe, Jill: East-Timor. Nationalism and Colonialism, St. Lucia (Queensland) 1978
Jones, F.C.: The Korean War, in: Survey of International Affairs, London 1949/ 50: 466-516
Joseph, Richard: Radical Nationalism in Cameroon. Social Origins of the UPC-Rebellion, Oxford 1977
Joseph, Richard: Le mouvement nationaliste au Cameroun. Les origines sociales de l'UPC, Paris 1986
Jung, Dietrich: Der Krieg im Libanon. Exemplarischer Versuch einer gesellschaftstheoretisch fundierten Kriegsursachenanalyse. in: Arbeitspapiere der Forschungsstelle Kriege, Rüstung und Entwicklung, Nr.61., Hamburg 1992
Jung, Harald: Soziale Kämpfe und Klassenstruktur in El Salvador, Bonn 1980 (1980a)
Jung, Harald: Nicaragua. Bereicherungsdiktatur und Volksaufstand, Frankfurt a.M. 1980 (1980b)
Kadritzke, Niels / Wagner, Wolf: Im Fadenkreuz der NATO, Ermittlungen am Beispiel Zypern, Berlin 1976
Kahin, G.M.T.: Nationalism and Revolution in Indonesia, Ithaca (N.Y.) 1959
Kalb, Marvin / Kalb, Bernhard: Kissinger, Frankfurt a.M. u.a. 1974
Kalmanovitz, Salomon: Economia y Nación. Una breve historia de Colombia, Bogotá 1988
Kaminski, G.: China-Taiwan, Frankfurt a.M. 1971
Kanza, Thomas: Conflict in the Congo, Harmondsworth 1972
Kaplan, Robert D.: Surrender or Starve. The Wars Behind the Famine, Boulder/ London 1988
Kappel, Robert / Korte, Werner: Human Rights Violations in Liberia 1980-1990. Dokumentation, Informationszentrum Afrika, Liberia Working Group, 1990
Kappel, Robert / Korte, Werner: Bürgerkrieg in Liberia, in: Jahrbuch Dritte Welt 1991, München 1991: 234-255
Kapur, Rajiv A.: Sikh Separatism: The Politics of Faith, London 1986
Kapur, Rajiv A.: Khalistan: India's Punjab Problem, in: Third World Quarterly, Vol.9, (Oct.) 4/1987: 1206-1224
Karibik-Informationszentrum: Grenada und die Probleme der "Wiederherstellung der Demokratie," Hamburg 1984
Karniol, Robert: Laotion Resistance emerges from Mist, in: International Defense Review, 3/1990: 269-272.
Kartveli,: Georgia's Narrow Path, in: European Journal of International Affairs 4/1989: 123-134
Karugire, Samwiri Rubazara: A Political History of Uganda, Nairobi/London 1980

Katjavivi, Peter: A History of Resistance in Namibia, London 1988
Kay, R,: Burundi since the genocide, London 1987
Kaye, G.D. / Grant, D.A. / Emond, E.J.: Major Armed Conflict: A Compendium of Interstate and Intrastate Conflict, 1720 to 1985.:(Hg.): Department of National Defence, Operational Research and Analysis Establishment (ORAE). Ottawa 1985
Kegel, Hannah: Die politischen Entwicklungen in Georgien, in: Osteuropa 7/ 1991
Kegel, Hannah: Georgien seit Herbst 1991, in: Osteuropa 3/1992 : 232-241
Keller, Walter (Hg.): Weg in den Ruin? Sri Lankas Minderheitenproblematik. Eine Dokumentation des Südasienbüros, Wuppertal 1986
Kelly, J. B.: Hadramaut, Oman, Dhufar: The Experience of Revolution, in: Middle Eastern Studies 1976: 213-230
Kempe, Frederick: Divorcing the Dictator, New York 1990
Kende, István: Twenty-five Years of Local Wars. in: Journal of Peace Research, Oslo, vol.8, no.1, 1971: 5-22.
Kende, István: Wars of Ten Years, 1967-76. in: Journal of Peace Research, Oslo, vol.15, no.3, 1978: 227-241.
Kende, István: Kriege nach 1945. Eine empirische Untersuchung. in: Militärpolitik Dokumentation, Heft 27. Frankfurt am Main 1982 (Habilitationsschrift, einzige aus dem Ungarischen übertragene Fassung.) (1982a).
Kende, István: Über die Kriege seit 1945. in: Deutsche Gesellschaft für Friedens- und Konfliktforschung (Hg.): DGFK-Hefte, Nr.16. Bonn 1982 (1982b)
Kende, István / Gantzel, Klaus Jürgen / Fabig, Kai : Die Kriege seit dem Zweiten Weltkrieg. Weltpolitik - Jahrbuch für internationale Beziehungen, Bd.2. Frankfurt am Main / New York 1982: 106-118.
Kennedy, Charles H.: The politics of ethnicity in Sindh, in: Asian Survey XXXI, 10, Oct. 1991: 938-955
Kent, Raymond K.: From Madagascar to the Malagasy Republic, Westport (Conn.) 1962
Kerr, Malcolm H.: Lebanese Views on the 1958 Crisis, in: Middle East International, Vol. 15, (Spring) 2/1961: 211-217
Kessler, Richard J.: Rebellion and Repression in the Philippines, New Haven/ London 1989
Khafaji, Isam al: Iraq's Seventh Year. Saddam's Quart d'Heure?, in: Middle East Report, 151/1988: 35-59
Khalid, M. (ed.): John Garang Speaks, London 1987
Khalidi, R.I.: The Uprising and the Palestine Question, in: World Policy Journal, Vol.5 (Summer) 3/1988: 497-518
Khalil, Fadel: Kurden heute, Wien/München/Zürich 1985

Khan, Kushi M.: Zur politischen Lage in Pakistan nach Zia, in: Aus Politik und Zeitgeschichte (24.Feb.) 9/1989: 15-24
Ki-Zerbo, Joseph: Die Geschichte Schwarz-Afrikas. Frankfurt a.M. 1981
Kiernan, Ben: Conflict in the Kampuchean Communist Movement, in: Journal of Contemporary Asia 1/2/1980
Kiernan, Ben: How Pol Pot Came to Power, London 1986
Kiernan, Ben / Boua, Chantou (Hg.): Peasants and Politics in Kampuchea 1942-81, London 1982
Kindermann, Gottfried-Karl: Pekings chinesische Gegenspieler, Düsseldorf 1977
Kindermann, Gottfried-Karl (Hg.): Chinas unbeendeter Bürgerkrieg. Wien/ München 1980
Kinnane, Derk: The Kurds and Kurdistan, London 1964
Kircheisen, I.: Der Einfluss des Kemalismus auf die Politik Afghanistans im ersten Jahrzehnt nach der Erringung der Unabhängigkeit, in: Barthel, Günther (Hg.): Der Kemalismus und die moderne Türkei, Berlin (Ost) 1979: 72-81
Kittel, Gabriele: Demokratische Außenpolitik als Voraussetzung der Friedensfähigkeit von Demokratien. Eine Untersuchung am Beispiel der "war powers" in den USA. in: Institut für Politikwissenschaft der Universität Tübingen, Arbeitsgruppe Friedensforschung (Hg.): Tübinger Arbeitspapiere zur internationalen Politik und Friedensforschung, Nr.20, Tübingen 1993
Kiwanuka, Samakula: Amin and the Tragedy of Uganda, München 1979
Klass, R.: Afghanistan: The Accords, in: Foreign Affairs 66/1988: 922-943
Klein, Herbert S.: Parties and Political Change in Bolivia 1880-1952, Cambridge 1969
Kliot, N.: The Collapse of the Lebanese State, in: MES 23/1987: 54-74
Klute, Georg: Die Revolte der Gastarbeiter: Die Auseinandersetzung zwischen Tuareg und Regierung in Mali und Niger, in: Blätter des iz3w 169/1990: 3-6
Kokeev, Alexander: Der Kampf um das Goldene Vlies. Zum Konflikt zwischen Georgien und Abchasien, in: HSFK-Report 8/1993
Kom, David: Le Cameroun. Essai d'analyse économique et politique, Paris 1971
Komitee Südliches Afrika (Hg.): Politisches Handbuch der PAIGC, Heidelberg 1974
Korbel, Josef: Danger in Kashmir, Princeton 1976
Kosiris, Hariton: Das politische System Griechenlands, Hersbruck 1980
Koszinowski, Thomas: Einigungsbestrebungen zwischen dem Nord- und Südjemen, in: Orient 1973: 29-32
Koszinowski, Thomas: Der Konflikt um Dhufar und die Aussichten auf seine Beilegung, in: Orient 53/1976: 66-87

Koszinowski, Thomas: Die Palästinenser und der arabisch-israelische Konflikt, in: Kaiser, Dieter / Steinbach, Udo (Hg.): Deutsch-Arabische Beziehungen, München/Wien 1981: 163-168
Koszinowski, Thomas: Die Krise der Ba'th-Herrschaft und die Rolle Assads bei der Sicherung der Macht, in: Orient 1985: 549-571
Koszinowski, Thomas: Nahostkonflikt: Die PLO in der Offensive, in: Jahrbuch Dritte Welt, München 1989: 153-168
Koszinowski, Thomas: Der Krieg in der Demokratischen Volksrepublik Jemen, in: Siegelberg, Jens (Red.): Die Kriege 1985 bis 1990. Analysen ihrer Ursachen, Münster/Hamburg 1991: 321-328
Koszinowski, Thomas / Menshaui, Magdi al / Meyer, Alois: Zur politischen und wirtschaftlichen Situation des Jemen. Einführung und Dokumentation, Hamburg 1980
Köhler, Wolfgang: Die Staatskrise im Libanon, in: Orient 4/1975: 68-104
Köhler, Wolfgang: Die Palästinenser in der arabischen Welt, in: Kaiser, Dieter / Steinbach, Udo (Hg.): Deutsch-arabische Beziehungen, München/Wien 1981: 77-95
Kreile, Renate: Zimbabwe: Von der Befreiungsbewegung zur Staatsmacht, Saarbrücken u.a. 1990
Krell, Gert / Kubbig, Bernd W. (Hg.): Krieg und Frieden am Golf, Frankfurt a.M. 1991
Krell, Gert / Müller, Harald (Hg.): Frieden und Konflikt in den internationalen Beziehungen. Festschrift für Ernst-Otto Czempiel. Frankfurt am Main / New York 1994
Krippendorff, Ekkehart: Staat und Krieg. Die historische Logik politischer Unvernunft. Frankfurt am Main 1985
Kroef, Justus M. v. d.: Indonesia and Irian Jaya. The enduring conflict, in: Asien, Deutsche Zeitschrift für Politik, Wirtschaft und Kultur, (Juli) 16/ 1985: 31-51
Krumwiede, Heinrich W.: Politik und katholische Kirche im gesellschaftlichen Modernisierungsprozeß. Tradition und Entwicklung in Kolumbien, Hamburg 1980
Krumwiede, Heinrich W.: Zur vergleichenden Analyse sozialrevolutionärer Prozesse: Zentralamerika, Ibero-Amerikanisches Archiv 10. Jg. (Sonderdruck) (Berlin) 4/1984
Kruse, Hans: Nordjemen, in: Ende, W. / Steinbach, Udo (Hg.): Der Islam in der Gegenwart, München 1989: 372-383
Kubitschek, Hans-Dieter / Wesser, Ingrid: Geschichte Indonesiens, Berlin (Ost) 1981
Kübler, H.: Ruanda, (hrsg. von der Deutschen Stiftung für Internationale Entwicklung, Deutsche Gesellschaft für Technische Zusammenarbeit), Bad Honnef 1988

Kühne, Winrich: Frieden im südwestlichen Afrika?, in: Europa-Archiv 4/1989
Kulke, H. / Rothermund, Dietmar: Geschichte Indiens, Stuttgart 1982
Kumar, V. Shiv: US Interventionism in Latin America. Dominican Crisis and the OAS, New York 1987
Kurtenbach, Sabine: Friedenssuche in Zentralamerika. Von der Contadora-Initiative über den Arias-Plan zum Abkommen von Guatemala, Lateinamerika. Analysen - Daten - Dokumentation, Beiheft 3, Hamburg (Nov.) 1987
Kurtenbach, Sabine: Staatliche Organisation und Krieg in Lateinamerika. Ein historisch-struktureller Vergleich der Entwicklung in Kolumbien und Chile. Kriege und militante Konflikte, Bd.3. Münster / Hamburg 1991
Kurzman, Dan: Santo Domingo. Revolt of the Damned, New York 1965
La Guerre d'Algérie et les Français: Colloque de l'Institut d'histoire du temps moderne, Paris 1990
LAB (Latin American Bureau) (ed.): Paraguay, Power Game, London 1980
LAB (Latin American Bureau) (ed.): Grenada - whose freedom ?, London 1984
LAB (Latin American Bureau) (ed.): Peru: Paths to Poverty, London 1985
LAB (Latin American Bureau) (ed.): Peru - a Time of Fear, London 1992
Lacar, L.: Neglected Dimensions in the Development of Muslim Mindanao, in: Journal Institute of Muslim Minority Affairs, Vol.9, (July) 1988: 286-310
Lachica, Eduardo: The HUKs: Philippine Agrarian Society in Revolt, New York/Washington/London 1971
Lackner, Helen: P.D.R. Yemen. Outpost of Socialist Development in Arabia, London 1985
LaFeber, Walter: The Panama Canal. The Crisis in Historical Perspective New York u.a. 1989
Lagerberg, Kees: West Irian and Jakarta Imperialism, London 1979
Laitin, David D. / Samatar, Said S.: Somalia. Nation in Search of a State, London 1987
Lamb, Alastair: The China-India Border. The Origins of the Disputed Boundaries, London 1964
Lamb, Alastair: The McMahon-Line. A Study in the relations between India, China and Tibet, London 1966 (1966a)
Lamb, Alastair: Crisis in Kashmir. 1947 to 1966, London 1966 (1966b)
Lamb, Alastair: Asian Frontiers. Studies in a Continuing Problem, London 1968
Lamb, Alastair: Kashmir: A Disputed Legacy 1846 - 1990, Hertingfordbury/ Herts 1991
Lamb, Christina: Waiting for Allah: Pakistan's Struggle for Democracy, London 1992

Lang; David Marshall: A Modern History of Georgia, London 1962
Lang, Josef: Das baskische Labyrinth. Unterdrückung und Widerstand in Euskadi 1890-1986, (2. Aufl.), Frankfurt a.M. 1988
Lao People's Democratic Republic / Ministry for Foreign Affairs: The Truth about Thai-Lao Relations, in : BBC (Hg.): Summary of World Broadcasts 25.9.84, FE/7757/C/1 - FE/7757/C/18, London 1984
Laqueur, Walter: Nahost - Vor dem Sturm. Die Vorgeschichte des Sechs-Tage-Krieges im Juni 1967, Frankfurt a.M. 1968
Lara, Patricia: Siembra Vientos y Recogerás Tempestades, Bogotá 1986
Lawson, Fred H.: Social Bases for the Hamah Revolt, in: Reip Reports, Nr. 110, 1982: 24-28
Lawson, Fred H.: South Yemen's troubles, in: Orient 1986, S. 441-449
LCHR (Lawyer's Committee for Human Rights) (ed.): Wages of War, Washington 1986
Leach, E. R.: Political Systems of Highland Burma. A Study of Kachin Social Structure, Boston 1972
Leal Buitrago, Francisco: Estado y Política en Colombia. Bogotá 1984
Lebow, R. N.: Der Krieg, den keiner wollte. Fehlwahrnehmungen im Falklandkonflikt, in: Beiträge zur Konfliktforschung, Frankfurt a.M. 1983
Leckie, Robert: Conflict. The History of the Korean War 1950-53, New York 1962
LeGrand, Catherine: Los Antecedentes Agrarios de la Violencia, in: Sánchez, Gonzalo / Peñaranda, Ricardo (Hg.): Pasado y Presente de la Violencia en Colombia, Bogotá 1986: 87-110
Legum, Colin (ed.): Africa Handbook, Harmondsworth 1969
Lehman, F.K.: Ethnic Categories in Burma and the Theory of Social Systems, in: Kunstadter, Peter (Hg.): Southeast Asian Tribes, Minorities and Nations, New Jersey 1967
Leifer, Walter (Hg.): Kenia, Tübingen/Basel 1977
Leitenberg, Milton: The Vietnam Conflict. Its Geographical Dimensions, PoliticalTraumas, and Military Developments, Santa Barbara (Cal.) 1973
Lemarchand, R.: Rwanda and Burundi, London 1970
Lesch, A. M.: Confrontation in the Southern Sudan, in: Middle East International, 3/1986
Levefer, Ernest W.: Crisis in the Congo. A United Nations Force in Action, Washington D.C. 1965
Levine, Daniel H. / Wilde, Alexander: The Catholic Church, "Politics", and Violence: The Colombian Case, in: The Review of Politics, Vol.39, 2/1977: 220-249
LeVine, Victor T.: The Cameroons from Mandate to Independence, Berkeley/ Los Angeles 1964

LeVine, Victor T.: Cameroon: Recent History, in: Africa South of the Sahara 1991, (hrsg. v. Europa Publications Ltd.), London 1990: 315-336
LeVine, Victor T. / Nye, Robert P.: Historical Dictionary of Cameroon, Metuchen (N.Y.) 1974
Lewis, B.: Die islamische Revolution, in: Kursbuch 93, Glauben 1988: 153-167
Lewis, I. M.: A Modern History of Somalia. Nation and State in the Horn of Africa, 3. Aufl., London/New York 1988
Libal, Wolfgang: Das Ende Jugoslawiens, Wien 1991
Liberation Support Movement (ed.): Namibia: SWAPO fights for freedom, Oakland 1978
Lider, Julian: Der Krieg. Deutungen und Doktrinen in Ost und West. Frankfurt am Main / New York 1983: (Engl. Original 1977.)
Liebermann, V.: Why the HUK Balahat Movement Failed, South East Asia Studies, New Haven (Conn.) 24/1968
Limia Peréz, E.: Reseña general del bandolerismo en España después de la guerra civil, Madrid 1957
Linde, Gerd: Dreiteilung des Subkontinents. Politische Konsequenzen des indisch-pakistanischen Krieges, in: Berichte des Bundesinstituts für Ostwissenschaftliche und internationale Studien (Nr.4) Köln 1972
Linde, Gerd: Bangla Desh, Indien und die Großmächte im pakistanischen Konflikt, Stuttgart 1982
Lindig, Wolfgang / Münzel, Mark: Die Indianer, in: Bd.2: Mittel- und Südamerika, München 1985
Lodge, Tom: Black Politics in South Africa since 1945, New York 1983
López, Julio / Núñez, Orlando / Chamorro, Carlos Fernando / Serres, Pascual: La caída del somocismo y la lucha sandinista en Nicaragua. EDUCA. Centroamérica, (2. Auflage) 1980
Lorenz, Konrad: Das sogenannte Böse. Zur Naturgeschichte der Aggression. Wien 1963
Lovelance, Daniel D.: China and People's War in Thailand 1964-1969, Berkeley 1971
Low, D.A.: Political Parties in Uganda 1949 - 62, London 1962
Lowenthal, Abraham F.: The Dominican Intervention in Retrospect, in: Public Policy (Vol.XVIII), 1969: 133-153
Lowenthal, Abraham F.: The Domincan Intervention, Cambridge (Mass.) 1972
Luard, Evan: War in International Society. A Study in International Sociology. London 1986
Lubeck, P.M.: Islamic Protest and Oil-based Capitalism: Agriculture, Rural Linkages and Urban Popular Movements in Northern Nigeria, in: Watts, M. (ed.): State, Oil, and Agriculture in Nigeria, Berkeley 1988: 268-290

Lucius, Robert von / Deutsche Afrikastiftung (Hg.): Südafrika: Schwarze Hoffnung, Weiße Reformbereitschaft, Bonn 1981

Ludwig, Klemens: Bedrohte Völker. Die Chittagong-Bergvölker in Bangla Desh, München 1985: 52-55

Ludwig, Klemens: Tibet, München 1989

Ludwig, K. / Horta, K.: Osttimor, Göttingen 1985

Luther, Hans U.: Der Mindanao Konflikt: "Interner Kolonialismus" und regionale Rebellion in den Südphilippinen, in: Khan, Kushi M./ Matthies, Volker (Hg.): Regionalkonflikte in der Dritten Welt, München 1981: 183-277

Maalmi, Abdelwahab: La politique de défense actuelle du Maroc: constance et perspectives d'évolution, Rabat 1993

Maaß, Citha D.: The Punjab Conflict: Failure of the Rajiv Ghandi-Longowal Accord, in: Kurzbibliographie: Der Konflikt in Punjab. (Hg: Deutsches Überseeinstitut, Übersee-Dokumentation, Referat Asien und Südpazifik), Hamburg 1988: 3-16

Maaß, Citha D.: Der Bangladesh-Krieg 1971. Ursachen und methodische Klassifizierungsprobleme, in: Arbeitskreis für Wehrforschung(Hg.): Unruhige Welt. Konflikt- und Kriegsursachen seit 1945, Koblenz 1989:139-158

Mackie, J. A. C.: "Konfrontation" - The Indonesia-Malaysia Dispute 1963-1966, Kuala Lumpur 1974

Mahajani, Usha: Soviet and American Aid to Indonesia 1949-1968. Ohio University Center for International Studies, Papers in International Studies, Southeast Asia Series, Athens (Ohio) 14/1970

Mahfoud, Bennoune: The making of contemporary Algeria, 1830-1987. Colonial upheaval and post-independence development, Cambridge 1988

Mahncke, Dieter: Konflikt in Südafrika, Paderborn 1989

Maibach, Marcia / Karajica, Zeljko: Desintegration und Krieg in Jugoslawien - Entstehung und Eskalation des serbisch-kroatischen Nationalitätenkonfliktes, in: Forschungsstelle Kriege, Rüstung und Entwicklung an der Universität Hamburg (Hg.): Arbeitspapiere, Nr. 57, Hamburg 1992

Malek, Mohammed H.: Kurdistan in the Middle East Conflict, in: New Left Review (May/June) 175/1989: 79-94

Malloy, James M.: Bolivia: The Uncompleted Revolution, Pittsburgh 1970

Malwal, Bona: People and Power in Sudan - The Struggle for National Stability, London 1981

Mamdami, Mahmood: Politics and Class Formation in Uganda, London 1976

Mamdami, Mahmood: Imperialism and Fascism in Uganda, Nairobi/Ibadan/ London 1983

Mandaza, Ibbo: Zimbabwe - The Political Economy of Transition 1980-86, Dakar 1986
Mangold, P.: Shaba I and Shaba II, in: Survival 21, 3/1979: 107-115
Manning, Peter: The United Nations Plan for Elections in Namibia, Southscan Occasional Paper, SWAPO, o.O. (June) 1989
Marcum, J.: The Angolan Revolution, Vol. I. - The Anatomy of an Explosion (1950-62), Cambridge /Mass 1969
Marcum, J.: The Angolan Revolution. Vol.I, The Anatomy of an Explosion (1950-62), Cambridge (Mass.) 1978
Markakis, John: National and Class Conflict in the Horn of Africa, Cambridge 1987
Marks, Thomas A.: The Karen Revolt in Burma, in: Issues and Studies, Taipei 14/1978: 48-84
Marquina, Antonio: Libya, the Maghreb and Mediterranean Security, Adelphi Papers 231/1988
Marsh, Jan: Stop the War against Angola and Mozambique, London 1981
Martin, Keith: Tajikistan: Civil War without End? in: Radio Free Europe / Radio Liberty 33/1993: 18-29
Martin, Pierre: Les chiites d'Irak de retour sur la scène politique, in: monde arabe. Maghreb - Machrek (avril-juin) 132/1991: 21-37
Martin, David / Johnson, Phyllis: The Struggle for Zimbabwe, London 1981
Martin, David / Johnson, Phyllis: Zimbabwe - Apartheid's Dilemma, in: Dies. (Hg.): Destructive Engagement, Harare 1986
Martín-Baró, Ignacio: El latino indolente. Carácter ideológico del fatalismo latinoamericano, in: Montero, Maritza (ed.): Psicología política latinoamericana, Caracas 1987a
Martín-Baró, Ignacio: Del opio religioso a la fe liberadora, in: Montero, Maritza (ed.): Psicología política latinoamericana, Caracas 1987b
Martz, John D.: Ecuador, in: Hopkins, Jack W. (Hg.): Latin American and Caribbean Contemporary Record, New York/London 1983: 301
Martz, John D.: Ecuador, in: Hopkins, Jack W. (Hg.): Latin American and Carribean Contemporary Record, New York/London 1985: 385
Mattes, Hans Peter: Sudan (Länderartikel), in: Deutsches Orient Institut (Hg.), Nahost-Jahrbuch 1989, Hamburg 1990: 264-270
Matthies, Volker: Der Shifta-Krieg in Kenia, in: Verfassung und Recht in Übersee 4/1973: 429-443
Matthies, Volker: Der Grenzkonflikt Somalias mit Äthiopien und Kenia, Hamburg 1977
Matthies, Volker: Der Eritrea-Konflikt - Ein "vergessener" Krieg am Horn von Afrika, Hamburg 1981
Matthies, Volker: Der Ogaden-Krieg zwischen Somalia und Äthiopien von

1977/78: Ursachen, Verlauf, Folgen, in: afrika spectrum, Vol.22, 3/1987 (Hamburg): 237-253.
Matthies, Volker: Geopolitik und Konfliktdynamik am Horn von Afrika: Umbruch der Allianzen, in: Senghaas, Dieter (Hg.): Regionalkonflikte in der Dritten Welt, Baden-Baden 1989: 121-147.
Matthies, Volker: Krieg am Horn von Afrika: Historische Aspekte von Gesellschaft, Rüstung und Verheerung, in: afrika spectrum 90/1 (Hamburg) 1990: 5-33 (1990a)
Matthies, Volker: Der Somalikonflikt: Nation ohne Staat, in: Brüne, Stefan / Matthies, Volker: Krisenregion Horn von Afrika, Hamburg 1990: 223-247 (1990b)
Matthies, Volker: Die Rivalität der Supermächte am Horn von Afrika, in: Brüne, Stefan / Matthies, Volker: Krisenregion Horn von Afrika, Hamburg 1990: 271-298 (1990c)
Mavrolan, Carlos: Marauding Warriors, in: Africa Events, April/1992: 21-31
Mawut, L. L.: The Southern Sudan: Why Back to Arms, Khartoum 1986
Maxwell, Neville: India's China War, Aylesbury/Bucks 1972
Maxwell, Neville: India and the Nagas, Minority Rights Group Report, No. 17, London 1973
Mayer, Thomas: The Islamic opposition in Syria 1961 - 1982, in: Orient 1983: 589-609
Mazrui, A.A.: Soldiers and Kinsmen in Uganda, Beverly Hills/London 1975
McClintock, Cynthia: Why Peasants Rebel: The Case of Peru's Sendero Luminoso, in: World Politics 1/1984: 48-84
McClintock, Cynthia: Peru's Sendero Luminoso Rebellion: Origins and Trajectory, in: Eckstein, Susan (ed.): Power and Popular Protest. Latin American Social Movements, Berkeley u.a. 1989: 61-101
Medina, Medófilo: La Resistencia Campesina en el Sur del Tolima, in: Sánchez, Gonzalo / Peñaranda, Ricardo (Hg.): Pasado y Presente de la Violencia en Colombia, Bogotá 1986: 233-265
Meinardus, Ronald: Die Türkei-Politik Griechenlands, Hamburg 1984
Meinicke-Kleint, Heinz: Algerien, Marokko, Tunesien. Unterjochung und Befreiung, Berlin 1965
Mejcher, Helmut Alexander.: Palästina in der Nahostpolitik europäischer Großmächte und der Vereinigten Staaten von Amerika 1918-1948, in: Mejcher, H. / Schölch: Die Palästinafrage 1917-1948, Paderborn 1981: 163-216
Menchú, Roberta: Klage der Erde. Der Kampf der Campesinos in Guatemala, Göttingen 1993
Mendler, Martin / Schwegler-Rohmeis, Wolfgang: Weder Drachentöter noch Sicherheitsingenieur. Bilanz und kritische Analyse der

sozialwissenschaftlichen Kriegsursachenforschung. in: Hessische Stiftung Friedens- und Konfliktforschung (Hg.): HSFK-Forschungsbericht, Nr.3/ 1989. Frankfurt am Main 1989

Menon, Vapal Pangunni: The Story of the Integration of the Indian States, London/New York 1956

Menzel, Ulrich / Senghaas, Dieter: Indikatoren zur Bestimmung von Schwellenländern. Ein Vorschlag zur Operationalisierung. in: Franz Nuscheler (Hg.): Politische Vierteljahresschrift, Sonderheft 16, 1985: 75-96.

Meschkat, Klaus / u.a.: Kolumbien. Geschichte eines Landes im Ausnahmezustand, Berlin 1980

Meschkat, Klaus / Oertzen, Eleonore v. / Richter, Ernesto / Rossbach, Lioba / Wünderich, Volker (Hg.): Mosquita. Die andere Hälfte Nicaraguas. Über Geschichte und Gegenwart an der Atlantikküste, Hamburg 1987

Mey, Wolfgang: Untersuchungen zur Wirtschaft in den Chittagong Hill Tracts, Bangla Desh, Bremen 1979

Mey, Wolfgang: Politische Systeme in den Chittagong Hill Tracts, Bangla Desh/ Bremen 1980

Mey, Wolfgang: Sie verbrennen jetzt ein Dorf nach dem anderen - Bengalischer Landraub und Völkermord in den Chittagong Hill Tracts, in: Nebelung, M. (Hg.): Bangla Desh im Schatten der Gewalt, Aachen 1986

Mey, Wolfgang: Wir wollen nicht euch - wir wollen euer Land, Göttingen/ Wien 1988

Mey, Wolfgang: Völkermord durch Entwicklung, in: blätter des iz3w, Nr. 184, Sept./Okt. 1992:47f

Meyer, E.: Ceylon / Sri Lanka, Paris 1977

Meyer, Gerd: Nationalitätenkonflikte in der Sowjetunion, Köln 1990

Meyer, Hektor: Die Entwicklung der kommunistischen Streitkräfte in China von 1927 bis 1949. Berlin/New York 1982

Meyer, Reinhold: Der Krieg im Tschad - Ursachen und Perspektiven, in: Gantzel, Klaus Jürgen (Hg.): Kriege in der Dritten Welt, Baden-Baden 1988: 329-350

Meyns, Peter: Befreiung und nationaler Wiederaufbau von Mozambique, Hamburg 1979

Meyns, Peter: Das südliche Afrika nach Nkomati, Hamburg 1987

Meza, Victor: Historia del movimiento obrero hondureño, Tegucigalpa 1980

Michler, Walter: Weißbuch Afrika, 2. Aufl., Bonn 1991

Millet, Richard: Guardianos de la Dinastía, San José 1979

Miners, N.J.: The Nigerian Army 1956 - 1966, London 1971

Ministre de l'Information (ed.): L'Information durant la guerre de libération. Dossiers Documentaires 42, Alger 1984.

MIPODOK - Militärpolitik Dokumentation (Hg.): Kontrollierte Intervention.

Destabilisierung, unerklärte Kriege und Militäraktionen gegen die Dritte Welt, Frankfurt a.M. 63/64/1988
Miranda, Carlos R.: The Stroessner Era, Westview Press 1990
Misra, Udayon: The Naga National Question, in: Economic and Political Weekly 8.4.1978: 618-624
Mittelmann, J.H.: Ideology and Politics in Uganda. From Obote to Amin, Ithaca 1975
Mohapatra, Srikant: The Siachen Imbroglio, in: IUMDA Newsletter, Vol.1, 2/1986: 5-12
Molano, Alfredo: Signiendo el Corte. Relato de Guerras y de Tierras, Bogotá 1989
Molano, Alfredo / Reyes, Alejandro: Los Bombardeos en el Pato. Controversia (Bogotá) 89/1978
Molloy, I.: The Decline of the Moro National Liberation Front in the Southern Philippines, in: Journal of Contemporary Asia, Vol.18, 1/1988: 59-76
Mondlane, Eduardo: Kampf um Mozambique, Frankfurt a.M. 1970
Moorsom, Richard: Walvis Bay: Namibia's Port, in : IDAF (ed.), London 1984
Morell, David / Samudavanija, Chai-anan: Political Conflict in Thailand: Reform, Reaction, Revolution, Cambridge (Mass.) 1981
Moreno, José A.: Barrios in Arms. Revolution in Santo Domingo, Pittsburgh 1970
Morgenthau, Hans Joachim: Macht und Frieden. Grundlegung einer Theorie der internationalen Politik. Gütersloh 1963 (US-amerik. Orig. 1948.)
Moser, Beat: Ethnischer Konflikt und Grenzkriege. Ursachen innen- und außenpolitischer Konflikte in Afrika. in: Zürcher Beiträge zur Politischen Wissenschaft, Bd.7. Diessenhofen 1983
Moss, Glenn / Obery, Ingrid: South African Review (Johannesburg) 5/1989
Mounir, Riad: Algérie: l'islamisme, sabre au clair, in: Le Novel Afrique Asie, Nr. 44, Mai 1993
Munro, Ross H.: Superpower Rising. Propelled by an Arms Buildup, India Asserts its Place on the World Stage, in: Time International (April 3) 14/1989: 10-17
Munslow, Barry: Mozambique: The Revolution and its Origin, London 1983
Münkler, Herfried: Krieg und Frieden. in: Fetscher / Münkler 1985: 279-325.
Münster, Arno: Guerrilla und Konterrevolution. Arbeiterklasse gegen oligarchische Diktatur und Gewerkschaftsbürokratie, München 1977
Mveng, Engelbert: Histoire du Cameroun, (Editions Présence Africaine) Paris 1963
Nabudere, Dan Wadada: Imperialism and Revolution in Uganda, London 1980
Nafziger, E.W.: The Economics of political instability: the Nigerian-Biafranwar, Boulder 1983
Nag, B.C.: Das Problem der Naga-Stämme, in: Indo-Asia 4/1965: 342-347

Nayar, Baldev Raj: Minority Politics in the Punjab, Princeton 1966
Nebelung, M.: Die stille Gewalt - Ländliche Armut und Landlosigkeit in Bangla Desh, in: Ders. (Hg.): Bangla Desh im Schatten der der Gewalt, Aachen 1986
Nebez, Jemal: Kurdistan und seine Revolution, Berlin 1972
Nelson, Harold D. (ed.): Area Handbook of the Malagasy Republic, Washington D.C. 1973
Nelson, Harold D. (ed.): Algeria - A Country Study. Area Handbook Series, Washington D.C. 1986.
Nemenzo, F.: Rectification Process in the Philippine Communist Movement, in: Joo-Jock, Lim / Vani, S. (ed.): Armed Communist Movements in Southeast Asia, Singapore 1984: 69-102
Neumann, Peter: Wirtschaft und materielle Kultur der Buschneger Surinams. Ein Beitrag zur Erforschung afroamerikanischer Probleme, Berlin (Ost) 1967.
Niblock, T.: Class and Power in Sudan, London/Basingstoke 1987
Niebling, Ursula: Der "nützliche" Krieg. El Salvador/Honduras 1969. in: Diskussionsbeiträge aus dem Institut für Politische Wissenschaften der Universität Hamburg, Nr.10, Hamburg 1984 (1984a)
Niebling, Ursula: Die Kriege zwischen Costa Rica und Nicaragua von 1948 und 1955 und ihre Vorgeschichte: Der Bürgerkrieg in Costa Rica 1948. in: Diskussionsbeiträge und Berichte aus dem Institut für Politische Wissenschaft der Universität Hamburg, Nr.3, Hamburg1984 (1984b)
Niebling, Ursula: Konfliktstrukturen und Kriegsursachen in Zentralamerika am Beispiel der Kriege zwischen Nicaragua und Costa Rica von 1948 und 1955, in: Friedensanalysen 21/1985
Niebling, Ursula: Esquipulas II in El Salvador. Friedensplan oder Befriedungsversuch?, in: Mittelamerika Magazin 65/66/1988 (Kiel) (1988)
Niebling, Ursula: Pax Americana versus Volkssouveränität. Volksaufstand und Contrakrieg in Nicaragua. Eine Ursachenanalyse, in: Forschungsstelle Kriege, Rüstung und Entwicklung an der Universität Hamburg (Hg.), Arbeitspaiere Nr. 46, Hamburg 1990
Niebling, Ursula: Der Bürgerkrieg in El Salvador. Herausbildung und Behauptungeiner Widerstandskultur, in: Forschungsstelle Kriege, Rüstung und Entwicklung an der Universität Hamburg (Hg.): Arbeitspapiere, Nr. 50, Hamburg 1991
Niebling, Ursula: Kriege in Zentralamerika seit 1945. Ein Beitrag zur vergleichenden Kriegsursachen- und Kulturforschung. Kriege und militante Konflikte, Bd.4. Münster / Hamburg 1992
Niebling, Ursula / Richter, Ernesto: El Salvador 1983. Kämpfen und verhandeln?, in: Mittelamerika Magazin (Kiel) 4/1983
Niedergang, Marcel: La Révolution de Saint-Domingue, Rennes 1966

Nieh, Yu-Hsi: Das indisch-chinesische Grenzproblem. Neue Gesichtspunkte, Mitteilungen des Instituts für Asienkunde, Nr. 43, Hamburg 1971
Nielebock, Thomas: Frieden zwischen Demokratien: Ein empirisches Gesetz der Internationalen Beziehungen auf der Sucher nach seiner Erklärung. in: Österreichische Zeitschrift für Politikwissenschaft, 22.Jg., Heft 2/1993: 179-193.
Nima, Ramy: The Wrath of Allah. Islamic Revolution and Reaction in Iran, London/Sydney 1983
Nöbel, Heinrich Wilhelm: Heer und Politik in Indonesien: Zielsetzung und Zielverwirklichung einer militärischen Organisation 1945- 1967, in: Militärgeschichtliches Forschungsamt (Hg.): Militärgeschichte seit 1945, Bd.2, Boppard am Rhein 1975
Noble, L.: Muslim Separatism in the Philippines 1972-81. The Making of a Stalement, in: Asian Survey, Vol.11, (Nov.) 1/1981: 1097-1114
Nohlen, Dieter: Bolivien, in: Nohlen, Dieter / Nuscheler, Franz (Hg.): Handbuch der Dritten Welt, Bd.2 - Südamerika, Hamburg 1982: 109-138
Nohlen, Dieter / Nuscheler, Franz (Hg.): Handbuch der Dritten Welt, 8 Bde, Hamburg 1982
Nohlen, Dieter / Nuscheler, Franz (Hg.): Handbuch der Dritten Welt, Bd.1: Grundprobleme, Theorien, Strategien. Bonn 1992
Nohlen, Dieter / Witzel, Renate: Guinea-Bissau, in: Nuschler, Franz / Ziemer, Klaus (Hg.): Politische Organisation und Repräsentation in Afrika, Berlin 1978
Noller, Jost F.: Theorie und Praxis der Apartheid, Frankfurt a.M. 1977
Nowacki, Helmut: Der Zypernkrieg 1974. Eine Analyse seiner Ursachen, Hamburg 1982
Nuscheler, Franz: Plädoyer für einen humanitären Interventionismus. Lehren aus der somalischen Apokalypse, in: Entwicklung und Zusammenarbeit, Jg 33, 11/1992:3-4
Nzongola-Ntalaja (ed.): The Crisis in Zaire: Myths and Realities, Trenton 1986
NZZ: Neue Zürcher Zeitung, Zürich
O'Ballance, Edgar: The Sinai Campaign, London 1959
O'Ballance, Edgar: The War in the Yemen, London 1971
O'Ballance, Edgar: The Third Arab-Israeli War, London 1972
O'Ballance, Edgar: The Kurdish Revolt: 1961-1970, London 1973 (1973a)
O'Ballance, Edgar: Arab Guerrilla Power 1967-1972, London 1973 (1973b)
O'Ballance, Edgar: The Arab-Israeli War 1948, Westport 1981
O'Ballance, Edgar: The Cyanide War. The Tamil Insurrection in Sri Lanka, London 1989
Obeyesekere, Gananath: The Origins and Institutionalization of Political Violence, in: Manor, J. (ed.): Sri Lanka in Change and Crisis, London 1984
ÖMZ: Österreichische Militärische Zeitschrift, Wien

Oertzen, Eleonore von: Man trägt wieder Unterschied: Der Rassismus der weißen Elite in Peru, in: Lateinamerika. Analysen und Berichte, Jahrbuch 1993, Münster/Hamburg 1992: 79-94
Offerman, Michael: Angola zwischen den Fronten, Pfaffenweiler 1988
Ohlson, Thomas: The Cuito Cuanavale Syndrome: Revealing SADF Vulnerabilities, in: South African Review (Johannesburg) 5/1979
Ojwando Abuor, C.: White Highlands No More - A Modern Political History of Kenya, Nairobi 1971
Ollier, Maria Matilda: Orden, poder y violencia 1966-1973, Buenos Aires 1983
Olschewski, Malte: Der Conducator Nicolae Ceausescu, Heidelberg 1990
Omara-Otunnu, Amii: Politics and the Military in Uganda 1890 - 1985, Houndmills 1987
Omer-Cooper, J.D.: History, in: Africa South of the Sahara, o.O. 1990
Opitz, Peter J.: Das Weltflüchtlingsproblem, München 1988
Oquist, Paul: Violencia, Conflicto y Politica en Colombia, Bogotá 1978
OCCHT (Organising Committee CHT Campaign): The Charge of Genocide - Human Rights in the Chittagong Hill Tracts of Bangla Desh, Amsterdam 1986
Orient-Institut (Deutsches Orient-Institut im Verbund der Stiftung Deutsches Übersee-Institut) (Hg.): Afghanistan seit dem Sturz der Monarchie: Dokumentation zur Politik, Wirtschaft und Bevölkerung, Hamburg 1981
Osborne, Robin: Indonesia's Secret War. The Guerrilla Struggle in Irian Jaya, Sydney 1985
O'Shaugnessy, Hugh: Grenada - Revolution, Invasion and Aftermath, London 1984
Osterud, Oyvind: War Termination in the Western Sahara, in: Bulletin of Peace Proposals, Vol. 20, (Sep) 3/1989: 309-317
Page, Stephen: The Soviet Union and the Yemens. Influence in Asymmetrical Relationships, New York 1985
Painter, James: Guatemala: False Hope, False Freedom. The Rich, the Poor and the Christian Democrats, London 1987
Palacios, Marco: El Café en Colombia (1859-1970). Una Historia Económica, Social y Política, Bogotá 1979
Palmer, David Scott: The Sendero Luminoso Rebellion in Rural Peru, in: Fauriol, Georges (Hg.): Latin American Insurgences, Georgetown Univ. 1985: 67-96
Papstein, R.: The Origins and Development of Tribalism in the Upper Zambezi Region of Zambia, 1830-1981, (Diss. paper Free University of Amsterdam) Amsterdam 1981
Pasricha, P.M.: The Siachen Glacier. An Orographic Offensive by Pakistan, in: Strategic Analysis, Vol.11, (Dec.) 9/1985: 851-860
Pathak, Angelika: Ethnische Konflikte und Flüchtlingsprobleme im Grenzgebiet

zwischen Indien und Bangla Desh, in: Jahrbuch Dritte Welt 1989, München 1988
Pathak, Angelika: Asien: Frieden in Südasien - Nur zu Indiens Bedingungen?, in: Jahrbuch Dritte Welt 1990, München 1989
Pazifik-Informationsstelle: Das Schweigen brechen. Osttimor. Bedrohtes Paradies im Pazifik, Dezember 1991
Pearce, Jenny: Under the Eagle. - U.S. Intervention in Central America and the Caribbean, London 1981
Pearce, Jenny: Colombia. Inside the Labyrinth, London 1990
Pécaut, Daniel: Orden y Violencia, Colombia 1930-1954, 2 Bde., Bogotá 1987
Pécaut, Daniel: Crónica de Dos Décadas de Politica Colombiana 1968-1988, Bogotá 1988
Peissel, Michael: Die Chinesen sind da!, Wien/Hamburg 1973
Peter, Marina: Ethnische Vielfalt als Herausforderung an staatliche Einheit im Sudan, (Unv. Dipl.Arb.) Hildesheim 1989
Peters, Christian: Zimbabwe: ein postkolonialer Machtkampf, in: Siegelberg, Jens (Red.): Die Kriege 1985-1990. Analysen ihrer Ursachen, Hamburg 1991: 139-148
Peters, Christian / Tetzlaff, Rainer / Wegemund, R.: Ethnisch-soziale Konflikte in Afrika, Hamburg 1991
Peterson, J. E.: Oman in the Twentieth Century. Political Foundations of an Emerging State, London/New York 1978
Peterson, N.: Südsudanesische Autonomie Schritt für Schritt zerstört, in: Pogrom, 107/1984
Pfeffer, K.H. (Hg.): Zeitschrift für Geopolitik, Darmstadt 1953
Pfetsch, Frank R. (Hg.): Konflikte seit 1945. Daten - Fakten - Hintergründe. 5 Bände. Freiburg / Würzburg 1991
Philipp, G.: Rwanda, in: Ders. (Hg.): Handbuch der Dritten Welt 2/II, Unterentwicklung und Entwicklung in Afrika, Hamburg 1976: 448-456.
Phillips, C.S.: Nigeria and Biafra, in: Shiels, F.L. (ed.), Ethnic separatism and world politics, Lanham 1984
Pieck, Werner: Algerien. Die wiedergewonnene Würde, Hildesheim/Zürich/ New York 1987
Piribauer, Karl: Hintergründe und Vorgeschichte des ungarischen Aufstandes, Osteuropa 1957
Pisani, A. du: Namibia: The Politics of Continuity and Change, Johannesburg 1985
Pizarro, Carlos: Guerra a la Guerra, Bogotá 1985
Pizarro, Eduardo: Los Origenes del Movimiento Armado Comunista en Colombia (1949-1966), in: Análisis Politico 7/1989: 7-31
Pizarro Leongómez, Eduardo: La Guerrilla Revolucionaria en Colombia, in:

Sánchez, Gonzalo / Peñaranda, Ricardo: Pasado y Presente de la Violencia en Colombia, Bogotá 1986
Pogrom (Hg.): Nationalitätenstaat Burma. Zur Situation der nicht-burmesischen Minderheitenvölker in Burma, Göttingen 12/1981: 32-37
Polimé, T. S. / Thoden van Velzen, H. U. E.: Vluchtelingen, Opstandelingen en andere Bosnegers van Oost-Suriname 1986-88, Utrecht 1988
Polley, Uwe: Indonesien. Der Kampf um ein freies Papua, in: Siegelberg, Jens (Red.): Die Kriege 1985-1990. Analysen ihrer Ursachen, Hamburg 1991
Pollock, David: Moscow and Aden: Coping with a Coup, in: Problems of Communism, May-June 1986: 50-70
Ponnambalam, S.: Sri Lanka - The National Question and the Tamil Liberation Struggle, London 1983
Pons Prades, Eduardo: Guerrillas españolas 1936-1960, Barcelona 1977
Posner, M.H.: Violations of Human Rights in Uganda, New York 1978
Prados, Luis / Rivas, Yavier: Ifni, la guerra prohibida, in: El País 24.3.1991
Premdas, Ralph R.: The Organisasi Papua Merdeka in Irian Jaya, in: Asian Survey, Vol. 25, Oct/1985
Pretzell, Klaus: Thailand. Neuer Grenzkonflikt mit Laos, in: SOAa (Jg.7) 2/1988: 98-99
Price, D. L.: Oman. Insurgency and Development, in: Conflict Studies, 53/1975
Prins, J.: Location, History, Forgotten Struggle: The South Moluccas - Rebellious Province or Occupied State, South Moluccan Information Service, Leyden 1959
Puertas, P.: Spagna, antifranchismo e lotta di classe 1939-1975, Milano 1975
Pye, Lucian W.: Guerrilla Communism in Malaya, Princeton 1956
Qubain, Fahim I.: Crisis in Lebanon, Washington D.C. 1961
Rabinovich, Itamar: The War for Lebanon 1970 - 1985, London 1985
Rabinovich, Itamar / Shaked, H. (Hg.): From June to October. The Middle East between 1967 and 1973, New Brunswick (N.Y.) 1978
Ramonet, Ignacio: Le labyrinthe algérien, in: Le Monde Diplomatique, Febr. 1993
Ramírez Tobon, William: La Liebre Mecánica y el Galgo Corredor. La Paz actual con el M-19, in: Análisis Politico 7/1989: 46-59
Ramírez V., Socorro / Restrepo M., Luis Alberto: Actores en el Conflicto por la Paz. El Proceso de Paz durante el Gobierno de Belisario Betancur, 1982-1986, Bogotá 1989
Razumkovski, D. Gräfin: Chaos Jugoslawien, München 1991
Rediske, Michael: Umbruch in Nicaragua. Die Entwicklung der Revolution aus dem Zerfall bürgerlicher Herrschaft, Berlin 1984
Reichel, Uwe: Militär und Gewalt im Südlichen Afrika, in: Militärpolitik Dokumentation, Frankfurt 19/1980

Reichert, Christoph: Das neue Zimbabwe, Bonn 1984
Reid Daly, Ron / Stiff, Peter: Selous Scouts, Top Secret War, Alberton 1982
Reinhard, Peter: Bedeutung des Baumwollanbaus für die tschadische Landwirtschaft und Analyse der Geldflüsse, die durch ihn ausgelöst werden, Bern/Frankfurt a.M. 1981
Reinknecht, Gottfried: Philippinen, in: Politisches Lexikon Asien und Südpazifik, München 1980 (1980a)
Reinknecht, Gottfried: Brunei (Sultanat Brunei), in: Politisches Lexikon Asien und Südpazifik, München 1980 (1980b)
Reyes, Alejandro / Bejarano, Ana Maria: Conflictos Agrarios y Luchas Armadas en la Colombia Contemporánea: Una Visión Geográfica, in: Análisis Politico 5/1988: 6-27
Richardson, Lewis F.: Statistics of Deadly Quarrels. Pittsburgh: Bookswood Press - Chicago 1960.
Richardson, Lewis F.: Der Rüstungswettlauf zum Ersten Weltkrieg. (Auszugsweise Übers. aus ders.: Arms and Insecurity, Pittsburgh / Chicago 1960.) Konflikt - in: K.J.Gantzel / G.Kress / V.Rittberger (Hg.): Eskalation - Krise, Sozialwissenschaftliche Studien zum Ausbruch des Ersten Weltkriegs, Düsseldorf 1972
Richter, Ernesto: Proceso de acumulación y dominación en la formación sociopolítica salvadoreña; CSUCA, Serie de Informes de Investigación, San José (Abril) 16/1976
Ridgway, Matthew B.: The Korean War, New York 1967
Ripken, Peter / Prein, Gabriele: Zimbabwe. Reiseführer mit Landeskunde, Buchschlag bei Frankfurt a.M. 1990
RIPS (Research Institute for Peace and Security, Tokyo): Asian Security 1990-91, London et al. 1990: 197-200
Risse-Kappen, Thomas: Demokratischer Frieden? Unfriedliche Demokratien? Überlegungen zu einem theoretischen Puzzle. In Krell / Müller (Hg.) 1994: 159-189.
Roberts, Michael: Elite Formation and Elites 1832-1931, in: Ders. (ed.): Collective Identities, Nationalisms and Protests in Modern Sri Lanka, Colombo 1979
Rondot, Philippe: Irak gegen Iran: Krieg ohne Entscheidung?, in: Europa-Archiv 36, 3/1981: 67-76
Rone, Jemera: Angola, Africa Watch (Hg.) London (April) 1989
Roosevelt, Archie: Die kurdische Republik Mahabad, in: Chaliand, Gérard (Hg.): Kurdistan und die Kurden, Göttingen 1984: 229-257
Rosberg, C. A. / Nottingham, J.: The Myth of "Mau-Mau." Nationalism in Kenya, London 1966
Rosett, Claudia: A Lonely Lao Fight for Freedom, in: The Wall Street Journal 13.6.1990.

Rotberg, Robert I.: Conflict and Compromise in South Africa, Boston 1980
Rothermund, Dietmar: Indien-Pakistan-Konflikt, in: Nohlen, Dieter / Boeckh, Andreas: Pipers Wörterbuch zur Politik. Bd. 5 - Internationale Beziehungen, München 1984: 207-209
Rotter, Gernot: Die Milizionarisierung des Libanon, in: Saeculum, Bd. 37, 1986: 192-198
Rowles, James: El conflicto Honduras-El Salvador y el orden jurídico internacional (1969), EDUCA, Centroamérica 1980
Roy, O.: Afghanistan vor einer ungewissen Zukunft, in: Europa-Archiv 9/1988: 233-240
Rösel, Jakob: Entwicklung und Dynamik des Tamilen-Konfliktes auf Sri Lanka. in: Arnold-Bergstraesser-Institut (Hg.): Informationspapiere zu Entwicklung und Politik, Nr.12, Freiburg 1987
Rösel, Jakob: Genese eines Bürgerkrieges - Ethnischer Bürgerkrieg, Jugendaufstand und Niedergang des Rechtsstaates in Sri Lanka, in: blätter des iz3w 1994/195: 45-50
Rubin, Barnett N.: The Fragmentation of Tajikistan, in: Survival, vol. 35 4/ 1993-94: 71-91
Rubin, Neville: Cameroon. An African Federation, London 1971
Rudebeck, Lars: Guinea-Bissau. A Study of Political Mobilisation, Uppsala 1974
Rudolph, Jörg Meinhard: Die Kommunistische Partei Chinas und Taiwans (1921-1981), München 1986
Rühland, J.: Die Philippinen. Zwischen Repression und Widerstand, in: Aktuelle Informationspapiere zu Entwicklung und Politik, Freiburg 5/1982
Ruiz, Hiram A. / US Committee for Refugees: Uprooted Liberans. Casualities of a Brutal War,. Issue Paper, American Council for Nationalities Service, 2/1992
Russel, Dominique Soulas de: Krisen und Konflikte im Tschad: Ursachen und Hintergründe, in: Veröffentlichungen des Instituts für Afrikakunde, Hamburg 39/1982
Saikal, A.: Die Schlußrunde im Afghanistan-Konflikt, in: Europa-Archiv 6/ 1989: 183-190
SAIRR (South African Institute of Race Relations) (ed.): Race Relations Survey 1989/90, Johannesburg 1990
Salibi, Kamal S.: The Modern History of Lebanon, London 1965
Salibi, Kamal S.: Cross Roads to Civil War, Lebanon 1958-1976, London 1976
Salinger, Pierre / Laurent, Eric: Guerre du Golfe. Le Dossier Secret, Paris 1990
Samatar, Ahmed I.: Socialist Somalia. Rhetoric and Reality, London 1988
Sánchez, Gonzalo (Hg.) / Peñaranda, Ricardo (Hg.): Pasado y Presente de la Violencia en Colombia, Bogotá 1986

Sandhu, Bhim: Unresolved Conflict: China and India, London 1988
Sandner, Gerhard: Zentralamerika und der Ferne Karibische Westen. Konjunkturen, Krisen und Konflikte 1503-1984, Stuttgart 1985
Sarasin, Viraphol: Reflections on Thai-Lao Relations, in: Asian Survey, Vol.25, 12/1985: 1260-1276
Saroyan, Mark: The "Karabakh Syndrome" and Azerbaijani Politics, in: Problems of Communism, No. 39, Sep/Okt 1990
Sathyamurthy, T.V.: The Political Development of Uganda 1900-1986, Oxford 1986
Savage, Peter: The National Liberation Struggle in West Irian. From Millenarianism to Socialist Revolution, in: Asian Survey 10/1978: 981-995
Schanberg, Sydney H.: Pakistan Devided, in: Foreign Affairs, Vol. 50, (Oct) 1/1971: 125-135
Scharpf, F.W.: Verhandlungssysteme, Verteilungskonflikte und Pathologien der politischen Steuerung, in: Politische Vierteljahreschrift 29, Sonderheft 19/1988: 61-87
Schermerhorn, Richard Alonzo: Ethnic Plurality in India, Tucson (Ariz.) 1978
Schiefer, Ulrich: Guiné-Bissau, Bonn 1986
Schier, Peter (Hg.): Zur Rolle der Vietnamesen in Kambodscha - Vietnamisierung oder brüderliche Hilfe?, in: Südostasien aktuell, Sept 1986: 496-499 (1986a)
Schier, Peter (Hg.): Kambodoscha 1985 - Zwischen Krokodilen und Tigern, in: SOAa (März), hrsg. Institut für Asienkunde, Hamburg 1986: 25-42 (1986b)
Schifter, Jacobo: La fase oculta de la guerra civil en Costa Rica, Educa, Costa Rica 1979
Schiller, A. Arthur: The Formation of Federal Indonesia 1945 - 1949, The Hague/Bandung 1955
Schipper, Edith: Spannungsfeld Südostasien: Die Konstellation der Beziehungen zwischen der Volksrepublik China, Kambodscha und Thailand in den Jahren 1975-1982, Frankfurt a.M. 1987
Schlesinger, Stephen / Kinzer, Stephen: Bananenkrieg. Das Exempel Guatemala, München 1986
Schlichte, Klaus: Krieg in Liberia. Eine kommentierte Pressedokumentation, in: Forschungsstelle Kriege, Rüstung und Entwicklung an der Universität Hamburg (Hg.): Arbeitspapiere Nr. 45, Hamburg 1990
Schlichte, Klaus: Der Krieg in Liberia seit 1989 - Eine Ursachenanalyse. in: Forschungsstelle Kriege, Rüstung und Entwicklung an der Universität Hamburg (Hg.): Arbeitspapiere Nr.59, Hamburg 1992.
Schlichte, Klaus: Is Ethnicity a Cause of War? in: Peace Review, vol.6, 1994/1: 59-65.

Schmitz, Jochen: Wurde die Stabilität in Surinam verspielt?, in: IIS-Auslandsinformationen (Bonn) 4/1982: 10-16
Schneider, Robin: Peshmerga - die dem Tode entgegengehen. Kurden zwischen den Fronten des Golfkrieges, in: Malanowski, Anja / Stern, Marianne (Hg.): Iran-Irak - "Bis die Gottlosen vernichtet sind," Reinbek 1987: 163-176
Schröder, Günter: Die EPLF-Führung ist nicht zu beneiden, in: Blätter des iz3w (Freiburg) 162/1989
Schubert, Alex: Panama. Geschichte eines Landes und eines Kanals, Berlin 1978
Schulz, Hermann: Nicaragua. Eine amerikanische Vision, Reinbek 1985
Schulze-Marmeling, Dietrich / Sotscheck, Ralf: Der lange Krieg. Macht und Menschen in Nordirland, Göttingen 1989
Schümer, M.: Zaïre und der Westen: Die amerikanische Zaïre-Politik 1975-1983, Bonn 1984
Seagrave, Sterling: Gelber Regen. Yellow Rain. Der Terror chemischer Kriegsführung, München 1983
Seddon, David: Morocco at War, in: Lawless, Richard / Monaham, Laila: War and Refugees. The Western Sahara Conflict, London/New York 1988
Seekins, D.: Historical Setting, in: Bunge, F.: Philippines, A Country Study, Washington 1984: 1-56
Seitz, Stefan: Historische Wurzeln der ethnischen Spannungen in Burundi, in: Afrika Jahrbuch 1988: 55-61
Sellin, Harald: Republik Sudan. Staat, Politik, Wirtschaft, in: Mitteilungen des Deutschen Orient-Instituts, Hamburg 1980
Sellin, Harald / Tetzlaff, Rainer: Sudan, in: Handbuch der Dritten Welt, Bd. 5, Hamburg 1982
Semin-Panzer, Ursula: Krieg gegen Mocambique, Institut für Politische Wissenschaft, Hamburg 1985
Senghaas, Dieter: Militärische Konflikte in der Dritten Welt. Quantitative Beobachtungen und politische Bewertung. in: Leviathan, 12.Jg., Heft 2/ 1984: 271-280.
Senghaas, Dieter: Wohin driftet die Welt? Über die Zukunft friedlicher Koexistenz. Frankfurt am Main 1994
Serajul Islam, S.: Bangladesh in 1986 - Entering a New Phase, in: Asian Survey, Vol.27, 2/1987: 163-172
Serajul Islam, S.: Bangladesh in 1987 - A Spectrum of Uncertainties, in: Asian Survey, Vol.28, 2/1988: 163-172
Shaha, R.: Nepali Politics. Retrospect and Prospect, Delhi/London/New York 1975
Shawcross, William: Schattenkrieg: Kissinger, Nixon und die Zerstörung Kambodschas, Berlin 1980
Short, Anthony: The Communist Insurrection in Malaya 1948-60, London 1975

Siegelberg, Jens: Schritte zu einer Theorie des Krieges. in: Forschungsstelle Kriege, Rüstung und Entwicklung an der Universität Hamburg (Hg.): Arbeitspapiere Nr. 42, Hamburg 1990

Siegelberg, Jens (Red): Die Kriege 1985 bis 1990 - Analyse ihrer Ursachen. Kriege und militante Konflikte, Bd.2. Münster / Hamburg 1991: Lit-Verlag.

Siegelberg, Jens (1994a): Kapitalismus und Krieg - Eine Theorie des Krieges in der Weltgesellschaft, Münster, Hamburg 1994

Siegelberg, Jens (1994b): Ethnizität als Kriegsursache: Realität oder Mythos? in: Birckenbach/Jäger/Wellmann (Hg.): Jahrbuch Frieden 1995 - Konflikte, Abrüstung, Friedensarbeit, München 1994: 29-41

Siemers, G.: Die "neue Volksarmee" in den Philippinen, in: Deutsche Gesellschaft für Asienkunde / Dahm, Bernhard (Hg.): Asien, Hamburg (April) 19/1986: 20-43

Sigrist, Christian: Probleme des demokratischen Neuaufbaus in Guinea-Bissau und auf den Kapverdischen Inseln, Bochum 1980

Silva, C. R. de: Singhala-Tamil Ethnic Rivalry. The Background, in: Goldmann, R.B. / Wilson, A.J. (ed.): From Independence to Stateshood, New York 1984: 111-124 (1984a)

Silva, C. R. de: Singhala-Tamil Relations and Education in Sri Lanka, in: Goldmann, R.B. / Wilson, A.J. (ed.): From Independence to Statehood, New York 1984 (1984b)

Silverstein, Josef: Burmese Politics - The Dilemma of National Unity, New Brunswick (N.Y.) 1980: 51-58

Silverstein, Josef: Minority Problems in Burma since 1962, in: Lehman, F.K.: Military Rule in Burma since 1962 - A Kaleidoscope of Views, Singapore 1981: 51-58

Sim, Richard: Kurdistan: The Search for Recognition, in: Conflict Studies 224/1980

Simon, Doris: Die Entwicklung des Grenzkonfliktes zwischen Indien und China bis zur militärischen Auseinandersetzung im Oktober 1961. Einstellung, Verhaltensweisen und Zielsetzungen der indischen und der chinesischen Regierung, (Diss.) Regensburg 1979

Simon, Gerhard: Die Unruhen in Armenien und Azerbaijan, in: Beiträge zur Konfliktforschung, 2/1988

Simon, Gerhard: Die Nationalitätenbewegungen und das Ende des Sowjetsystems, in: Osteuropa 8/1991

Simonitsch, Pierre: Ein eigener Weg mit vielen Windungen, in: Frankfurter Rundschau 11.4.1988.

Singelenberg, Richard: Surinam, in: Nohlen, Dieter/ Nuscheler, Franz (Hg.): Handbuch der Dritten Welt, Bd. 3, Ludwigsburg 1982: 413-424

SIPRI (ed.): SIPRI Yearbook 1968. World Armaments and Disarmament, Stockholm / London / New York 1968ff

Slater, Jerome: Intervention and Negotiation. The United States and the Dominican Revolution, New York/Evanston/London 1970
Small, Melvin / Singer, J. David: Resort to Arms. International and Civil Wars, 1816-1980. Beverly Hills / London / New Delhi: Sage. (1st ed.: J.D.Singer / M.Small, The Wages of War, 1816-1965. A Statistical Handbook. New York / London / Sydney / Toronto ²1982
Small Arms World Report (seit 1990). Vierteljährlicher Informationsbrief, (Hg.): Institute for Research on Small Arms in International Security. Washington, D.C.
Snow, Peter: Hussein. König und Soldat, Düsseldorf/Köln 1973
Somerville, K.: Angola: Politics, Economics and Society, London 1986
Sorel, Andrés: Guerrilla española del siglo XX, Paris 1970
Sorokin, Pitirim A. (1937). Social and Cultural Dynamics, vol.III: Fluctuation of Social Relationships, War, and Revolution. New York: American Book.
Spencer, W.: Historical Dictionary of Morocco, London 1980
Spencer, Jonathan (ed.): Sri Lanka. History and Roots of the Conflict, London/New York 1990
Stadler, Klaus: Nordirland - Analyse eines Bürgerkrieges, München 1979
Steinbach, Udo: Die arabischen Staaten und der Konflikt mit Israel, in: Kaiser, Dieter / Steinbach, Udo (Hg.): Deutsch-arabische Beziehungen, München/Wien 1981: 261-281
Steinberg, David I.: Constitutional and Political Base of Minority Insurrection in Burma, in: Lim Joo-Jock (Hg.): Armed Separatism in Southeast Asia, Singapore 1984
Stoneman, Colin: Zimbabwe's Prospects, London 1988
Stoneman, Colin / Cliffe, Lionel: Zimbabwe. Politics, Economics and Society, London u.a. 1989
Stookey, Robert W.: Yemen. The Politics of the Yemen Arab Republic, Boulder 1978
Stölting, Erhard: Eine Weltmacht zerbricht. Nationalitäten und Religionen der UdSSR, Frankfurt a.M. 1990
Stubbs, Richard: Hearts and Minds in Guerrilla Warfare: The Malayan Emergency 1948-60, New York 1989
Study Commission on U.S. Policy towards South Africa: South Africa: Time Running Out, Berkeley 1981
Sullivan, J. L.: ETA and the Basque Nationalism: The Fight for Euskadi 1890 - 1986, Croom Helm 1988
Suny, Ronald: The Making of the Georgian Nation, Bloomington/Indianapolis 1988
Suny, Ronald: The Revenge of the Past: Socialism and Ethnic Conflict in Transcaucasia, in: New Left Review, 1990: 5-34

Suter, Keith: West Irian, East Timor and Indonesia, in: Minority Rights Group Report No. 42, London 1980
SWAPO: The Constitution and the Political Program, Lusaka 1976
SWAPO: To be Born a Nation: The Liberation Struggle for Namibia, London 1981
Syed, A.H.: Political Parties and the Nationality Question in Pakistan, in: Journal of South Asian and Middle Eastern Studies 12/1988
Szulc, Tad: Dominican Diary, New York 1965
Talbot, Phillips: The Subcontinent. Ménage à trois, in: Foreign Affairs (July) 4/1972: 698-710
Tan, Samuel K.: The Filipino Muslim Armed Struggle 1900 - 1972, Filipinas 1977
Taylor, Robert H.: Burma's National Unity Problem and 1974 Constitution, in: Contemporary Southeast Asia (Singapore) 1/1979: 247-248
Taylor, David: The Kashmir Crisis, in: Asian Affairs Oct. 1991: 303-313
taz: Tageszeitung
Tellez, A.: Sabaté. Stadtguerrilla in Spanien nach dem Bürgerkrieg 1945-1960, München 1975
Tenekides, George: Zypern. Jüngste Geschichte und politische Perspektive, Genf 1966
Tetzlaff, Rainer: Sudan: Der zweite Bürgerkrieg, in: Internationales Afrikaforum, 4/1984
Tetzlaff, Rainer: Sudan (Länderanalyse), in: Steinbach, Udo / Robert, R. (Hg.): Der Nahe und Mittlere Osten, Bd. 2, Lerverkusen 1988:371-384
Tetzlaff, Rainer, u.a.: Politisierte Ethnizität - eine unterschätzte Realität im nachkolonialen Afrika. in: Afrika Spectrum, Deutsche Zeitschrift für moderne Afrikaforschung, 26.Jg. 1/1991: 5-28.
Thee, Marek: Die Indochina-Kriege: Die Verwicklung der Großmächte - Eskalation und Disengagement, in: Friedensanalysen 8/1978: 12-36
Thoden van Velzen, H.U.E.: De Brunswijk-opstand: Antropologischen kanttekeningen bij de Surinaamse burgerroorlog; in: Sociologische Gids 35/3/1988: 212-236
Thomas, Raju G. C.: Indian Security Policy, Princeton 1986
Thomas-Hope, Elizabeth: Surinam. In: EP (Europa Publications) (ed.): South America, Central America and the Caribbean 1988, London 1987: 609-621
Thompson, Carol B.: Challenge to Imperialism, Boulder/London 1985
Thorton, Richard C.: The Struggle for Power 1917-1972, Bloomington (Ind.) 1973
Thürk, H.: Nachts weint die Sampaguita. Kampf und Niederlage der HUK auf den Philippinen, Berlin (Ost) 1980
Tibi, Bassam: Konfliktregion Naher Osten, München 1989

Tillman, Robert O.: The Non-Lessons of the Malayan Emergency, Yale 1967
Torres, Camilo: Revolutionary Writings, New York 1972
Torres-Rivas, Edelberto: Guatemala - Crisis and Political Violence, in: NACLA (North American Report on the Americas) Vol. XIV, 1/1980: 16-27
Trautner, Bernhard J.: Der Afghanistan-Konflikt, in: Pfetsch, Frank R. (Hg.): Konflikte seit 1945. Daten - Fakten - Hintergründe. Bd. b): Die arabisch-islamische Welt, Freiburg/Würzburg 1991: 143-156
Tremayne, Penelope: Sri Lanka. The Problem of the Tamils, in: RUSI-Yearbook (Brassey's Defence Yearbook), London 1987: 217-237
Tremayne, Penelope: Sri Lanka - Chronology of a Conflict, in: RUSI-Yearbook (Brassey's Defense Yearbook), London 1988: 387-431
Triebswetter, Jutta: Probleme der jemenitischen Politik, in: Jemen-Report, Nr. 10, Schwaig 1979
Trout, Frank E.: Morocco's Saharan Frontiers, Genf 1969
Tuchman, Barbara: Die Torheit der Regierenden. Von Troja bis Vietnam. Frankfurt am Main 1984: S. Fischer.
Uhlig, H. (Hg.): Südasien, in: Fischer-Länderkunde, Frankfurt a.M. 1977
Urban, Gisela / Urban, Ingo: Moskaus Krieg in Südostasien: Mit Giftgas gegen die Bevölkerung, in: Die Welt v. 7.12.1982.
URNG: Manifiesto al Pueblo de Guatemala, in: El Día, México D.F. 22.01.1982
Vanderlinden, J.: Du Congo au Zaïre 1960-1980. Essai du bilan, Brüssel 1981
Vanly, Ismet Sherif: Kurdistan im Irak, in: Chaliand, Gérard (Hg.): Kurdistan und die Kurden, Bd. 1, Göttingen 1984: 259-394
Vanly, Ismet Sherif: Kurdistan und die Kurden, Bd. 2, Türkei und Irak - Fortsetzung, Göttingen/Wien 1986
Veliz, Claudio: The Centralist Tradition of Latin America, Princeton (New Jersey) 1980
Vergara, M. Raul / et al.: Centroamérica. La Guerra de Baja Intersidad, San José 1987
Verhaegen, Benoit: Rebellions au Congo, 2 Bde., Brüssel 1969
Vickery, Michael: Cambodia 1975-82, Boston (Mass.) 1984
Vickery, Michael: Kampuchea - Politics, Economics and Society, London 1986
Vilas, Carlos M. (Hg.): Perfiles de la revolución sandinista, La Habana 1984
Wachendorfer, Achim: Madagaskar, in: Nuscheler, Franz / Ziemer, Klaus (Hg.): Politische Organisation und Repräsentation in Afrika, Berlin 1978: 1153-1196
Waldmann, Peter: Ursachen der Guerrilla in Argentinien, in: Jahrbuch für Geschichte von Staat, Wirtschaft und Gesellschaft Lateinamerikas Bd.15, Köln 1978: 295-348
Waldmann, Peter: Mitgliederstruktur, Sozialisationsmethoden und gesellschaftlicher Rückhalt der baskischen ETA, in Politische Vierteljahresschrift, Jg. 22, 1/1981: 45-69

Waldmann, Peter: Die Bedeutung der ETA für die Gesellschaft und Politik im spanischen Baskenland, in: Aus Politik und Zeitgeschichte (4.Nov.) 45/1988: 3-20

Waldmann, Peter: Guerrillabewegungen in Lateinamerika: Das Beispiel des Sendero Luminoso, in: Langewiesche, Dieter (Hg.): Revolution und Krieg: zur Dynamik historischen Wandels seit dem 18. Jahrhundert, Paderborn 1989: 171-187 (1989a)

Waldmann, Peter: Ethnischer Radikalismus. Ursachen und Folgen gewaltsamer Minderheitenkonflikte am Beispiel des Baskenlandes, Nordirlands und Quebecs, Opladen 1989 (1989b)

Wallensteen, Peter: Structure and War - On International Relations, 1920-1968. Stockholm 1973

Watson, Murray: Vom Wunder des Überlebens in Somalia, in: Frankfurter Rundschau, Nr. 168, 22.07.1992

Weber, Max: (1916). Richtungen und Stufen religiöser Weltablehnung. In derselbe (1956): 441-483.

Weber, Max: (1919). Vom inneren Beruf zur Wissenschaft. In derselbe (1956): 311-339.

Weber, Max: Soziologie - Weltgeschichtliche Analysen - Politik. (Hg.): Johannes Winckelmann. Stuttgart ^{2}1956

Weber, Max: Wirtschaft und Gesellschaft. Grundriß der verstehenden Soziologie. Studienausgabe, (Hg.): Johannes Winckelmann. Köln / Berlin 1964

Weeks, John / Gunson, Phil: Panama. Made in the USA, London 1991

Wegemund, Regina: Der Casamance-Konflikt, in: Afrika spectrum, 26. Jg., Nr. 3, 1991

Weggel, Oskar: Beobachtung eines Ritts auf dem Tiger (China und die Philippinen), in: China aktuell, Hamburg 1973: 515-537

Weggel, Oskar: Indochina im Berichtszeitraum (Oktober/November 1980), in: China aktuell, 9. Jg., 11/1980: 856-972.

Weggel, Oskar: Indochina im Berichtszeitraum (Januar 1981), in: China aktuell, 10. Jg., 1/1981: 56-66.

Weggel, Oskar: Die “Laotische Volksarmee” (LVA), in: SOAa 5/1983: 527-554

Weggel, Oskar: Minderheiten und Minoritätenpolitik in Laos, in: SOAa 3/1986: 289-299.

Weggel, Oskar: Indochina - Vietnam, Kambodscha, Laos, München 1987

Weggel, Oskar: Gesamtbericht Vietnam, Kambodscha, Laos, in: SOAa 2/1988: 128-142 (1988a)

Weggel, Oskar: Gesamtbericht Vietnam, Kambodscha, Laos. in: SOAa 1988 (1988b)

Weggel, Oskar: Gesamtbericht Vietnam, Kambodscha, Laos, in: SOAa 1/1990: 44-85 (1990a)
Weggel, Oskar: Gesamtbericht Vietnam, Kambodscha, Laos, in SOAa 4/1990: 319-346 (1990b)
Weggel, Oskar: Gesamtbericht: Vietnam, Kambodscha, Laos. In: SOAa März 1992 (1992a)
Weggel, Oskar: Gesamtbericht: Vietnam, Kambodscha, Laos. In: SOAa Nov. 1992 (1992b)
Weidemann, Heiko: Birma: Krieg im Dschungel, in: Jens Siegelberg (Red.): Die Kriege 1985 bis 1990 - Analyse ihrer Ursachen, Münster / Hamburg 1991: 397-410
Weinstein, W.: Historical Dictionary of Burundi, New York 1976
Weinstein, W.: Military constituties in the Rwanda state, in: Mazrui, Ali (Ed.): The warrior tradition in modern Africa, Leiden 1977
Weinstein, W. / Schrire, R.: Political Conflict and Ethnic Strategies. A Case Study of Burundi, New York 1976
Weisenfeld, Ernst: Frankreichs Geschichte seit dem Krieg: Ereignisse, Gestalten, Hintergründe 1944-1980, München 1980
Weiss, Marianne: Rwanda, in: Hofmeier, Rolf/Institut für Afrika-Kunde (Hg.): Afrika Jahrbuch 1987, Opladen 1988: 224-246
Weiss, Marianne: Rwanda, in: Hofmeier, Rolf/Institut für Afrika-Kunde (Hg.): Afrika Jahrbuch 1988, Opladen 1989: 238-240
Weiss, Marianne: Rwanda, in: Hofmeier, Rolf/Institut für Afrika-Kunde (Hg.): Afrika-Jahrbuch 1989, Opladen 1990: 249-252
Weiss, Marianne: Senegal, in: Hofmeier, Rolf/Institut für Afrika-Kunde (Hg.): Afrika Jahrbuch 1990, Opladen 1991: 168-174
Wele, Patrick: Kaunda and the Mushala-Rebellions. The Untold Story, Lusaka 1987
Wenner, Manfred W.: Yemen. A Selected Annotated Bibliography of Literature since 1960, Washington 1965
Wenner, Manfred W.: Modern Yemen: 1918-1966, Baltimore 1967
Werle, Rainer: Marxismus contra Tribalismus, in: Blätter des iz3w, 5/1986: 3-8
Werning, Rainer (Hg.): US-Imperialismus auf den Philippinen. Der Mindanao-Konflikt, Münster 1983
Werobel-La Rochelle, J.M.: Madagaskar, in: Hofmeier / Schönborn (Hg.): München 1978
Werz, Nikolaus: Parteien, Staat und Entwicklung in Venezuela, München 1983
Wheelock, Jaime: Imperialismo y Dictatura, México D.F. (5. Auflage) 1980
Wiant, John A.: Insurgency in the Shan State, in: Lim Joo-Jock (ed.): Armed Separatism in Southeast Asia, Singapore 1984
Wiarda, Howard J.: The United States and the Dominican Republic. Interven-

tion, Dependency, and Tyrannicide, in: Journal of Interamerican Studies and World Affairs, Vol. 22 No.2 (Mai 1980): 247-260

Wickham-Crowley, Timothy P.: Winners, Losers and Also-Rans: Toward a Comparative Sociology of Latin American Guerrilla Movements, in: Eckstein, Susan (ed.): Power and Popular Protest. Latin American Social Movements, Berkeley u.a. 1989: 132-181

Wiebe, D.: Afghanistan. Ein mittelasiatisches Entwicklungsland im Umbruch, Stuttgart 1984

Wilkinson, David: Deadly Quarrels. Lewis F.Richardson and the Statistical Study of War. Berkeley / Los Angeles / London 1980

Will, Gerhardt: Vietnam 1975-79. Von Krieg zu Krieg, Hamburg 1987

Will, Gerhardt: Sieg ohne Ende. Die Entfaltung der Konflikte in Indochina seit 1975 in: Senghaas, Dieter (Hg.): Regionalkonflikte in der Dritten Welt, Baden-Baden 1989: 85-120

Wilson, A. Jeyaratnam: Politics in Sri Lanka 1947-79, London 1979

Wilson, A. Jeyaratnam: The Break-up of Sri Lanka. The Singhalese-Tamil Conflict, London 1988

Wilson, Monica / Thompson, Leonard: The Oxford History of South Africa, Vol.I und II, Oxford 1971

Wirsing, R.G.: Ethnicity and Political Reform in Pakistan, in: Asian Affairs 15/1988

Wirz, A.: Krieg in Afrika: Die nachkolonialen Konflikte in Nigeria, Sudan, Tschad und Kongo, Wiesbaden 1982

Wit, Daniel: Thailand: Another Vietnam?, New York 1969

Wolf, Christa (1983). Kassandra. Darmstadt / Neuwied.

Wolfers, Michael / Bergerol, Jane: Angola in the Frontline, London 1983

Wood, Hugh Garland: American Reaction to Limited War in Asia: Korea and Vietnam, Ann Arbor (Mich.) 1974

Woods, Donald: Biko, London 1978

World Watch Institute Report (WWIR): Zur Lage der Welt 87/88, (Kap.1) Bevölkerung: Ursachen der demographischen Falle, Frankfurt a.M. 1987: 13-44

Wright, Quincy: A Study of War. 2nd Ed., with a Commentary on War since 1942, and a Preface by Karl W.Deutsch. Chicago / London: 1965 (1st ed. 1942.)

Wrobèl-Leipold, Andreas: Konflikt und Massenflucht in Tropisch-Afrika. Die Fallstudien Äthiopien und Tschad, Bern/Frankfurt a.M. 1986

Wyatt, David K.: Thailand. A Short History, New Haven 1984

Young, Crawford: Politics in the Congo. Decolonization and Independence, Princeton (N.J.) 1965

Young, Crawford: Zaire: The Unending Crisis, in: Foreign Affairs 57, 1, 1978: 169-185

Young, Thomas: Angola. Recent History, in: Africa South of the Sahara 1990, London 1990: 234-239

Young, Crawford / Turner, Thomas: The Rise and the Decline of the Zairian State, London/Madison (Wisc.) 1985

Zabih, Sepehr: The Left in Contemporary Iran. Ideology, Organisation and the Soviet Connection, London/Sydney 1986

Zamir, Meir: The Lebanese Presidential Elections of 1970 and their Impact on the Civil War of 1975-1976, in: MES, Vol.16, (Jan.) 1/1980: 49-70

Ziegler, Jean: Afrika. Die neue Kolonisation, Darmstadt 1980

Ziem, K.: Ein Zwischenbericht: Zur Lage in Afghanistan, in: Außenpolitik, Zeitschrift für internationale Fragen 1989: 43-58

Ziemer, Klaus: Kamerun, in: Nuscheler, Franz / Ziemer, Klaus (Hg.): Politische Organisation und Repräsentation in Afrika, 1. Halbband, Berlin/New York 1978: 188-222

Zingel, W.-P.: Bangladesh, in: Nohlen, Dieter / Nuschler, Franz (Hg.): Handbuch der Dritten Welt, Bd.7, Hamburg 1983

Zirnig, L.: Domestic Politics and the Regional Security Perspective of Pakistan, in: Strategic Analysis 12(4)/1988

Zitelmann, Thomas: Die Konstruktion kollektiver Identität im Prozeß der Flüchtlingsbewegungen am Horn von Afrika - eine sozialanthropologische Studie am Beispiel der 'saba oromoo' (Nation der Oromo), Berlin 1990

Zürrer, Werner: Iran/Irak - der unerklärte Krieg am Golf, in: Weltgeschehen 4/1981: 421-448

Authors

Key to abbreviations

AAB	Anton Andreas Bloom
AB	Annette Bitter
AH	Andrea Hilgers
BHu	Bertold Hunger
BM	Bernd Möller
BMu	Bernd Musch
CF	Claudia Fregien
CN	Claus Neukirch
CL	Claudius Lieven
CP	Wolff-Christian Peters-Berries
CS	Christian Schaudwet
DJ	Dietrich Jung
DS	Daniel Schwarz
DSf	Dirk Scheffel
HJ	Heike Janßen
HSa	Helena Sabbagh
HW	Heiko Weidemann
JH	Jörg Helmedach
JK	Jürgen Kolk
JM	Joachim Müller
JPF	Jens Peter Franke
JS	Jens Siegelberg
KJG	Klaus Jürgen Gantzel
KS	Klaus Schlichte
MB	Marianne Bleckmann
ML	Matthias Lentze
MMT	Marcus M. Tetz laff
MS	Matthias Schmitt
MV	Marc Vogel
NN	Nicole Nagel
PK	Peter Körner
PM	Petra Marx
PT	Peter Tautkus
RtH	Reinhardt te Heesen

SA	Sebastian Asmus
SG	Susanne Grossmann
SiS	Siegfried Schröder
SK	Sabine Kurtenbach
SJ	Susanne Jesche
ST	Silja Teege
SW	Susanne Wiesemann
TD	Thomas Dorenwendt
TF	Tim Felder
TK	T. Köllmann
TR	Thomas Rabehl
TS	Torsten Schwinghammer
UB	Ulrike Borchardt
U.N.	Ursula Niebling
UP	Uwe Polley
US	Ute Spangenberg
VL	Volkmar Lass
WR	Wiebke Reimers
WW	Werner Wilbert
ZK	Zeljko Karajica

For Product Safety Concerns and Information please contact our EU
representative GPSR@taylorandfrancis.com
Taylor & Francis Verlag GmbH, Kaufingerstraße 24, 80331 München, Germany

www.ingramcontent.com/pod-product-compliance
Ingram Content Group UK Ltd.
Pitfield, Milton Keynes, MK11 3LW, UK
UKHW022113190926
475702UK00013B/1416

* 9 7 8 1 1 3 8 5 1 7 7 9 0 *